Data-Driven Modeling & Scientific Computation

Data-Driven Modeling
&
Scientific Computation

Methods for Complex Systems & Big Data

J. NATHAN KUTZ

Department of Applied Mathematics
University of Washington

UNIVERSITY PRESS

OXFORD
UNIVERSITY PRESS

Great Clarendon Street, Oxford, OX2 6DP,
United Kingdom

Oxford University Press is a department of the University of Oxford.
It furthers the University's objective of excellence in research, scholarship,
and education by publishing worldwide. Oxford is a registered trade mark of
Oxford University Press in the UK and in certain other countries

© J. Nathan Kutz 2013

The moral rights of the author have been asserted

First Edition published in 2013
Impression: 2

Published in the United States of America by Oxford University Press
198 Madison Avenue, New York, NY 10016, United States of America

British Library Cataloguing in Publication Data

Data available

Library of Congress Control Number: 2013937977

ISBN 978-0-19-966033-9 (hbk.)
ISBN 978-0-19-966034-6 (pbk.)

Printed and bound by
CPI Group (UK) Ltd, Croydon, CR0 4YY

Links to third party websites are provided by Oxford in good faith and
for information only. Oxford disclaims any responsibility for the materials
contained in any third party website referenced in this work.

Dedication

For Kristy, Lauren and Katie

Acknowledgments

The idea of the first part of this book began as a series of conversations with Dave Muraki. It then grew into the primary set of notes for a scientific computing course whose ambition was to provide a truly versatile and useful course for students in the engineering, biological and physical sciences. And over the last couple of years, the book expanded to included methods for data analysis, thus bolstering the intellectual scope of the book significantly. Unbeknownst to them, much of the data analysis portion of the book was heavily inspired by the fantastic works of Emmanuel Candés, Yannis Kevrekidis and Clancy Rowley and various conversations I had with each of them. I've also benefitted greatly from early discussions with James Rossmanith, and with implementation ideas with Peter Blossey and Sorin Mitran; and more recently on dimensionality reduction methods with Steven Brunton, Edwin Ding, Joshua Proctor, Peter Schmid, Eli Shlizerman, Jonathan Tu and Matthew Williams. Leslie Butson, Sarah Hewitt and Jennifer O'Neil have been very helpful in editing the book so that it is more readable, useful and error-free. A special thanks should also be given to all the many wonderful students who have provided so much critical commentary and vital feedback for improving the delivery, style and correctness of the book. Of course, all errors in this book are the fault of my daughters' hamsters Fluffy and Quickles.

Contents

PART I Basic Computations and Visualization

PART IV   Scientific Applications

26 Applications of Data Analysis   620

Prolegomenon

Scientific computing is ubiquitous in the physical, biological and engineering sciences. Today, proficiency with computational methods, or lack thereof, can have a major impact on a researcher's ability to effectively analyze a given problem. Although a host of numerical analysis courses are traditionally offered in the mathematical sciences, the typical audience is the professional mathematician. Thus the emphasis is on establishing proven techniques and working through rigorous stability arguments, for instance. No doubt, this vision of numerical analysis is essential and provides the basic groundwork for most numerical analysis courses and books. This more traditional approach to the teaching of numerical methods generally requires more than a year in coursework to achieve a level of proficiency necessary for solving practical problems since the focus is on establishing rigor versus implementation of the methods with a high-level programming language.

The goal of this book is to embark on a new tradition: establishing computing proficiency as the first and foremost priority above rigorous analysis. Thus the major computational methods established over the past few decades are considered with emphasis on their use and implementation versus their rigorous analytic framework. A terse time-frame is also necessary in order to effectively augment the education of students from a wide variety of scientific departments. The three major techniques for solving partial differential equations are all considered: finite differences, finite elements and spectral methods. And the addition in this manuscript of data analysis techniques represents a fairly radical departure from the standard curriculum in mathematics and applied mathematics departments.

MATLAB has established itself as the leader in scientific computing software. The built-in algorithms developed by MATLAB allow the computational focus to shift from technical details to overall implementation and solution techniques. Heavy and repeated use is made of MATLAB's linear algebra packages, fast Fourier transform routines, and finite element (partial differential equations) package. These routines are the workhorses for most solution techniques and are treated to a large extent as blackbox operations. Of course, cursory explanations are given of the underlying principles in any of the routines utilized, but it is largely left as reference material in order to focus on the application of the routine. It is assumed that when necessary, one could implement these techniques by using standard libraries from, for instance, LAPACK.

The end goal is for the student to develop a sense of confidence about implementing computational techniques. Specifically, at the end of the book, the student should be able to solve almost any 1D, 2D or 3D problem of the elliptic, hyperbolic or parabolic type. Or at the least, they should have a great deal of knowledge about how to solve the problem and should have enough information and references at their disposal to circumvent any implementation difficulties. Likewise, with the data analysis framework presented, the key concepts of statistics, time–frequency analysis and low dimensional reduction (SVD) should allow one to have an excellent starting point for moving forward in a given problem where data plays a critical role.

Overall, the combination of methods for solving complex spatio-temporal systems along with understanding how to integrate data into the analysis provides an excellent framework for integrating dynamics of complex systems with big data. This is especially true in the later chapters (Chapters 18–23) where a clear integration of data methods and dynamics is advocated. Given the growing importance of big data methods, it is essential that scientific computing keep pace with some of the exciting developments in this field, especially as it relates to more traditional scientific computing applications.

How to Use This Book

There are a number of intended audiences for this book as is shown in the division of the book into three numerical methods parts and one applications portion. Indeed, the various parts of the book were developed with different student audiences in mind. In what follows, the specific target audiences are highlighted along with the portions of the book appropriate for their use. Ultimately, this gives a roadmap on how to use this book for one's desired ends. All portions of the book have been heavily vetted by undergraduate and graduate students alike, thus improving the overall readability and usefulness. What is unique about this book is the applications portion which attempts to allow students to solve real problems of broad scientific interest. My own personal frustration with the myriad of MATLAB computing books is that many of the problems designed or used for illustration of key concepts are fairly uninspiring and devoid of real-world applicability. More will be said about this in the following paragraphs.

Undergraduate course in beginning scientific computing

The first part of this book reflects the topical coverage of material taught at the University of Washington campus for freshman–sophomore level engineering and physical science students. Given that MATLAB has become the programming language of choice in our engineering programs, the coverage begins by considering basic concepts and theoretical ideas in the context of programming language infrastructure. Thus programming and algorithm development becomes an integral part of developing practical routines for computing, for instance, least-square fits, derivatives or integrals. Moreover, simple things like making nice plots, generating movies and importing/exporting data files with MATLAB are incorporated into the pedagogical structure. Thus Chapters 1 through 6 give a basic introduction to MATLAB, to programming infrastructure (**if** and **for** loops) and to problem solving development. In addition to these first five chapters, Chapter 6 on differential equations is covered towards the end of the book. The ability to quickly and efficiently solve differential equations is critical for many junior and senior level courses in the engineering sciences such as aeronautics/astronautics (the three-body problem) or electrical engineering (circuit theory). Thus the first part of this book has a well-defined, beginning scientific computing audience.

Graduate course in scientific computing

In addition to beginning students, there is a great deal of effort in providing an educational infrastructure and high-level overview of scientific computing methods to graduate students (or advanced undergraduates) from a broad range of scientific disciplines. A traditional graduate course in numerical analysis, while extremely relevant, often is focused on the mathematical infrastructure versus practical implementation. Further, many numerical analysis sequences are quite devoid of engineering, biological and physical science students. The second part of this book, Chapters 7 through 11, provides a high-level introduction to the computational methods used for solving differential and partial differential equations. It certainly is the case that such systems provide an underlying theoretical framework for many applications of broad scientific interest. Thus the key elements of finite difference, spectral and finite elements are all considered. The starting point for the graduate students is in Chapter 7 with stepping techniques for differential equations. On the one hand, one can envision skipping all of the first part of the book to begin the graduate level treatment of scientific computing. But on the other hand, the inclusion of Part I of this book allows graduate students to have a refresher section for simple things such as plotting, constructing movies, importing/exporting data, or simply reviewing programming architecture. Thus the graduate student can use the first part of this book as a reference for their intended studies.

Graduate course in computational methods for data analysis

Parts I and II of this book offer a fairly standard treatment, although slanted heavily towards implementation here, of beginning and advanced numerical methods leading to the construction of numerical solutions of partial differential equations. In contrast, Part III of this book (Chapters 12 through 23) offers a unique perspective on data analysis methods. Indeed, one would be hard pressed to find such a treatment in any current textbook in the mathematical sciences. The aim of this third part is to introduce graduate students (or advanced undergraduates) in the sciences to the burgeoning field of data analysis. This area of research is expanding at an incredible pace in the sciences due to the proliferation of data collection in almost every field of science. The enormous data sets routinely encountered in the sciences now certainly give enormous incentive to their synthesis, interpretation and conjectured meaning. This portion of the book attempts to bring together in a self-consistent fashion the key ideas from (i) statistics, (ii) time–frequency analysis and (iii) low dimensional reductions in order to provide meaningful insight into the data sets one is faced with in any scientific field today. This is a tremendously exciting area and much of this part of the book is driven by intuitive examples of how the three areas (i)–(iii) can be used in combination to give critical insight into the fundamental workings of various problems. As with Part II of this book, access to the introductory material in Part I allows students from various backgrounds to supplement their background where appropriate, thus making Part I an indispensable part of the overall architecture of the book.

Computational methods reference guide

In addition to its primary use as a textbook for either a graduate or undergraduate course in scientific computing or data analysis, the book also serves as a helpful reference guide. As a reference guide, its strength lies in either giving the appropriate high-level overview necessary along with key pieces of MATLAB code for implementation, or the scientific applications portion of the book provides example ideas and techniques for solving a broad class of problems. Using either the applications or theory sections, or both in combination, provides an effective and quick way to refresh one's skills and/or analytic understanding of a solutions technique. In practice, the most common comment I hear concerning this book is that students have found them useful long after their course work has been completed. I believe this is attributed to the low entry threshold for obtaining both practical theoretical knowledge and key snippets of MATLAB code for use in developing a piece of code beyond what is covered in the text. Further, since many high-level

MATLAB subroutines are introduced, a person referencing the book can be assured of exposure to techniques well beyond the simple and trivial methods that might be at first considered.

Scientific applications—Bringing it all together

The most critical aspect of this book is the scientific applications part (Chapters 24 through 26). The philosophy here is simple: solve real problems. There is something a bit disingenuous to me about developing sophisticated methodology and then applying it to either (i) highly contrived examples, or (ii) greatly oversimplified problems that are constructed for theoretical/analytical convenience. It is well understood that each of these serve their purpose in pedagogy. However, if the aim is to make one proficient in building real code, then real problems, with all their complexities, must be considered. The selection of problems is broad enough that an instructor should be able to pick something of interest to themselves. Further, each problem has a well laid out background so that the problem is developed in its appropriate context. The level of difficulty of each problem increases as the problem is developed so that the same problem can serve for undergraduates, advanced undergraduates or graduate students alike, the only difference being in where the student was asked to stop. This allows students the ability to push beyond the basic formulation if they desire. Moreover, there is a context in which their computations are placed, making connection to important historical problems such as the three-body problem, quantum mechanics, 2D advection–diffusion, etc. It is hoped that such problem formulation is much more exciting for both students and faculty alike. This is a unique way in which to think about implementing the theoretical and computational tools learned in the book.

About MATLAB

MATLAB (**MAT**rix **LAB**oratory) has become the *tool of choice* for the teaching of and rapid prototyping of a myriad of problems arising in the physical, engineering and biological sciences. As a high-level language rooted in matrix and vector mathematics, it provides an exceptional integrated programming environment for algorithm development, data analysis, visualization and numerical computation. Using MATLAB, you can solve technical computing problems at a fraction of the programming effort required with traditional languages such as C, C++ or Fortran.

The objective of this book is to provide methods for computationally solving problems, and then to actually solve them. As such, a high-level scientific language is ideal for generating small, yet extremely powerful algorithms. Moreover, full advantage is taken of professionally developed algorithms that few students could match in terms of efficiency, accuracy and/or cutting-edge relevance. Armed with such high-level program tools, rapid and significant progress can be made in solving significant problems that arise in the sciences. Thus the focus is on solving the scientific problem versus the algorithm implementation and its nuances in accuracy, stability and speed.

MATLAB is not free. It is developed by Mathworks Inc. based outside of Boston, MA. It is a privately held company that today has over one million users from all walks of life and industries. A student version is available online and/or from select bookstores, and is highly recommended, for those studying at any institution of learning (www.mathworks.com).

Given its dominance and importance, it is only natural that there should be a MATLAB-styled version through open source development. GNU Octave is a high-level interpreted language, primarily intended for numerical computations. Much like MATLAB, it provides capabilities for the numerical solution of linear and nonlinear problems, and for performing other numerical experiments. It also provides extensive graphics capabilities for data visualization and manipulation. Octave is normally used through its interactive command line interface, but it can also be used to write noninteractive programs. The Octave language is quite similar to MATLAB so that most programs are easily portable. Octave is free, although there is a suggested donation.

Whether it be MATLAB or Octave, a high-level interpreted language is assumed to be available for your use in going through this book. The availability of such high-level languages is transformative in one's ability to rapidly implement and develop the sophisticated computing algorithms developed herein. It is my view that such high-level languages let one much more clearly see the forest through the trees.

PART I

Basic Computations and Visualization

MATLAB Introduction

The first five chapters of this book deal with the preliminaries necessary for manipulating data, constructing logical statements and plotting data. The remainder of the book relies heavily on proficiency with these basic skills.

1.1 Vectors and Matrices

The manipulation of matrices and vectors is of fundamental importance in MATLAB proficiency. This section deals with the construction and manipulation of basic matrices and vectors. We begin by considering the construction of row and column vectors. Vectors are simply a subclass of matrices which have only a single column or row. A row vector can be created by executing the command structure

```
>>x=[1  3  2]
```

which creates the row vector

$$\vec{x} = (1\ 2\ 3). \tag{1.1.1}$$

Row vectors are oriented in a horizontal fashion. In contrast, column vectors are oriented in a vertical fashion and are created with either of the two following command structures:

```
>>x=[1; 3; 2]
```

where the semicolons indicate advancement to the next row. Otherwise a return can be used in MATLAB to initiate the next row. Thus the following command structure is equivalent

```
>>x=[1
    3
    2]
```

Either one of these creates the column vector

$$\vec{x} = \begin{pmatrix} 1 \\ 3 \\ 2 \end{pmatrix}.$$
(1.1.2)

All row and column vectors can be created in this way.

Vectors can also be created by use of the colon. Thus the following command line

```
>>x=0:1:10
```

creates a row vector which goes from 0 to 10 in steps of 1 so that

$$\vec{x} = (0\ 1\ 2\ 3\ 4\ 5\ 6\ 7\ 8\ 9\ 10).$$
(1.1.3)

Note that the command line

```
>>x=0:2:10
```

creates a row vector which goes from 0 to 10 in steps of 2 so that

$$\vec{x} = (0\ 2\ 4\ 6\ 8\ 10).$$
(1.1.4)

Steps of integer size need not be used. This allows

```
>>x=0:0.2:1
```

to create a row vector which goes from 0 to 1 in steps of 0.2 so that

$$\vec{x} = (0\ 0.2\ 0.4\ 0.6\ 0.8\ 1).$$
(1.1.5)

Thus the basic structure associated with the colon operator for forming vectors is as follows:

```
>>x=a:h:b
```

where $a = $ start value, $b = $ end value, and $h = $ step-size. One comment should be made about the relation of b and h. It is possible to choose a step-size h so that starting with a, you will not end on b. As a specific example, consider

```
>>x=0:2:9
```

This command creates the vector

$$\vec{x} = (0\ 2\ 4\ 6\ 8). \tag{1.1.6}$$

The end value of $b = 9$ is not generated in this case because starting with $a = 0$ and taking steps of $h = 2$ is not commensurate with ending at $b = 9$. Thus the value of b can be thought of as an upper bound rather than the ending value. Finally, if no step-size h is included, it is assumed to be 1 so that

```
>>x=0:4
```

produces the vector

$$\vec{x} = (0\ 1\ 2\ 3\ 4). \tag{1.1.7}$$

Not specifying the step-size h is particularly relevant in integer operations and loops where it is assumed that the integers are advanced from the value a to $a + 1$ in successive iterations or loops.

Matrices are just as simple to generate. Now there are a specified number of rows (N) and columns (M). This matrix would be referred to as an $N \times M$ matrix. The 3×3 matrix

$$\mathbf{A} = \begin{pmatrix} 1 & 3 & 2 \\ 5 & 6 & 7 \\ 8 & 3 & 1 \end{pmatrix} \tag{1.1.8}$$

can be created by use of the semicolons

```
>>A=[1 3 2; 5 6 7; 8 3 1]
```

or by using the return key so that

```
>>A=[1 3 2
     5 6 7
     8 3 1]
```

In this case, the matrix is square with an equal number of rows and columns ($N = M$).

Accessing components of a given matrix or vector is a task that relies on knowing the row and column of a certain piece of data. The coordinate of any data item in a matrix is found from its row and column location so that the elements of a matrix $\mathbf{A}$ are given by

$$\mathbf{A}(i, j) \tag{1.1.9}$$

where i denotes the row and j denotes the column. To access the second row and third column of (1.1.8), which takes the value of 7, use the command

```
>>x=A(2,3)
```

This will return the value $x = 7$. To access an entire row, the use of the colon is required

```
>>x=A(2,:)
```

This will access the entire second row and return the row vector

$$\vec{x} = (5 \ 6 \ 7). \tag{1.1.10}$$

Columns can be similarly extracted. Thus

```
>>x=A(:,3)
```

will access the entire third column to produce

$$\vec{x} = \begin{pmatrix} 2 \\ 7 \\ 1 \end{pmatrix}. \tag{1.1.11}$$

The colon operator is used here in the same operational mode as that presented for vector creation. Indeed, the colon notation is one of the most powerful shorthand notations available in MATLAB.

More sophisticated data extraction and manipulation can be performed with the aid of the colon operational structure. To show examples of these techniques we consider the 4×4 matrix

$$\mathbf{B} = \begin{pmatrix} 1 & 7 & 9 & 2 \\ 2 & 3 & 3 & 4 \\ 5 & 0 & 2 & 6 \\ 6 & 1 & 5 & 5 \end{pmatrix}. \tag{1.1.12}$$

The command

```
>>x=B(2:3,2)
```

removes the second through third row of column 2. This produces the column vector

$$\vec{x} = \begin{pmatrix} 3 \\ 0 \end{pmatrix}. \tag{1.1.13}$$

The command

```
>>x=B(4,2:end)
```

removes the second through last columns of row 4. This produces the row vector

$$\vec{x} = (1\ 5\ 5). \tag{1.1.14}$$

We can also remove a specified number of rows and columns. The command

```
>>C=B(1:end-1,2:4)
```

removes the first row through the next to last row along with the second through fourth columns. This then produces the matrix

$$C = \begin{pmatrix} 7 & 9 & 2 \\ 3 & 3 & 4 \\ 0 & 2 & 6 \end{pmatrix}. \tag{1.1.15}$$

As a last example, we make use of the transpose symbol which turns row vectors into column vectors and vice versa. In this example, the command

```
>>D=[B(1,2:4);   B(1:3,3).']
```

makes the first row of **D** the second through fourth columns of the first row of **B**. The second row of **D**, which is initiated with the semicolon, is made from the transpose (. ′) of the first three rows of the third column of **B**. This produces the matrix

$$D = \begin{pmatrix} 7 & 9 & 2 \\ 9 & 3 & 2 \end{pmatrix}. \tag{1.1.16}$$

An important comment about the transpose function is in order. In particular, when transposing a vector with complex numbers, the period must be put in before the ′ symbol. Specifically, when considering the transpose of the column vector

$$\vec{x} = \begin{pmatrix} 3+2i \\ 1 \\ 8 \end{pmatrix}. \tag{1.1.17}$$

where i is the complex (imaginary) number $i = \sqrt{-1}$, the command

```
>>y=x.'
```

produces the row vector

$$\vec{y} = (3+2i \ 1 \ 8), \tag{1.1.18}$$

whereas the command

```
>>y=x'
```

produces the row vector

$$\vec{y} = (3-2i \ 1 \ 8). \tag{1.1.19}$$

Thus the use of the ′ symbol alone also conjugates the vector, i.e. it changes the sign of the imaginary part.

As a final example of vector and/or matrix manipulation, consider the following vector

```
x=[-1 2 3 5 -2];
```

This produces a simple row vector with both positive and negative numbers. The following commands can be quite useful in practice for manipulating such a vector and setting, for example, threshold rules for the vector.

```
y1=(x>0);
y2=(x>0).*x;
y3=(x>0)*3;
```

These commands produce the vectors

$$\vec{y}_1 = (0\ 1\ 1\ 1\ 0) \tag{1.1.20a}$$

$$\vec{y}_2 = (0\ 2\ 3\ 5\ 0) \tag{1.1.20b}$$

$$\vec{y}_3 = (0\ 3\ 3\ 3\ 0) \tag{1.1.20c}$$

respectively. This provides a very convenient way to quickly sweep through a vector or matrix while applying a logical statement.

1.2   Logic, Loops and Iterations

The basic building blocks of any MATLAB program, or any other mathematical software or programming languages, are **for** loops and **if** statements. They form the background for carrying out the complicated manipulations required of sophisticated and simple codes alike. This section will focus on the use of these two ubiquitous logic structures in programming.

To illustrate the use of the **for** loop structure, we consider some very basic programs which revolve around its implementation. We begin by constructing a loop which will recursively sum a series of numbers. The basic format for such a loop is the following:

```
a=0
for j=1:5
    a=a+j
end
```

This program begins with the variable sum, **a**, being zero. It then proceeds to go through the **for** loop five times, i.e. the counter j takes the value of one, two, three, four and five in succession. In the loop, the value of sum is updated by adding the current value of j. Thus starting with an initial value of sum equal to zero, we find that the variable a is equal to 1 ($j = 1$), 3 ($j = 2$), 6 ($j = 3$), 10 ($j = 4$), and 15 ($j = 5$).

The default incremental increase in the counter j is 1. However, the increment can be specified as desired. The program

```
a=0
for j=1:2:5
    a=a+j
end
```

is similar to the previous program. But for this case the incremental steps in j are specified to be 2. Thus starting with an initial value of sum equal to zero, we find that the variable a is equal to

1 ($j = 1$), 4 ($j = 3$), and 9 ($j = 5$). And even more generally, the **for** loop counter can be simply given by a row vector. As an example, the program

```
a=0
for j=[1 5 4]
    a=a+j
end
```

will go through the loop three times with the values of j being 1, 5, and 4 successively. Thus starting with an initial value of a equal to zero, we find that a is equal to 1 ($j = 1$), 6 ($j = 5$), and 10 ($j = 4$).

The **if** statement is similarly implemented. However, unlike the **for** loop, a series of logic statements is usually considered. The basic format for this logic is as follows:

```
if (logical statement)
      (expressions to execute)
elseif (logical statement)
      (expressions to execute)
elseif (logical statement)
      (expressions to execute)
else
      (expressions to execute)
end
```

In this logic structure, the last set of expressions is executed if the three previous logical statements do not hold. The basic logic architecture in MATLAB is given in Table 1.1. Here, some of the

Table 1.1 Common logic expressions in MATLAB used in if statement architectures

Logic	MATLAB expression
equal to	==
not equal	~=
greater than	>
less than	<
greater than or equal to	>=
less than or equal to	<=
AND	&
OR		

common mathematical logic statements are given such as, for example, equal to, greater than, less than, AND and OR.

In practice, the logical **if** may only be required to execute a command if something is true. Thus there would be no need for the *else* logic structure. Such a logic format might be as follows:

```
if (logical statement)
      (expressions to execute)
elseif (logical statement)
      (expressions to execute)
end
```

In such a structure, no commands would be executed if the logical statements do not hold.

Bisection method for root solving

To make a practical example of the use of **for** and **if** statements, we consider the bisection method for finding zeros of a function. In particular, we will consider the transcendental function

$$\exp(x) - \tan(x) = 0 \tag{1.2.1}$$

for which the values of x which make this true must be found computationally. Figure 1.1 plots (a) the functions $f(x) = \exp(x)$ and $f(x) = \tan(x)$ and (b) the function $f(x) = \exp(x) - \tan(x)$.

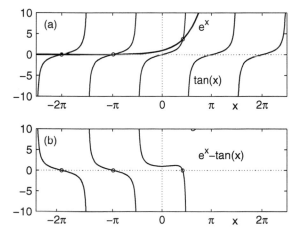

Figure 1.1: (a) Plot of the functions $f(x) = \exp(x)$ (bold) and $f(x) = \tan(x)$. The intersection points (circles) represent the roots $\exp(x) - \tan(x) = 0$. In (b), the function of interest, $f(x) = \exp(x) - \tan(x)$, is plotted with the corresponding zeros (circles).

The intersection points of the two functions (circles) represent the roots of the equation. We can begin to get an idea of where the relevant values of x are by plotting this function. The following MATLAB script will plot the function over the interval $x \in [-10, 10]$.

```
clear all  % clear all variables
close all  % close all figures

x=-10:0.1:10;     % define plotting range
y=exp(x)-tan(x);  % define function to consider
plot(x,y)         % plot the function
```

It should be noted that $\tan(x)$ takes on the value of $\pm\infty$ at $\pi/2 + n\pi$ where $n = \ldots, -2, -1, 0, 1, 2, \ldots$. By zooming in to smaller values of the function, one can find that there are a large number (infinite) of roots to this equation. In particular, there is a root located at $x \in [-4, 2.8]$. At $x = -4$, the function $\exp(x) - \tan(x) > 0$ while at $x = -2.8$ the function $\exp(x) - \tan(x) < 0$. Thus, in between these points lies a root.

The bisection method simply cuts the given interval in half and determines if the root is on the left or right side of the cut. Once this is established, a new interval is chosen with the midpoint now becoming the left or right end of the new domain, depending of course on the location of the root. This method is repeated until the interval has become so small and the function considered has come to within some tolerance of zero. The following algorithm uses an outside **for** loop to continue cutting the interval in half while the imbedded **if** statement determines the new interval of interest. A second **if** statement is used to ensure that once a certain tolerance has been achieved, i.e. the absolute value of the function $\exp(x) - \tan(x)$ is less than 10^{-5}, then the iteration process is stopped.

bisection.m

```
clear all  % clear all variables

xr=-2.8;  % initial right boundary
xl=-4;    % initial left boundary

for j=1:1000  % j cuts the interval

    xc=(xl+xr)/2;        % calculate the midpoint
    fc=exp(xc)-tan(xc);  % calculate function

    if fc>0
        xl=xc;  % move left boundary
```

Continued

```
    else
        xr=xc;   % move right boundary
    end

    if abs(fc)<10^(-5)
        break    % quit the loop
    end

end
xc    % print value of root
fc    % print value of function
```

Table 1.2 Convergence of the bisection iteration scheme to an accuracy of 10^{-5} given the end points of $x_l = -4$ and $x_r = -2.8$

j	xc	fc
1	−3.4000	−0.2977
2	−3.1000	0.0034
3	−2.9500	−0.1416
⋮	⋮	⋮
10	−3.0965	−0.0005
⋮	⋮	⋮
14	−3.0964	0.0000

Note that the **break** command ejects you from the current loop. In this case, that is the j loop. This effectively stops the iteration procedure for cutting the intervals in half. Further, extensive use has been made of the semicolons at the end of each line. The semicolon simply represses output to the computer screen, saving valuable time and clutter. The progression of the root finding algorithm is given in Table 1.2. An accuracy of 10^{-5} is achieved after 14 iterations.

1.3   Iteration: The Newton–Raphson Method

Iteration methods are of fundamental importance, not only as a mathematical concept, but as a practical tool for solving many problems that arise in the engineering, physical and biological sciences. In a large number of applications, the convergence or divergence of an iteration scheme is of critical importance. Iteration schemes also help illustrate the basic functionality of

for loops and **if** statements. In later chapters, iteration schemes become important for solving many problems which include:

- root finding $f(x) = 0$ (Newton's method);
- linear systems $\mathbf{A}\vec{x} = \vec{b}$ (system of equations);
- differential equations $dy/dt = f(\mathbf{y}, t)$.

In the first two applications listed above, the convergence of an iteration scheme is of fundamental importance. For applications in differential equations, the iteration predicts the future state of a given system provided the iteration scheme is stable. The stability of iteration schemes will be discussed in later sections on differential equations.

The generic form of an iteration scheme is as follows:

$$p_{k+1} = g(p_k), \tag{1.3.1}$$

where $g(p_k)$ determines the iteration rule to be applied. Thus starting with an initial value p_0 we can construct the hierarchy

$$p_1 = g(p_0) \tag{1.3.2a}$$
$$p_2 = g(p_1) \tag{1.3.2b}$$
$$p_3 = g(p_2) \tag{1.3.2c}$$
$$\vdots \tag{1.3.2d}$$

Such an iteration procedure is trivial to implement with a **for** loop structure. Indeed, an example code will be given shortly. Of interest is the values taken by successive iterations p_k. In particular, the values can converge, diverge, become periodic, or vary randomly (chaotically).

Fixed points

Fixed points are fundamental to understanding the behavior of a given iteration scheme. Fixed points are points defined such that the input is equal to the output in Eq. (1.3.1):

$$p_k = g(p_k). \tag{1.3.3}$$

The fixed points determine the overall behavior of the iteration scheme. Effectively, we can consider two functions

$$y_1 = x \tag{1.3.4a}$$
$$y_2 = g(x) \tag{1.3.4b}$$

and the fixed point occurs for

$$y_1 = y_2 \quad \rightarrow \quad x = g(x). \tag{1.3.5}$$

Defining these quantities in a continuous fashion allows for an insightful geometrical interpretation of the iteration process. Figure 1.2 demonstrates graphically the location of the fixed point

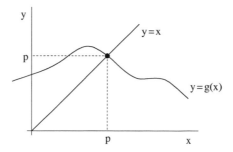

Figure 1.2: Graphical representation of the fixed point. The value p is the point at which $p = g(p)$, i.e. where the input value of the iteration is equal to the output value.

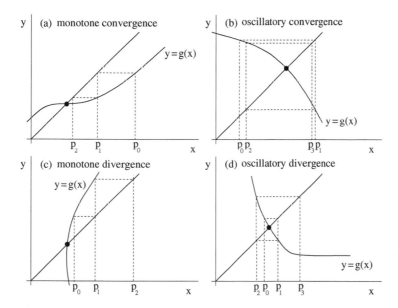

Figure 1.3: Geometric interpretation of the iteration procedure $p_{k+1} = g(p_k)$ for four different functions $g(x)$. Some of the possible iteration behaviors near the fixed point included monotone convergence or divergence and oscillatory convergence or divergence.

for a given $g(x)$. The intersection of $g(x)$ with the line $y = x$ gives the fixed point locations of interest. Note that for a given $g(x)$, multiple fixed points may exist. Likewise, there may be no fixed points for certain iteration functions $g(x)$.

From this graphical interpretation, the concept of convergence or divergence can be easily understood. Figure 1.3 shows the iteration process given by (1.3.1) for four prototypical iteration functions $g(x)$. Illustrated is convergence to and divergence from (both in a monotonic and oscillatory sense) the fixed point where $x = g(x)$. Note that in practice, when iterating towards a convergent solution, the computations should be stopped once a desired accuracy is achieved. It is not difficult to imagine from Figs. 1.3(b) and (d) that the iteration may also result in periodicity of the iteration scheme, thus the ability of iteration to produce periodic orbits and solutions.

The Newton–Raphson method

One of the classic uses of iteration is for finding roots of a given function or polynomial. We therefore develop the basic ideas of the Newton–Raphson iteration method, commonly known as a Newton's method. We begin by considering a single nonlinear equation

$$f(x_r) = 0 \tag{1.3.6}$$

where x_r is the root to the equation and the value being sought. We would like to develop a scheme which systematically determines the value of x_r. The Newton–Raphson method is an iterative scheme which relies on an initial guess, x_0, for the value of the root. From this guess, subsequent guesses are determined until the scheme either converges to the root x_r or the scheme diverges and another initial guess is used. The sequence of guesses $(x_0, x_1, x_2, \ldots)$ is generated from the slope of the function $f(x)$. The graphical procedure is illustrated in Fig. 1.4. In essence, everything relies on the slope formula as illustrated in Fig. 1.4(a):

$$\text{slope} = \frac{df(x_n)}{dx} = \frac{\text{rise}}{\text{run}} = \frac{0 - f(x_n)}{x_{n+1} - x_n}. \tag{1.3.7}$$

Rearranging this gives the Newton–Raphson iterative relation

$$x_{n+1} = x_n - \frac{f(x_n)}{f'(x_n)}. \tag{1.3.8}$$

A graphical example of how the iteration procedure works is given in Fig. 1.4(b) where a sequence of iterations is demonstrated. Note that the scheme fails if $f'(x_n) = 0$ since then the slope line never intersects $y = 0$. Further, for certain guesses the iterations may diverge. Provided the initial guess is sufficiently close, the scheme usually will converge. Conditions for convergence can be found in Burden and Faires [11].

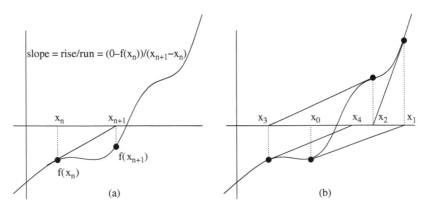

Figure 1.4: Construction and implementation of the Newton–Raphson iteration formula. In (a), the slope is the determining factor in deriving the Newton–Raphson formula. In (b), a graphical representation of the iteration scheme is given.

To implement this method, consider once again the example problem of the last section for which we are to determine the roots of $f(x) = \exp(x) - \tan(x) = 0$. In this case, we have

$$f(x) = \exp(x) - \tan(x) \tag{1.3.9a}$$

$$f'(x) = \exp(x) - \sec^2(x) \tag{1.3.9b}$$

so that the Newton iteration formula becomes

$$x_{n+1} = x_n - \frac{\exp(x_n) - \tan(x_n)}{\exp(x_n) - \sec^2(x_n)}. \tag{1.3.10}$$

This results in the following algorithm for searching for a root of this function.

newton.m

```
clear all   % clear all variables

x(1)=-4;   % initial guess

for j=1:1000   % j is the iteration variable

  x(j+1)=x(j)-(exp(x(j))-tan(x(j)))/(exp(x(j))-sec(x(j))^2);
  fc=exp(x(j+1))-tan(x(j+1));

    if abs(fc)<10^(-5)
        break   % quit the loop
    end

end
x(j+1)   % print value of root
fc   % print value of function
```

Table 1.3 Convergence of the Newton iteration scheme to an accuracy of 10^{-5} given the initial guess $x_0 = -4$

j	x(j)	fc
1	−4	1.1761
2	−3.4935	0.3976
3	−3.1335	0.0355
4	−3.0964	2.1946×10^{-6}

Note that the **break** command ejects you from the current loop. In this case, that is the j loop. This effectively stops the iteration procedure from proceeding to the next value of $f(x_n)$. The progression of the root finding algorithm is given in Table 1.3. An accuracy of 10^{-5} is achieved after only four iterations. Compare this to 14 iterations for the bisection method of the last section.

 ## 1.4 Function Calls, Input/Output Interactions and Debugging

A fairly standard way to program in MATLAB is to write most algorithms as functions and to call them through other functions or directly in the MATLAB dialog box. Thus building functions is one of the most common techniques for code implementation.

   In what follows, a number of examples will be used to illustrate how to build a functions in MATLAB. The goal is to start with simple functions and to build in sophistication from there. Thus consider building a function that evaluates the simple function

$$f(x) = x^3 + \sin(x). \tag{1.4.1}$$

Obviously, the function requires a value of x (input) which must be used to evaluate the function. This *input* must be specified in order for the function call to make sense. In MATLAB, this function can be easily implemented by building the following MATLAB script:

myfun.m

```
function y=myfun(x)
y=x.^3+sin(x);
```

The dot in the cube of x is for the case that an initial vector is input into the function call. Access to this function comes from a simple command line call such as the following

```
>> y=myfun(1);
```

This returns the value of y evaluated at $x = 1$ so that $y = 1^3 + \sin(1) = 1.8415$. Alternatively, one can evaluate the function at multiple values of x by defining the input as a vector as follows:

```
>> x=(0:1:10).';
>> y=myfun(x);
```

This yields

```
y =

         0
    1.8415
    8.9093
   27.1411
   63.2432
  124.0411
  215.7206
  343.6570
  512.9894
  729.4121
  999.4560
```

An important task often associated with function calls is the passing of variables or parameters through to the function itself. The following illustrates how to pass a variable along with a parameter. Specifically, assume that we wish to evaluate the function $f(x) = \sin Bx$ where the parameter is a function of the input parameter A so that $B = A^2 + \cos A$. The following code evaluates the function for a given input vector **x** and parameter A.

funcA.m

```
function [fA B]=funcA(x,A)

B=A^2 + cos(A);
fA=sin(B*x);
```

Note that `function` returns both the value of the function and the value of the parameter B. Specifically, the code can be executed as

```
[fA B]=funcA([3 4 5],2)
```

which produces the output

```
fA =

    -0.9704     0.9804    -0.8018


B =

     3.5839
```

Finally, functions can be called within functions, which is often convenient in executing complex codes. The following function **funcC.m** uses two parameters A and B to evaluate a third variable C which is then used to evaluate the final function. Thus the function has two embedded functions within the called function. There is no limit on how many functions can be embedded within another.

funcC.m

```
function [f A B C]=funcC(x)
A=2;  B=3;
C=f_param(A,B);
f=func(x,C);


function C=f_param(A,B)
C=A^2+B^2;

function f=func(x,C)
f=C*sin(C*x);
```

This function can be called by the command line

```
x=0:1:10;
[f a b c]=funcC(x);
```

This returns the vector **f** evaluated at the vector **x** values along with the parameters A, B and C.

Input and output to MATLAB codes

It is also easy to make MATLAB interactive in nature. Thus MATLAB can ask for user input as the algorithm progresses through its execution. Consider the function

func.m

```
function f=func(x)
A=input('A=')
f=sin(A*x);
```

This code will pause at the second line and display the following to the screen

```
A=
```

The user then inputs any numerical value of A desired and the function will then continue the function evaluation with this user specified (and interactive) value of A. Text strings can also be input into the system. The following is an example

```
reply = input('Do you want more? Y/N [Y]: ', 's');
if isempty(reply)
    reply = 'Y';
end
```

The option 's' in the input command allows for the acceptance of a string. The **isempty** line is in case no input is generated and a preassigned value is given to the variable **reply**.

Output to the screen can also be easily generated in a variety of ways. The **disp** (for display) is perhaps the easiest to use. So, for instance, one could use the command

```
disp('hello world')
```

to display the test string: *hello world*. Alternatively, one could also simply put a text string in quotes to produce a text string

```
'hello world'
```

Alternatively, one could define a variable that can be viewed at any time by defining, for instance, the following:

```
x='hello world'
```

Thus the variable **x** takes on the value of the text string.

Debugging

MATLAB provides a convenient debugging environment for tracking the flow of your code on a step-by-step basis. The debugger is fairly straightforward to use and can greatly aid in tracking down coding errors (see Fig. 1.5). The debugger allows you to (i) set/remove breakpoints, (ii) clear all breakpoints, (iii) step forward, (iv) step in to loops, (v) step out of loops, (vi) continue with the code and (vii) exit the debugging mode. Thus the code is executed line by line to see if it actually can perform the slated task. In addition to manually stepping through the code,

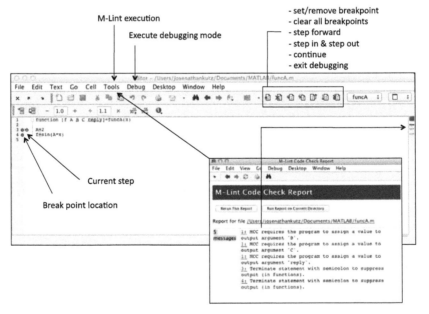

Figure 1.5: Debugging options in MATLAB. The **tools** and **debug** buttons allow for full functionality of the debugging tools, including the M-lint debugging report (bottom right panel). The debugger allows you to (i) set/remove breakpoints, (ii) clear all breakpoints, (iii) step forward, (iv) step in to loops, (v) step out of loops, (vi) continue with the code and (vii) exit the debugging mode. Breakpoints can be set (red dots) and the current execution location is denoted by a green arrow.

the **M-lint** report allows for an overview of potentially problematic portions of the code. Thus a simple and efficient method is generated in MATLAB for making a first pass at the debugging process. Experience in debugging code is perhaps one of the most important aspects of developing into a strong programmer.

1.5   Plotting and Importing/Exporting Data

The graphical representation of data is a fundamental aspect of any technical scientific communication. MATLAB provides an easy and powerful interface for plotting and representing a large variety of data formats. Like all MATLAB structures, plotting is dependent on the effective manipulation of vectors and matrices.

To begin the consideration of graphical routines, we first construct a set of functions to plot and manipulate. To define a function, the plotting domain must first be specified. This can be done by using a routine which creates a row vector

```
x1=-10:0.1:10;
y1=sin(x1);
```

Here the row vector $\vec{x}$ spans the interval $x \in [-10, 10]$ in steps of $\Delta x = 0.1$. The second command creates a row vector $\vec{y}$ which gives the values of the sine function evaluated at the corresponding values of $\vec{x}$. A basic plot of this function can then be generated by the command

```
plot(x1,y1)
```

Note that this graph lacks a title or axis labels. These are important in generating high quality graphics.

The preceding example considers only equally spaced points. However, MATLAB will generate functions values for any specified coordinate values. For instance, the two lines

```
x2=[-5 -sqrt(3) pi];
y2=sin(x2);
```

will generate the values of the sine function at $x = -5, -\sqrt{3}$ and π. The **linspace** command is also helpful in generating a domain for defining and evaluating functions. The basic procedure for this function is as follows

```
x3=linspace(-10,10,64);
y3=x3.*sin(x3);
```

This will generate a row vector **x** which goes from -10 to 10 in 64 steps. This can often be a helpful function when considering intervals where the number of discretized steps gives a complicated Δx. In this example, we are considering the function $x \sin x$ over the interval $x \in [-10, 10]$. By doing so, the period must be included before the multiplication sign. This will then perform a *component-by-component* multiplication, thus creating a row vector with the values of $x \sin x$.

To plot all of the above data sets in a single plot, we can do either one of the following routines.

```
figure(1)
plot(x1,y1), hold on
plot(x2,y2)
plot(x3,y3)
```

In this case, the **hold on** command is necessary after the first plot so that the second plot command will not overwrite the first data plotted. This will plot all three data sets on the same graph with a default blue line. This graph will be *figure 1*. Any subsequent plot commands will plot the new data on top of the current data until the **hold off** command is executed. Alternatively, all the data sets can be plotted on the same graph with

```
figure(2)
plot(x1,y1,x2,y2,x3,y3)
```

For this case, the three pairs of vectors are prescribed within a single plot command. This figure generated will be *figure 2*. An advantage to this method is that the three data sets will be of different colors, which is better than having them all the default color of blue.

This is only the beginning of the plotting and visualization process. Many aspects of the graph must be augmented with specialty commands in order to more accurately relay the information. Of significance is the ability to change the line colors and styles of the plotted data. By using the **help plot** command, a list of options for customizing data representation is given. In the following, a new figure is created which is customized as follows

```
figure(3)
plot(x1,y1,x2,y2,'g*',x3,y3,'mo:')
```

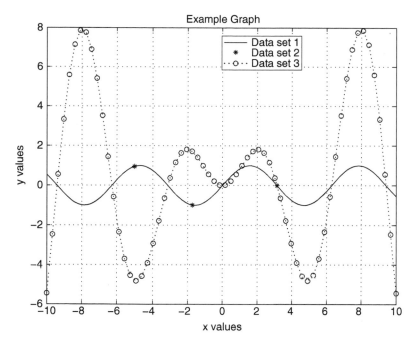

Figure 1.6: Plot of the functions $f(x) = \sin(x)$ and $f(x) = x\sin(x)$. The stars (*****) are the value of $\sin(x)$ at the locations $x = \{-5, -\sqrt{3}, \pi\}$. The **grid on** and **legend** command have both been used. Note that the default font size is smaller than would be desired. More advanced plot settings and adjustments are considered in Section 5.5 on visualization.

This will create the same plot as in *figure 2*, but now the second data set is represented by green stars (*****) and the third data set is represented by a magenta dotted line with the actual data points given by a magenta hollow circle. This kind of customization greatly helps distinguish the various data sets and their individual behaviors. An example of this plot is given in Fig. 1.6. Extra features, such as axis label and titles, are discussed in the following paragraphs. Table 1.4 gives a list of options available for plotting different line styles, colors and symbol markers.

Labeling the axis and placing a title on the figure is also of fundamental importance. This can be easily accomplished with the commands

```
xlabel('x values')
ylabel('y values')
title('Example Graph')
```

The strings given within the ' sign are now printed in a centered location along the x-axis, y-axis and title location, respectively.

Table 1.4   Plotting options which can be used in a standard **plot** command. The format would typically be, for instance, **plot(x,y,'r:h')** which would put a red dotted line through the data points marked with a hexagram

Line style	Color	Symbol
- = solid	k = black	. = point
: = dotted	b = blue	o = circle
-. = dashed-dot	r = red	x = x-mark
– = dashed	c = cyan	+ = plus
(none) = no line	m = magenta	* = star
	y = yellow	s = square
	g = green	d = diamond
		v = triangle (down)
		$\wedge$ = triangle (up)
		< = triangle (left)
		> = triangle (right)
		p = pentagram
		h = hexagram

A legend is then important to distinguish and label the various data sets which are plotted together on the graph. Within the **legend** command, the position of the legend can be easily specified. The command **help legend** gives a summary of the possible legend placement positions, one of which is to the right of the graph and does not interfere with the graph data. To place a legend in the above graph in an *optimal* location, the following command is used.

```
legend('Data set 1','Data set 2','Data set 3','Location','Best')
```

Here the strings correspond to the three plotted data sets in the order they were plotted. The location command at the end of the **legend** command is the option setting for placement of the legend box. In this case, the option *Best* tries to pick the best possible location which does not interfere with any of the data. Table 1.5 gives a list of the location options that can be used in MATLAB. And in addition to these specified locations, a 1×4 vector may be placed in this slot with the specific location desired for the legend.

Subplots

Another possible method for representing data is with the **subplot** command. This allows for multiple graphs to be arranged in a row and column format. In this example, we have three data sets which are under consideration. To plot each data set in an individual subplot requires three

Table 1.5 Legend position placement options. The default is to place the legend on the inside top right of the graph. One can also place a 1×4 vector in the specification to manually give a placement of the legend

MATLAB position specification	Legend position
'North'	inside plot box near top
'South'	inside bottom
'East'	inside right
'West'	inside left
'NorthEast'	inside top right (default)
'NorthWest'	inside top left
'SouthEast'	inside bottom right
'SouthWest'	inside bottom left
'NorthOutside'	outside plot box near top
'SouthOutside'	outside bottom
'EastOutside'	outside right
'WestOutside'	outside left
'NorthEastOutside'	outside top right
'NorthWestOutside'	outside top left
'SouthEastOutside'	outside bottom right
'SouthWestOutside'	outside bottom left
'Best'	least conflict with data in plot
'BestOutside'	least unused space outside plot

plot frames. Thus we construct a plot containing three rows and a single column. The format for this command structure is as follows:

```
figure(4)
subplot(3,1,1), plot(x1,y1), axis([-10 10 -10 10])
subplot(3,1,2), plot(x2,y2,'g*'), axis([-10 10 -10 10])
subplot(3,1,3), plot(x3,y3,'mo:')
```

Note that the **subplot** command is of the form **subplot(row, column, graph number)** where the graph number refers to the current graph being considered. In this example, the axis command has also been used to make all the graphs have the same x and y ranges. The command structure for the axis command is **axis([xmin xmax ymin ymax])**. For each subplot, the use of the **legend**, **xlabel**, **ylabel** and **title** command can be used. An example of this plot is given in Fig. 1.7.

Remark 1 All the graph customization techniques discussed can be performed directly within the MATLAB graphics window. Simply go to the edit button and choose to edit axis properties. This will give full control of all the axis properties discussed so far and more. However, once

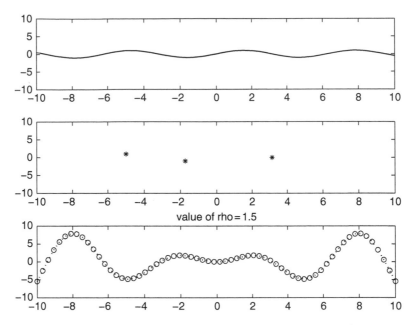

Figure 1.7: Subplots of the functions $f(x) = \sin(x)$ and $f(x) = x\sin(x)$. The stars (*) in the middle plot are the value of $\sin(x)$ at the locations $x = \{-5, -\sqrt{3}, \pi\}$. The **num2str** command has been used for the title of the bottom plot.

MATLAB is shut down or the graph is closed, all the customization properties are lost and you must start again from scratch. This gives an advantage to developing a nice set of commands in a **.m** file to customize your graphics.

Remark 2 To put a grid on the a graph, simply use the **grid on** command. To take it off, use **grid off**. The number of grid lines can be controlled from the editing options in the graphics window. This feature can aid in illustrating more clearly certain phenomena and should be considered when making plots.

Remark 3 To put a variable value in a string, the **num2str** (number to string) command can be used. For instance, the code

```
rho=1.5;
title(['value of rho=' num2str(rho)])
```

creates a title which reads "value of rho=1.5". Thus this command converts variable values into useful string configurations.

Remark 4 More advanced plot settings, adjustments and modifications are considered in Section 6.

Load, save and print

Once a graph has been made or data generated, it may be useful to save the data or print the graphs for inclusion in a write-up or presentation. The **save** and **print** commands are the appropriate commands for performing such actions.

The **save** command will write any data out to a file for later use in a separate program or for later use in MATLAB. To save workspace variables to a file, the following command structure is used

```
save filename
```

This will save all the current workspace variables to a binary file named **filename.mat**. In the preceding example, this will save all the vectors created along with any graphics settings. This is an extremely powerful and easy command to use. However, you can only access the data again through MATLAB. To recall this data at a future time, simply load it back into MATLAB with the command

```
load filename
```

This will reload into the workspace all the variables saved in **filename.mat**. This command is ideal when closing down operations on MATLAB and resuming at a future time.

Alternatively, it may be advantageous to save data for use in a different software package. In this case, data needs to be saved in an ASCII format in order to be read by other software engines. The **save** command can then be modified for this purpose. The command

```
save x1.dat x1 -ASCII
```

saves the row vector **x1** generated previously to the file **x1.dat** in ASCII format. This can then be loaded back into MATLAB with the command

```
load x1.dat
```

This saving option is advantageous when considering the use of other software packages to manipulate or analyze data generated in MATLAB.

It is also necessary at times to save files names according to a loop variable. For instance, you may create a loop variable which executes a MATLAB script a number of times with a different key parameter which is controlled by the loop variable. In this case, it is often imperative to save

Table 1.6 Some figure export options for MATLAB. Most common output options are available which can then be imported to Word, PowerPoint or Latex

Figure output format	Print option
encapsulated postscript	-deps
encapsulated color postscript	-depsc
JPEG image	-djpeg
TIFF image	-dtiff
portable network graphics	-dpng

the variables according to the loop name as repeated passage through the loop will overwrite the variables you wish to save. The following example illustrates the saving procedure

```
for loop=1:5
    (expressions to execute)
    save(['loopnumber' num2str(loop)])
end
```

Subsequent passes through the loop will save files called **loopnumber1.mat, loopnumber2.mat**, etc. This can often be helpful for running scripts which generate the same common variable names in a loop.

If you desire to print a figure to a file, the **print** command needs to be utilized. There a large number of graphics formats which can be used. By typing **help print**, a list of possible options will be listed. A common graphics format would involve the command

```
print -djpeg fig.jpg
```

which will print the current figure as a jpeg file named **fig.jpg**. Note that you can also print or save from the figure window by pressing the **file** button and following the links to the print or export option, respectively. A list of some of the common figure export options can be found in Table 1.6. As with saving according to a loop variable name, batch printing can also be easily performed. The command structure

```
for j=1:5
    (plot expressions)
    print ('-djpeg',['figure' num2str(j) '.jpg'])
end
```

prints the plots generated in the *j* loop as **figure1.jpg**, **figure2.jpg**, etc. This is a very useful command for generating batch printing.

Linear Systems

2

The solution of linear systems is one of the most basic aspects of computational science. In many applications, the solution technique often gives rise to a system of linear equations which need to be solved as efficiently as possible. In addition to Gaussian elimination, there are a host of other techniques which can be used to solve a given problem. This chapter offers an overview of these methods and techniques. Ultimately the goal is to solve large linear systems of equations as quickly as possible. Thus consideration of the operation count is critical.

2.1   Direct Solution Methods for Ax = b

A central concern in almost any computational strategy is a fast and efficient computational method for achieving a solution of a large system of equations $\mathbf{Ax} = \mathbf{b}$. In trying to render a computation tractable, it is crucial to minimize the operations it takes in solving such a system. There are a variety of direct methods for solving $\mathbf{Ax} = \mathbf{b}$: Gaussian elimination, LU decomposition and inverting the matrix $\mathbf{A}$. In addition to these direct methods, iterative schemes can also provide efficient solution techniques. Some basic iterative schemes will be discussed in what follows.

The standard beginning to discussions of solution techniques for $\mathbf{Ax} = \mathbf{b}$ involves Gaussian elimination. We will consider a very simple example of a 3×3 system in order to understand the operation count and numerical procedure involved in this technique. As a specific example, consider the three equations and three unknowns

$$x_1 + x_2 + x_3 = 1 \tag{2.1.1a}$$

$$x_1 + 2x_2 + 4x_3 = -1 \tag{2.1.1b}$$

$$x_1 + 3x_2 + 9x_3 = 1. \tag{2.1.1c}$$

In matrix algebra form, we can rewrite this as $\mathbf{Ax} = \mathbf{b}$ with

$$\mathbf{A} = \begin{pmatrix} 1 & 1 & 1 \\ 1 & 2 & 4 \\ 1 & 3 & 9 \end{pmatrix} \qquad \mathbf{x} = \begin{pmatrix} x \\ y \\ z \end{pmatrix} \qquad \mathbf{b} = \begin{pmatrix} 1 \\ -1 \\ 1 \end{pmatrix}. \qquad (2.1.2)$$

The Gaussian elimination procedure begins with the construction of the augmented matrix

$$[\mathbf{A}|\mathbf{b}] = \begin{bmatrix} \mathbf{1} & 1 & 1 & 1 \\ 1 & 2 & 4 & -1 \\ 1 & 3 & 9 & 1 \end{bmatrix}$$

$$= \begin{bmatrix} 1 & 1 & 1 & 1 \\ 0 & 1 & 3 & -2 \\ 0 & 2 & 8 & 0 \end{bmatrix}$$

$$= \begin{bmatrix} 1 & 1 & 1 & 1 \\ 0 & \mathbf{1} & 3 & -2 \\ 0 & 1 & 4 & 0 \end{bmatrix}$$

$$= \begin{bmatrix} 1 & 1 & 1 & 1 \\ 0 & 1 & 3 & -2 \\ 0 & 0 & 1 & 2 \end{bmatrix} \qquad (2.1.3)$$

where we have underlined and bolded the pivot of the augmented matrix. Back-substituting then gives the solution

$$x_3 = 2 \qquad \rightarrow \quad x_3 = 2 \qquad (2.1.4a)$$

$$x_2 + 3x_3 = -2 \rightarrow \quad x_2 = -8 \qquad (2.1.4b)$$

$$x_1 + x_2 + x_3 = 1 \rightarrow \quad x_1 = 7. \qquad (2.1.4c)$$

This procedure can be carried out for any matrix $\mathbf{A}$ which is nonsingular, i.e. $\det \mathbf{A} \neq 0$. In this algorithm, we simply need to avoid these singular matrices and occasionally shift the rows to avoid a zero pivot. Provided we do this, it will always yield a unique answer.

In scientific computing, the fact that this algorithm works is secondary to the concern of the time required in generating a solution. The operation count for the Gaussian elimination can easily be estimated from the algorithmic procedure for an $N \times N$ matrix:

1   Movement down the N pivots.

2   For each pivot, perform N additions/subtractions across the columns.

3   For each pivot, perform the addition/subtraction down the N rows.

In total, this results in a scheme whose operation count is $O(N^3)$. The back-substitution algorithm can similarly be calculated to give an $O(N^2)$ scheme.

LU decomposition

Each Gaussian elimination operation costs $O(N^3)$ operations. This can be computationally prohibitive for large matrices when repeated solutions of $\mathbf{Ax} = \mathbf{b}$ must be found. When working with the same matrix $\mathbf{A}$, however, the operation count can easily be brought down to $O(N^2)$ using LU factorization which splits the matrix $\mathbf{A}$ into a lower triangular matrix L, and an upper triangular matrix U. For a 3×3 matrix, the LU factorization scheme splits $\mathbf{A}$ as follows:

$$\mathbf{A} = \mathbf{LU} \rightarrow \begin{pmatrix} a_{11} & a_{12} & a_{13} \\ a_{21} & a_{22} & a_{23} \\ a_{31} & a_{32} & a_{33} \end{pmatrix} = \begin{pmatrix} 1 & 0 & 0 \\ m_{21} & 1 & 0 \\ m_{31} & m_{32} & 1 \end{pmatrix} \begin{pmatrix} u_{11} & u_{12} & u_{13} \\ 0 & u_{22} & u_{23} \\ 0 & 0 & u_{33} \end{pmatrix}. \tag{2.1.5}$$

Thus the **L** matrix is lower triangular and the **U** matrix is upper triangular. This then gives

$$\mathbf{Ax} = \mathbf{b} \rightarrow \mathbf{LUx} = \mathbf{b} \tag{2.1.6}$$

where by letting $\mathbf{y} = \mathbf{Ux}$ we find the coupled system

$$\mathbf{Ly} = \mathbf{b} \quad \text{and} \quad \mathbf{Ux} = \mathbf{y}. \tag{2.1.7}$$

The system $\mathbf{Ly} = \mathbf{b}$

$$y_1 = b_1 \tag{2.1.8a}$$

$$m_{21}y_1 + y_2 = b_2 \tag{2.1.8b}$$

$$m_{31}y_1 + m_{32}y_2 + y_3 = b_3 \tag{2.1.8c}$$

can be solved by $O(N^2)$ forward-substitution and the system $\mathbf{Ux} = \mathbf{y}$

$$u_{11}x_1 + u_{12}x_2 + u_{13}x_3 = y_1 \tag{2.1.9a}$$

$$u_{22}x_2 + u_{23}x_3 = y_2 \tag{2.1.9b}$$

$$u_{33}x_3 = y_3 \tag{2.1.9c}$$

can be solved by $O(N^2)$ back-substitution. Thus once the factorization is accomplished, the LU results in an $O(N^2)$ scheme for arriving at the solution. The factorization itself is $O(N^3)$, but you only have to do this once. Note, you should **always** use LU decomposition if possible. Otherwise, you are doing far more work than necessary in achieving a solution.

As an example of the application of the LU factorization algorithm, we consider the 3×3 matrix

$$\mathbf{A} = \begin{pmatrix} 4 & 3 & -1 \\ -2 & -4 & 5 \\ 1 & 2 & 6 \end{pmatrix}. \tag{2.1.10}$$

The factorization starts from the matrix multiplication of the matrix $\mathbf{A}$ and the identity matrix $\mathbf{I}$

$$\mathbf{A} = \mathbf{IA} = \begin{pmatrix} 1 & 0 & 0 \\ 0 & 1 & 0 \\ 0 & 0 & 1 \end{pmatrix} \begin{pmatrix} \mathbf{4} & 3 & -1 \\ -2 & -4 & 5 \\ 1 & 2 & 6 \end{pmatrix}. \tag{2.1.11}$$

The factorization begins with the pivot element. To use Gaussian elimination, we would multiply the pivot by $-1/2$ to eliminate the first column element in the second row. Similarly, we would multiply the pivot by $1/4$ to eliminate the first column element in the third row. These multiplicative factors are now part of the first matrix above:

$$\mathbf{A} = \begin{pmatrix} 1 & 0 & 0 \\ -1/2 & 1 & 0 \\ 1/4 & 0 & 1 \end{pmatrix} \begin{pmatrix} 4 & 3 & -1 \\ 0 & \underline{-2.5} & 4.5 \\ 0 & \underline{1.25} & 6.25 \end{pmatrix}. \tag{2.1.12}$$

To eliminate on the third row, we use the next pivot. This requires that we multiply by $-1/2$ in order to eliminate the second column, third row. Thus we find

$$\mathbf{A} = \begin{pmatrix} 1 & 0 & 0 \\ -1/2 & 1 & 0 \\ 1/4 & -1/2 & 1 \end{pmatrix} \begin{pmatrix} 4 & 3 & -1 \\ 0 & -2.5 & 4.5 \\ 0 & 0 & 8.5 \end{pmatrix}. \tag{2.1.13}$$

Thus we find that

$$\mathbf{L} = \begin{pmatrix} 1 & 0 & 0 \\ -1/2 & 1 & 0 \\ 1/4 & -1/2 & 1 \end{pmatrix} \quad \text{and} \quad \mathbf{U} = \begin{pmatrix} 4 & 3 & -1 \\ 0 & -2.5 & 4.5 \\ 0 & 0 & 8.5 \end{pmatrix}. \tag{2.1.14}$$

It is easy to verify by direct multiplication that indeed $\mathbf{A} = \mathbf{LU}$. Just like Gaussian elimination, the cost of factorization is $O(N^3)$. However, once $\mathbf{L}$ and $\mathbf{U}$ are known, finding the solution is an $O(N^2)$ operation.

The permutation matrix

As will often happen with Gaussian elimination, following the above algorithm will at times result in a zero pivot. This is easily handled in Gaussian elimination by shifting rows in order to find a nonzero pivot. However, in LU decomposition, we must keep track of this row shift since it will effect the right-hand side vector $\mathbf{b}$. We can keep track of row shifts with a row permutation matrix $\mathbf{P}$. Thus if we need to permute two rows, we find

$$\mathbf{Ax} = \mathbf{b} \rightarrow \mathbf{PAx} = \mathbf{Pb} \rightarrow \mathbf{LUx} = \mathbf{Pb} \tag{2.1.15}$$

thus $\mathbf{PA} = \mathbf{LU}$ where $\mathbf{PA}$ no longer has any zero pivots. To shift rows one and two, for instance, we would have

$$\mathbf{P} = \begin{pmatrix} 0 & 1 & 0 & \cdots \\ 1 & 0 & 0 & \cdots \\ 0 & 0 & 1 & \cdots \\ \vdots & & & \end{pmatrix}. \tag{2.1.16}$$

Thus the permutation matrix starts with the identity matrix. If we need to shift rows j and k, then we shift these corresponding rows in the permutation matrix $\mathbf{P}$. If permutation is necessary, MATLAB can supply the permutation matrix associated with the LU decomposition.

MATLAB: **A** \ **b**

Given the alternatives for solving the linear system $\mathbf{Ax} = \mathbf{b}$, it is important to know how the MATLAB command structure for $\mathbf{A} \setminus \mathbf{b}$ works. Indeed, it is critical to note that MATLAB first attempts to use the most efficient algorithm to solve a given matrix problem. The following is an outline of the algorithm performed.

1   It first checks to see if $\mathbf{A}$ is triangular, or some permutation thereof. If it is, then all that is needed is a simple $O(N^2)$ substitution routine.

2   It then checks if $\mathbf{A}$ is symmetric, i.e. Hermitian or self-adjoint. If so, a Cholesky factorization is attempted. If $\mathbf{A}$ is positive definite, the Cholesky algorithm is always succesful and takes half the run time of LU factorization.

3   It then checks if $\mathbf{A}$ is Hessenberg. If so, it can be written as an upper triangular matrix and solved by a substitution routine.

4   If all the above methods fail, then LU factorization is used and the forward- and backward-substitution routines generate a solution.

5   If $\mathbf{A}$ is not square, a QR (Householder) routine is used to solve the system.

6   If $\mathbf{A}$ is not square and sparse, a least-squares solution using QR factorization is performed.

Note that solving by $\mathbf{x} = \mathbf{A}^{-1}\mathbf{b}$ is the slowest of all methods, taking 2.5 times longer or more than $\mathbf{A}\setminus\mathbf{b}$. It is not recommended. However, just like LU factorization, once the inverse is known for a given matrix $\mathbf{A}$, it need not be calculated again since the inverse does not change as $\mathbf{b}$ changes. Care must also be taken when det $\mathbf{A} \approx 0$, i.e. the matrix is ill-conditioned. In this case, the determination of the inverse matrix $\mathbf{A}^{-1}$ becomes suspect.

MATLAB commands

The commands for executing the linear system solve are as follows

▪   **A****b:** Solve the system in the order above.

▪ $[L, U] = lu(A)$**:** Generate the L and U matrices.

▪ $[L, U, P] = lu(A)$**:** Generate the L and U factorization matrices along with the permutation matrix P.

2.2   Iterative Solution Methods for Ax = b

In addition to the standard techniques of Gaussian elimination or LU decomposition for solving $\mathbf{Ax} = \mathbf{b}$, a wide range of iterative techniques is available. These iterative techniques can often go under the name of Krylov space methods [1]. The idea is to start with an initial guess for the solution and develop an iterative procedure that will converge to the solution. The simplest example of this method is known as a Jacobi iteration scheme. The implementation of this scheme is best illustrated with an example. We consider the linear system

$$4x - y + z = 7 \tag{2.2.1a}$$

$$4x - 8y + z = -21 \tag{2.2.1b}$$

$$-2x + y + 5z = 15. \tag{2.2.1c}$$

We can rewrite each equation as follows

$$x = \frac{7 + y - z}{4} \tag{2.2.2a}$$

$$y = \frac{21 + 4x + z}{8} \tag{2.2.2b}$$

$$z = \frac{15 + 2x - y}{5}. \tag{2.2.2c}$$

To solve the system iteratively, we can define the following Jacobi iteration scheme based on the above

$$x_{k+1} = \frac{7 + y_k - z_k}{4} \tag{2.2.3a}$$

$$y_{k+1} = \frac{21 + 4x_k + z_k}{8} \tag{2.2.3b}$$

$$z_{k+1} = \frac{15 + 2x_k - y_k}{5}. \tag{2.2.3c}$$

An algorithm is then easily implemented computationally. In particular, we would follow the structure:

1 Guess initial values: (x_0, y_0, z_0).
2 Iterate the Jacobi scheme: $\mathbf{x}_{k+1} = \mathbf{D}^{-1}((\mathbf{D} - \mathbf{A})\mathbf{x}_k + \mathbf{b})$.
3 Check for convergence: $\| \mathbf{x}_{k+1} - \mathbf{x}_k \| <$tolerance.

Note that the choice of an initial guess is often critical in determining the convergence to the solution. Thus the more that is known about what the solution is supposed to look like, the higher the chance of successful implementation of the iterative scheme. Although there is no reason *a priori* to believe this iteration scheme would converge to the solution of the corresponding system $\mathbf{Ax} = \mathbf{b}$, Table 2.1 shows that indeed this scheme convergences remarkably quickly to the solution for this simple example.

Given the success of this example, it is easy to conjecture that such a scheme will always be effective. However, we can reconsider the original system by interchanging the first and last set of equations. This gives the system

$$-2x + y + 5z = 15 \tag{2.2.4a}$$

$$4x - 8y + z = -21 \tag{2.2.4b}$$

$$4x - y + z = 7. \tag{2.2.4c}$$

Table 2.1   Convergence of the Jacobi iteration scheme to the solution value of $(x, y, z) = (2, 4, 3)$ from the initial guess $(x_0, y_0, z_0) = (1, 2, 2)$

k	x_k	y_k	z_k
0	1.0	2.0	2.0
1	1.75	3.375	3.0
2	1.84375	3.875	3.025
⋮	⋮	⋮	⋮
15	1.99999993	3.99999985	2.9999993
⋮	⋮	⋮	⋮
19	2.0	4.0	3.0

To solve the system iteratively, we can define the following Jacobi iteration scheme based on this rearranged set of equations

$$x_{k+1} = \frac{y_k + 5z_k - 15}{2} \tag{2.2.5a}$$

$$y_{k+1} = \frac{21 + 4x_k + z_k}{8} \tag{2.2.5b}$$

$$z_{k+1} = y_k - 4x_k + 7. \tag{2.2.5c}$$

Of course, the solution should be exactly as before. However, Table 2.2 shows that applying the iteration scheme leads to a set of values which grow to infinity. Thus the iteration scheme quickly fails.

Strictly diagonal dominant

The difference in the two Jacobi schemes above involves the iteration procedure being strictly diagonal dominant. We begin with the definition of strict diagonal dominance. A matrix **A** is

Table 2.2   Divergence of the Jacobi iteration scheme from the initial guess $(x_0, y_0, z_0) = (1, 2, 2)$

k	x_k	y_k	z_k
0	1.0	2.0	2.0
1	−1.5	3.375	5.0
2	6.6875	2.5	16.375
3	34.6875	8.015625	−17.25
⋮	⋮	⋮	⋮
	±∞	±∞	±∞

strictly diagonal dominant if for each row, the sum of the absolute values of the off-diagonal terms is less than the absolute value of the diagonal term:

$$|a_{kk}| > \sum_{j=1, j \neq k}^{N} |a_{kj}|. \tag{2.2.6}$$

Strict diagonal dominance has the following consequence; given a strictly diagonal dominant matrix $\mathbf{A}$, then an iterative scheme for $\mathbf{Ax} = \mathbf{b}$ converges to a unique solution $\mathbf{x} = \mathbf{p}$. Jacobi iteration produces a sequence $\mathbf{p}_k$ that will converge to $\mathbf{p}$ for any $\mathbf{p}_0$. For the two examples considered here, this property is crucial. For the first example (2.2.1), we have

$$\mathbf{A} = \begin{pmatrix} 4 & -1 & 1 \\ 4 & -8 & 1 \\ -2 & 1 & 5 \end{pmatrix} \rightarrow \begin{matrix} \text{row 1: } |4| > |-1| + |1| = 2 \\ \text{row 2: } |-8| > |4| + |1| = 5 \\ \text{row 3: } |5| > |2| + |1| = 3, \end{matrix} \tag{2.2.7}$$

which shows the system to be strictly diagonal dominant and guaranteed to converge. In contrast, the second system (2.2.4) is not stricly diagonal dominant as can be seen from

$$\mathbf{A} = \begin{pmatrix} -2 & 1 & 5 \\ 4 & -8 & 1 \\ 4 & -1 & 1 \end{pmatrix} \rightarrow \begin{matrix} \text{row 1: } |-2| < |1| + |5| = 6 \\ \text{row 2: } |-8| > |4| + |1| = 5 \\ \text{row 3: } |1| < |4| + |-1| = 5. \end{matrix} \tag{2.2.8}$$

Thus this scheme is not guaranteed to converge. Indeed, it diverges to infinity.

Modification and enhancements: Gauss–Seidel

It is sometimes possible to enhance the convergence of a scheme by applying modifications to the basic Jacobi scheme. For instance, the Jacobi scheme given by (2.2.3) can be enhanced by the following modifications

$$x_{k+1} = \frac{7 + y_k - z_k}{4} \tag{2.2.9a}$$

$$y_{k+1} = \frac{21 + 4x_{k+1} + z_k}{8} \tag{2.2.9b}$$

$$z_{k+1} = \frac{15 + 2x_{k+1} - y_{k+1}}{5}. \tag{2.2.9c}$$

Here use is made of the supposedly improved value x_{k+1} in the second equation and x_{k+1} and y_{k+1} in the third equation. This is known as the Gauss–Seidel scheme. Table 2.3 shows that the Gauss–Seidel procedure converges to the solution in half the number of iterations used by the Jacobi scheme in later sections.

Like the Jacobi scheme, the Gauss–Seidel method is guaranteed to converge in the case of strict diagonal dominance. Further, the Gauss–Seidel modification is only one of a large number of possible changes to the iteration scheme which can be implemented in an effort to enhance convergence. It is also possible to use several previous iterations to achieve convergence. Krylov space methods [1] are often high-end iterative techniques especially developed for rapid convergence.

Table 2.3 Convergence of the Jacobi iteration scheme to the solution value of $(x, y, z) = (2, 4, 3)$ from the initial guess $(x_0, y_0, z_0) = (1, 2, 2)$

k	x_k	y_k	z_k
0	1.0	2.0	2.0
1	1.75	3.75	2.95
2	1.95	3.96875	2.98625
$\vdots$	$\vdots$	$\vdots$	$\vdots$
10	2.0	4.0	3.0

Included in these iteration schemes are conjugate gradient methods and generalized minimum residual methods which we will discuss and implement [1].

2.3   Gradient (Steepest) Descent for Ax = b

The Jacobi and Gauss–Seidel iteration schemes are only two potential iteration schemes that can be developed for solving $\mathbf{Ax} = \mathbf{b}$. But when it comes to high-performance computing and solving extremely large systems of equations, more efficient algorithms are needed to overcome the computational costs of the methods discussed so far. In fact, modern iterative techniques are perhaps the most critical algorithms developed to date for solving large systems which would otherwise be computationally intractable. In this section, the gradient descent, or steepest descent, algorithm is developed. It illustrates how one might engineer an algorithm that continually looks to iterate, in an efficient and intuitive manner, towards the solution of the linear problem $\mathbf{Ax} = \mathbf{b}$. Moreover, it hints at the kind of underlying algorithms and techniques that dominate many high-performance computing applications today including the bi-conjugate gradient descent method (**bicgstab**) and generalized method of residuals (**gmres**) mentioned in the last section.

To begin, we will consider the concept of the *quadratic form*

$$f(\mathbf{x}) = \frac{1}{2}\mathbf{x}^T\mathbf{Ax} - \mathbf{b}^T\mathbf{x} + \mathbf{c} \tag{2.3.1}$$

where $\mathbf{A}$ is a matrix, $\mathbf{b}$ and $\mathbf{x}$ are vectors, and c is a scalar constant. Note that the quadratic form produces a scalar value $f(\mathbf{x})$.

The gradient of the quadratic form is defined as

$$\nabla f(\mathbf{x}) = \begin{bmatrix} \frac{\partial}{\partial x_1} f(\mathbf{x}) \\ \frac{\partial}{\partial x_2} f(\mathbf{x}) \\ \vdots \\ \frac{\partial}{\partial x_n} f(\mathbf{x}) \end{bmatrix}. \tag{2.3.2}$$

The gradient is a standard quantity from vector calculus that produces a vector field that points in the direction of greatest increase of the function $f(\mathbf{x})$.

The gradient of the function (2.3.1) can be easily computed to be

$$\nabla f(\mathbf{x}) = \frac{1}{2}\mathbf{A}^T\mathbf{x} + \frac{1}{2}\mathbf{A}\mathbf{x} - \mathbf{b}. \tag{2.3.3}$$

If the matrix $\mathbf{A}$ is symmetric, then $\mathbf{A}^T = \mathbf{A}$ and the gradient has a critical point (maximum or minimum) when $\nabla f = 0$ and $\mathbf{A}\mathbf{x} = \mathbf{b}$:

$$\nabla f(\mathbf{x}) = \mathbf{A}\mathbf{x} - \mathbf{b} = 0. \tag{2.3.4}$$

Thus the solution to $\mathbf{A}\mathbf{x} = \mathbf{b}$ is a critical point of the gradient. If, in addition, the matrix $\mathbf{A}$ is *positive definite* ($\mathbf{A}^T\mathbf{x} > 0$), then the solution is a *minimum* of $f(\mathbf{x})$. These observations concerning positive definite and symmetric matrices are critical in the development of the gradient descent technique, i.e. a technique that will try to take advantage of knowing the greatest direction of increase (decrease) of the quadratic form. If $\mathbf{A}$ is not symmetric, then the gradient descent algorithm finds the solution to the linear system $(1/2)(\mathbf{A}^T + \mathbf{A})\mathbf{x} = \mathbf{b}$.

To develop an intuitive feel for the gradient descent technique and the importance of positive definite matrices, consider the following illustrative example system that is both positive definite and symmetric [2]

$$\mathbf{A} = \begin{bmatrix} 3 & 2 \\ 2 & 6 \end{bmatrix}, \quad \mathbf{b} = \begin{bmatrix} 2 \\ -8 \end{bmatrix}, \quad c = 0. \tag{2.3.5}$$

With this $\mathbf{A}$ and $\mathbf{b}$, solving for $\mathbf{x}$ is a fairly trivial exercise which yields the solution $\mathbf{x} = [x_1 \ x_2]^T = [2 \ -2]^T$. Our objective is to develop an iterative scheme which moves towards this solution in an efficient manner given an initial guess. First, we calculate the quadratic form for this case to be

$$f(\mathbf{x}) = \frac{1}{2}[x_1 \ x_2]\begin{bmatrix} 3 & 2 \\ 2 & 6 \end{bmatrix}\begin{bmatrix} x_1 \\ x_2 \end{bmatrix} - [2 \ -8]\begin{bmatrix} x_1 \\ x_2 \end{bmatrix} = \frac{3}{2}x_1^2 + 2x_1x_2 + 3x_2^2 - 2x_1 + 8x_2 \tag{2.3.6}$$

with the gradient

$$\nabla f(\mathbf{x}) = \begin{bmatrix} 3x_1 + 2x_2 - 2 \\ 2x_1 + 6x_2 + 8 \end{bmatrix}. \tag{2.3.7}$$

Figure 2.1 shows the quadratic form surface $f(\mathbf{x})$ along with its contour plot and its gradient. The solution to $\mathbf{A}\mathbf{x} = \mathbf{b}$ is shown as the circle at $\mathbf{x} = [2 \ -2]^T$. As a result of the matrix being positive definite, the surface is seen to be a paraboloid whose minimum is at the desired solution. Thus the idea will be use to construct an algorithm that marches to the bottom of the paraboloid in an efficient manner. If the matrix is not positive definite, then the gradient algorithm is unlikely to work. Figure 2.2 shows four examples that are important to consider: a positive definite matrix, a negative definite matrix, an indefinite (saddle) matrix and a singular matrix. For the negative definite matrix, the gradient descent algorithm would *fall away* from the maximum point. Similarly for the saddle or singular matrix, the algorithm would either converge to a line of solutions (singular case), or drop away along the saddle.

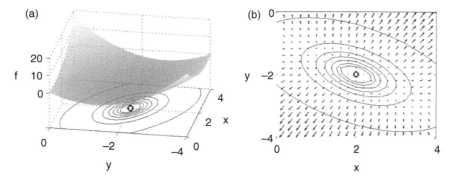

Figure 2.1: Graphical depiction of the gradient descent algorithm. The quadratic form surface $f = (3/2)x_1^2 + 2x_1x_2 + 3x_2^2 - 2x_1 + 8x_2$ is plotted along with its contour lines. In (a), the surface is plotted along with the contour plot beneath. The objective of gradient descent is to find the bottom of the paraboloid. In (b), the gradient is calculated and shown with a quiver plot.

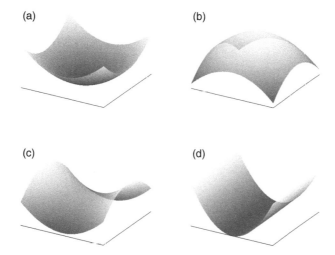

Figure 2.2: Four characteristic matrix forms: (a) positive definite, (b) negative definite, (c) saddle and (d) singular matrix. Intuition would suggest that a descent algorithm will only work for a positive definite matrix.

Given this graphical interpretation of the gradient descent algorithm, one can see how it may generalize to higher dimensions and larger systems. In particular, for a high dimensional system, one would construct a hyper-paraboloid for a positive definite matrix where the algorithm would force, in an optimal way, the solution to the bottom of the paraboloid and the solution. The positive definite restriction is similar to the Jacobi algorithm only being guaranteed to work for strictly diagonal dominant matrices.

To build an algorithm that takes advantage of the gradient, an initial guess point is first given and the gradient $\nabla f(\mathbf{x})$ computed. This gives the steepest descent towards the minimum point of $f(\mathbf{x})$, i.e. the minimum is located in the direction given by $-\nabla f(\mathbf{x})$. Note that the gradient does not point at the minimum, but rather gives the steepest path for minimizing $f(\mathbf{x})$. The geometry of the steepest descent suggests the construction of an algorithm whereby the next point of iteration is picked by following the steepest descent so that

$$\xi(\tau) = \mathbf{x} - \tau \nabla f(\mathbf{x}) \qquad (2.3.8)$$

where the parameter τ dictates how far to move along the gradient descent curve. In gradient descent, it is crucial to determine when this bottom is reached so that the algorithm is always *going downhill* in an optimal way. This requires the determination of the correct value of τ in the algorithm.

To compute the value of τ, consider the construction of a new function

$$F(\tau) = f(\xi(\tau)) \qquad (2.3.9)$$

which must be minimized now as a function of τ. This is accomplished by computing $dF/d\tau = 0$. Thus one finds

$$\frac{\partial F}{\partial \tau} = -\nabla f(\xi) \nabla f(\mathbf{x}) = 0. \qquad (2.3.10)$$

The geometrical interpretation of this result is the following: $\nabla f(\mathbf{x})$ is the gradient direction of the current iteration point and $\nabla f(\xi)$ is the gradient direction of the future point, thus τ is chosen so that the two gradient directions are orthogonal.

The following MATLAB code performs a gradient descent search on the linear problem given by (2.3.5). Note that we have already calculated the quadratic form to be minimized along with its gradient.

```
x(1)=1; y(1)=-0.2;  % initial guess and function value
f(1)=(3/2)*x(1)^2+2*x(1)*y(1)+3*y(1)^2-2*x(1)+8*y(1);

for j=1:100  % iteration loop
   tau=fminsearch('tausearch',0.2,[],x(end),y(end));  % optimal tau
   x(j+1)=x(j)-tau*(3*x(j)+2*y(j)-2);   % update x, y, and f
   y(j+1)=y(j)-tau*(2*x(j)+6*y(j)+8);
   f(j+1)=(3/2)*x(j+1)^2+2*x(j+1)*y(j+1)+3*y(j+1)^2-2*x(j+1)+8*y(j+1);

   if abs(f(j+1)-f(j))<10^(-6)   % check convergence
      break
   end
end
```

Note that an initial guess of $\tau = 0.2$ is used in **fminsearch**. This function calls on the following subroutine that is minimized as a function of τ.

tausearch.m

```
function mintau=tausearch(tau,x,y)
x0=x-tau*(3*x+2*y-2); y0=y-tau*(2*x+6*y+8);
mintau= (3/2)*x0^2+2*x0*y0+3*y0^2-2*x0+8*y0;
```

Figure 2.3(a) shows how this algorithm converges to the solution of $\mathbf{Ax} = \mathbf{b}$ for the values of τ computed from (2.3.10). Indeed, the algorithm converges quickly (11 iterations are required) to the correct value of $\mathbf{x} = [2 \ -2]^T$. Note that if a simple radially symmetric function is considered, then the gradient descent converges in a single iteration since the gradient descent would point directly at the minimum. This figure assumes a line search algorithm to find an optimal value of τ. In particular, the value of τ picked here is optimal in the sense that a given line search is conducted so that the minimum of the gradient direction is picked as the next iteration point. However, this is not a requirement. In fact, one can simply choose a fixed value of τ for stepping forward along the gradient direction. Figure 2.3(b) demonstrates this case for $\tau = 0.1$. This method also converges to the solution, however at a much slower rate. Such a method may be favorable in a case where the steepest descent algorithm *zig-zags* a large amount in trying to make the projective steps orthogonal. This can happen in cases where *long-valley* type structures exist in the function we are trying to minimize.

Gradient descent is a first example of how to use a quadratic form along with a derivate (gradient) for building an iterative scheme for solving $\mathbf{Ax} = \mathbf{b}$. It can also be used as the basis for an optimization routine. As might be guessed, there are a myriad of improvements to such a scheme. Indeed, the gradient descent algorithm is the core algorithm of advanced iterative solvers such as the bi-conjugate gradient descent method (**bicgstab**) and generalized method of residuals (**gmres**), both of which are essentially improvements on the gradient descent considered here. For

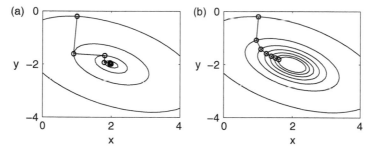

Figure 2.3: Gradient descent algorithm applied to the function (quadratic form) $f = (3/2)x_1^2 + 2x_1x_2 + 3x_2^2 - 2x_1 + 8x_2$. In both panels, the contours are plotted for each successive value (x, y) in the iteration algorithm given the initial guess $(x, y) = (1, -0.2)$. For (a), the optimal τ value is calculated so that each successive gradient in the steepest descent algorithm is orthogonal to the previous step. In (b), a prescribed value of $\tau = 0.1$ is used. In this case, the iteration still converges to the solution, but at a much slower rate since the descent steps are not chosen in an optimal way.

a more detailed analysis of these schemes, please see the excellent overview by J. R. Shewchuk [2]. The implementation of **gmres** and **bicgstab** in MATLAB is mentioned in the previous section.

 ## 2.4 Eigenvalues, Eigenvectors and Solvability

Another class of linear systems of equations which are of fundamental importance are known as eigenvalue problems. Unlike the system $\mathbf{Ax} = \mathbf{b}$ which has the single unknown vector $\vec{x}$, eigenvalue problems are of the form

$$\mathbf{Ax} = \lambda\mathbf{x} \tag{2.4.1}$$

which have the unknowns $\mathbf{x}$ and λ. The values of λ are known as the *eigenvalues* and the corresponding $\mathbf{x}$ are the *eigenvectors*.

Eigenvalue problems often arise from differential equations. Specifically, we consider the example of a linear set of coupled differential equations

$$\frac{d\mathbf{y}}{dt} = \mathbf{Ay}. \tag{2.4.2}$$

By attempting a solution of the form

$$\mathbf{y}(t) = \mathbf{x}\exp(\lambda t), \tag{2.4.3}$$

where all the time-dependence is captured in the exponent, the resulting equation for $\mathbf{x}$ is

$$\mathbf{Ax} = \lambda\mathbf{x} \tag{2.4.4}$$

which is just the eigenvalue problem. Once the full set of eigenvalues and eigenvectors of this equation is found, the solution of the differential equation is written as the linear superposition

$$\vec{y} = c_1\mathbf{x}_1\exp(\lambda_1 t) + c_2\mathbf{x}_2\exp(\lambda_2 t) + \cdots + c_N\mathbf{x}_N\exp(\lambda_N t) \tag{2.4.5}$$

where N is the number of linearly independent solutions to the eigenvalue problem for the matrix $\mathbf{A}$ which is of size $N \times N$. Thus solving a linear system of differential equations relies on the solution of an associated eigenvalue problem.

The question remains: how are the eigenvalues and eigenvectors found? To consider this problem, we rewrite the eigenvalue problem as

$$\mathbf{Ax} = \lambda\mathbf{I}x \tag{2.4.6}$$

where a multiplication by unity has been performed, i.e. $\mathbf{I}x = x$ where $\mathbf{I}$ is the identity matrix. Moving the right-hand side to the left side of the equation gives

$$\mathbf{Ax} - \lambda\mathbf{I}x = 0. \tag{2.4.7}$$

Factoring out the vector **x** then gives the desired result

$$(\mathbf{A} - \lambda\mathbf{I})\mathbf{x} = \mathbf{0}. \tag{2.4.8}$$

Two possibilities now exist.

Option I

The determinant of the matrix $(\mathbf{A} - \lambda\mathbf{I})$ is not zero. If this is true, the matrix is *nonsingular* and its inverse, $(\mathbf{A} - \lambda\mathbf{I})^{-1}$, exists. The solution to the eigenvalue problem (2.4.8) is then

$$\mathbf{x} = (\mathbf{A} - \lambda\mathbf{I})^{-1}\mathbf{0} \tag{2.4.9}$$

which implies that

$$\mathbf{x} = \mathbf{0}. \tag{2.4.10}$$

This trivial solution could have been guessed from (2.4.8). However, it is not relevant as we require nontrivial solutions for **x**.

Option II

The determinant of the matrix $(\mathbf{A} - \lambda\mathbf{I})$ is zero. If this is true, the matrix is *singular* and its inverse, $(\mathbf{A} - \lambda\mathbf{I})^{-1}$, cannot be found. Although there is no longer a guarantee that there is a solution, it is the only scenario which allows for the possibility of $\mathbf{x} \neq \mathbf{0}$. It is this condition which allows for the construction of eigenvalues and eigenvectors. Indeed, we choose the eigenvalues λ so that this condition holds and the matrix is singular.

To illustrate how the eigenvalues and eigenvectors are computed, an example is shown. Consider the 2×2 matrix

$$\mathbf{A} = \begin{pmatrix} 1 & 3 \\ -1 & 5 \end{pmatrix}. \tag{2.4.11}$$

This gives the eigenvalue problem

$$\mathbf{Ax} = \begin{pmatrix} 1 & 3 \\ -1 & 5 \end{pmatrix}\mathbf{x} = \lambda\mathbf{x} \tag{2.4.12}$$

which when manipulated to the form $(\mathbf{A} - \lambda\mathbf{I})\mathbf{x} = \mathbf{0}$ gives

$$\left[\begin{pmatrix} 1 & 3 \\ -1 & 5 \end{pmatrix} - \lambda\begin{pmatrix} 1 & 0 \\ 0 & 1 \end{pmatrix}\right]\mathbf{x} = \begin{pmatrix} 1-\lambda & 3 \\ -1 & 5-\lambda \end{pmatrix}\mathbf{x} = \mathbf{0}. \tag{2.4.13}$$

We now require that the determinant is zero:

$$\det\begin{vmatrix} 1-\lambda & 3 \\ -1 & 5-\lambda \end{vmatrix} = (1-\lambda)(5-\lambda) + 3 = \lambda^2 - 6\lambda + 8 = (\lambda-2)(\lambda-4) = 0. \tag{2.4.14}$$

The *characteristic equation* for determining λ in the 2×2 case reduces to finding the roots of a quadratic equation. Specifically, this gives the two eigenvalues

$$\lambda = 2, 4. \tag{2.4.15}$$

For an $N \times N$ matrix, the characteristic equation is an N degree polynomial that has N roots.

The eigenvectors are then found from (2.4.13) as follows:

$$\lambda = 2 : \quad \begin{pmatrix} 1-2 & 3 \\ -1 & 5-2 \end{pmatrix} \mathbf{x} = \begin{pmatrix} -1 & 3 \\ -1 & 3 \end{pmatrix} \mathbf{x} = 0. \tag{2.4.16}$$

Given that $\mathbf{x} = (x_1 \ x_2)^T$, this leads to the single equation

$$-x_1 + 3x_2 = 0. \tag{2.4.17}$$

This is an underdetermined system of equations. Thus we have freedom in choosing one of the values. Choosing $x_2 = 1$ gives $x_1 = 3$ and determines the first eigenvector to be

$$\mathbf{x}_1 = \begin{pmatrix} 3 \\ 1 \end{pmatrix}. \tag{2.4.18}$$

The second eigenvector comes from (2.4.13) as follows:

$$\lambda = 4 : \quad \begin{pmatrix} 1-4 & 3 \\ -1 & 5-4 \end{pmatrix} \mathbf{x} = \begin{pmatrix} -3 & 3 \\ -1 & 1 \end{pmatrix} \mathbf{x} = 0. \tag{2.4.19}$$

Given that $\mathbf{x} = (x_1 \ x_2)^T$, this leads to the single equation

$$-x_1 + x_2 = 0. \tag{2.4.20}$$

This again is an underdetermined system of equations. Thus we have freedom in choosing one of the values. Choosing $x_2 = 1$ gives $x_1 = 1$ and determines the second eigenvector to be

$$\mathbf{x}_2 = \begin{pmatrix} 1 \\ 1 \end{pmatrix}. \tag{2.4.21}$$

These results can be found from MATLAB by using the **eig** command. Specifically, the command structure

```
[V,D]=eig(A)
```

gives the matrix $\mathbf{V}$ containing the eigenvectors as columns and the matrix $\mathbf{D}$ whose diagonal elements are the corresponding eigenvalues.

For very large matrices, it is often only required to find the largest or smallest eigenvalues. This can easily be done with the **eigs** command. Thus to determine the K largest in magnitude eigenvalues, the following command is used

```
[V,D]=eigs(A,K,'LM')
```

Instead of all N eigenvectors and eigenvalues, only the first K will now be returned with the largest magnitude. The option LM denotes the largest magnitude eigenvalues. Table 2.4 lists the various options for searching for specific eigenvalues. Often the smallest, largest or smallest real, largest real eigenvalues are required and the **eigs** command allows you to construct these easily.

Matrix powers

Another important operation which can be performed with eigenvalues and eigenvectors is the evaluation of

$$\mathbf{A}^M \tag{2.4.22}$$

where M is a large integer. For large matrices $\mathbf{A}$, this operation is computationally expensive. However, knowing the eigenvalues and eigenvectors of $\mathbf{A}$ allows for a significant reduction in computational expense. Assuming we have all the eigenvalues and eigenvectors of $\mathbf{A}$, then

$$\begin{aligned}
\mathbf{A}\mathbf{x}_1 &= \lambda_1\mathbf{x}_1 \\
\mathbf{A}\mathbf{x}_2 &= \lambda_2\mathbf{x}_2 \\
&\vdots \\
\mathbf{A}\mathbf{x}_n &= \lambda_n\mathbf{x}_n.
\end{aligned}$$

Table 2.4 Eigenvalue search options for the **eigs** command. For the BE option, it will return one more from the high end if K is odd. If no options are specified, **eigs** will return the largest six (magnitude) eigenvalues

Option	Eigenvalues
LM	largest magnitude
SM	smallest magnitude
LA	largest algebraic
SA	smallest algebraic
LR	largest real
SR	smallest real
LI	largest imaginary
SI	smallest imaginary
BE	both ends

This collection of eigenvalues and eigenvectors gives the matrix system

$$\mathbf{AS} = \mathbf{S\Lambda} \qquad (2.4.23)$$

where the columns of the matrix $\mathbf{S}$ are the eigenvectors of $\mathbf{A}$,

$$\mathbf{S} = (\mathbf{x}_1 \;\; \mathbf{x}_2 \;\; \cdots \;\; \mathbf{x}_n), \qquad (2.4.24)$$

and $\mathbf{\Lambda}$ is a matrix whose diagonals are the corresponding eigenvalues

$$\mathbf{\Lambda} = \begin{pmatrix} \lambda_1 & 0 & \cdots & & 0 \\ 0 & \lambda_2 & 0 & \cdots & 0 \\ \vdots & & \ddots & & \vdots \\ 0 & & \cdots & 0 & \lambda_n \end{pmatrix}. \qquad (2.4.25)$$

By multiplying (2.4.24) on the right by $\mathbf{S}^{-1}$, the matrix $\mathbf{A}$ can then be rewritten as

$$\mathbf{A} = \mathbf{S\Lambda S}^{-1}. \qquad (2.4.26)$$

The final observation comes from

$$\mathbf{A}^2 = (\mathbf{S\Lambda S}^{-1})(\mathbf{S\Lambda S}^{-1}) = \mathbf{S\Lambda}^2\mathbf{S}^{-1}. \qquad (2.4.27)$$

This then generalizes to

$$\mathbf{A}^M = \mathbf{S\Lambda}^M\mathbf{S}^{-1} \qquad (2.4.28)$$

where the matrix $\mathbf{\Lambda}^M$ is easily calculated as

$$\mathbf{\Lambda}^M = \begin{pmatrix} \lambda_1^M & 0 & \cdots & & 0 \\ 0 & \lambda_2^M & 0 & \cdots & 0 \\ \vdots & & \ddots & & \vdots \\ 0 & & \cdots & 0 & \lambda_n^M \end{pmatrix}. \qquad (2.4.29)$$

Since raising the diagonal terms to the Mth power is easily accomplished, the matrix $\mathbf{A}^M$ can then be easily calculated by multiplying the three matrices in (2.4.28).

Solvability and the Fredholm-alternative theorem

It is easy to ask under what conditions the system

$$\mathbf{Ax} = \mathbf{b} \qquad (2.4.30)$$

can be solved. Aside from requiring the $\det \mathbf{A} \neq 0$, we also have a solvability condition on $\mathbf{b}$. Consider the adjoint problem

$$\mathbf{A}^{\dagger}\mathbf{y} = 0 \qquad (2.4.31)$$

where $\mathbf{A}^{\dagger} = \mathbf{A}^{*T}$ is the adjoint which is the transpose and complex conjugate of the matrix $\mathbf{A}$.

The definition of the adjoint is such that

$$\mathbf{y} \cdot \mathbf{A}\mathbf{x} = \mathbf{A}^{\dagger}\mathbf{y} \cdot \mathbf{x}. \qquad (2.4.32)$$

Since $\mathbf{A}\mathbf{x} = \mathbf{b}$, the left side of the equation reduces to $\mathbf{y} \cdot \mathbf{b}$ while the right side reduces to $\mathbf{0}$ since $\mathbf{A}^{\dagger}\mathbf{y} = \mathbf{0}$. This then gives the condition

$$\mathbf{y} \cdot \mathbf{b} = 0 \qquad (2.4.33)$$

which is known as the *Fredholm-alternative* theorem, or a *solvability* condition. In words, the equation states that in order for the system $\mathbf{A}\mathbf{x} = \mathbf{b}$ to be solvable, the right-hand side forcing $\mathbf{b}$ must be orthogonal to the null space of the adjoint operator $\mathbf{A}^{\dagger}$.

2.5   Eigenvalues and Eigenvectors for Face Recognition

Normally, to motivate the concept of what eigenvalues and eigenvectors mean, examples would be given from some physical and/or engineering system. Often, however, the insight given into the physical system assumes that you actually know the workings of the physical system fairly well. Thus eigenvalues and eigenvectors often fail to impress students as they should. As a generic statement, it is difficult to refute the claim that if you know the eigenvectors and eigenvalues of a given system (or matrix), then you know everything there is to know about that system. Given then their importance, some measure of intuition will be developed about their meaning and use.

To motivate the importance of eigenvalues and eigenvectors, they will be considered in the context of the computer vision problem of human face recognition. The approach of using so-called *eigenfaces* for recognition was developed by Sirovich and Kirby [3, 4] and used by Turk and Pentland [5] in face classification. It is considered the first successful example of facial recognition technology. In what follows, some very basic ideas of eigenfaces will be given.

First, consider a set of face images. Figure 2.4 will be our so-called *training set* of images. This will be a highly limited example as we will consider only faces from four different people representing the political, entertainment and sports arenas, namely George Clooney, Barack Obama, Margaret Thatcher and Matt Damon. For each celebrity, five images are considered that are roughly cropped the same way. If you were a computer scientist, you would probably also make sure to subtract the background from the pictures, make sure all faces are looking directly at the camera without a head tilt, have photos with identical (or nomarlized) lighting, crop in a more systematic way, etc. However, this is only a rough guide on how the techniques work and

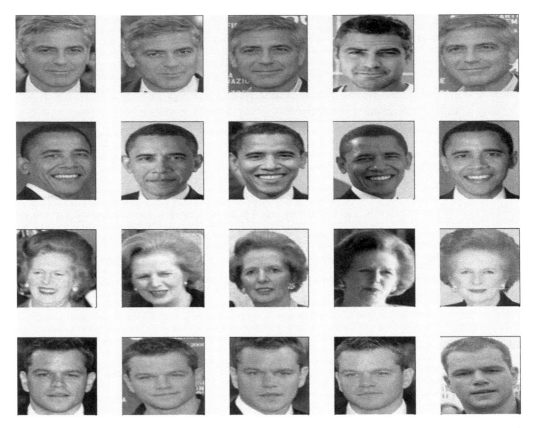

Figure 2.4: Five image samples of head shots of four different faces including George Clooney (top row), Barack Obama (second row), Margaret Thatcher (third row) and Matt Damon (bottom row). Each image was cropped, turned to gray-scale, and resized to 120 vertical and 80 horizontal pixels. All images are public domain, licensed under Creative Commons Attribution-Share Alike 2.0, or licensed under Creative Commons Attribution 3.0.

we will not concern ourselves with all of the extras that can make the eigenface methodology perform better.

To begin, all images are imported, turned into gray-scale, and resized to the same number of pixels in the vertical (120 pixels) and horizontal (80 pixels) directions. The following code takes an image called **pic.jpg**, converts it into gray-scale and turns the image format into double precision numbers for manipulation. Resizing is also accomplished.

```
A=imresize(double(rgb2gray(imread('pic','jpeg'))),[120 80]);
```

This preprocessing is done for each picture to create an image matrix **A**. A convenient way to view the image is with the **pcolor** command

```
pcolor(flipud(A)), shading interp, colormap(gray)
set(gca,'Xtick',[],'Ytick',[])
```

Note that the **flipud** command flips the image to be right-side-up for visualization. Further, the **imresize** command is part of the image processing toolbox.

To start understanding the face recognition process, we can begin by trying to understand the concept of an *average* face. In particular, if one takes the five images of a single person represented in Fig. 2.4 and averages over them, one would arrive at an average face for that person. Figure 2.5 demonstrates the average face associated with the five images of the four considered celebrities. The average faces are trivially computed by the following MATLAB code

```
Ave=(A1+A2+A3+A4+A5)/5;
```

where **A1** through **A5** are the five images of a particular celebrity. Note that although each celebrity image has a different background and is slightly misaligned from the others, the average still represents the basic look of the celebrity. In some sense, this is their *average* face. Interestingly enough, each picture can be thought of as the average plus corrections, or variance from the average. And such is the fundamental philosophy of the approach of eigenfaces: there exists an average face and all other faces are simply variances (perhaps small or large) from the average face. Indeed, it is conjectured that each person's variance from the average face is unique so that

Figure 2.5: Average faces generated from the five images of each celebrity of Fig. 2.4. Despite the varying backgrounds and image misalignments, the images retain the fundamental aspects of the individuals considered.

identification of the face can be made from them. To get a good average face for human beings, or even individuals or select groups of individuals, one would need very large data sets to improve the eigenface methodology.

What the average face actually emphasizes is the common features among a group of pictures, i.e. the pixels that are correlated. This gives a clue on how to pursue a face recognition algorithm: look for the correlations and common features of faces. Alternatively, two faces that are highly uncorrelated are most likely different. Correlations between data sets can be computed using the *covariance*. To do this, all the images in Fig. 2.4 are arranged into a matrix so that

$$\mathbf{B} = \begin{pmatrix} image1 \\ image2 \\ \vdots \\ imageN \end{pmatrix} \tag{2.5.1}$$

where each image has been reshaped into a vector using the command

```
a=reshape(A,1,120*80)
```

The new matrix **B** now has individual images that comprise each row. In our case, this will be a matrix with 20 rows with 120×80 columns.

To compute a correlation matrix, it only remains to multiply each row vector by each other row vector. Specifically, the inner product of any two rows gives the correlation between those two images. One can also normalize this if desired, but we will not do so here. Further, most eigenface methodologies subtract the average face of the entire data set so that everything is about the deviation from this average face. Again, this is not necessary to demonstrate the method. To correlate all the pixels and images, i.e. to multiply all rows by all other rows, we can simply compute the correlation matrix

$$C = B^T B \tag{2.5.2}$$

where the superscript T denotes the transpose of the data matrix. This results in a very large correlation matrix (9600×9600).

Our objective is to then compute the eigenvalues and eigenvectors of this matrix in order to gain some insight into the face recognition algorithm. To compute the eigenvalues and eigenvectors, simply apply the code from the last section

```
C=B' * B;
[V,D]=eigs(C,20,'lm');
```

where the first 20 eigenvalues and eigenvectors are generated with the largest magnitude. The matrix **V** contains the first eigenvectors as the columns and the vector **D** contains the first

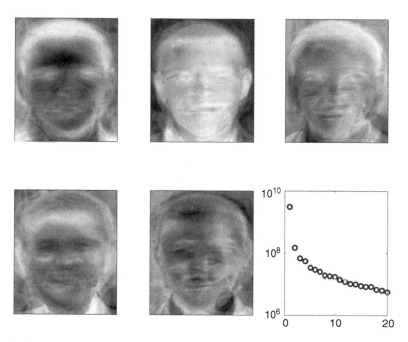

Figure 2.6: The first five eigenvectors (or eigenface components) of the correlation matrix **C** arranged from top left to bottom right. In the bottom right panel the first 20 eigenvalues are shown. Note the dominance of the first mode on the log scale which suggests it captures the average face in the data matrix. The additional eigenvectors are added in a weighted fashion to the average mode in order to produce the individual faces (or variances) of each image.

20 eigenvalues as its diagonals. Figure 2.6 shows the first five eigenvectors (reshaped back to matrix form for viewing) of our data matrix. Additionally, the first 20 eigenvalues are demonstrated showing the decrease in their value. The magnitude of the eigenvalues is directly related to how important its eigenvector is in composing the images. In the log plot of Fig. 2.6, the first mode dominates the *energy* content of all the images. This is effectively the average face within the 25 selected images. The other modes are shown as a function of decreasing importance. The idea is that one can reconstruct any of the faces considered as a weighted sum of the eigenvectors shown. All that remains is a demonstration on how the weighting of the eigenvectors is achieved. Given the dominance of the first few eigenvectors and their importance in the weighting, they are often called *principal components*, or a principal component analysis (PCA). As the name implies, one can think of them as the principal quantities (or most important quantities) of interest in the face recognition algorithm. A great deal more will be discussed concerning PCA in the third part of this book on data analysis. What is important here is that the eigenvalue/eigenvector decomposition of the matrix highlights the most essential features of the data matrix and also ranks, or prioritizes, through the eigenvalues the importance of each eigenvector.

To represent an individual in terms of the eigenvectors of the full data matrix **C**, it remains to understand the weighting of each image on the eigenvectors. This can be done by simply

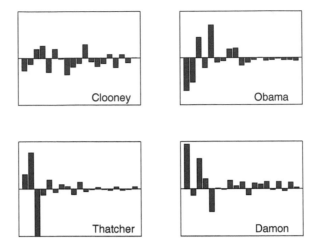

Figure 2.7: Projection of the average faces in Fig. 2.5 onto the eigenvector space computed from the full correlation matrix **C**. Note that each celebrity has a different, and hopefully unique, set of coefficients that can be used for face recognition.

computing the inner product of each image on each eigenvector. This can be done very simply with the MATLAB command

```
proj_a=a*V;
```

where it should be recalled that **a** is the reshaped picture in vector form of a specific image. Multiplying by the eigenvector matrix **V** gives a vector of inner products of the image onto each eigenvector. This operation can be performed for the average image (see Fig. 2.5) of the four celebrities considered in Fig. 2.4. This gives a *unique* representation of each celebrity on the eigenvectors. Figure 2.7 shows the projection coefficients of the four average faces of Fig. 2.5. Note that each celebrity has a different set of coefficients in this representation. It is these differences that can be used as the basis for face recognition.

To demonstrate how the eigenvectors can be used in practice, we consider reconstruction of a *new* image using the eigenvectors formed from our *training set*. Figure 2.8 shows three new images that are not part of the training set. In the first, a new picture of Margaret Thatcher (Fig. 2.8 top row) is introduced and projected onto the first 20 eigenvalues. The weighting coefficients are computed and a 20 mode representation is then constructed in the third panel of the top row. This shows that some semblance of Margaret Thatcher can indeed be reconstructed from our training set which included five different pictures of Margaret Thatcher. In the last panel of the top row, the difference between the projection vector of the new image and the projection vector of the original five Margaret Thatcher images is given. Thus the following quantity is computed

$$E_j = \frac{\|c_j - c_{\text{new}}\|}{\|c_j\|} \tag{2.5.3}$$

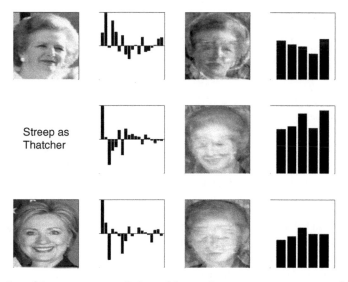

Figure 2.8: Projection of three new images (left panels) onto the eigenvector space created from Fig. 2.4. The coefficients of the projection are shown in the second column of data panels while the reconstruction using 20 eigenvectors is shown in the third column of data. The fourth column shows the distance in the coefficient measure (2.5.3) for each new picture against the original five Margaret Thatcher images. The new Margaret Thatcher image is able to produce a reasonable reconstruction while Meryl Streep (as Thatcher) and Hillary Clinton both perform poorly under reconstruction. All images are public domain, licensed under Creative Commons Attribution-Share Alike 2.0, or licensed under Creative Commons Attribution 3.0.

where the double bar denotes the L^2 norm, c_j is the projection of the jth Margaret Thatcher image in the original set of Fig. 2.4 and c_{new} is the new image presented. In MATLAB, this can be accomplished with

```
E=norm(proj_a-proj_new)/norm(proj_a);
```

This gives some measure of the difference between the new images and the original images already used in the correlation matrices. The conjecture is that if the image is similar to those in the training set, then this error, or difference, should be small. In the second row of Fig. 2.8, the results are reproduced with a new picture (not shown) of Meryl Streep who plays Margaret Thatcher in the movie *Iron Lady* (2012). We can compare her Margaret Thatcher look to the real Margaret Thatcher to conclude that with a 20 mode reconstruction (second row, third panel), she does not look much like Margaret Thatcher from the perspective of the training set used. The last panel in the second row shows the distance measure from the new picture to the five originals in terms of the coefficients of the first 20 eigenvalues. Note that if a totally different person is projected onto the data set (Hillary Clinton in the last row of Fig. 2.8), then no recognizable face is produced in the 20 mode reconstruction. Interestingly, however, Hillary Clinton does appear to be closer in the coefficient space to Margaret Thatcher than Meryl Streep.

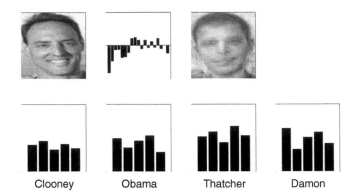

Clooney   Obama   Thatcher   Damon

Figure 2.9: Projection of the author onto the eigenvector space created from Fig. 2.4. The coefficients of the projection are shown in the second panel of the top row while the reconstruction using 20 eigenvectors is shown in the third column. It seems like there is some resemblance to Obama and Damon. The bottom row shows the distance in the coefficient measure (2.5.3) for the author's picture against the original 25 images, showing that the author looks most like Obama than the others.

Finally, the author wishes to see how he might project onto this data set. This projection is shown in Fig. 2.9. In the top row, the original picture is shown along with the projection onto the coefficients of the eigenvectors and the reconstruction using 20 eigenvectors. What seems to be the case is that the author kind of looks like a cross between Barack Obama and Matt Damon. In fact, when comparing in the eigenvector coefficient space against all the images of Fig. 2.4, the distance calculation (2.5.3) shows the author to be quite similar to Obama/Damon in a couple of his pictures. The author was hoping to look perhaps more like George Clooney.

As a final note, the technique used in creating eigenfaces and using them for recognition is also used outside of facial recognition. This technique is also used for handwriting analysis, lip reading, voice recognition, sign language/hand gestures interpretation and medical imaging analysis. Therefore, some do not use the term eigenface, but prefer to use *eigenimage*.

2.6 Nonlinear Systems

The preceding sections deal with a variety of techniques to solve linear systems of equations. Provided the matrix **A** is nonsingular, there is one unique solution to be found. Often, however, it is necessary to consider a *nonlinear* system of equations. This presents a significant challenge since a nonlinear system may have no solutions, one solution, five solutions, or an infinite number of solutions. Unfortunately, there is no general theorem concerning nonlinear systems which can narrow down the possibilities. Thus even if we are able to find a solution to a nonlinear system, it may be simply one of many. Furthermore, it may not be a solution we are particularly interested in.

To help illustrate the concept of nonlinear systems, we can once again consider the Newton–Raphson method of Section 1.3. In this method, the roots of a function $f(x)$ are determined by the Newton–Raphson iteration method. As a simple example, Fig. 2.10 illustrates a simple cubic

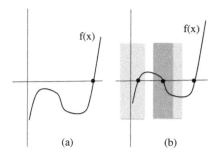

Figure 2.10: Two example cubic functions and their roots. In (a), there is only a single root to converge on with a Newton iteration routine. Aside from guessing a location which has $f'(x) = 0$, the solution will always converge to this root. In (b), the cubic has three roots. Guessing in the dark gray region will converge to the middle root. Guessing in the light gray regions will converge to the left root. Guessing anywhere else will converge to the right root. Thus the initial guess is critical to finding a specific root location.

function which has either one or three roots. For an initial guess, the Newton solver will converge to the one root for the example shown in Fig. 2.10(a). However, Fig. 2.10(b) shows that the initial guess is critical in determining which root the solution finds. In this simple example, it is easy to graphically see there are either one or three roots. For more complicated systems of nonlinear equations, we in general will not be able to visualize the number of roots or how to guess. This poses a significant problem for the Newton iteration method.

The Newton method can be generalized for solving systems of nonlinear equations. The details will not be discussed here, but the Newton iteration scheme is similar to that developed for the single-function case. Given a system:

$$\mathbf{F}(\mathbf{x}) = \begin{bmatrix} f_1(x_1, x_2, x_3, \ldots, x_N) \\ f_2(x_1, x_2, x_3, \ldots, x_N) \\ \vdots \\ f_N(x_1, x_2, x_3, \ldots, x_N) \end{bmatrix} = 0, \tag{2.6.1}$$

the iteration scheme is

$$\mathbf{x}_{n+1} = \mathbf{x}_n + \Delta\mathbf{x}_n \tag{2.6.2}$$

where

$$\mathbf{J}(\mathbf{x}_n)\Delta\mathbf{x}_n = -\mathbf{F}(\mathbf{x}_n) \tag{2.6.3}$$

and $\mathbf{J}(\mathbf{x}_n)$ is the Jacobian matrix

$$\mathbf{J}(\mathbf{x}_n) = \begin{bmatrix} \frac{\partial f_1}{\partial x_1} & \frac{\partial f_1}{\partial x_2} & \cdots & \frac{\partial f_1}{\partial x_N} \\ \frac{\partial f_2}{\partial x_1} & \frac{\partial f_2}{\partial x_2} & \cdots & \frac{\partial f_2}{\partial x_N} \\ \vdots & \vdots & & \vdots \\ \frac{\partial f_N}{\partial x_1} & \frac{\partial f_N}{\partial x_2} & \cdots & \frac{\partial f_N}{\partial x_N} \end{bmatrix}. \tag{2.6.4}$$

This algorithm relies on initially guessing values for $x_1, x_2, \ldots, x_N$. As before, the algorithm is guaranteed to converge for an initial iteration value which is sufficiently close to a solution of (2.6.1). Thus a good initial guess is critical to its success. Further, the determinant of the Jacobian cannot equal zero, $\det \mathbf{J}(\mathbf{x}_n) \neq 0$, in order for the algorithm to work. Indeed, significant problems can occur any time the determinant is nearly zero. A simple check of the Jacobian can be made with the condition number command: **cond(J)**. If this condition number is large, i.e. greater than 10^6, then problems are likely to occur in the iteration algorithm. This is equivalent to having an almost zero derivative in the Newton algorithm displayed in Fig. 1.4. Specifically, a large condition number or nearly zero derivative will move the iteration far away from the objective fixed point.

To show the practical implementation of this method, consider the following nonlinear system of equations:

$$f_1(x_1, x_2) = 2x_1 + x_2 + x_1^3 = 0 \tag{2.6.5a}$$

$$f_2(x_1, x_2) = x_1 + x_1 x_2 + \exp(x_1) = 0. \tag{2.6.5b}$$

Figure 2.11 shows a surface plot of the functions $f_1(x_1, x_2)$ and $f_2(x_1, x_2)$ with bolded lines for demarcation of the lines where these functions are zero. The intersection of the zero lines of f_1 and f_2 are the solutions of the example 2×2 system (2.6.5).

To apply the Newton iteration algorithm, the Jacobian must first be calculated. The partial derivatives required for the Jacobian are first evaluated:

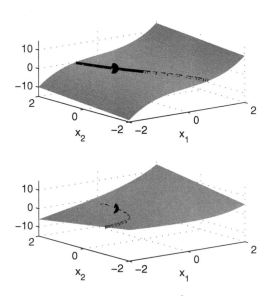

Figure 2.11: Plot of the two surfaces: $f_1(x_1, x_2) = 2x_1 + x_2 + x_1^3$ (top) and $f_2(x_1, x_2) = x_1 + x_1 x_2 + \exp(x_1)$ (bottom). The solid lines show where the crossing at zero occurs for both surfaces. The dot is the intersection of zero surfaces and the solution of the 2×2 system (2.6.5).

$$\frac{\partial f_1}{\partial x_1} = 2 + 3x_1^2 \tag{2.6.6a}$$

$$\frac{\partial f_1}{\partial x_2} = 1 \tag{2.6.6b}$$

$$\frac{\partial f_2}{\partial x_1} = 1 + x_2 + \exp(x_1) \tag{2.6.6c}$$

$$\frac{\partial f_2}{\partial x_2} = x_1 \tag{2.6.6d}$$

and the Jacobian is constructed

$$\mathbf{J} = \begin{pmatrix} 2 + 3x_1^2 & 1 \\ 1 + x_2 + \exp(x_1) & x_1 \end{pmatrix}. \tag{2.6.7}$$

A simple iteration procedure can then be developed for MATLAB. The following MATLAB code illustrates the ease of implementation of this algorithm.

```
x=[0; 0]
for j=1:1000

    J=[2+3*x(1)^2  1; 1+x(2)+exp(x(1))  x(1)];
    f=[2*x(1)+x(2)+x(1)^3; x(1)+x(1)*x(2)+exp(x(1))];

    if norm(f)<10^(-6)
        break
    end

    df=-J\f;
    x=x+df;

end
x
```

In this algorithm, the initial guess for the solution was taken to be $(x_1, x_2) = (0, 0)$. The Jacobian is updated at each iteration with the current values of x_1 and x_2. Table 2.5 shows the convergence to an accuracy of 10^{-6} of this simple algorithm to the solution of (2.6.5). It only requires four iterations to calculate the solution of (2.6.5). Recall, however, that if there were multiple roots in such a system, then the difficulties would arise from trying to guess initial solutions so that one can find all the solution roots.

The built-in MATLAB command which solves nonlinear system of equations is **fsolve**. The **fsolve** algorithm is a more sophisticated version of the Newton method presented here. As with

Table 2.5 Convergence of the Newton iteration algorithm to the roots of (2.6.5). A solution to an accuracy of 10^{-6} is achieved, with only four iterations with a starting guess of $(x_1, x_2) = (0, 0)$

Iteration	x_1	x_2
1	−0.50000000	1.00000000
2	−0.38547888	0.81006693
3	−0.37830667	0.81069593
4	−0.37830658	0.81075482

any iteration scheme, it requires that the user provide an initial starting guess for the solution. The basic command structure is as follows

```
x=fsolve('system',[0 0])
```

where the vector [0 0] is the initial guess values for x_1 and x_2. This command calls upon the function **system.m** that contains the function whose roots are to be determined. To follow the previous example, the following function would need to be constructed

system.m

```
function f=system(x)
f=[2*x(1)+x(2)+x(1)^3; x(1)+x(1)*x(2)+exp(x(1))];
```

Execution of the **fsolve** command with this function **system.m** once again yields the solution to the 2 × 2 nonlinear system of equations (2.6.5).

A nice feature of the **fsolve** command is that it has many options concerning the search algorithm and its accuracy settings. Thus one can search for more or less accurate solutions as well as set a limit on the number of iterations allowed in the **fsolve** function call.

Curve Fitting

Analyzing data is fundamental to any aspect of science. Often data can be noisy in nature and only the trends in the data are sought. A variety of curve fitting schemes can be generated to provide simplified descriptions of data and its behavior. The least-squares fit method is explored along with fitting methods of polynomial fits and splines.

3.1 Least-Square Fitting Methods

One of the fundamental tools for data analysis and recognizing trends in physical systems is curve fitting. The concept of curve fitting is fairly simple: use a simple function to describe a trend by minimizing the error between the selected function to fit and a set of data. The mathematical aspects of this are laid out in this section.

Suppose we are given a set of n data points

$$(x_1, y_1), \ (x_2, y_2), \ (x_3, y_3), \ \cdots , (x_n, y_n).$$ (3.1.1)

Further, assume that we would like to fit a best fit line through these points. Then we can approximate the line by the function

$$f(x) = Ax + B$$ (3.1.2)

where the constants A and B are chosen to minimize some error. Thus the function gives an approximation to the true value which is off by some error so that

$$f(x_k) = y_k + E_k$$ (3.1.3)

where y_k is the true value and E_k is the error from the true value.

Various error measurements can be minimized when approximating with a given function $f(x)$. Three standard possibilities are given as follows

   I. MaximumError :

$$E_\infty(f) = \max_{1<k<n} |f(x_k) - y_k|. \tag{3.1.4a}$$

  II. AverageError :

$$E_1(f) = \frac{1}{n} \sum_{k=1}^{n} |f(x_k) - y_k|. \tag{3.1.4b}$$

 III. Root-meanSquare :

$$E_2(f) = \left(\frac{1}{n} \sum_{k=1}^{n} |f(x_k) - y_k|^2 \right)^{1/2}. \tag{3.1.4c}$$

In practice, the root-mean square error is most widely used and accepted. Thus when fitting a curve to a set of data, the root-mean square error is chosen to be minimized. This is called a *least-square fit*. Figure 3.1 depicts three line fits for the errors E_∞, E_1 and E_2 listed above. The E_∞ error line fit is strongly influenced by the one data point which does not fit the trend. The E_1 and E_2 line fit nicely through the bulk of the data.

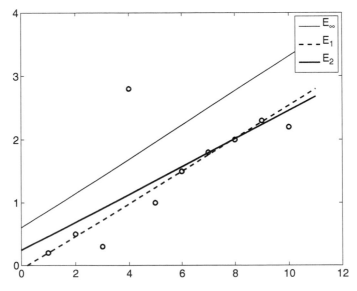

Figure 3.1: Linear fit $Ax + B$ to a set of data (open circles) for the three error definitions given by E_∞, E_1 and E_2. The weakness of the E_∞ fit results from the one stray data point at $x = 4$.

Least-squares line

To apply the least-square fit criteria, consider the data points $\{x_k, y_k\}$, where $k = 1, 2, 3, \ldots, n$. To fit the curve

$$f(x) = Ax + B \tag{3.1.5}$$

to this data, the E_2 is found by minimizing the sum

$$E_2(f) = \sum_{k=1}^{n} |f(x_k) - y_k|^2 = \sum_{k-1}^{n} (Ax_k + B - y_k)^2. \tag{3.1.6}$$

Minimizing this sum requires differentiation. Specifically, the constants A and B are chosen so that a minimum occurs; thus we require: $\partial E_2/\partial A = 0$ and $\partial E_2/\partial B = 0$. Note that although a zero derivative can indicate either a minimum or maximum, we know this must be a minimum to the error since there is no maximum error, i.e. we can always choose a line that has a larger error. The minimization condition gives:

$$\frac{\partial E_2}{\partial A} = 0 : \quad \sum_{k=1}^{n} 2(Ax_k + B - y_k)x_k = 0 \tag{3.1.7a}$$

$$\frac{\partial E_2}{\partial B} = 0 : \quad \sum_{k=1}^{n} 2(Ax_k + B - y_k) = 0. \tag{3.1.7b}$$

Upon rearranging, the 2×2 system of linear equations is found for A and B:

$$\begin{pmatrix} \sum_{k=1}^{n} x_k^2 & \sum_{k=1}^{n} x_k \\ \sum_{k=1}^{n} x_k & n \end{pmatrix} \begin{pmatrix} A \\ B \end{pmatrix} = \begin{pmatrix} \sum_{k=1}^{n} x_k y_k \\ \sum_{k=1}^{n} y_k \end{pmatrix}. \tag{3.1.8}$$

This equation can be easily solved using the backslash command in MATLAB.

This method can be easily generalized to higher polynomial fits. In particular, a parabolic fit to a set of data requires the fitting function

$$f(x) = Ax^2 + Bx + C \tag{3.1.9}$$

where now the three constants A, B and C must be found. These can be found from the 3×3 system which results from minimizing the error $E_2(A, B, C)$ by taking .

$$\frac{\partial E_2}{\partial A} = 0 \tag{3.1.10a}$$

$$\frac{\partial E_2}{\partial B} = 0 \tag{3.1.10b}$$

$$\frac{\partial E_2}{\partial C} = 0. \tag{3.1.10c}$$

Data linearization

Although a powerful method, the minimization procedure can result in equations which are nontrivial to solve. Specifically, consider fitting data to the exponential function

$$f(x) = C \exp(Ax).$$ (3.1.11)

The error to be minimized is

$$E_2(A, C) = \sum_{k=1}^{n} (C \exp(Ax_k) - y_k)^2.$$ (3.1.12)

Applying the minimizing conditions leads to

$$\frac{\partial E_2}{\partial A} = 0: \quad \sum_{k=1}^{n} 2(C \exp(Ax_k) - y_k) C x_k \exp(Ax_k) = 0$$ (3.1.13a)

$$\frac{\partial E_2}{\partial C} = 0: \quad \sum_{k=1}^{n} 2(C \exp(Ax_k) - y_k) \exp(Ax_k) = 0.$$ (3.1.13b)

This in turn leads to the 2×2 system

$$C \sum_{k=1}^{n} x_k \exp(2Ax_k) - \sum_{k=1}^{n} x_k y_k \exp(Ax_k) = 0$$ (3.1.14a)

$$C \sum_{k=1}^{n} \exp(Ax_k) - \sum_{k=1}^{n} y_k \exp(Ax_k) = 0.$$ (3.1.14b)

This system of equations is nonlinear and cannot be solved in a straightforward fashion. Indeed, a solution may not even exist. Or many solution may exist. Section 2.6 describes a possible iterative procedure for solving this nonlinear system of equations.

To avoid the difficulty of solving this nonlinear system, the exponential fit can be *linearized* by the transformation

$$Y = \ln(y)$$ (3.1.15a)

$$X = x$$ (3.1.15b)

$$B = \ln C.$$ (3.1.15c)

Then the fit function

$$f(x) = y = C \exp(Ax)$$ (3.1.16)

can be linearized by taking the natural log of both sides so that

$$\ln y = \ln(C \exp(Ax)) = \ln C + \ln(\exp(Ax)) = B + Ax \rightarrow Y = AX + B.$$ (3.1.17)

So by fitting to the natural log of the y-data

$$(x_i, y_i) \rightarrow (x_i, \ln y_i) = (X_i, Y_i)$$ (3.1.18)

the curve fit for the exponential function becomes a linear fitting problem which is easily handled.

General fitting

Given the preceding examples, a theory can be developed for a general fitting procedure. The key idea is to assume a form of the fitting function

$$f(x) = f(x, C_1, C_2, C_3, \cdots, C_M) \tag{3.1.19}$$

where the C_i are constants used to minimize the error and $M < n$. The root-mean square error is then

$$E_2(C_1, C_2, C_3, \cdots, C_m) = \sum_{k=1}^{n} (f(x_k, C_1, C_2, C_3, \cdots, C_M) - y_k)^2 \tag{3.1.20}$$

which can be minimized by considering the $M \times M$ system generated from

$$\frac{\partial E_2}{\partial C_j} = 0 \quad j = 1, 2, \cdots, M. \tag{3.1.21}$$

In general, this gives the *nonlinear* set of equations

$$\sum_{k=1}^{n} (f(x_k, C_1, C_2, C_3, \cdots, C_M) - y_k) \frac{\partial f}{\partial C_j} = 0 \quad j = 1, 2, 3, \cdots, M. \tag{3.1.22}$$

Solving this set of equations can be quite difficult. Most attempts at solving nonlinear systems arc based upon iterative schemes which require good initial guesses to converge to the solution. Regardless, the general fitting procedure is straightforward and allows for the construction of a best fit curve to match the data. Section 2.6 gives an example of a specific algorithm used to solve such nonlinear systems. In such a solution procedure, it is imperative that a reasonable initial guess be provided for by the user. Otherwise, the desired rapid convergence or convergence to the desired root may not be achieved.

3.2 Polynomial Fits and Splines

One of the primary reasons for generating data fits from polynomials, splines or least-square methods is to *interpolate* or *extrapolate* data values. In practice, when considering only a finite number of data points

$$(x_0, y_0)$$
$$(x_1, y_1)$$
$$\vdots$$
$$(x_n, y_n)$$

the value of the curve at points other than the x_i are unknown. *Interpolation* uses the data points to predict values of $y(x)$ at locations where $x \neq x_i$ and $x \in [x_0, x_n]$. *Extrapolation* is similar, but it predicts values of $y(x)$ for $x > x_n$ or $x < x_0$, i.e. outside the range of the data points.

Interpolation and extrapolation are easy to do given a least-squares fit. Once the fitting curve is found, it can be evaluated for any value of x, thus giving an interpolated or extrapolated value. Polynomial fitting is another method for getting these values. With polynomial fitting, a polynomial is chosen to go through all data points. For the $n + 1$ data points given above, an nth degree polynomial is chosen

$$p_n(x) = a_n x^n + a_{n-1} x^{n-1} + \cdots + a_1 x + a_0 \tag{3.2.1}$$

where the coefficients a_j are chosen so that the polynomial passes through each data point. Thus we have the resulting system

$$(x_0, y_0): \quad y_0 = a_n x_0^n + a_{n-1} x_0^{n-1} + \cdots + a_1 x_0 + a_0$$
$$(x_1, y_1): \quad y_1 = a_n x_1^n + a_{n-1} x_1^{n-1} + \cdots + a_1 x_1 + a_0$$
$$\vdots$$
$$(x_n, y_n): \quad y_n = a_n x_n^n + a_{n-1} x_n^{n-1} + \cdots + a_1 x_n + a_0 .$$

This system of equations is nothing more than a linear system $\mathbf{Ax} = \mathbf{b}$ which can be solved by using the backslash command in MATLAB. Note that unlike the least-square fitting method, the error of this fit is identically zero since the polynomial goes through each individual data point. However, this does not necessarily mean that the resulting polynomial is a good fit to the data. This will be shown in what follows.

Although this polynomial fit method will generate a curve which passes through all the data, it is an expensive computation since we have to first set up the system and then perform an $O(n^3)$ operation to generate the coefficients a_j. A more direct method for generating the relevant polynomial is the *Lagrange polynomials* method. Consider first the idea of constructing a line between the two points (x_0, y_0) and (x_1, y_1). The general form for a line is $y = mx + b$ which gives

$$y = y_0 + (y_1 - y_0) \frac{x - x_0}{x_1 - x_0} . \tag{3.2.2}$$

Although valid, it is hard to continue to generalize this technique to fitting higher order polynomials through a larger number of points. Lagrange developed a method which expresses the line through the above two points as

$$p_1(x) = y_0 \frac{x - x_1}{x_0 - x_1} + y_1 \frac{x - x_0}{x_1 - x_0} \tag{3.2.3}$$

which can be easily verified to work. In a more compact and general way, this first degree polynomial can be expressed as

$$p_1(x) = \sum_{k=0}^{1} y_k L_{1,k}(x) = y_0 L_{1,0}(x) + y_1 L_{1,1}(x) \tag{3.2.4}$$

where the *Lagrange coefficients* are given by

$$L_{1,0}(x) = \frac{x - x_1}{x_0 - x_1} \tag{3.2.5a}$$

$$L_{1,1}(x) = \frac{x - x_0}{x_1 - x_0}. \tag{3.2.5b}$$

Note the following properties of these cofficients:

$$L_{1,0}(x_0) = 1 \tag{3.2.6a}$$

$$L_{1,0}(x_1) = 0 \tag{3.2.6b}$$

$$L_{1,1}(x_0) = 0 \tag{3.2.6c}$$

$$L_{1,1}(x_0) = 1. \tag{3.2.6d}$$

By design, the Lagrange coefficients take on the binary values of zero or unity at the given data points. The power of this method is that it can be easily generalized to consider the $n + 1$ points of our original data set. In particular, we fit an nth degree polynomial through the given data set of the form

$$p_n(x) = \sum_{k=0}^{n} y_k L_{n,k}(x) \tag{3.2.7}$$

where the Lagrange coefficient is

$$L_{n,k}(x) = \frac{(x - x_0)(x - x_1)\cdots(x - x_{k-1})(x - x_{k+1})\cdots(x - x_n)}{(x_k - x_0)(x_k - x_1)\cdots(x_k - x_{k-1})(x_k - x_{k+1})\cdots(x_k - x_n)}, \tag{3.2.8}$$

so that

$$L_{n,k}(x_j) = \begin{cases} 1 & j = k \\ 0 & j \neq k. \end{cases} \tag{3.2.9}$$

Thus there is no need to solve a linear system to generate the desired polynomial. This is the preferred method for generating a polynomial fit to a given set of data and is the core algorithm employed by most commercial packages such as MATLAB for polynomial fitting.

Figure 3.2 shows the polynomial fit that in theory has zero error. In this example, 15 data points were used and a comparison is made between the polynomial fit, least-square line fit and a spline. The polynomial fit does an excellent job in the interior of the fitting regime, but generates a phenomenon known as polynomial wiggle at the edges. This makes a polynomial fit highly suspect for any application due to its poor ability to accurately capture interpolations or extrapolations.

Splines

Although MATLAB makes it a trivial matter to fit polynomials or least-square fits through data, a fundamental problem can arise as a result of a polynomial fit: *polynomial wiggle* (see Fig. 3.2). Polynomial wiggle is generated by the fact that an nth degree polynomial has, in general, $n - 1$ turning points from up to down or vice versa. One way to overcome this is to use a *piecewise*

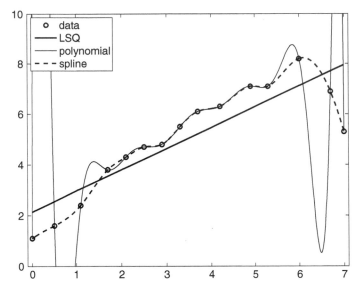

Figure 3.2: Fitting methods to a given data set (open circles) consisting of 15 points. Represented is a least-square (LSQ) line fit, a polynomial fit and a spline. The polynomial fit has the drawback of generating polynomial wiggle at its edges making it a poor candidate for interpolation or extrapolation.

polynomial interpolation scheme. Essentially, this simply draws a line between neighboring data points and uses this line to give interpolated values. This technique is rather simple minded, but it does alleviate the problem generated by polynomial wiggle. However, the interpolating function is now only a piecewise function. Therefore, when considering interpolating values between the points (x_0, y_0) and (x_n, y_n), there will be n linear functions each valid only between two neighboring points.

The data generated by a piecewise linear fit can be rather crude and it tends to be choppy in appearance. *Splines* provide a better way to represent data by constructing cubic functions between points so that the first and second derivatives are continuous at the data points. This gives a smooth looking function without polynomial wiggle problems. The basic assumption of the spline method is to construct a cubic function between data points:

$$S_k(x) = S_{k,0} + S_{k,1}(x - x_k) + S_{k,2}(x - x_k)^2 + S_{k,3}(x - x_k)^3 \qquad (3.2.10)$$

where $x \in [x_k, x_{k+1}]$ and the coefficients $S_{k,j}$ are to be determined from various constraint conditions. Four constraint conditions are imposed:

$$S_k(x_k) = y_k \qquad (3.2.11a)$$

$$S_k(x_{k+1}) = S_{k+1}(x_{k+1}) \qquad (3.2.11b)$$

$$S'_k(x_{k+1}) = S'_{k+1}(x_{k+1}) \qquad (3.2.11c)$$

$$S''_k(x_{k+1}) = S''_{k+1}(x_{k+1}). \qquad (3.2.11d)$$

This allows for a smooth fit to the data since the four constraints correspond to fitting the data, continuity of the function, continuity of the first derivative, and continuity of the second derivative, respectively.

To solve for these quantities, a large system of equations $\mathbf{Ax} = \mathbf{b}$ is constructed. The number of equations and unknowns must first be calculated.

- $S_k(x_k) = y_k \rightarrow$ Solution fit: $n + 1$ equations;
- $S_k = S_{k+1} \rightarrow$ Continuity: $n - 1$ equations;
- $S'_k = S'_{k+1} \rightarrow$ Smoothness: $n - 1$ equations;
- $S''_k = S''_{k+1} \rightarrow$ Smoothness: $n - 1$ equations.

This gives a total of $4n - 2$ equations. For each of the n intervals, there are four parameters which gives a total of $4n$ unknowns. Thus two extra constraint conditions must be placed on the system to achieve a solution. There are a large variety of options for assumptions which can be made at the edges of the spline. It is usually a good idea to use the default unless the application involved requires a specific form. The spline problem is then reduced to a simple solution of an $\mathbf{Ax} = \mathbf{b}$ problem which can be solved with the backslash command.

As a final remark on splines, splines are heavily used in computer graphics and animation. The primary reason is for their relative ease in calculating, and for their smoothness properties. There is an entire spline toolbox available for MATLAB which attests to the importance of this technique for this application. Further, splines are also commonly used for smoothing data before differentiating. This will be considered further in upcoming chapters.

3.3 Data Fitting with MATLAB

This section will discuss the practical implementation of the curve fitting schemes presented in the preceding two sections. The schemes to be explored are least-square fits, polynomial fits, line interpolation and spline interpolation. Additionally, a nonpolynomial least-square fit will be considered which results in a nonlinear system of equations. This nonlinear system requires additional insight into the problem and sophistication in its solution technique.

To begin the data fit process, we first import a relevant data set into the MATLAB environment. To do so, the **load** command is used. The file *linefit.dat* is a collection of x and y data values put into a two-column format separated by spaces. The file is the following:

linefit.dat

```
0.0   1.1
0.5   1.6
1.1   2.4
```

Continued

```
1.7   3.8
2.1   4.3
2.5   4.7
2.9   4.8
3.3   5.5
3.7   6.1
4.2   6.3
4.9   7.1
5.3   7.1
6.0   8.2
6.7   6.9
7.0   5.3
```

This data set is just an example of what you may want to import into MATLAB. The command structure to read this data is as follows

```
load linefit.dat
x=linefit(:,1);
y=linefit(:,2);
figure(1), plot(x,y,'o:')
```

After reading in the data, the two vectors **x** and **y** are created from the first and second column of the data, respectively. It is this set of data which will be explored with line fitting techniques. The code will also generate a plot of the data in *figure 1* of MATLAB. Note that this is the data considered in Fig. 3.2 of the preceding section.

Least-squares fitting

The least-squares fit technique is considered first. The **polyfit** and **polyval** commands are essential to this method. Specifically, the **polyfit** command is used to generate the $n + 1$ coefficients a_j of the nth degree polynomial

$$p_n(x) = a_n x^n + a_{n-1} x^{n-1} + \cdots + a_1 x + a_0 \tag{3.3.1}$$

used for fitting the data. The basic structure requires that the vectors **x** and **y** be submitted to **polyval** along with the desired degree of polynomial fit n. To fit a line ($n = 1$) through the data, use the command

```
pcoeff=polyfit(x,y,1);
```

The output of this function call is a vector **pcoeff** which includes the coefficients a_1 and a_0 of the line fit $p_1(x) = a_1x + a_0$. To evaluate and plot this line, values of x must be chosen. For this example, the line will be plotted for $x \in [0, 7]$ in steps of $\Delta x = 0.1$.

```
xp=0:0.1:7;
yp=polyval(pcoeff,xp);
figure(2), plot(x,y,'O',xp,yp,'m')
```

The **polyval** command uses the coefficients generated from **polyfit** to generate the y-values of the polynomial fit at the desired values of x given by **xp**. *Figure 2* in MATLAB depicts both the data and the best line fit in the least-square sense. Figure 3.2 of the last section demonstrates a least-squares line fit through the data.

To fit a parabolic profile through the data, a second degree polynomial is used. This is generated with

```
pcoeff2=polyfit(x,y,2);
yp2=polyval(pcoeff2,xp);
figure(3), plot(x,y,'O',xp,yp2,'m')
```

Here the vector **yp2** contains the parabolic fit to the data evaluated at the x-values **xp**. These results are plotted in MATLAB *figure 3*. Figure 3.3 shows the least-square parabolic fit to the data. This can be contrasted to the linear least-square fit in Fig. 3.2. To find the least-square error, the sum of the squares of the differences between the parabolic fit and the actual data must be evaluated. Specifically, the quantity

$$E_2(f) = \left(\frac{1}{n} \sum_{k=1}^{n} |f(x_k) - y_k|^2 \right)^{1/2} \tag{3.3.2}$$

is calculated. For the parabolic fit considered in the last example, the polynomial fit must be evaluated at the x-values for the given data **linefit.dat**. The error is then calculated

```
yp3=polyval(pcoeff2,x);
E2=sqrt( sum( ( abs(yp3-y) ).^2  )/n )
```

This is a quick and easy calculation which allows for an evaluation of the fitting procedure. In general, the error will continue to drop as the degree of the polynomial is increased. This is because every extra degree of freedom allows for a better least-squares fit to the data. Indeed, a degree $n - 1$ polynomial has zero error since it goes through all the data points.

Interpolation

In addition to least-square fitting, interpolation techniques can be developed which go through all the given data points. The error in this case is zero, but each interpolation scheme must be evaluated for its accurate representation of the data. The first interpolation scheme is a polynomial fit to the data. Given $n + 1$ points, an nth degree polynomial is chosen. The **polyfit** command is again used for the calculation

```
n=length(x)-1;
pcoeffn=polyfit(x,y,n);
ypn=polyval(pcoeffn,xp);
figure(4), plot(x,y,'O',xp,ypn,'m')
```

The MATLAB script will produce an nth degree polynomial through the data points. But as always there is the danger with polynomial interpolation: *polynomial wiggle* can dominate the behavior. This is indeed the case as illustrated in *figure 4*. The strong oscillatory phenomenon at the edges is a common feature of this type of interpolation. Figure 3.2 has already illustrated the pernicious behavior of the polynomial wiggle phenomenon. Indeed, the idea of using a high-degree polynomial for a data fit becomes highly suspect upon seeing this example.

In contrast to a polynomial fit, a piecewise linear fit gives a simple minded connect-the-dot interpolation to the data. The **interp1** command gives the piecewise linear fit algorithm

```
yint=interp1(x,y,xp);
figure(5), plot(x,y,'O',xp,yint,'m')
```

The linear interpolation is illustrated in *figure 5* of MATLAB. There are a few options available with the **interp1** command, including the *nearest* option and the *spline* option. The *nearest* option gives the nearest value of the data to the interpolated value while the *spline* option gives a cubic spline fit interpolation to the data. The two options can be compared with the default *linear* option.

```
yint=interp1(x,y,xp);
yint2=interp1(x,y,xp,'nearest')
yint3=interp1(x,y,xp,'spline')
figure(5), plot(x,y,'O',xp,yint,'m',xp,int2,'k',xp,int3,'r')
```

Note that the spline option is equivalent to using the **spline** algorithm supplied by MATLAB. Thus a smooth fit can be achieved with either **spline** or **interp1**. Figure 3.3 demonstrates the nearest neighbor and linear interpolation fit schemes. The most common (default) scheme is the simple linear interpolation between data points.

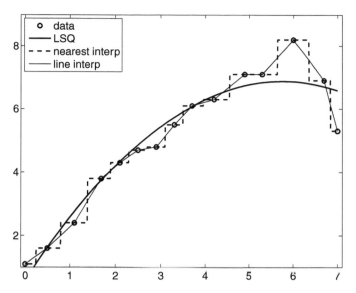

Figure 3.3: Fitting methods to a given data set (open circles) consisting of 15 points. Represented is a least-square (LSQ) quadratic fit, a nearest neighbor interpolation fit and a linear interpolation between data points.

The **spline** command is used by giving the x and y data along with a vector **xp** for which we desire to generate corresponding y-values.

```
yspline=spline(x,y,xp);
figure(6), plot(x,y,'O',xp,yspline,'k')
```

The generated spline is depicted in *figure 6*. This is the same as that using **interp1** with the *spline* option. Note that the data is smooth as expected from the enforcement of continuous smooth derivatives. The spline fit was already demonstrated in the last section in Fig. 3.2.

Nonpolynomial least-square fitting

To consider more sophisticated least-square fitting routines, consider the following data set which looks like it could be nicely fitted with a Gaussian profile.

gaussfit.dat

```
-3.0   -0.2
-2.2    0.1
-1.7    0.05
```

Continued

```
 -1.5     0.2
 -1.3     0.4
 -1.0     1.0
 -0.7     1.2
 -0.4     1.4
 -0.25    1.8
 -0.05    2.2
  0.07    2.1
  0.15    1.6
  0.3     1.5
  0.65    1.1
  1.1     0.8
  1.25    0.3
  1.8    -0.1
  2.5     0.2
```

To fit the data to a Gaussian, we indeed assume a Gaussian function of the form

$$f(x) = A \exp(-Bx^2).$$

(3.3.3)

Following the procedure to minimize the least-square error leads to a set of nonlinear equations for the coefficients A and B. In general, solving a nonlinear set of equations can be a difficult task. A solution is not guaranteed to exist. And in fact, there may be many solutions to the problem. Thus the nonlinear problem should be handled with care.

To generate a solution, the least-square error must be minimized. In particular, the sum

$$E_2 = \sum_{k=0}^{n} |f(x_k) - y_k|^2$$

(3.3.4)

must be minimized. The command **fminsearch** in MATLAB minimizes a function of several variables. In this case, we minimize (3.3.4) with respect to the variables A and B. Thus the minimum of

$$E_2 = \sum_{k=0}^{n} |A \exp(-Bx_k^2) - y_k|^2$$

(3.3.5)

must be found with respect to A and B. The **fminsearch** algorithm requires a function call to the quantity to be minimized along with an initial guess for the parameters which are being used for minimization, i.e. A and B.

```
coeff=fminsearch('gauss_fit',[1 1]);
```

This command structure uses as initial guesses $A = 1$ and $B = 1$ when calling the file **gauss_fit.m**. After minimizing, it returns the vector **coeff** which contains the appropriate values of A and B which minimize the least-square error. The function called is then constructed as follows:

gauss_fit.m

```
function E=gauss_fit(x0)

load gaussfit.dat
x=gaussfit(:,1);
y=gaussfit(:,2);
E=sum(  ( x0(1)*exp(-x0(2)*x.^2)-y ).^2  )
```

The vector x_0 accepts the initial guesses for A and B and is updated as A and B are modified through an iteration procedure to converge to the least-square values. Note that the sum is simply a statement of the quantity to be minimized, namely (3.3.5). The results of this calculation can be illustrated with MATLAB *figure 7*:

```
xga=-3:0.1:3;
a=coeff(1); b=coeff(2)
yga=a*exp(-b*xga.^2);
figure(7), plot(x2,y2,'O',xga,yga,'m')
```

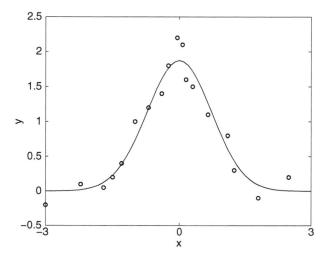

Figure 3.4: Gaussian fit $f(x) = A\exp(-Bx^2)$ to a given data set (open circles). The **fminsearch** command performs an iterative search to minimize the associated, nonlinear least-squares fit problem. It determines the values to be $A = 1.8733$ and $B = 0.9344$.

Note that for this case, the initial guess is extremely important. For any given problem where this technique is used, an educated guess for the values of parameters like A and B can determine if the technique will work at all. The results should also be checked carefully since there is no guarantee that a minimization near the desired fit can be found. Figure 3.4 shows the results of the least-square fit for the Gaussian example considered here. In this case, MATLAB determines that $A = 1.8733$ and $B = 0.9344$ minimizes the least-square error.

Numerical Differentiation and Integration

Differentiation and integration form the backbone of the mathematical techniques required to describe and analyze physical systems. These two mathematical concepts describe how certain quantities of interest change with respect to either space and time or both. Understanding how to evaluate these quantities numerically is essential to understanding systems beyond the scope of analytic methods.

4.1 Numerical Differentiation

Given a set of data or a function, it may be useful to differentiate the quantity considered in order to determine a physically relevant property. For instance, given a set of data which represents the position of a particle as a function of time, then the derivative and second derivative give the velocity and acceleration, respectively. From calculus, the definition of the derivative is given by

$$\frac{df(t)}{dt} = \lim_{\Delta t \to 0} \frac{f(t + \Delta t) - f(t)}{\Delta t}. \tag{4.1.1}$$

Since the derivative is the slope, the formula on the right is nothing more than a rise-over-run formula for the slope. The general idea of calculus is that as $\Delta t \to 0$, then the rise-over-run gives the instantaneous slope. Numerically, this means that if we take Δt sufficiently small, than the approximation should be fairly accurate. To quantify and control the error associated with approximating the derivative, we make use of *Taylor series* expansions.

To see how the Taylor expansions are useful, consider the following two Taylor series:

$$f(t + \Delta t) = f(t) + \Delta t \frac{df(t)}{dt} + \frac{\Delta t^2}{2!} \frac{d^2 f(t)}{dt^2} + \frac{\Delta t^3}{3!} \frac{d^3 f(c_1)}{dt^3} \qquad (4.1.2a)$$

$$f(t - \Delta t) = f(t) - \Delta t \frac{df(t)}{dt} + \frac{\Delta t^2}{2!} \frac{d^2 f(t)}{dt^2} - \frac{\Delta t^3}{3!} \frac{d^3 f(c_2)}{dt^3} \qquad (4.1.2b)$$

where $c_1 \in [t, t + \Delta t]$ and $c_2 \in [t, t - \Delta t]$. Subtracting these two expressions gives

$$f(t + \Delta t) - f(t - \Delta t) = 2\Delta t \frac{df(t)}{dt} + \frac{\Delta t^3}{3!} \left(\frac{d^3 f(c_1)}{dt^3} + \frac{d^3 f(c_2)}{dt^3} \right). \qquad (4.1.3)$$

By using the intermediate value theorem of calculus, we find $f'''(c) = (f'''(c_1) + f'''(c_2))/2$. Upon dividing the above expression by $2\Delta t$ and rearranging, we find the following expression for the first derivative:

$$\frac{df(t)}{dt} = \frac{f(t + \Delta t) - f(t - \Delta t)}{2\Delta t} - \frac{\Delta t^2}{6} \frac{d^3 f(c)}{dt^3} \qquad (4.1.4)$$

where the last term is the truncation error associated with the approximation of the first derivative using this particular Taylor series generated expression. Note that the truncation error in this case is $O(\Delta t^2)$. Figure 4.1 demonstrates graphically the numerical procedure and approximation for the second-order slope formula. Here, nearest neighbor points are used to calculate the slope.

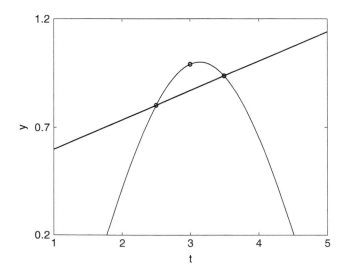

Figure 4.1: Graphical representation of the second-order accurate method for calculating the derivative with finite differences. The slope is simply rise over run where the nearest neighbors are used to determine both quantities. The specific function considered is $y = -\cos(t)$ with the derivative being calculated at $t = 3$ with $\Delta t = 0.5$.

We could improve on this by continuing our Taylor expansion and truncating it at higher orders in Δt. This would lead to higher accuracy schemes. Specifically, by truncating at $O(\Delta t^5)$, we would have

$$f(t + \Delta t) = f(t) + \Delta t \frac{df(t)}{dt} + \frac{\Delta t^2}{2!} \frac{d^2 f(t)}{dt^2}$$
$$+ \frac{\Delta t^3}{3!} \frac{d^3 f(t)}{dt^3} + \frac{\Delta t^4}{4!} \frac{d^4 f(t)}{dt^4} + \frac{\Delta t^5}{5!} \frac{d^5 f(c_1)}{dt^5} \tag{4.1.5a}$$

$$f(t - \Delta t) = f(t) - \Delta t \frac{df(t)}{dt} + \frac{\Delta t^2}{2!} \frac{d^2 f(t)}{dt^2}$$
$$- \frac{\Delta t^3}{3!} \frac{d^3 f(t)}{dt^3} + \frac{\Delta t^4}{4!} \frac{d^4 f(t)}{dt^4} - \frac{\Delta t^5}{5!} \frac{d^5 f(c_2)}{dt^5} \tag{4.1.5b}$$

where $c_1 \in [t, t + \Delta t]$ and $c_2 \in [t, t - \Delta t]$. Again subtracting these two expressions gives

$$f(t + \Delta t) - f(t - \Delta t) = 2\Delta t \frac{df(t)}{dt} + \frac{2\Delta t^3}{3!} \frac{d^3 f(t)}{dt^3} + \frac{\Delta t^5}{5!} \left(\frac{d^5 f(c_1)}{dt^5} + \frac{d^5 f(c_2)}{dt^5} \right). \tag{4.1.6}$$

In this approximation, there is a third derivative term left over which needs to be removed. By using two additional points to approximate the derivative, this term can be removed. Thus we use the two additional points $f(t + 2\Delta t)$ and $f(t - 2\Delta t)$. Upon replacing Δt by $2\Delta t$ in (4.1.6), we find

$$f(t + 2\Delta t) - f(t - 2\Delta t) = 4\Delta t \frac{df(t)}{dt} + \frac{16\Delta t^3}{3!} \frac{d^3 f(t)}{dt^3} + \frac{32\Delta t^5}{5!} \left(\frac{d^5 f(c_3)}{dt^5} + \frac{d^5 f(c_4)}{dt^5} \right) \tag{4.1.7}$$

where $c_3 \in [t, t + 2\Delta t]$ and $c_4 \in [t, t - 2\Delta t]$. By multiplying (4.1.6) by 8 and subtracting (4.1.7) and using the intermediate value theorem on the truncation terms twice, we find the expression:

$$\frac{df(t)}{dt} = \frac{-f(t + 2\Delta t) + 8f(t + \Delta t) - 8f(t - \Delta t) + f(t - 2\Delta t)}{12\Delta t} + \frac{\Delta t^4}{30} f^{(5)}(c) \tag{4.1.8}$$

where $f^{(5)}$ is the fifth derivative and the truncation is of $O(\Delta t^4)$.

Approximating higher derivatives works in a similar fashion. By starting with the pair of equations (4.1.2) and adding, this gives the result

$$f(t + \Delta t) + f(t - \Delta t) = 2f(t) + \Delta t^2 \frac{d^2 f(t)}{dt^2} + \frac{\Delta t^4}{4!} \left(\frac{d^4 f(c_1)}{dt^4} + \frac{d^4 f(c_2)}{dt^4} \right). \tag{4.1.9}$$

By rearranging and solving for the second derivative, the $O(\Delta t^2)$ accurate expression is derived

$$\frac{d^2 f(t)}{dt^2} = \frac{f(t + \Delta t) - 2f(t) + f(t - \Delta t)}{\Delta t^2} + O(\Delta t^2) \tag{4.1.10}$$

where the truncation error is of $O(\Delta t^2)$ and is found again by the intermediate value theorem to be $-(\Delta t^2/12) f''''(c)$. This process can be continued to find any arbitrary derivative. Thus, we could also approximate the third, fourth and higher derivatives using this technique. It is also possible to generate backward- and forward-difference schemes by using points only behind or in front of the current point, respectively. Tables 4.1–4.3 summarize the second-order

Table 4.1   Second-order accurate center-difference formulas

$O(\Delta t^2)$ center-difference schemes.

$f'(t) = [f(t + \Delta t) - f(t - \Delta t)]/2\Delta t$
$f''(t) = [f(t + \Delta t) - 2f(t) + f(t - \Delta t)]/\Delta t^2$
$f'''(t) = [f(t + 2\Delta t) - 2f(t + \Delta t) + 2f(t - \Delta t) - f(t - 2\Delta t)]/2\Delta t^3$
$f''''(t) = [f(t + 2\Delta t) - 4f(t + \Delta t) + 6f(t) - 4f(t - \Delta t) + f(t - 2\Delta t)]/\Delta t^4$

Table 4.2   Fourth-order accurate center-difference formulas

$O(\Delta t^4)$ center-difference schemes.

$f'(t) = [-f(t + 2\Delta t) + 8f(t + \Delta t) - 8f(t - \Delta t) + f(t - 2\Delta t)]/12\Delta t$
$f''(t) = [-f(t + 2\Delta t) + 16f(t + \Delta t) - 30f(t) + 16f(t - \Delta t) - f(t - 2\Delta t)]/12\Delta t^2$
$f'''(t) = [-f(t + 3\Delta t) + 8f(t + 2\Delta t) - 13f(t + \Delta t) + 13f(t - \Delta t) - 8f(t - 2\Delta t) + f(t - 3\Delta t)]/8\Delta t^3$
$f''''(t) = [-f(t + 3\Delta t) + 12f(t + 2\Delta t) - 39f(t + \Delta t) + 56f(t) - 39f(t - \Delta t) + 12f(t - 2\Delta t)$
$\qquad - f(t - 3\Delta t)]/6\Delta t^4$

Table 4.3   Second-order accurate forward- and backward-difference formulas

$O(\Delta t^2)$ forward- and backward-difference schemes.

$f'(t) = [-3f(t) + 4f(t + \Delta t) - f(t + 2\Delta t)]/2\Delta t$
$f'(t) = [3f(t) - 4f(t - \Delta t) + f(t - 2\Delta t)]/2\Delta t$
$f''(t) = [2f(t) - 5f(t + \Delta t) + 4f(t + 2\Delta t) - f(t + 3\Delta t)]/\Delta t^2$
$f''(t) = [2f(t) - 5f(t - \Delta t) + 4f(t - 2\Delta t) - f(t - 3\Delta t)]/\Delta t^2$

and fourth-order central difference schemes along with the forward- and backward-difference formulas which are accurate to second order.

A final remark is in order concerning these differentiation schemes. The central difference schemes are an excellent method for generating the values of the derivative in the interior points of a data set. However, at the end points, forward- and backward-difference methods must be used since they do not have neighboring points to the left and right, respectively. Thus special care must be taken at the end points of any computational domain.

It may be tempting to deduce from the difference formulas that as $\Delta t \to 0$, the accuracy only improves in these computational methods. However, this line of reasoning completely neglects the second source of error in evaluating derivatives: numerical round-off.

Round-off and optimal step-size

An unavoidable consequence of working with numerical computations is round-off error. When working with most computations, *double precision* numbers are used. This allows for 16-digit accuracy in the representation of a given number. This round-off has a significant impact upon numerical computations and the issue of time-stepping.

As an example of the impact of round-off, we consider the approximation to the derivative

$$\frac{dy}{dt} = \frac{y(t + \Delta t) - y(t - \Delta t)}{2\Delta t} + \epsilon(y(t), \Delta t) \tag{4.1.11}$$

where $\epsilon(y(t), \Delta t)$ measures the truncation error. Upon evaluating this expression in the computer, round-off error occurs so that

$$y(t + \Delta t) = Y(t + \Delta t) + e(t + \Delta t) \tag{4.1.12a}$$

$$y(t - \Delta t) = Y(t - \Delta t) + e(t - \Delta t), \tag{4.1.12b}$$

where $Y(\cdot)$ is the approximated value given by the computer and $e(\cdot)$ measures the error from the true value $y(t)$. Thus the combined error between the round-off and truncation gives the following expression for the derivative:

$$\frac{dy}{dt} - \frac{Y(t + \Delta t) - Y(t - \Delta t)}{2\Delta t} + F(y(t), \Delta t) \tag{4.1.13}$$

where the total error, E, is the combination of round-off and truncation such that

$$E = E_{\text{round}} + E_{\text{trunc}} = \frac{e(t + \Delta t) - e(t - \Delta t)}{2\Delta t} - \frac{\Delta t^2}{6}\frac{d^3 y(c)}{dt^3}. \tag{4.1.14}$$

We now determine the maximum size of the error. In particular, we can bound the maximum value of the round-off error and the value of the second derivative to be

$$|e(t + \Delta t)| \le e_r \tag{4.1.15a}$$

$$|-e(t - \Delta t)| \le e_r \tag{4.1.15b}$$

$$M = \max_{c \in [t_n, t_{n+1}]} \left\{ \left| \frac{d^3 y(c)}{dt^3} \right| \right\}. \tag{4.1.15c}$$

This then gives the maximum error to be

$$|E| \le \frac{e_r + e_r}{2\Delta t} + \frac{\Delta t^2}{6}M = \frac{e_r}{\Delta t} + \frac{\Delta t^2 M}{6}. \tag{4.1.16}$$

Note that as Δt gets large, the error grows linearly due to the truncation error. However, as Δt decreases to zero, the error is dominated by round-off which grows like $1/\Delta t$. The error as a function of the step-size Δt for this second-order scheme is represented in Fig. 4.2. Note the dominance of the error on numerical round-off as $\Delta t \to 0$.

To minimize the error, we require that $\partial |E| / \partial(\Delta t) = 0$. Calculating this derivative gives

$$\frac{\partial |E|}{\partial(\Delta t)} = -\frac{e_r}{\Delta t^2} + \frac{\Delta t M}{3} = 0, \tag{4.1.17}$$

so that

$$\Delta t = \left(\frac{3e_r}{M} \right)^{1/3}. \tag{4.1.18}$$

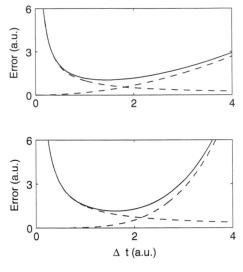

Figure 4.2: Graphical representation of the error which is made up of two components (dotted lines): numerical round-off and truncation. The total error in arbitrary units is shown for both a second-order scheme (top panel) and a fourth-order scheme (bottom panel). For convenience, we have taken $M = 1$ and $e_r = 1$. Note the dominance of numerical round-off as $\Delta t \to 0$.

This gives the step-size resulting in a minimum error. Thus the smallest step-size is not necessarily the most accurate. Rather, a balance between round-off error and truncation error is achieved to obtain the optimal step-size. For $e_r \approx 10^{-16}$, the optimal $\Delta t \approx 10^{-5}$. Below this value of Δt, numerical round-off begins to dominate the error.

A similar procedure can be carried out for evaluating the optimal step-size associated with the $O(\Delta t^4)$ accurate scheme for the first derivative. In this case

$$\frac{dy}{dt} = \frac{-f(t+2\Delta t) + 8f(t+\Delta t) - 8f(t-\Delta t) + f(t-2\Delta t)}{12\Delta t} + E(y(t), \Delta t) \qquad (4.1.19)$$

where the total error, E, is the combination of round-off and truncation such that

$$E = \frac{-e(t+2\Delta t) + 8e(t+\Delta t) - 8e(t-\Delta t) + e(t-2\Delta t)}{12\Delta t} + \frac{\Delta t^4}{30}\frac{d^5 y(c)}{dt^5}. \qquad (4.1.20)$$

We now determine the maximum size of the error. In particular, we can bound the maximum value of round-off to e as before and set $M = \max\left\{\left|y''''' (c)\right|\right\}$. This then gives the maximum error to be

$$|E| \le \frac{3e_r}{2\Delta t} + \frac{\Delta t^4 M}{30}. \qquad (4.1.21)$$

Note that as Δt gets large, the error grows like a quartic due to the truncation error. However, as Δt decreases to zero, the error is again dominated by round-off which grows like $1/\Delta t$. The error as a function of the step-size Δt for this fourth-order scheme is represented in Fig. 4.2. Note the dominance of the error on numerical round-off as $\Delta t \to 0$.

To minimize the error, we require that $\partial |E|/\partial(\Delta t) = 0$. Calculating this derivative gives

$$\Delta t = \left(\frac{45e_r}{4M}\right)^{1/5}.$$
(4.1.22)

Thus in this case the optimal step $\Delta t \approx 10^{-3}$. This shows that the error can be quickly dominated by numerical round-off if one is not careful to take this significant effect into account.

 ## 4.2   Numerical Integration

Numerical integration simply calculates the area under a given curve. The basic ideas for performing such an operation come from the definition of integration

$$\int_a^b f(x)dx = \lim_{h \to 0} \sum_{j=0}^{N} f(x_j)h$$
(4.2.1)

where $b - a = Nh$. Thus the area under the curve, from the calculus standpoint, is thought of as a limiting process of summing up an ever-increasing number of rectangles. This process is known as numerical quadrature. Specifically, any sum can be represented as follows

$$Q[f] = \sum_{j=0}^{N} w_k f(x_k) = w_0 f(x_0) + w_1 f(x_1) + \cdots + w_N f(x_N)$$
(4.2.2)

where $a = x_0 < x_1 < x_2 < \cdots < x_N = b$. Thus the integral is evaluated as

$$\int_a^b f(x)dx = Q[f] + E[f]$$
(4.2.3)

where the term $E[f]$ is the error in approximating the integral by the quadrature sum (4.2.2). Typically, the error $E[f]$ is due to truncation error. To integrate, use will be made of polynomial fits to the y-values $f(x_j)$. Thus we assume the function $f(x)$ can be approximated by a polynomial

$$P_n(x) = a_n x^n + a_{n-1} x^{n-1} + \cdots + a_1 x + a_0$$
(4.2.4)

where the truncation error in this case is proportional to the $(n+1)$th derivative $E[f] = Af^{(n+1)}(c)$ and A is a constant. This process of polynomial fitting the data gives the *Newton–Cotes formulas*. Figure 4.3 gives the standard representation of the integration process and the division of the integration interval into a finite set of integration intervals.

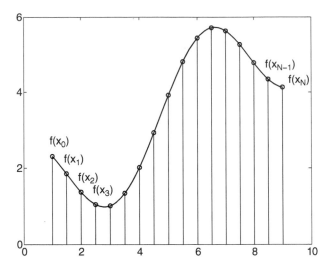

Figure 4.3: Graphical representation of the integration process. The integration interval is broken up into a finite set of points. A quadrature rule then determines how to sum up the area of a finite number of rectangles.

Newton–Cotes formulas

The following integration approximations result from using a polynomial fit through the data to be integrated. It is assumed that

$$x_k = x_0 + hk \qquad\qquad f_k = f(x_k). \tag{4.2.5}$$

This gives the following integration algorithms:

$$\text{Trapezoid rule} \int_{x_0}^{x_1} f(x)dx = \frac{h}{2}(f_0 + f_1) - \frac{h^3}{12}f''(c) \tag{4.2.6a}$$

$$\text{Simpson's rule} \int_{x_0}^{x_2} f(x)dx = \frac{h}{3}(f_0 + 4f_1 + f_2) - \frac{h^5}{90}f''''(c) \tag{4.2.6b}$$

$$\text{Simpson's 3/8 rule} \int_{x_0}^{x_3} f(x)dx = \frac{3h}{8}(f_0 + 3f_1 + 3f_2 + f_3) - \frac{3h^5}{80}f''''(c) \tag{4.2.6c}$$

$$\text{Boole's rule} \int_{x_0}^{x_4} f(x)dx = \frac{2h}{45}(7f_0 + 32f_1 + 12f_2 + 32f_3 + 7f_4) - \frac{8h^7}{945}f^{(6)}(c). \tag{4.2.6d}$$

These algorithms have varying degrees of accuracy. Specifically, they are $O(h^2)$, $O(h^4)$, $O(h^4)$ and $O(h^6)$ accurate schemes, respectively. The accuracy condition is determined from the truncation terms of the polynomial fit. Note that the *trapezoid rule* uses a sum of simple trapezoids to approximate the integral. *Simpson's rule* fits a quadratic curve through three points and calculates the area under the quadratic curve. *Simpson's 3/8 rule* uses four points and a cubic polynomial

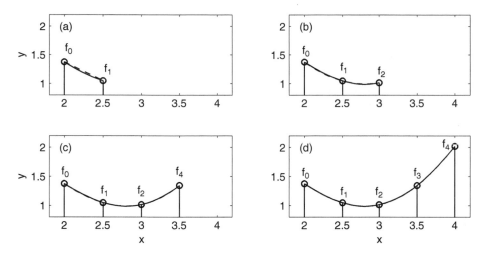

Figure 4.4: Graphical representation of the integration process for the quadrature rules in (4.2.6). The quadrature rules represented are (a) trapezoid rule, (b) Simpson's rule, (c) Simpson's 3/8 rule and (d) Boole's rule. In addition to the specific function considered, the dotted line represents the polynomial fit to the actual function for the four different integration rules. Note that the dotted line and function line are essentially identical for (c) and (d).

to evaluate the area, while *Boole's rule* uses five points and a quartic polynomial fit to generate an evaluation of the integral. Figure 4.4 represents graphically the different quadrature rules and its approximation (dotted line) to the actual function.

The derivation of these integration rules follows from simple polynomial fits through a specified number of data points. To derive Simpson's rule, consider a second degree Lagrange polynomial through the three points (x_0, f_0), (x_1, f_2) and (x_2, f_2):

$$p_2(x) = f_0 \frac{(x - x_1)(x - x_2)}{(x_0 - x_1)(x_0 - x_2)} + f_1 \frac{(x - x_0)(x - x_2)}{(x_1 - x_0)(x_1 - x_2)} + f_2 \frac{(x - x_0)(x - x_1)}{(x_2 - x_0)(x_2 - x_1)}. \tag{4.2.7}$$

This quadratic fit is derived by using Lagrange coefficients. The truncation error could also be included, but we neglect it for the present purposes. By plugging in (4.2.7) into the integral

$$\int_{x_0}^{x_2} f(x)dx \approx \int_{x_0}^{x_2} p_2(x)dx = \frac{h}{3}(f_0 + 4f_1 + f_2). \tag{4.2.8}$$

The integral calculation is easily performed since it only involves integrating powers of x^2 or less. Evaluating at the limits then causes many terms to cancel and drop out. Thus Simpson's rule is recovered. The trapezoid rule, Simpson's 3/8 rule and Boole's rule are all derived in a similar fashion. To make connection with the quadrature rule (4.2.2), $Q = w_0 f_0 + w_1 f_1 + w_2 f_2$, Simpson's rule gives $w_0 = h/3$, $w_1 = 4h/3$ and $w_2 = h/3$ as weighting factors.

Composite rules

The integration methods (4.2.6) give values for the integrals over only a small part of the integration domain. The trapezoid rule, for instance, only gives a value for $x \in [x_0, x_1]$. However, our fundamental aim is to evaluate the integral over the entire domain $x \in [a, b]$. Assuming once again that our interval is divided as $a = x_0 < x_1 < x_2 < \cdots < x_N = b$, then the trapezoid rule applied over the interval gives the total integral

$$\int_a^b f(x)dx \approx Q[f] = \sum_{j=1}^{N-1} \frac{h}{2} \left(f_j + f_{j+1} \right). \tag{4.2.9}$$

Writing out this sum gives

$$\sum_{j=1}^{N-1} \frac{h}{2} \left(f_j + f_{j+1} \right) = \frac{h}{2}(f_0 + f_1) + \frac{h}{2}(f_1 + f_2) + \cdots + \frac{h}{2}(f_{N-1} + f_N)$$

$$= \frac{h}{2}(f_0 + 2f_1 + 2f_2 + \cdots + 2f_{N-1} + f_N) \tag{4.2.10}$$

$$= \frac{h}{2} \left(f_0 + f_N + 2 \sum_{j=1}^{N-1} f_j \right).$$

The final expression no longer double counts the values of the points between f_0 and f_N. Instead, the final sum only counts the intermediate values once, thus making the algorithm about twice as fast as the previous sum expression. These are computational savings which should always be exploited if possible.

Recursive improvement of accuracy

Given an integration procedure and a value of h, a function or data set can be integrated to a prescribed accuracy. However, it may be desirable to improve the accuracy without having to disregard previous approximations to the integral. To see how this might work, consider the trapezoidal rule for a step-size of $2h$. Thus, the even data points are the only ones of interest and we have the basic one-step integral

$$\int_{x_0}^{x_2} f(x)dx \approx \frac{2h}{2}(f_0 + f_2) = h(f_0 + f_2). \tag{4.2.11}$$

The composite rule associated with this is then

$$\int_a^b f(x)dx \approx Q[f] = \sum_{j=0}^{N/2-1} h\left(f_{2j} + f_{2j+2} \right). \tag{4.2.12}$$

Writing out this sum gives

$$\sum_{j=0}^{N/2-1} h(f_{2j} + f_{2j+2}) = h(f_0 + f_2) + h(f_2 + f_4) + \cdots + h(f_{N-2} + f_N)$$

$$= h(f_0 + 2f_2 + 2f_4 + \cdots + 2f_{N-2} + f_N) \tag{4.2.13}$$

$$= h\left(f_0 + f_N + 2\sum_{j=1}^{N/2-1} f_{2j}\right).$$

Comparing the middle expressions in (4.2.10) and (4.2.13) gives a great deal of insight into how recursive schemes for improving accuracy work. Specifically, we note that the more accurate scheme with step-size h contains all the terms in the integral approximation using step-size $2h$. Quantifying this gives

$$Q_h = \frac{1}{2}Q_{2h} + h(f_1 + f_3 + \cdots + f_{N\ 1}), \tag{4.2.14}$$

where Q_h and Q_{2h} are the quadrature approximations to the integral with step-size h and $2h$, respectively. This then allows us to cut the value of h in half and improve accuracy without jettisoning the work required to approximate the solution with the accuracy given by a step-size of $2h$. This recursive procedure can be continued so as to give higher accuracy results. Further, this type of procedure holds for Simpson's rule as well as any of the integration schemes developed here. This recursive routine is used in MATLAB's integration routines in order to generate results to a prescribed accuracy.

4.3 Implementation of Differentiation and Integration

This section focuses on the implementation of differentiation and integration methods. Since they form the backbone of calculus, accurate methods to approximate these calculations are essential. To begin, we consider a specific function to be differentiated, namely a hyperbolic secant. Part of the reason for considering this function is that the exact values of its first two derivatives are known. This allows us to make a comparison of our approximate methods with the actual solution. Thus, we consider

$$u = \text{sech}(x) \tag{4.3.1}$$

whose derivative and second derivative are

$$\frac{du}{dx} = -\text{sech}(x)\tanh(x) \tag{4.3.2a}$$

$$\frac{d^2u}{dx^2} = \text{sech}(x) - \text{sech}^3(x). \tag{4.3.2b}$$

Differentiation

To begin the calculations, we define a spatial interval. For this example we take the interval $x \in [-10, 10]$. In MATLAB, the spatial discritization, Δx, must also be defined. This gives

```
dx=0.1;         % spatial discritization
x=-10:dx:10;    % spatial domain
```

Once the spatial domain has been defined, the function to be differentiated must be evaluated

```
u=sech(x);
ux_exact=-sech(x).*tanh(x);
uxx_exact=sech(x)-sech(x).^3;
figure(1), plot(x,u,x,ux_exact,x,uxx_exact)
```

MATLAB *figure 1* produces the function and its first two derivatives.

To calculate the derivative numerically, we use the center-, forward-, and backward-difference formulas derived for differentiation. Specifically, we will make use of the following four first-derivative approximations from Tables 4.1–4.3:

$$\text{center-difference } O(h^2) : \frac{y_{n+1} - y_{n-1}}{2h} \tag{4.3.3a}$$

$$\text{center-difference } O(h^4) : \frac{-y_{n+2} + 8y_{n+1} - 8y_{n-1} + y_{n-2}}{12h} \tag{4.3.3b}$$

$$\text{forward-difference } O(h^2) : \frac{-3y_n + 4y_{n+1} - y_{n+2}}{2h} \tag{4.3.3c}$$

$$\text{backward-difference } O(h^2) : \frac{3y_n - 4y_{n-1} + y_{n-2}}{2h}. \tag{4.3.3d}$$

Here we have used $h = \Delta x$ and $y_n = y(x_n)$.

Second-order accurate derivative

To calculate the second-order accurate derivative, we use the first, third and fourth equations of (4.3.3). In the interior of the domain, we use the center-difference scheme. However, at the left and right boundaries, there are no left and right neighboring points, respectively, to calculate the derivative with. Thus we require the use of forward- and backward-difference schemes, respectively. This gives the basic algorithm

```
n=length(x)
% 2nd-order accurate
ux(1)=(-3*u(1)+4*u(2)-u(3))/(2*dx);
for j=2:n-1
    ux(j)=(u(j+1)-u(j-1))/(2*dx);
end
ux(n)=(3*u(n)-4*u(n-1)+u(n-2))/(2*dx);
```

The values of $ux(1)$ and $ux(n)$ are evaluated with the forward- and backward-difference schemes.

Fourth-order accurate derivative

A higher degree of accuracy can be achieved by using a fourth-order scheme. A fourth-order center-difference scheme such as the second equation of (4.3.3) relied on two neighboring points. Thus the first two and last two points of the computational domain must be handled separately. In what follows, we use a second-order scheme at the boundary points, and a fourth-order scheme for the interior. This gives the algorithm

```
% 4th-order accurate
ux2(1)=(-3*u(1)+4*u(2)-u(3))/(2*dx);
ux2(2)=(-3*u(2)+4*u(3)-u(4))/(2*dx);
for j=3:n-2
    ux2(j)=(-u(j+2)+8*u(j+1)-8*u(j-1)+u(j-2))/(12*dx);
end
ux2(n-1)=(3*u(n-1)-4*u(n-2)+u(n-3))/(2*dx);
ux2(n)=(3*u(n)-4*u(n-1)+u(n-2))/(2*dx);
```

For the $\Delta x = 0.1$ considered here, the second-order and fourth-order schemes should result in errors of 10^{-2} and 10^{-4}, respectively.

To view the accuracy of the first derivative approximations, the results are plotted together with the analytic solution for the derivative

```
figure(2), plot(x,ux_exact,'o',x,ux,'c',x,ux2,'m')
```

To the naked eye, these results all look to be on the exact solution. However, by zooming in on a particular point, it quickly becomes clear that the errors for the two schemes are indeed 10^{-2} and 10^{-4}, respectively. The failure of accuracy can also be observed by taking large values of Δx. For instance, $\Delta x = 0.5$ and $\Delta x = 1$ illustrate how the derivatives fail to be properly determined with large Δx values. Figure 4.5 shows the function and its first two numerically calculated

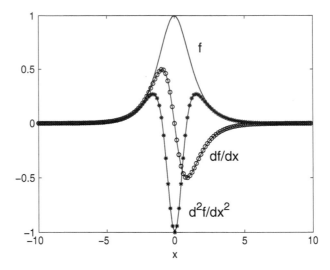

Figure 4.5: Calculation of the first and second derivatives of $f(x) = \text{sech}(x)$ using the finite difference method. The approximate solutions for the first and second derivatives are compared with the exact analytic solutions (circles and stars, respectively).

derivatives for the simple function $f(x) = \text{sech}(x)$. A more careful comparison of the accuracy of both schemes is given in Fig. 4.6. This figures shows explicitly that for $\Delta x = 0.1$ the second-order and fourth-order schemes produce errors on the order of 10^{-2} and 10^{-4}, respectively.

As a final note, if you are given a set of data to differentiate, it is always recommended that you first run a spline through the data with an appropriately small Δx and then differentiate the spline. This will help to give smooth differentiated data. Otherwise, the data will tend to be highly inaccurate and choppy.

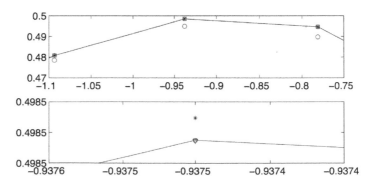

Figure 4.6: Accuracy comparison between second- and fourth-order finite difference methods for calculating the first derivative. Note that by using the *axis* command, the exact solution (line) and its approximations can be magnified from Fig. 4.5 near an arbitrary point of interest. Here, the top figure shows that the second-order finite difference (circles) method is within $O(10^{-2})$ of the exact derivative (box). The fourth-order finite difference (star) is within $O(10^{-5})$ of the exact derivative.

Integration

There are a large number of integration routines built into MATLAB. So unlike the differentiation routines presented here, we will simply make use of the built-in MATLAB functions. The most straightforward integration routine is the trapezoidal rule. Given a set of data, the **trapz** command can be used to implement this integration rule. For the x and u data defined previously for the hyperbolic secant, the command structure is

```
int_sech=trapz(x,u.^2);
```

where we have integrated $\text{sech}^2(x)$. The value of this integral is exactly 2. The **trapz** command gives a value that is within 10^{-7} of the true value. To generate the cumulative values, the **cumtrapz** command is used

```
int_sech2=cumtrapz(x,u.^2);
figure(3), plot(x,int_sech2)
```

MATLAB *figure 3* gives the value of the integral as a function of x. Thus if the function u represented the velocity of an object, then the vector **int_sech2** generated from **cumtrapz** would represent the position of the object.

Alternatively, a function can be specified for integration over a given domain. The **quad** command is implemented by specifying a function and the range of integration. This will give a value of the integral using a recursive Simpson's rule that is accurate to within 10^{-6}. The command structure for this evaluation is as follows

```
int_quad=quad(@fun,a,b)
```

where **fun** specifies the function to integrate and a and b are the lower and upper limits of integration, respectively. To integrate the function we have been considering thus far, the specific command structure is as follows:

```
A=1;
int_quad=quad(inline('A*sech(x).^2','x','A'),-10,10,[],[],A)
```

Here the **inline** command allows us to circumvent a traditional function call. The **quad** command can, however, be used with a function call. This command executes the integration of $\text{sech}^2(x)$ over $x \in [-10, 10]$ with the parameter A. Often in computations, it is desirable to pass in

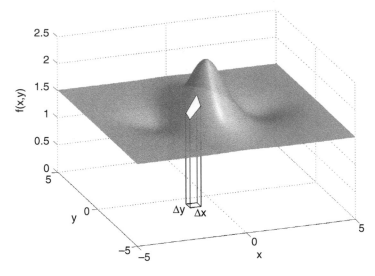

Figure 4.7: Two-dimensional trapezoidal rule integration procedure. As with one dimension, the domain is broken down into a grid of width and height Δx and Δy. The area of each parallelepiped is then easily calculated and summed up.

these parameters. Thus this example is particularly helpful in illustrating the full use of the **quad** command. The two empty brackets are for tolerance and trace settings.

Double and triple integrals over two-dimensional rectangles and three-dimensional boxes can be performed with the **dblquad** and **triplequad** commands. Note that no version of the **quad** command exists that produces cumulative integration values. However, this can be easily handled in a **for** loop. Double integrals can also be performed with the **trapz** command using a **trapz** imbedded within another **trapz**. As an example of a two-dimensional integration, we consider the **dblquad** command for integration of the following two-dimensional function

$$f(x, y) = \cos(x)\text{sech}(x)\text{sech}(y) + 1.5 \qquad (4.3.4)$$

on the box $x \in [-5, 5]$ and $y \in [-5, 5]$. The **trapz** MATLAB code for solving this problem is the following

```
Area=dblquad(inline('cos(x).*sech(x).*sech(y)+1.5'),-5,5,-5,5)
```

This integrates the function over the box of interest. Figure 4.7 demonstrates the trapezoidal rule that can be used to handle the two-dimensional integration procedure. A similar procedure is carried out in three dimensions.

Basic Optimization 5

Optimization methods are prevalent throughout a vast range of applications. In its simplest form, an optimization routine simply attempts to find the maximum or minimum of a real-valued function, i.e. the *objective function*, by developing an algorithm that systematically chooses input values from an allowed set, or *feasible* set, and computes the value of the function. Typically, the optimization algorithm is built upon an iteration scheme that continues to choose new input values so that the maximum or minimum of the objective function is achieved. The generalization of optimization theory and techniques to other formulations comprises a large area of applied mathematics. More generally, optimization includes finding *best available* values of some objective function given a defined domain, including a variety of different types of objective functions and different types of domains. In the following sections, a brief introduction to these techniques will be given along with their MATLAB function calls.

5.1 Unconstrained Optimization (Derivative-Free Methods)

The concept of optimization is fairly simple: find the minimum or maximum of a function. In optimization, the function is real-valued and called the *objective function*. Moreover, it is often the case that you would like to find the maximum or minimum under some *constraints* on the input values to the function. Thus there is a concept of *feasibility* of the solution. To begin, we will consider optimization without constraints, thus the aptly named *unconstrained optimization* problem. With a background in calculus, the mention of minimization or maximization of a function automatically evokes the concept of the derivative, i.e. setting the derivative of a function to zero implies it is either a minimum or maximum. In this section, however, derivative free methods for computing the minimum or maximum will be considered. Such methods are often employed when computing the derivatives is either very expensive computationally, or simply intractable in practice.

Before proceeding forward with the development of the methods, abstracting the unconstrained optimization problem is necessary. Thus consider minimizing the objective function

$$\min f(\mathbf{x}) \tag{5.1.1}$$

where the objective function depends generally on a number of variables

$$x_1, x_2, \cdots , x_n . \tag{5.1.2}$$

These are called the *control variables* because we can choose their values. Indeed, our objective is to choose these values so that the objective function (5.1.1) is minimized. It is also these control variables that are often constrained in practice so that it becomes a *constrained optimization* problem.

By definition, $f(\mathbf{x})$ has a minimum at a point $\mathbf{x} = \mathbf{x_0}$ in a region R if

$$f(\mathbf{x}) \geq f(\mathbf{x_0}) \quad \text{for } \mathbf{x} \in R. \tag{5.1.3}$$

A maximum can be defined in a similar way. However, a maximum of $f(\mathbf{x})$ becomes a minimum of $-f(\mathbf{x})$. Thus it suffices from here forward to only consider the minimization problem. Note that in this definition of the minimum, it is not specified wether this is a local or global minimum. For highly nontrivial functions $f(\mathbf{x})$, it may be that there exists many local minima, thus a search algorithm that converges to a minimum solution may not be the actual desired solution. This will be illustrated shortly in practice.

For simplicity, we will consider derivative-free optimization methods for a function of a single variable $(\min f(x))$. The methods here are very much like the root solving techniques demonstrated in the earlier chapters of the book, i.e. they are simple iterative schemes that progressively produce better and better minimization values. In both methods shown, an initial interval is required in which the minimum lies.

Golden section search

Given an interval in which the minimum is known to exist, the golden section search algorithm is an efficient method for finding the minimum of the function. To be more specific, the function must be *unimodal* in the interval $x \in [a, b]$, meaning there is only a single relative minimum in the interval chosen. The algorithm is much like the bisection method for root solving. Specifically, within the interval where the minimum exists, two points (x_1 and x_2) are chosen so that $a < x_1 < x_2 < b$. Evaluation of the function $f(x)$ is then performed at x_1 and x_2. The following observations hold:

$$\text{if } f(x_1) \leq f(x_2) \quad \text{then retain interval } x \in [a, x_2]$$
$$\text{if } f(x_1) > f(x_2) \quad \text{then retain interval } x \in [x_1, b].$$

This ensures that the minimum is now in a smaller subinterval of the original $x \in [a, b]$.

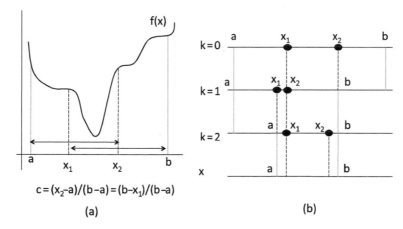

Figure 5.1: Graphical depiction of the golden section search. (a) Two points, x_1 and x_2, are used for the iteration process. The points are chosen so as to keep the ratio of $(b - x_1)/(b - a)$ and $(x_2 - a)/(b - a)$ the same. (b) The first three iterations are depicted.

The golden section search gives a simple procedure for selecting the evaluation points x_1 and x_2. Specifically, they are chosen so that (i) we want a constant reduction factor, say c, for the size of the interval, and (ii) they can be reused at the next iteration, thus saving a function evaluation. From the observations above, criterion (i) is enforced mathematically with the conditions:

$$c = \frac{x_2 - a}{b - a} \quad \rightarrow \quad x_2 = (1 - c)a + cb \qquad (5.1.4a)$$

$$c = \frac{b - x_1}{b - a} \quad \rightarrow \quad x_1 = ca + (1 - c)b \qquad (5.1.4b)$$

where c is the reduction factor to the new intervals $x \in [a, x_2]$ and $x \in [x_1, b]$, respectively (see Fig. 5.1). The key observation is the following: if $f(x_1) < f(x_2)$ then the new interval is $x \in [a, x_2]$. Moreover, a new x_1 must be evaluated while the old x_2 is the x_1. Thus we have the following upon making use of (5.1.4):

$$x_2^{\text{new}} = x_1^{\text{old}} = ca + (1 - c)b \qquad (5.1.5a)$$

$$x_2^{\text{new}} = (1 - c)a + cx_2^{\text{old}} = (1 - c)a + c[(1 - c)a + cb]. \qquad (5.1.5b)$$

But these two new values of x_2 must be equivalent, thus we have

$$ca + (1 - c)b = (1 - c)a + c[(1 - c)a + cb]$$
$$c^2(b - a) + c(b - a) - (b - a) = 0$$
$$c^2 + c - 1 = 0$$
$$c = (-1 + \sqrt{5})/2 \approx 0.6180 \qquad (5.1.6)$$

where only the positive root of c is kept since c must be positive. Note that the resulting ratio is simply unity minus the golden ratio, i.e. $\phi = 1 + c$, thus the name *golden section search*. Now that c is determined, x_1 and x_2 can be found from (5.1.4). Such a routine can be easily implemented in MATLAB. The following finds the minimum of the function $f(x) = x^4 + 10x\sin(x^2)$ on the interval $x \in [-2, 1]$.

goldensearch.m

```
a=-2; b=1;  % initial interval
c=(-1+sqrt(5))/2;   % golden section

x1=c*a + (1-c)*b;
x2=(1-c)*a + c*b;
f1=x1.^4+10*x1.*sin(x1.^2);
f2=x2.^4+10*x2.*sin(x2.^2);

for j=1:100
  if f1<f2    % move right boundary
     b=x2; x2=x1; f2=f1;
     x1=c*a+(1-c)*b;
     f1=x1.^4+10*x1.*sin(x1.^2);
  else    % move left boundary
     a=x1; x1=x2; f1=f2;
     x2=(1-c)*a + c*b;
     f2=x2.^4+10*x2.*sin(x2.^2);
  end

  if (b-a)<10^(-6)   % break if close
     break
  end
end
```

The above algorithm converges in 31 iterations to the minimum $f = -10.0882$ at $x = -1.2742$ with an accuracy of 10^{-6}. A theorem regarding the golden search algorithm states that after k iterations, starting from the interval $x \in [a, b]$, the midpoint of this final interval is within $c^k(b - a)/2$ of the minimum. Thus a guaranteed convergence rate can be established.

Successive parabolic interpolation

In the golden section search, no information was used about the values of $f(x_1)$ and $f(x_2)$ in selecting a new subinterval. Thus if $f(x_1) \ll f(x_2)$, it would be judicious to assume that the minimum might be closer to the point x_1 than x_2 and the interval should cut accordingly. A technique that makes use of the function evaluation in choosing how to refine the interval is the method of successive parabolic interpolation.

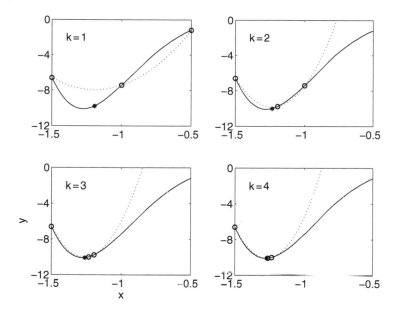

Figure 5.2: Graphical depiction of the successive parabolic interpolation algorithm. The three data points are depicted (circles) along with the point evaluated at the minimum of the parabola (star). The solid line is the function and the dotted line is the parabola generated. In this case, the guesses give a rapid convergence (14 iterations for 10^{-6} accuracy) to the minimum. However, the method is not guaranteed to converge and can, in fact, easily diverge. Thus it is important to have a good local starting point.

The idea behind successive parabolic interpolation is to choose three points near the vicinity of the minimum: x_1, x_2 and x_3. These points correspond to left, middle and right points, respectively (see Fig. 5.2). With the choice of these three points, one can evaluate the function values associated with each point, $f(x_1)$, $f(x_2)$ and $f(x_3)$ and fit a parabola through them. Using the Lagrange polynomial coefficients, this gives a parabolic function

$$p(x) = f(x_1)\frac{(x-x_2)(x-x_3)}{(x_1-x_2)(x_1-x_3)} + f(x_2)\frac{(x-x_1)(x-x_3)}{(x_2-x_1)(x_2-x_3)} + f(x_3)\frac{(x-x_1)(x-x_2)}{(x_3-x_1)(x_3-x_2)} \tag{5.1.7}$$

where $p(x)$ is now locally approximating the function $f(x)$. The minimum of the parabola now serves as a temporary proxy for the minimum of the actual function $f(x)$. The minimum of the parabola can be found by setting the first derivative to zero so that we evaluate $p'(x_0) = 0$. This gives, after some algebra, the following minimum:

$$x_0 = \frac{x_1 + x_2}{2} - \frac{(f_2 - f_1)(x_3 - x_1)(x_3 - x_2)}{2[(x_2 - x_1)(f_3 - f_2) - (f_2 - f_1)(x_3 - x_2)]}. \tag{5.1.8}$$

The idea now is to use this new point x_0 as our new middle point x_2. There are two cases of interest:

$$x_0 < x_2 : \quad x_1^{\text{new}} = x_1^{\text{old}}$$
$$x_2^{\text{new}} = x_0$$
$$x_3^{\text{new}} = x_2^{\text{old}}$$

$$x_0 > x_2 : \quad x_1^{\text{new}} = x_2^{\text{old}}$$
$$x_2^{\text{new}} = x_0$$
$$x_3^{\text{new}} = x_3^{\text{old}}.$$

This gives a simple algorithm that progressively converges to the minimum by using information about the function values. Moreover, it only requires a single function evaluation per iterative step.

Figure 5.2 gives a graphical depiction of this local iteration process. The convergence is typically extremely fast once you can find a good neighborhood to work in. However, the method is not guaranteed to converge, unfortunately. Thus great care should be used with this method. Alternatively, very good starting points must always be used. The following code implements the successive parabolic approximation.

goldensearch.m

```
x1=-1.5; x2=-1; x3=-.5; % initial guesses
f1=x1.^4+10*x1.*sin(x1.^2);
f2=x2.^4+10*x2.*sin(x2.^2);
f3=x3.^4+10*x3.*sin(x3.^2);

for j=1:100
  x0 =(x1+x2)/2 - ( (f2-f1)*(x3-x1)*(x3-x2) ) ...
       /( 2*( (x2-x1)*(f3-f2)-(f2-f1)*(x3-x2) ) );

  if x0>x2
    x1=x2; f1=f2;
    x2=x0; f2=x0.^4+10*x0.*sin(x0.^2);
  else
    x3=x2; f3=f2;
    x2=x0; f2=x0.^4+10*x0.*sin(x0.^2);
  end

  if abs(x2-x3)<10^(-6) | abs(x2-x1)<10^(-6)
    break
  end

end
```

This algorithm converges to the solution in less than half the iterations of the golden section search. However, it is easy to show that if the initial guesses are changed, then the minimization will simply not work.

fminbnd

MATLAB has a built-in one-dimensional search algorithm where the function and the interval are specified. The function **fminbnd** is based upon a combination of the golden section search and successive parabolic search. Integrated together they form an effective technique for finding minima. The following code gives an example of how to execute the function:

```
x=fminbnd('x^2*cos(x)',3,4)
```

Here the left and right values of the search interval are given by $x = 3$ and $x = 4$, respectively. In this case, the function $f(x) = x^2 \cos(x)$ was found to have a minimum at $x = 3.6436$.

5.2   Unconstrained Optimization (Derivative Methods)

The methods of the previous section do not utilize any derivative information about the objective function of interest. However, in many cases the explicit functional form to be considered for minimization is known, thus suggesting that derivatives may help in finding optimization solutions. Indeed, in simple one-dimensional problems for finding the minimum of $f(x) = 0$, it is well known that a minimum is found when $f'(x) = 0$ and $f''(x) > 0$. A maximum can be found when $f'(x) = 0$ and $f''(x) < 0$. Such ideas are easily integrated into an optimization algorithm.

To begin, we generalize the concept of a minimum or maximum, i.e. an extremum for a multi-dimensional function $f(\mathbf{x})$. At an extremum, the gradient must be zero so that

$$\nabla f(\mathbf{x}) = 0.$$ (5.2.1)

Unlike the one-dimensional case, there is no simple second derivative test to apply to determine if the extremum point is a minimum or maximum. The idea behind gradient descent, or steepest descent, is to use the derivative information as the basis of an iterative algorithm that progressively moves closer and closer to the minimum point $f(\mathbf{x}) = 0$.

To illustrate how to proceed in practice, consider the simple example two-dimensional surface

$$f(x, y) = x^2 + 3y^2$$ (5.2.2)

which has the minimum located at the origin $(x, y) = 0$. The gradient for this function can be easily computed

$$\nabla f(\mathbf{x}) = \frac{\partial f}{\partial x}\hat{\mathbf{x}} + \frac{\partial f}{\partial y}\hat{\mathbf{y}} = 2x\hat{\mathbf{x}} + 6y\hat{\mathbf{y}} \tag{5.2.3}$$

where $\hat{\mathbf{x}}$ and $\hat{\mathbf{y}}$ are unit vectors in the x- and y-directions, respectively.

Figure 5.3 illustrates the steepest descent (gradient descent) algorithm. At the initial guess point, the gradient $\nabla f(\mathbf{x})$ can be computed. This gives the steepest descent towards the minimum point of $f(\mathbf{x})$, i.e. the minimum is located in the direction given by $-\nabla f(\mathbf{x})$. Note that the gradient does not point at the minimum, but rather gives the steepest path for minimizing $f(\mathbf{x})$. The geometry of the steepest descent suggests the construction of an algorithm whereby the next point of iteration is picked by following the steepest descent so that

$$\boldsymbol{\xi}(\tau) = \mathbf{x} - \tau \nabla f(\mathbf{x}) \tag{5.2.4}$$

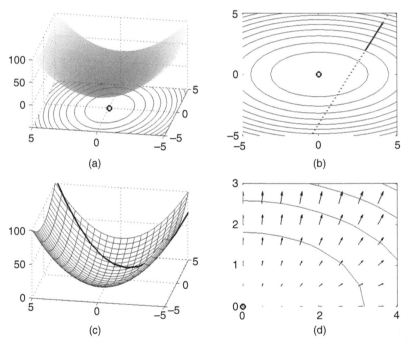

(a)   (b)

(c)   (d)

Figure 5.3: Graphical depiction of the gradient descent algorithm. The surface $f(x, y) = x^2 + 3y^2$ is plotted along with its contour lines. In (a), the surface is plotted along with the contour plot beneath. In (b), the gradient is calculated at the point $(x, y) = (3, 2)$. The gradient, which points away from the mininum, is plotted with the dark bolded line while the gradient line through $(x, y) = (3, 2)$ is plotted with a dotted line. The gradient descent moves along the steepest line of descent as shown in panel (c). Once the bottom of the descent curve is reached, a new descent path is picked. Panel (d) shows the overall gradient of the surface in the upper right quadrant.

where the parameter τ dictates how far to move along the gradient descent curve. Figure 5.3(c) shows that the gradient descent curves gives a descent path that eventually *reaches bottom* and starts to go back up again. In gradient descent, it is crucial to determine when this bottom is reached so that the algorithm is always *going downhill* in an optimal way. This requires the determination of the correct value of τ in the algorithm.

To compute the value of τ, consider the construction of a new function

$$F(\tau) = f(\boldsymbol{\xi}(\tau)) \tag{5.2.5}$$

which must be minimized now as a function of τ. This is accomplished by computing $dF/d\tau = 0$. Thus one finds

$$\frac{\partial F}{\partial \tau} = -\nabla f(\boldsymbol{\xi}) \nabla f(\mathbf{x}) = 0. \tag{5.2.6}$$

The geometrical interpretation of this result is the following: $\nabla f(\mathbf{x})$ is the gradient direction of the current iteration point and $\nabla f(\boldsymbol{\xi})$ is the gradient direction of the future point, thus τ is chosen so that the two gradient directions are orthogonal.

For the example given above with $f(x, y) = x^2 + 3y^2$, we can easily compute this conditions as follows:

$$\boldsymbol{\xi} = \mathbf{x} - \tau \nabla f(\mathbf{x}) = (1 - 2\tau)x\,\hat{\mathbf{x}} + (1 - 6\tau)y\,\hat{\mathbf{y}}. \tag{5.2.7}$$

This is then used to compute

$$F(\tau) = f(\boldsymbol{\xi}(\tau)) = (1 - 2\tau)^2 x^2 + 3(1 - 6\tau)^2 y^2 \tag{5.2.8}$$

whereby its derivative with respect to τ gives

$$F'(\tau) = -4(1 - 2\tau)x^2 - 36(1 - 6\tau)y^2. \tag{5.2.9}$$

Setting $F'(\tau) = 0$ then gives

$$\tau = \frac{x^2 + 9y^2}{2x^2 + 54y^2} \tag{5.2.10}$$

as the optimal descent step length. This gives us all the information necessary to perform the steepest descent search for the minimum of the given function. As is clearly evident, this descent search algorithm based upon derivative information is very much like Newton's method for root finding both in one dimension as well as higher dimensions. Moreover, the gradient descent algorithm is the core algorithm of advanced iterative solvers such as the bi-conjugate gradient descent method (**bicgstab**) and generalized method of residuals (**gmres**).

In what follows, we develop a MATLAB code to perform the gradient descent search for the function $f(x, y) = x^2 + 3y^2$.

```
x(1)=3; y(1)=2;  % initial guess
f(1)=x(1)^2+3*y(1)^2; % initial function value

for j=1:100
  tau=(x(j)^2 +9*y(j)^2)/(2*x(j)^2 + 54*y(j)^2);
  x(j+1)=(1-2*tau)*x(j);  % update values
  y(j+1)=(1-6*tau)*y(j);
  f(j+1)=x(j+1)^2+3*y(j+1)^2;

  if abs(f(j+1)-f(j))<10^(-6)  % check convergence
    break
  end
end
```

The above algorithm converges in only 11 iteration steps to the minimal solution (see Fig. 5.4). Interestingly enough, if a simple radially symmetric function is considered, then the gradient descent converges in a single iteration since the gradient descent would point directly at the minimum. As with other iterative schemes of this sort, including the root finding algorithms based upon the Newton method, convergence to the solution often depends on a user's ability to provide a good initial guess for the minimal value.

The above algorithm assumes a line search algorithm to find an optimal value of τ. In particular, the value of τ picked here is optimal in the sense that a given line search is conducted so

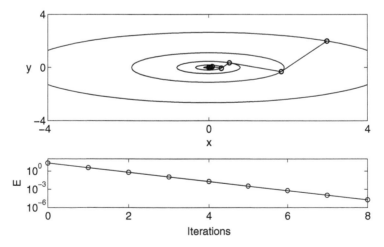

Figure 5.4: Gradient descent algorithm applied to the function $f(x,y) = x^2 + 3y^2$. In the top panel, the contours are plotted for each successive value (x, y) in the iteration algorithm given the initial guess $(x, y) = (3, 2)$. Note the orthogonality of each successive gradient in the steepest descent algorithm. The bottom panel demonstrates the rapid convergence and error (E) to the minimum (optimal) solution.

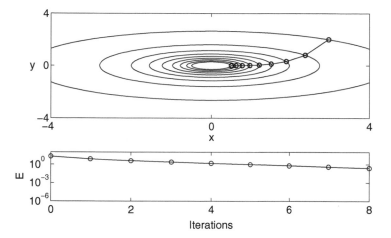

Figure 5.5: Gradient descent algorithm applied to the function $f(x, y) = x^2 + 3y^2$ with a fixed $\tau = 0.1$. In the top panel, the contours are plotted for each successive value (x, y) in the iteration algorithm given the initial guess $(x, y) = (3, 2)$. In this case, successive gradients are no longer orthogonal. The convergence and error (E) to the minimum (optimal) solution is slower with this line search method of a fixe value of τ.

that the minimum of the gradient direction is picked as the next iteration point. However, this is not a requirement. In fact, one can simply choose a fixed value of τ for stepping forward along the gradient direction. Figure 5.5 demonstrates this case for $\tau = 0.1$. This method also converges to the solution, however at a much slower rate. Such a method may be favorable in a case where the steepest descent algorithm *zig-zags* a large amount in trying to make the projective steps orthogonal. This can happen in cases where *long-valley* type structures exist in the function we are trying to minimize.

fminsearch

Although not based upon gradient descent algorithms, the **fminsearch** algorithm in MATLAB is a generic, nonlinear unconstrained optimization method based upon the Nelder–Mead simplex method [6]. We have already used this method as a means of doing nonlinear curve fitting. In that case, the objective function was the E_2 error which was to be minimized. As a second example of this technique, consider once again a set of data that we wish to fit with the function $f(x) = A \cos(Bx) + C$ where A, B and C are the variables to be chosen for minimizing the error. Our objective function in this case is the least-square error $E_2 = \sqrt{(1/N) \sum |f(x_j) - y_j|^2}$. Thus we only need to consider minimizing $\sum |f(x_j) - y_j|^2$ with respect to A, B and C to achieve our goal.

The following code performs the optimization process with initial guesses given by $(A, B, C) = (12, \pi/12, 63)$.

```
c=fminsearch('datafit',[12 pi/12 63]);   % optimization
```

The function **temp fit** is given by

datafit.m

```
function e2=tempfit(c)

x=1:24;
y=[75 77 76 73 69 68 63 59 57 55 54 52 50 ...
    50 49 49 49 50 54 56 59 63 67 72];

e2=sqrt(sum((c(1)*cos(c(2)*x)+c(3)-y).^2)/24);
```

The algorithm will rapidly converge to new values of the vector **c** which contains the updated and optimal value of A, B and C. To plot the results and compare the fit (see Fig. 5.6), the following code is used:

```
t=1:24;    % raw data
tem=[75 77 76 73 69 68 63 59 57 55 54 52 ...
  50 50 49 49 49 50 54 56 59 63 67 72];

tt=1:0.01:24;
yfit=(c(1)*cos(c(2)*tt)+c(3)).';

plot(t,tem,'ko',tt,yfit,'k-')
```

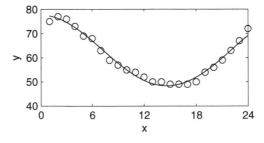

Figure 5.6: Minimization algorithm **fminsearch** used for curve fitting to a nonlinear function. The dots are the original data points and the solid line is the least-square fit. In this case, the least-square error E_2 is the objective function.

This example illustrates both the construction of an objective function as well as the implementation of one of the most important unconstrained optimization tools that is available in MATLAB. Critical to success in his algorithm is the initial guess used for the optimal (minimal) solution.

5.3   Linear Programming

We now come to perhaps the most import aspect in terms of application: optimization with constraint. This is still a highly active area of research and many methods exist which exploit the underlying nature of the problem being considered. Here, we will limit our discussion to a classic problem known as a *linear program*. A linear program is an optimization problem in which the objective function is linear in the unknown and the constraints consist of linear inequalities and equalities.

To illustrate the linear programming concept, the so-called *standard form* will first be considered.

$$
\begin{aligned}
\text{minimize} \quad & c_1 x_1 + c_2 x_2 + \cdots + c_n x_n \\
\text{subject to} \quad & a_{11} x_1 + a_{12} x_2 + \cdots + a_{1n} x_n = b_1 \\
& a_{21} x_1 + a_{22} x_2 + \cdots + a_{2n} x_n = b_2 \\
& \quad \vdots \\
& a_{m1} x_1 + a_{m2} x_2 + \cdots + a_{mn} x_n = b_m \\
\text{and} \quad & x_1 \geq 0, x_2 \geq 0, \cdots, x_n \geq 0
\end{aligned}
\tag{5.3.1}
$$

which can be written in a much more elegant form via vector and matrix notation

$$
\begin{aligned}
\text{minimize} \quad & \mathbf{c}^T \mathbf{x} \\
\text{subject to} \quad & \mathbf{A}\mathbf{x} = \mathbf{b} \text{ and } \mathbf{x} \geq 0.
\end{aligned}
\tag{5.3.2}
$$

Thus given the matrix $\mathbf{A}$ and the vectors $\mathbf{b}$ and $\mathbf{c}$, the goal is to find the vector $\mathbf{x}$ that minimizes the linear objective function given by $\mathbf{c}$.

Of course, not all linear optimization problems come directly in this form. But they can be transformed to the standard form by simple techniques.

Slack variables

Consider instead the following related problem which has inequality constraints instead of equality constraints.

$$\text{minimize} \quad c_1x_1 + c_2x_2 + \cdots + c_nx_n$$

$$\text{subject to} \quad a_{11}x_1 + a_{12}x_2 + \cdots + a_{1n}x_n \leq b_1$$

$$a_{21}x_1 + a_{22}x_2 + \cdots + a_{2n}x_n \leq b_2$$

$$\vdots$$

$$a_{m1}x_1 + a_{m2}x_2 + \cdots + a_{mn}x_n \leq b_m$$

$$\text{and} \quad x_1 \geq 0, x_2 \geq 0, \cdots, x_n \geq 0.$$

(5.3.3)

This problem is no longer in the standard form. However, it can be easily put into the standard form by introducing *slack variables* so that the inequalities can be made into equalities. Thus we transform the problem to the following:

$$\text{minimize} \quad c_1x_1 + c_2x_2 + \cdots + c_nx_n$$

$$\text{subject to} \quad a_{11}x_1 + a_{12}x_2 + \cdots + a_{1n}x_n + y_1 = b_1$$

$$a_{21}x_1 + a_{22}x_2 + \cdots + a_{2n}x_n + y_2 = b_2$$

$$\vdots$$

$$a_{m1}x_1 + a_{m2}x_2 + \cdots + a_{mn}x_n + y_m = b_m$$

$$\text{and} \quad x_1 \geq 0, x_2 \geq 0, \cdots, x_n \geq 0$$

$$\text{and} \quad y_1 \geq 0, y_2 \geq 0, \cdots, y_n \geq 0.$$

(5.3.4)

The introduction of the new m variables given by $\mathbf{y}$ now sets the problem to be in standard form. In particular, the new matrix $\bar{\mathbf{A}}$ associated with the problem is now of the special form $\bar{\mathbf{A}} = [\mathbf{A}, \mathbf{I}]$ where $\mathbf{I}$ is the identify matrix, and the new vector $\bar{\mathbf{x}}$ to be solved for is $\bar{\mathbf{x}} = [\mathbf{x}, \mathbf{y}]$.

Other techniques exist to transform a linear optimization problem. If the inequalities are the opposite to the above, then *surplus* variables are introduced. If some of the unknown variables are actually not required to be positive, then they can be transformed using *free variables* [12]. MATLAB's own built-in linear programming subroutine accepts a different form than the standard form, saving you the work of transforming it to this specific form.

Any vector $\mathbf{x}$ that satisfies the constraints of (5.3.2) is a *feasible solution*. A feasible solution is called an *optimal solution* if, in addition, the objective function in (5.3.2), i.e. $\mathbf{c}^T\mathbf{x}$, is minimal in comparison with all other feasible solutions. A *basic feasible solution* is one for which $m - n$ of the variables $\mathbf{x}$ are zero, i.e. the number of nonzero solution elements is commensurate with the number of constraints. This leads to an important theorem of linear programming [12]:

Fundamental theorem of linear programming: *Given a linear program in the standard form (5.3.2) where $\mathbf{A}$ is an $m \times n$ matrix of rank m,*

(i)   *if there exists a feasible solution, there is a basic feasible solution.*

(ii)   *if there is an optimal feasible solution, there is an optimal basic feasible solution.*

The goal of linear programming is to find the optimal basic feasible solution of (5.3.2). As one might imagine, there have been a great number of mathematical techniques developed to solve this critically important problem [12]. Here we will consider how to think about (5.3.2) graphically and then MATLAB's linear programming function will be introduced.

A graphical interpretation

To illustrate the idea of *feasible solutions*, **basic feasible solutions** and transforming to the *standard form*, consider the following simple example

$$\begin{aligned} \text{minimize} \quad & -2x_1 - x_2 \\ \text{subject to} \quad & x_1 + (8/3)x_2 \leq 4 \\ & x_1 + x_2 \leq 2 \\ & 2x_1 \leq 3 \\ \text{and} \quad & x_1 \geq 0, x_2 \geq 0. \end{aligned}$$
(5.3.5)

The idea is to first write this in the standard form by introducing three slack variables to handle the three constraint inequalities. Thus we have the new problem in standard form:

$$\begin{aligned} \text{minimize} \quad & -2x_1 - x_2 \\ \text{subject to} \quad & x_1 + (8/3)x_2 + x_3 = 4 \\ & x_1 + x_2 + x_4 = 2 \\ & 2x_1 + x_5 = 3 \\ \text{and} \quad & x_1 \geq 0, x_2 \geq 0, x_3 \geq 0, x_4 \geq 0, x_5 \geq 0 \end{aligned}$$
(5.3.6)

where the slack variables are x_3, x_4 and x_5. Given the three constraints, it is ideal to find a *basic feasible solution* that has two of the five variables set to zero.

To begin discussing the solution of this problem, we first consider the region of feasibility solutions. Thus we consider $x_3 = 0$ in the first constraint, $x_4 = 0$ in the second constraint and $x_5 = 0$ in the third constraint. Figure 5.7 demonstrates the feasibility region associated with

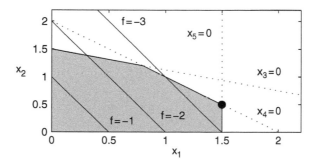

Figure 5.7: Graphical representation of the feasible region (shaded) given the constraints (5.3.5). The constraints are represented in terms of the slack variables. The objective function $f = \mathbf{c}^T\mathbf{x}$ is evaluated along contour lines. Thus the linear program seeks to minimize f while satisfying the constraints, i.e. the linear program would identify the point $(x_1, x_2) = (1.5, 0.5)$ as the optimal basic feasible solution.

this example. Once the feasibility solution is found, our objective is to minimize the objective function

$$\min f(x_1, x_2) = \min \mathbf{c}^T \mathbf{x} = -2x_1 - x_2 . \tag{5.3.7}$$

Figure 5.7 also demonstrates the lines of constant f. Note that the value of f decreases as the line of constant f is pushed to the right. The point $(x_1, x_2) = (1.5, 0.5)$ is the furthest point in the feasible region that one can push to the right, thus it is the optimal solution. Moreover, it is a basic optimal solution since $x_4 = x_5 = 0$ at this solution point.

linprog

Of course, what is desired is a systematic way to find the basic optimal solution. In the example given previously, it was simple to see from plotting alone where the optimal solution would be. However, in higher dimensional problems, the aid of such graphical techniques is rarely available. Thus algorithmic constructs for finding feasible solutions, and then iterating towards the optimal feasible solution, are of primary importance. Such methods have been developed; for example, the simplex method and/or interior point methods. These fall outside the scope of this book, but they can be followed up on in the literature [12].

   Here, MATLAB's linear program subroutine, **linprog**, will be considered. This is an extremely powerful tool for solving linear programming problems. The form of the linear program used by MATLAB is slightly different from the standard form. In particular, MATLAB will solve the following problem:

$$\begin{aligned} \text{minimize} \quad & \mathbf{c}^T \mathbf{x} & (5.3.8) \\ \text{subject to} \quad & \mathbf{A}\mathbf{x} \le \mathbf{b} \\ & \bar{\mathbf{A}}\mathbf{x} = \bar{\mathbf{b}} \\ & \mathbf{x}_- \le \mathbf{x} \le \mathbf{x}_+ & (5.3.9) \end{aligned}$$

where $\mathbf{x}_-$ and $\mathbf{x}_+$ are lower and upper bounds on the values of $\mathbf{x}$, respectively. Note that in this formulation, the equality and inequality constraints are separated. MATLAB automatically formulates the slack/surplus variables for you.

   The example given previously can be rewritten as

$$\begin{aligned} \text{minimize} \quad & -2x_1 - x_2 \\ \text{subject to} \quad & x_1 + (8/3)x_2 \le 4 \\ & x_1 + x_2 \le 2 \\ & 2x_1 \le 3 & (5.3.10) \\ & -x_1 \le 0 & (5.3.11) \\ & -x_2 \le 0 . \end{aligned}$$

In the matrix form as required by (5.3.8), one would then have

$$\mathbf{A} = \begin{bmatrix} 1 & 8/3 \\ 1 & 1 \\ 2 & 0 \\ -1 & 0 \\ 0 & -1 \end{bmatrix}, \quad \mathbf{b} = \begin{bmatrix} 4 \\ 2 \\ 3 \\ 0 \\ 0 \end{bmatrix}, \quad \mathbf{c} = \begin{bmatrix} -2 \\ -1 \end{bmatrix}. \tag{5.3.12}$$

Note that in this case, there are no equality constraints so that $\bar{\mathbf{A}}$ and $\bar{\mathbf{b}}$ do not need to be defined. Nor do we need to define bounds on the solution. The MATLAB code for implementing this linear program is as follows:

```
c=[-2 -1];
A=[1 8/3; 1 1; 2 0; -1 0; 0 -1];
b=[4; 2; 3; 0; 0];

x = linprog(c,A,b)
```

This produces the optimal solution $(x_1, x_2) = (1.5, 0.5)$.

More generally, the **linprog** is of the following form

```
[x,fval,exitflag]= linprog(c,A,b,Abar,bbar,xl,xu,x0,options)
```

where **xl** and **xu** correspond to the lower and upper bound vectors, **Abar** and **bbar** are the matrices corresponding to the equality constraints, **x0** is an initial guess for the solution if available, and the **options** allow for toggling of the error tolerance, for instance. Upon return, the variable **exitflag** describes if the optimization routine converged (it is equal to unity), or wether the maximum number of iterations were performed without converging, or if something else went wrong in the linear programming procedure.

Open source optimization packages: cvx

Of course, linear programming can be quite restrictive since it is, in fact, limited to linear objective functions and linear constraints. There are methods available for *nonlinear programming* [12], however, they are beyond the scope of this book. Thankfully, there are a number of open source convex optimization codes that can be downloaded from the Internet. In the compressive sensing chapter to come, a convex optimization package is used that can be directly implemented with MATLAB: **http://cvxr.com/cvx/**. This is one of several codes that can be downloaded that use state-of-the-art optimization techniques that go far beyond both the constraints of linear programming.

5.4 Simplex Method

Before moving on from the linear programming method, a key issue must be addressed: From a feasible solution, how can new feasible solutions be generated that are more optimal? Indeed, how can one find the optimal solution, which is the solution of the linear programming algorithm. Here, the *simplex method* is discussed which was developed by G.B. Dantzig in 1948. As is expected, the simplex method is a systematic iterative technique which aims to take a given *basic feasible solution* to another *basic feasible solution* for which the objective function is smaller.

Consider once again the linear program in standard form:

$$
\begin{aligned}
\text{minimize} \quad & f(\mathbf{x}) = c_1 x_1 + c_2 x_2 + \cdots + c_n x_n \\
\text{subject to} \quad & a_{11} x_1 + a_{12} x_2 + \cdots + a_{1n} x_n = b_1 \\
& a_{21} x_1 + a_{22} x_2 + \cdots + a_{2n} x_n = b_2 \\
& \quad\quad \vdots \\
& a_{m1} x_1 + a_{m2} x_2 + \cdots + a_{mn} x_n = b_m \\
\text{and} \quad & x_1 \geq 0, x_2 \geq 0, \cdots, x_n \geq 0
\end{aligned}
\tag{5.4.1}
$$

where we have now represented the objective function as $f(\mathbf{x})$.

First, we can easily consider the constraint conditions and feasibility. Specifically, the constraint equations are simply $\mathbf{Ax} = \mathbf{b}$ where $\mathbf{A}$ is an $m \times n$ matrix where $m < n$. Thus the constraint system is underdetermined and there are an infinite number of possible solutions (see Fig. 5.7 which shows the entire (shaded) region of infinite solutions). Thus since we are guaranteed a solution to the underdetermined system, we are guaranteed a feasible solution. But once this feasible solution is found, we can easily put it into the form of a *basic feasible solution* by converting it (via Gaussian elimination type techniques) to the canonical form:

$$
\begin{aligned}
x_1 + y_{1,m+1} x_{m+1} + y_{1,m+2} x_{m+2} + \cdots + y_{1,n} x_n &= y_{10} \\
x_2 + y_{2,m+1} x_{m+1} + y_{2,m+2} x_{m+2} + \cdots + y_{2,n} x_n &= y_{20} \\
\vdots \quad\quad\quad\quad & \\
x_m + y_{m,m+1} x_{m+1} + y_{m,m+2} x_{m+2} + \cdots + y_{m,n} x_n &= y_{m0}.
\end{aligned}
\tag{5.4.2}
$$

Once in the canonical form, a basic feasible solution is found where

$$
x_1 = y_{10}, x_2 = y_{20}, x_m = y_{m0}, \quad \text{and} \quad x_{m+1} = 0, x_{m+2} = 0, \cdots, x_n = 0.
\tag{5.4.3}
$$

This canonical solution is also a basic feasible solution. The variables $x_1, x_2, \cdots, x_m$ are called *basic* and the variables $x_{m+1}, x_{m+2}, \cdots, x_n$ are called *nonbasic*.

Here is the fundamental question to ask: Do we have the right basic and nonbasic variables? Specifically, what if there is a variable x_p, where p is from somewhere in $m + 1$ to n, such that it would be a better choice as a basic variable, i.e. it would give a more optimal solution where the objective function is smaller. The simplex method fundamentally is concerned with making basic those variables that, in fact, give an optimal solution. Thus an iteration procedure must be created to perform such an action.

To move forward, the simplex *tableau* is created for the above basic feasible solution. Thus we can write this in a more shorthand notation as:

$$
\begin{array}{ccccccccc}
1 & 0 & \cdots & 0 & y_{1,m+1} & y_{1,m+2} & \cdots & y_{1,n} & y_{1,0} \\
0 & 1 & \cdots & 0 & y_{2,m+1} & y_{2,m+2} & \cdots & y_{2,n} & y_{2,0} \\
\cdot & \cdot & & \cdot & \cdot & \cdot & & \cdot & \cdot \\
\cdot & \cdot & & \cdot & \cdot & & \cdot & & \cdot \\
0 & 0 & \cdots & 1 & y_{m,m+1} & y_{m,m+2} & \cdots & y_{m,n} & y_{m,0}.
\end{array}
\tag{5.4.4}
$$

The purpose in writing it in this form is that the operations which will be performed for the simplex method are much like those in Gaussian elimination. In particular, it is often advantageous to switch basic and nonbasic variables, thus necessitating column and row reductions to achieve this goal.

Here is the critical observation, and the fundamental point, of the simplex method. Although it is natural to use the basic solutions from the computed tableau above, it is also clear that arbitrary values of $x_{m+1}, x_{m+2}, \cdots, x_n$ can be chosen. Recall that this is an underdetermined system, so an infinite number of solutions are allowed, including those with nontrivial nonbasic variables. If these nontrivial basic variables are chosen arbitrarily, then the above tableau gives the following values of the basic variables:

$$
\begin{aligned}
x_1 &= y_{10} - \sum_{j=m+1}^{n} y_{1,j} x_j \\
x_2 &= y_{20} - \sum_{j=m+1}^{n} y_{2,j} x_j \\
&\quad \cdot \quad \cdot \quad \cdot \\
&\quad \cdot \quad \cdot \quad \cdot \\
x_m &= y_{m0} - \sum_{j=m+1}^{n} y_{m,j} x_j .
\end{aligned}
\tag{5.4.5}
$$

Of course, this is a trivial observation. But it has a profound impact when considering the objective function

$$
\begin{aligned}
f &= c_1 x_1 + c_2 x_2 + \cdots + c_n x_n \\
&= f_0 + (c_{m+1} - f_{m+1}) x_{m+1} + (c_{m+2} - f_{m+2}) x_{m+2} + \cdots + (c_n - f_n) x_n
\end{aligned}
\tag{5.4.6}
$$

where

$$
f_j = y_{1,j} c_1 + y_{2,j} c_2 + \cdots + y_{m,j} c_m
\tag{5.4.7}
$$

with $m + 1 \leq j \leq n$. The critical observation is that this formulation gives the value of the objective function $f(\mathbf{x})$ in terms of the nonbasic variables $x_{m+1}, x_{m+2}, \cdots, x_n$. Thus from this we can determine if there is an advantage to switching basic to nonbasic variables in order to minimize the objective function. Specifically, if any $(c_j - f_j) < 0$ in the above formula, then the objective function will be lowered. The following theorem then applies:

Theorem (Improvement of basic feasible solution): *Given a nondegenerate basic feasible solution with corresponding objective function f_0, suppose that for some j there holds $(c_j - z_j) < 0$. Then there is a feasible solution with objective function value $f < f_0$. If the column $\mathbf{a}_j$ can be substituted* *for some vector in the original basis to yield a new basic feasible solution, this new solution will have $f < f_0$. If $\mathbf{a}_j$ cannot be substituted to yield a basic feasible solution, then the solution set is unbounded and the objective function can be made arbitrarily small (toward minus infinity).*

The above theorem is the basis for the simplex method. Thus from a given basic feasible solution, it only remains to identify $(c_j - z_j) < 0$ and use pivoting and row reduction techniques to swap basic and nonbasic solutions so that a new, and smaller, objective function is achieved. This process is continued until no $(c_j - z_j) < 0$ remain. In fact, the following optimality theorem then holds:

Theorem (Optimality condition theorem): *If for some basic feasible solution $(c_j - z_j) \geq 0$ for all j, then the solution is optimal.*

Armed with the above two theorems, the simplex method can be constructed and a termination point reached in the iteration method. Note that just like Gaussian elimination, which involves the same basic procedures in row and column reductions and manipulations, the larger the matrix, the greater the time in computation.

To illustrate how the actual process is achieved, consider the objective function $f(\mathbf{x}) = \sum x_j$ and the following simplex tableau:

$$
\begin{array}{ccccccc}
1 & 0 & 0 & 2 & 4 & 6 & 4 \\
0 & 1 & 0 & 1 & 2 & 3 & 3 \\
0 & 0 & 1 & -1 & 2 & 1 & 1
\end{array}
\tag{5.4.8}
$$

where the first six columns correspond to the coefficients of x_1, x_2, x_3, x_4, x_5 and x_6. The final column is the coefficients of the constraint vector $\mathbf{b}$. The basic feasible solution in this case is

$$
\mathbf{x} = (4, 3, 1, 0, 0, 0) \quad \text{with} \quad f = \sum x_j = 8.
\tag{5.4.9}
$$

Now suppose we elect to bring the fourth column $\mathbf{a}_4$ into the basis, i.e. make it a basic versus nonbasic variable. Then it is necessary to determine which element in the fourth column is the appropriate pivot. The following three ratios are computed

$$
b(1)/y_{1,4} = 4/2 = 2, \quad b(2)/y_{2,4} = 3/1 = 3, \quad b(3)/y_{3,4} = 1/-1 = -1.
\tag{5.4.10}
$$

The idea is to choose the smallest positive pivot, thus the pivot point will be the $b(1)/y_{1,4} = 2$ term and the pivoting happens about the first row, fourth column. As with Gaussian elimination, the goal is to make the other elements of the fourth column zero which can be achieved by adding and subtracting appropriately scaled rows. This yields the new simplex tableau:

$$
\begin{array}{ccccccc}
1/2 & 0 & 0 & 1 & 2 & 3 & 2 \\
-1/2 & 1 & 0 & 0 & 0 & 0 & 1 \\
1/2 & 0 & 1 & 0 & 4 & 4 & 3
\end{array}
\tag{5.4.11}
$$

which has the basic feasible solution

$$
\mathbf{x} = (0, 1, 3, 2, 0, 0) \quad \text{with} \quad f = \sum x_j = 6.
\tag{5.4.12}
$$

This simple example shows that the objective function was reduced from 8 to 6 simply by switching a basic with nonbasic variable.

More generally, the simplex algorithm proceeds as follows:

(i)   Form a simplex tableau from the initial basic feasible solution and compute the $(c_j - f_j)$.

(ii)   If each $(c_j - f_j) \geq 0$, stop the algorithm since the basic feasible solution is optimal.

(iii)   Select the jth column for which $(c_j - f_j) < 0$ is the least negative. This column will be made into a basic variable.

(iv)   Determine all the potential pivot values by evaluating y_{k0}/y_{kj} for $y_{kj} > 0$ and $k = 1, 2, \cdots, m$. If no $y_{kj} > 0$, then stop as the problem is unbounded. Otherwise, select p as the index k corresponding to the minimum ratio.

(v)   Pivot on the pjth element, updating all rows including the last. Return to the first step (i).

This gives the basic outline of the technique. Of course, just like Gaussian elimination, certain problems can arise in the pivoting process, including if there is degeneracy in the system. There are numerous techniques and algorithm improvements for the simplex method, and one is encouraged to follow these up in the literature [12].

5.5   Genetic Algorithms

Other methods developed for optimization problems are the so-called *genetic algorithms* which are a subset of evolutionary algorithms. The principle is quite simple and mirrors what is perceived to occur in evolution and/or genetic mutations. In particular, given a set of feasible trial solutions (either constrained or unconstrained), the objective function is evaluated. In the language of genetic algorithms, the objective function is now called the *fitness function*. The idea is to keep those solutions that give the minimal value of the objective function and mutate them in order to try and do even better. Thus beneficial mutations, in the sense of giving a better minimization, are kept while those that perform poorly are thrown away, i.e. survival of the fittest. This process is repeated through a prescribed number of iterations, or *generations*, with the idea that better and better fitness function values are generated via the mutation process.

To be more precise about the genetic algorithm structure, consider the unconstrained optimization problem with the objective function

$$\min f(\mathbf{x}) \tag{5.5.1}$$

where $\mathbf{x}$ is an n-dimensional vector. Suppose that m initial guesses are given for the values of $\mathbf{x}$ so that

$$\text{guess } j \text{ is } \mathbf{x}_j . \tag{5.5.2}$$

Thus m solutions are evaluated and compared with each other in order to see which of the solutions generate the smallest objective function since our goal is to minimize it. We can order the guesses so that the first $p < m$ give the smallest values of $f(\mathbf{x})$. Arranging our data, we then have

$$\begin{aligned} \text{keep} \quad & \mathbf{x}_j \, j = 1, 2, \cdots, p \\ \text{discard} \quad & \mathbf{x}_j \, j = p + 1, p + 2, \cdots, m. \end{aligned} \qquad (5.5.3)$$

Since the first p solutions are the best, these are kept in the next generation. In addition, we now generate $m - p$ new trial solutions that are randomly mutated from the p best solutions. This process is repeated through a finite number of iterations with the hope that convergence to the optimal solution is achieved.

Interestingly, there are really no theorems about the convergence of such a technique, so one may wonder why it should be considered at all when guaranteed convergence can be achieved with alternative algorithms. The use of such algorithms is due to a few key advantages that are difficult to find elsewhere. First, many iteration schemes can *get stuck* in the iteration process for a variety of reasons. Thus convergence is extremely slow, or only a local minimum can be found. In the mutation process of the genetic algorithm, there exists the possibility of moving well beyond these pernicious points so that the iteration can continue moving towards the optimal solution. Second, in all that has been considered thus far, an optimization problem can be neatly packaged as a set of constraints with an objective function. However, suppose the problem is sufficiently complex so that nonlinear constraints exist and the methods developed previously simply no longer hold. Alternatively, what if the objective function to be minimized is only computable after a larger simulation has been performed? Thus the idea is to choose the parameters of this larger simulation based upon the genetic algorithm itself and its ability to minimize the objective function.

To demonstrate the concept, consider the example in Fig. 5.6 which was solved using the **fminsearch** algorithm. In this case, the function form $f(x) = A \cos(Bx) + C$ is assumed and the fitness function (objective function) is the E_2 error. Here, a genetic algorithm will be developed that will search for the optimal solution using a set of initial guesses followed by mutations of the best solutions. We begin by defining some initial parameters for the genetic algorithm. In particular, 200 generations will be run with 50 trial solutions. Only the top 10 best solutions will be kept and mutated at the next generation. As before with **fminsearch**, an initial guess of $A = 12$, $B = pi/12$ and $C = 60$ will be used.

```
m=200; % number of generations
n=50; % number of trials
n2=10; % number of trials to be kept
A=12+randn(n,1); B=pi/12+randn(n,1); C=60+randn(n,1);
```

The main loop of the genetic algorithm is now ready to be performed. Below, the objective function is evaluated for each trial solution. The trials are then ordered from smallest to largest and the best 10 are kept. Of these 10, four mutations are made of each and the process repeated.

```
for jgen=1:m

  for j=1:n  % evaluate objective function
    E(j)= sum((A(j)*cos(B(j)*x)+C(j)-y).^2);
  end

  [Es,Ej]=sort(E);  % sort from small to large

  Ak1=A(Ej(1:n2)); % best 10 solutions
  Bk1=B(Ej(1:n2));
  Ck1=C(Ej(1:n2));

  Ak2=Ak1+randn(n2,1)/jgen; % 10 new mutations
  Bk2=Bk1+randn(n2,1)/jgen;
  Ck2=Ck1+randn(n2,1)/jgen;

  Ak3=Ak1+randn(n2,1)/jgen; % 10 new mutations
  Bk3=Bk1+randn(n2,1)/jgen;
  Ck3=Ck1+randn(n2,1)/jgen;

  Ak4=Ak1+randn(n2,1)/jgen; % 10 new mutations
  Bk4=Bk1+randn(n2,1)/jgen;
  Ck4=Ck1+randn(n2,1)/jgen;

  Ak5=Ak1+randn(n2,1)/jgen; % 10 new mutations
  Bk5=Bk1+randn(n2,1)/jgen;
  Ck5=Ck1+randn(n2,1)/jgen;

  A=[Ak1; Ak2; Ak3; Ak4; Ak5]; % group new 50
  B=[Bk1; Bk2; Bk3; Bk4; Bk5];
  C=[Ck1; Ck2; Ck3; Ck4; Ck5];

end
```

Note that the algorithm takes progressively smaller mutations as the generations progress, thus the divide by **jgen**. Although a contrived example, this genetic algorithm converges nicely to the least-square solution. It should be noted, however, that this is an extremely slow method for doing curve fitting. So this example should be thought of as illustrative only. The convergence of the scheme and the data fit can be found in Fig. 5.8. Figure 5.9 shows the error of the 50 trials at various generations of the algorithm.

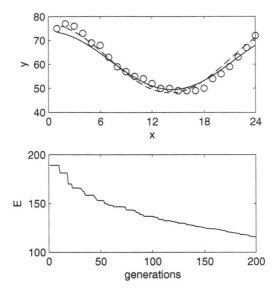

Figure 5.8: The top panel shows the data and curve fit from the genetic algorithm as developed here (solid line) and MATLAB's genetic algorithm (dotted line). The bottom panel shows the error of the best solution at each successive generation. The error slowly converges to the same solution as **fminsearch**.

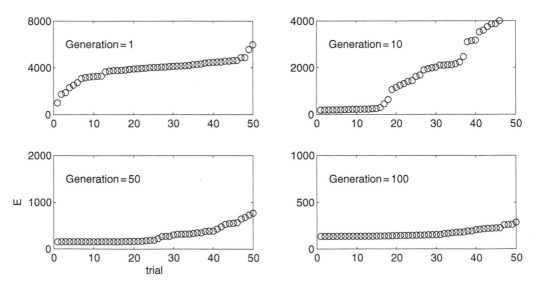

Figure 5.9: Error of the 50 trial solutions at generation 1, 10, 50 and 100. Note the convergence to the optimal solution as generations progress forward.

ga

MATLAB also has a built-in genetic algorithm code that is easy to use and implement. Moreover, it can be set up to do constrained optimization problems, even with nonlinear objective functions. To begin, the **ga** algorithm is illustrated with the simple example of the curve fit just illustrated. The code for solving this problem is given by

```
lower=[10 pi/20 50];
upper=[15 pi/4 70];
x=ga(@(x)fit_line(x),3,[],[],[],[],lower,upper)
```

where the objective function is given by

fit_line.m

```
function E=fit_line(x)

xx=1:24;
yy=[75 77 76 73 69 68 63 59 57 55 54 52 ...
    50 50 49 49 49 50 54 56 59 63 67 72];

E=sum((x(1)*cos(x(2)*xx)+x(3)-yy).^2);
```

Note that in the code, a lower and upper bound on the solution has been provided. This is equivalent to providing a good initial guess. If upper and lower bounds are not provided, then the algorithm fails completely.

More generally, the genetic algorithm as developed by MATLAB allows for both equality and inequality constraints. Additionally, nonlinear constraints can be imposed on the problem, thus making the **ga** algorithm extremely powerful. To be specific, the following problem can be solved

$$\text{minimize} \quad f(\mathbf{x}) \tag{5.5.4}$$
$$\text{subject to} \quad \mathbf{A}\mathbf{x} \leq \mathbf{b}$$
$$\bar{\mathbf{A}}\mathbf{x} = \bar{\mathbf{b}}$$
$$g(\mathbf{x}) \leq 0$$
$$\bar{g}(\mathbf{x}) = 0$$
$$\mathbf{x}_- \leq \mathbf{x} \leq \mathbf{x}_+ . \tag{5.5.5}$$

The generic optimization thus allows for a nonlinear objective function $f(\mathbf{x})$ along with a set of linear equality and inequality constraints, $\mathbf{A}\mathbf{x} \leq \mathbf{b}$ and $\bar{\mathbf{A}}\mathbf{x} = \bar{\mathbf{b}}$, respectively, a set of nonlinear equality and inequality constraints, $g(\mathbf{x}) \leq 0$ and $\bar{g}(\mathbf{x}) = 0$, respectively, and upper and lower bounds, $\mathbf{x}_+$ and $\mathbf{x}_-$, respectively. A function call to the **ga** algorithm is given by

```
x = ga('fit',n,A,b,Abar,bbar,xl,xu,nonlin,options)
```

where the above constraints are placed one by one into the genetic algorithm. The options may become important as the maximum number of generations and tolerance, for instance, are set within this variable space.

Visualization

Visualization is one of the most important aspects of MATLAB that needs to be considered when communicating your results with others. The default settings on MATLAB are often lacking when professional looking graphs and plots need to be produced. In the following subsections, both full customization of graphs are considered as well as the high-end two- and three-dimensional graphic capabilities of MATLAB. The final subsection will then consider how to use these plotting algorithms to produce movie and animation files.

6.1 Customizing Plots and Basic 2D Plotting

Section 1.5 illustrates many of the basic features of the plotting algorithms of MATLAB. However, it is rarely the case that a plot created with the default line thicknesses and font sizes of MATLAB can be used for professional purposes. Thus customization of the plot is important to do. It should be noted that all the customization can be performed directly from the figure by following the links from the **Edit** button on the graphs. However, this is not recommended since once the figure is closed, all the customization settings are lost. Further, if one wishes to use the same custom setting for a large number of plots, it is much better to have a customized MATLAB code that can be applied to all the data sets. All the figures in this book have been customized in order to get the font sizes and line thicknesses correct. Indeed, the default line thickness and text font sizes typically show up poorly when printed out in a report or manuscript unless they are modified.

To begin the consideration of the customization process, the following specific functions will be plotted

$$f(x) = \cos(x) \tag{6.1.1a}$$
$$g(x) = \sin(0.2x^2)\exp(-0.02x^2) \tag{6.1.1b}$$

on the interval $x \in [-10, 10]$. These two functions can easily be made by using the MATLAB commands

```
L=10;
x=linspace(-L,L,100);
f=cos(x);
g=sin(0.2*x.^2).*exp(-0.02*x.^2);
```

Note that in this example, the functions have been discretized into 100 points.

Plotting these functions can be done trivially in MATLAB with the commands

```
figure(1)
subplot(2,1,1)
plot(x,f,'m',x,g,'k')
```

The top panel of Fig. 6.1 shows the default image generated from MATLAB. Although a subjective opinion, the default font size and line thickness are simply not appropriately chosen for the purpose of insertion into a manuscript or report.

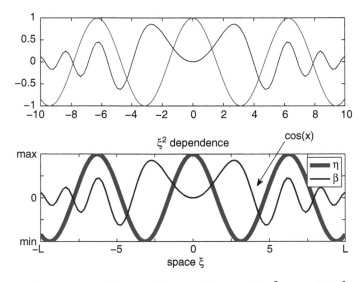

Figure 6.1: Plots of the functions $f(x) = \cos(x)$ and $g(x) = \sin(0.2x^2)\exp(-0.02x^2)$ over the interval $x \in [-10, 10]$. The top graph is plotted with the default settings of MATLAB. The bottom graph is a fully customized plot with font size and line thickness adjusted to more appropriate settings.

To customize the plot, access to its various properties must be obtained. The line width can easily be adjusted with the command structure:

```
subplot(2,1,2)
plot(x,f,'m','Linewidth',6),hold on
plot(x,g,'k','Linewidth',2)
```

This sets the thickness to a prescribed value. Generally, it is best to have a line thickness of a least value 2. The default value is unity.

The other properties can be altered with the **set(gca)** command in MATLAB. This **cga** stands for *get current axis*. The following lines of code set the axis limits, where tick marks are to be displayed, and what is to be displayed at each tick mark. Further, the font size is increased from the default value of 12 to 15.

```
set(gca,'Xlim',[-10 10], ...
        'Xtick',[-10 -7.5 -5 -2.5 0 2.5 5 7.5 10], ...
        'Xticklabel',{'-L',' ','-5',' ','0',' ','5',' ','L'}, ...
        'Ytick',[-1.5 -1 -0.5 0 0.5 1.0 1.5], ...
        'Yticklabel',{'','min',' ','0',' ','max',''}, ...
        'FontSize',[15])
```

The results of this modification can be seen in the bottom panel of Fig. 6.1. Thus full control of the graph can be established and all the defaults can be overridden.

In addition to customizing the font sizes, ticks and line thicknesses, Latex commands can be used in conjunction with MATLAB. Latex is a professional typesetting software package that is freely available. Further, it has become the dominant typesetting software in the scientific community. Most Latex commands are initiated with the backslash command. The following lines of codes show the implementation of Latex commands in the label, title and legend.

```
xlabel('space \xi','Fontsize',[20])
title('\xi^2 dependence','Fontsize',[20])
legend('\eta','\beta')
```

The ability to implement Latex structures in a plot is instrumental for many scientifically oriented plots where Greek characters typically represent the quantities of interest.

In addition to customizing the properties of the graph, annotations (arrows) can be added to a plot. An arrow is specified by defining its starting (x, y) location and ending (x, y) location. The entire figure space is defined with $(0, 0)$ as the bottom left corner and $(1, 1)$ as the top right corner. All annotation locations are defined with these reference points in mind. The following command annotates one of the plot lines for the bottom graph of Fig. 6.1.

```
A=annotation('textarrow',[0.75 0.68],[0.45 0.3]);
set(A,'String','cos(x)','Fontsize',15)
```

It should be noted that the string specified by the annotation does not handle Latex input like the axis label, legend and title commands.

Basic 2D plotting

In addition to modifying and adjusting all the default plotting parameters, there is an obvious desire to consider plotting higher dimensional functions. Our starting point is a function of two spatial dimensions x and y. For instance, we can consider the spiral wave defined as the following:

$$u(x, y) = \tanh\left[\sqrt{x^2 + y^2} \cos\left(m\angle (x + iy) - \sqrt{x^2 + y.^2}\right)\right] \tag{6.1.2}$$

where the $\angle$ denotes the phase angle of the quantity $(x + iy)$. In MATLAB, such a function can be defined as the following

```
L=10; x=linspace(-L,L,50); y=x;
[X,Y]=meshgrid(x,y);
m=1; % number of spirals
u=tanh(sqrt(X.^2+Y.^2)).*cos(m*angle(X+i*Y)-(sqrt(X.^2+Y.^2)));
```

Here the parameter m is the number of spiral arms in the function. There are a variety of techniques for producing a graphical representation of this 2D function. Figure 6.2 displays the first set of visualizations of the spiral wave. The four panels are given by the MATLAB code:

```
subplot(2,2,1), surf(x,y,u)
  set(gca,'Xtick',[],'Ytick',[],'Ztick',[])
subplot(2,2,2), surfc(x,y,u)   % with contour
  set(gca,'Xtick',[],'Ytick',[],'Ztick',[])
subplot(2,2,3), surfl(x,y,u)   % with lighting
  set(gca,'Xtick',[],'Ytick',[],'Ztick',[])
subplot(2,2,4), mesh(x,y,u)
  set(gca,'Xtick',[],'Ytick',[],'Ztick',[])
```

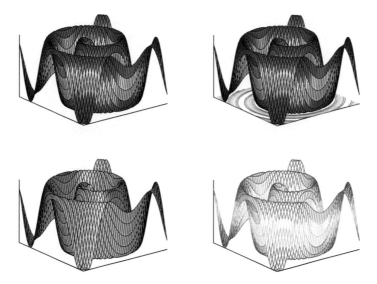

Figure 6.2: Some of the common options for plotting of data including **surf** (surface plot), **surfc** (surface plot with a contour), **surfl** (lighted surface plot) and **mesh** (just the mesh of the data).

The ticks and labels have all been removed for visualization purposes. The **surf** command is the standard surface plot and is displayed in the top left of Fig. 6.2. To combine both a surface plot and a contour plot the **surfc** command is used (top right of Fig. 6.2). The **surfl** command is a lighted surface (bottom left of Fig. 6.2) while the **mesh** command is simply the mesh of the data (bottom right of Fig. 6.2).

Other options exist for plotting the data. One of the most important is the **pcolor** command which plots a color scale projection of the data. This topographical plot is often a very insightful way to visualize the data, especially if it has a complicated structure which can be hidden in one of the other plotting options of Fig. 6.2. The following plotting options produce the topographical visualization:

```
subplot(2,2,1), pcolor(x,y,u),
  set(gca,'Xtick',[],'Ytick',[],'Ztick',[])
subplot(2,2,2), pcolor(x,y,u), shading interp
  set(gca,'Xtick',[],'Ytick',[],'Ztick',[])
```

Note that the **shading interp** command interpolates over the matrix values given by **u**. This is often a nice option for visualization. Figure 6.3 shows the results of the **pcolor**.

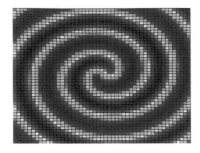 

Figure 6.3: Topographical plot of the spiral wave. Note that the **shading interp** command smooths out the data by interpolation (right panel).

Different color schemes can also be considered in the plotting. In Fig. 6.4, a variety of color schemes are considered with the **surfl** command. The following code generates this figure:

```
figure(8), surfl(x,y,u), shading interp
figure(9), surfl(x,y,u), shading interp, colormap(hot)
figure(10), surfl(x,y,u), shading interp, colormap(gray)
figure(11), surfl(x,y,u), shading interp, colormap(copper)
```

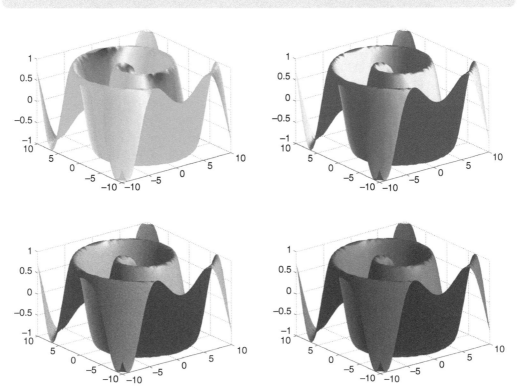

Figure 6.4: Lighted surface plots in differing colormaps and gray-scale.

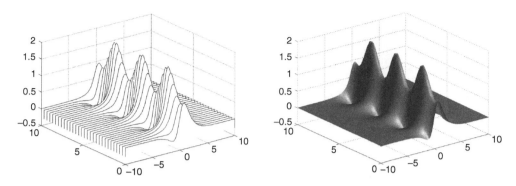

Figure 6.5: Waterfall and lighted surface plots in black-and-white and gray-scale.

Note that each of the figures generated are different figure numbers. This is due to the fact that only one colormap can be implemented per figure.

Finally, the **waterfall** command is considered for plotting. In Fig. 6.5 the waterfall command is used with a black-and-white colormap. The two plots generated in this figure show some nice options for gray-scale plotting. Often for journals, black-and-white is an important option due to the cost of printing in color. The code for this is the following:

```
x=linspace(-10,10,100);
t=linspace(0,10,30);
[X,T]=meshgrid(x,t);
f=sech(X).*(1-0.5*cos(2*T))+(sech(X).*tanh(X)).*(1-0.5*sin(2*T));
figure(4), waterfall(X,T,f), colormap([0 0 0])
  set(gca,'Fontsize',[15])
figure(5), surfl(X,T,f), shading interp, colormap(gray)
  set(gca,'Fontsize',[15])
```

The **waterfall** command is especially useful for showing spatio-temporal dynamics of a system in a simple way and in black-and-white.

 ## 6.2   More 2D and 3D Plotting

The aim of this section will be to outline some of the plot combining methods that can be used for 2D functions as well as develop visualization methods for 3D functions. 3D functions are particularly difficult because all of the spatial dimensions x, y and z already fill up the visualization domain. Thus methods must be available that allow for essentially 4D data, i.e. the value of some function $u(x, y, z)$ expressed as a function of all three spatial dimensions.

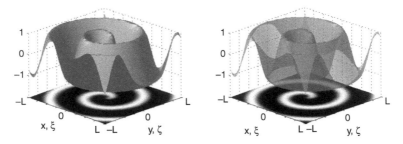

Figure 6.6: Combination plot of the spiral wave using both the **surfl** and **pcolor** plotting commands. The z-axes have been relabeled in order to accurately reflect the z-values. Further, the right figure has been made semi-transparent by setting **alpha(0.8)**. The **fontweight** option has been used to bold the x- and y-axis labels.

To begin, 2D combination plots are considered. Specifically, the spiral wave of the last section is revisited. Thus the function to be plotted is generated from the code:

```
L=10; x=linspace(-L,L,50); y=x;
[X,Y]=meshgrid(x,y);
m=1; % number of spirals
u=tanh(sqrt(X.^2+Y.^2)).*cos(m*angle(X+i*Y)-(sqrt(X.^2+Y.^2)));
```

As before, only a single spiral wave will be considered as a function of the spatial variables x and y.

A combination plot can be extremely helpful in visualizing a solution. Specifically, when considering the visualization of the spiral in the previous section, the spiral was well represented by both the **surfl** (lighted surface) and **pcolor** (topographical) plots. However, each of these plots has drawbacks. In the case of the lighted surface, portions of the spiral wave block out what is behind it. The topographical plot generated from **pcolor**, on the other hand, gives a complete representation of the solution without blocking anything out. But in this case, it is difficult to see what the heights correspond to in terms of units. One can certainly add a **colorbar** to the plot which will map the colors to physical values, making it a more user friendly plot. Alternatively, one could combine both the **surfl** and **pcolor** together to give an excellent representation of the solution.

For combination plots, the **hold on** command must be used in order to keep the various plot commands from overwriting the previous plot. Here the aim will be to plot both a **surfl** and **pcolor** figure together in combination. Figure 6.6 depicts the net effect of this process. The code for doing this is as follows:

```
subplot(2,2,1)
surfl(x,y,u+2), shading interp
colormap(gray)
view(45,30)
```

Continued

```
hold on
pcolor(x,y,u), shading interp
xlabel('x, \xi','FontSize',[12],'FontWeight','bold');
ylabel('y, \zeta','FontSize',[12],'FontWeight','bold');
set(gca,'Xlim',[-10 10], ...
        'Xtick',[-10 -7.5 -5 -2.5 0 2.5 5 7.5 10], ...
        'Xticklabel',{'-L',' ',' ',' ','0',' ',' ',' ','L'}, ...
        'Ytick',[-10 -7.5 -5 -2.5 0 2.5 5 7.5 10], ...
        'Yticklabel',{'-L',' ',' ',' ','0',' ',' ',' ','L'}, ...
        'FontSize',[12],'Ztick',[0 1 2 3], ...
        'Zticklabel',{'','-1','0','1'})

subplot(2,2,2)
surfl(x,y,u+2), shading interp
colormap(gray)
alpha(0.5)
view(45,30)
hold on
pcolor(x,y,u), shading interp
xlabel('x, \xi','FontSize',[12],'FontWeight','bold');
ylabel('y, \zeta','FontSize',[12],'FontWeight','bold');
set(gca,'Xlim',[-10 10], ...
        'Xtick',[-10 -7.5 -5 -2.5 0 2.5 5 7.5 10], ...
        'Xticklabel',{'-L',' ',' ',' ','0',' ',' ',' ','L'}, ...
        'Ytick',[-10 -7.5 -5 -2.5 0 2.5 5 7.5 10], ...
        'Yticklabel',{'-L',' ',' ',' ','0',' ',' ',' ','L'}, ...
        'FontSize',[12],'Ztick',[0 1 2 3], ...
        'Zticklabel',{'','-1','0','1'})
```

There are three key features of this plot. First, when plotting, the command

```
surfl(x,y,u+2)
```

is used to depict the **surfl**. The reason that the solution is lifted up by 2 units is that the **pcolor** only plots the solution at $z = 0$. Thus the **color** plot would slice directly through the **surfl** plot and create a visually unappealing plot. Instead, the solution is lifted up by 2 so that the **pcolor** plot no longer slices through the **surfl** plot. This creates a problem, however, since the z-scale now shows the data goes from $z = 1$ to $z = 3$ instead of the correct $z = -1$ to $z = 1$. This can be easily fixed by overwriting the axis labels so that $z = 1$ ($z = 3$) is labeled as $z = -1$ ($z = 1$).

A second interesting feature of these plots is that the **surfl** plot can be made semi-transparent. In the left panel of Fig. 6.6, the plot is not transparent. However, in the right panel, the figure is made semi-transparent with the parameter **alpha**. The value of alpha can be chosen from 0 (fully transparent or invisible) to 1 (nontransparent). Here the value was chosen to be 0.8 allowing for partial transparency. The transparency can often be used for favorable viewing of an image. As a final note, the **fontweight** command has been used to bold the axis labels.

3D plots

Three-dimensional plots are difficult to produce due to the fact that all three spatial dimensions are used. Thus representing the value of the solution must be done with some other method. The simple function to be considered here is a 3D periodic function

$$u(x, y, z) = \cos(x) \cos(y) \cos(z). \tag{6.2.1}$$

This is an *egg-carton* type structure in 3D space. The following lines of MATLAB can be used to produce the function itself.

```
x=linspace(-3,3,20); y=x; z=x;
[X,Y,Z]=meshgrid(x,y,z);
u=cos(X).*cos(Y).*cos(Z);
```

The **meshgrid** command can be easily generalized to 3D (or even higher), thus producing 3D matrices **X, Y, Z** and **U**. For each component of the matrix $U(i, j, k)$, a value is assigned based upon the function chosen.

Two nice options are available for plotting this function, the **isosurface** command and the **slice** command. The **isosurface** command uses predefined isosurface values in order to plot surfaces of constant value. As with the combination plots, the **hold on** command must be used. Figure 6.7 demonstrates the implementation of the **isosurface**. The following lines of MATLAB are used to generate this plot:

```
isosurface(x,y,z,u,0.5), grid on
alpha(0.8), hold on
isosurface(x,y,z,u,-0.5)
isosurface(x,y,z,u,0.25)
isosurface(x,y,z,u,-0.25)
set(gca,'Xlim',[-3 3],'Ylim',[-3 3],'Zlim',[-3 3],'Fontsize',[18])
```

The isosurface values chosen in this case are for the surface height of 0.5, 0.25, −0.25 and −0.5. Thus four isosurfaces are represented in the plot with the colors being related to the isosurface height.

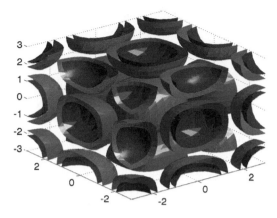

Figure 6.7: Isosurfaces generated from the **isosurface** command. Four surfaces are generated with values of −0.5, −0.25, 0.25 and 0.5. The plot has been made semi-transparent by setting alpha to 0.8.

As with the combination plots, two or more plot types can be incorporated together into one figure. Here, the **isosurface** is combined with the **slice** command to produce a composite figure. The **slice** command allows the user to specify **pcolor** planes on which the data can be projected. Thus it is really a generalization of the **pcolor** plotting routine for 3D. In the simple implementation here, slices are placed at the planes $x = 3$, $y = 3$ and $z = -3$. The following lines of code have been used to plot this.

```
isosurface(x,y,z,u,0.5), grid on
alpha(0.8), hold on
isosurface(x,y,z,u,-0.5)
isosurface(x,y,z,u,0.25)
isosurface(x,y,z,u,-0.25)
slice(x,y,z,u,3,3,-3)
set(gca,'Xlim',[-3 3],'Ylim',[-3 3],'Zlim',[-3 3],'Fontsize',[18])
colormap(hot)
```

In this case, the hot colormap has been used in Fig. 6.8 for the visualization. This gives a combination representation which in some visualizations can be extremely useful.

As a final example, the **slice** command is used exclusively for the 3D visualization. The following lines of code are used.

```
subplot(2,2,1)
slice(x,y,z,u,[0 3],3,[0 -3])
colormap(hot)
set(gca,'Xlim',[-3 3],'Ylim',[-3 3],'Zlim',[-3 3],'Fontsize',[14])
```
Continued

```
subplot(2,2,2)
slice(x,y,z,u,[0 3],3,[0 -3])
colormap(hot)
shading interp
set(gca,'Xlim',[-3 3],'Ylim',[-3 3],'Zlim',[-3 3],'Fontsize',[14])
```

In this visualization, several slice planes are used. Specifically the slice planes are at $x = 0, 3, y = 3$ and $z = -3, 0$. Figure 6.9 left, shows the plot without the shading interpolation. The grid lines can often be a nice guide for the eye for generating nice plots. On the right, the shading interpolation is used. Depending on the data, one may want to use either one.

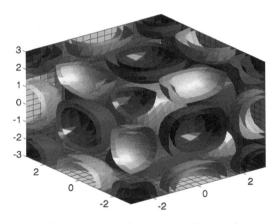

Figure 6.8: Isosurfaces generated from the **isosurface** command in combination with the **slice** command for projecting the solution on the surfaces at the edge of the domain. Four surfaces are generated with values of $-0.5, -0.25, 0.25$ and 0.5. The plot has been made semi-transparent by setting alpha to 0.8.

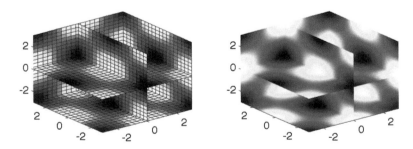

Figure 6.9: Implementation of the **slice** plotting routine for the slice planes at $x = 0, 3, y = 3$ and $z = -3, 0$. The only difference between the plots is that the **shading interp** command has been applied to the right figure.

6.3   Movies and Animations

Making movies in MATLAB is a fairly easy proposition. It also affords an excellent opportunity to highlight the hard work accomplished via simulations or experiments. Essentially, the movie making is based upon using the figures that are generated in MATLAB and using them as frames in a movie.

Two movie files will be exhibited here, one which is a simple implementation that generates a movie file to be read by MATLAB, and a second file which generates a more standard output, namely an AVI file. The example to be considered is the spiral wave of Section 6.1 but with a time-changing phase. Thus the spiral wave rotates in time. Each frame of the movie is a plot of the spiral wave at a different time-frame.

Two key commands are used in the basic implementation of the movie making technique: **getframe** and **movie**. The **getframe** command takes the currently plotted figure and takes this to be a frame of the movie. The collection of frames make up the movie. The **movie** command can then run this collection of frames as a movie via MATLAB.

The following code develops a movie by creating frames and saving them into a movie matrix **M**.

```
L=10; x=linspace(-L,L,50); y=x;
[X,Y]=meshgrid(x,y);

for j = 1:30
    u=tanh(sqrt(X.^2+Y.^2)).*cos((angle(X+i*Y)-(sqrt(X.^2+Y.^2)))+j);
    pcolor(X,Y,u)
    shading interp
    drawnow
    M(j)=getframe;
end

movie(M,3)
```

At the end of the code, the **movie(M, 3)** command plays the movie file **M** a total number of three times. The file **M** can be saved for later viewing in MATLAB. Note that the only command in the plot loop that has not been covered previously is the **getframe** command that saves each frame into the *j*th frame of **M**.

AVI files

The movie files saved as in the previous example are good for viewing in MATLAB only. A more general way to save files is in the **.avi** file format. The second example code is a more general, and essentially preferable, way of saving the movie in a format that can be read by many media players outside of MATLAB. The following MATLAB file produces the equivalent of the first code but now in the AVI format.

```
L=10; x=linspace(-L,L,50); y=x;
[X,Y]=meshgrid(x,y);

fh = figure(1)
mov = avifile('test.avi','fps',10)

for j = 1:30
  u=tanh(sqrt(X.^2+Y.^2)).*cos((angle(X+i*Y)-(sqrt(X.^2+Y.^2)))+j);
  pcolor(X,Y,u)
  shading interp
  drawnow
  M=getframe(fh);
  mov=addframe(mov,M);
end

mov = close(mov)
```

Figure 6.10: The first six frames of the movie file generated from MATLAB. In this movie, the spiral wave rotates in time. Note that for the plot, the tick marks have been removed.

Note that if the movie is not closed at the end, an error will occur in trying to read the AVI file. In this case, the **mov** file is defined at the onset as an AVI file. The command **'fps'** controls the frames-per-second for the movie. The **addframe** command takes the frames procured from the **getframe** command and adds them to the AVI file MOV. The output of this code is a file called **test.avi**.

The first six frames of the movie generated from either code is illustrated in Fig. 6.10. The tick marks have been removed for convenience. It is observed in this movie that the spiral wave rotates in time. This is very nicely observed from the generated **test.avi** movie file.

PART II

Differential and Partial Differential Equations

Initial and Boundary Value Problems of Differential Equations

Our ultimate goal is to solve very general nonlinear partial differential equations of elliptic, hyperbolic, parabolic or mixed type. However, a variety of basic techniques are required from the solutions of ordinary differential equations. By understanding the basic ideas for computationally solving initial and boundary value problems for differential equations, we can solve more complicated partial differential equations. The development of numerical solution techniques for initial and boundary value problems originates from the simple concept of the Taylor expansion. Thus the building blocks for scientific computing are rooted in concepts from freshman calculus. Implementation, however, often requires ingenuity, insight and clever application of the basic principles. In some sense, our numerical solution techniques reverse our understanding of calculus. Whereas calculus teaches us to take a limit in order to define a derivative or integral, in numerical computations we take the derivative or integral of the governing equation and go backwards to define it as the difference.

7.1 Initial Value Problems: Euler, Runge–Kutta and Adams Methods

The solutions of general partial differential equations rely heavily on the techniques developed for ordinary differential equations. Thus we begin by considering systems of differential equations of the form

$$\frac{d\mathbf{y}}{dt} = f(t, \mathbf{y}) \tag{7.1.1}$$

where $\mathbf{y}$ represents the solution vector of interest and the general function $f(t, \mathbf{y})$ models the specific system of interest. Indeed, the function $f(t, \mathbf{y})$ is the primary quantity required for calculating the dynamical evolution of a given system. As such, it will be the critical part of building

MATLAB codes for solving differential equations. In addition to the evolution dynamics, the initial conditions are given by

$$\mathbf{y}(0) = \mathbf{y}_0 \tag{7.1.2}$$

with $t \in [0, T]$. Although very simple in appearance, this equation cannot be solved analytically in general. Of course, there are certain cases for which the problem can be solved analytically, but it will generally be important to rely on numerical solutions for insight. For an overview of analytic techniques, see Boyce and DiPrima [7]. Note that the function $f(\mathbf{y}, t)$ is what is ultimately required by MATLAB to solve a given differential equation system.

The simplest algorithm for solving this system of differential equations is known as the *Euler method*. The Euler method is derived by making use of the definition of the derivative:

$$\frac{d\mathbf{y}}{dt} = \lim_{\Delta t \to 0} \frac{\Delta \mathbf{y}}{\Delta t}. \tag{7.1.3}$$

Thus over a time-span $\Delta t = t_{n+1} - t_n$ we can approximate the original differential equation by

$$\frac{d\mathbf{y}}{dt} = f(t, \mathbf{y}) \quad \Rightarrow \quad \frac{\mathbf{y}_{n+1} - \mathbf{y}_n}{\Delta t} \approx f(t_n, \mathbf{y}_n). \tag{7.1.4}$$

The approximation can easily be rearranged to give

$$\mathbf{y}_{n+1} = \mathbf{y}_n + \Delta t \cdot f(t_n, \mathbf{y}_n). \tag{7.1.5}$$

Thus the Euler method gives an iterative scheme by which the future values of the solution can be determined. Generally, the algorithm structure is of the form

$$\mathbf{y}(t_{n+1}) = F(\mathbf{y}(t_n)) \tag{7.1.6}$$

where $F(\mathbf{y}(t_n)) = \mathbf{y}(t_n) + \Delta t \cdot f(t_n, \mathbf{y}(t_n))$. The graphical representation of this iterative process is illustrated in Fig. 7.1 where the slope (derivative) of the function is responsible for generating each subsequent approximation to the solution $\mathbf{y}(t)$. Note that the Euler method is exact as the step-size decreases to zero: $\Delta t \to 0$.

The Euler method can be generalized to the following iterative scheme:

$$\mathbf{y}_{n+1} = \mathbf{y}_n + \Delta t \cdot \phi \tag{7.1.7}$$

where the function ϕ is chosen to reduce the error over a single time-step Δt and $\mathbf{y}_n = \mathbf{y}(t_n)$. The function ϕ is no longer constrained, as in the Euler scheme, to make use of the derivative at the left end point of the computational step. Rather, the derivative at the midpoint of the time-step and at the right end of the time-step may also be used to possibly improve accuracy. In particular, by generalizing to include the slope at the left and right ends of the time-step Δt, we can generate an iteration scheme of the following form:

$$\mathbf{y}(t + \Delta t) = \mathbf{y}(t) + \Delta t \left[Af(t, \mathbf{y}(t)) + Bf\left(t + P \cdot \Delta t, \mathbf{y}(t) + Q\Delta t \cdot f(t, \mathbf{y}(t))\right) \right] \tag{7.1.8}$$

where A, B, P and Q are arbitrary constants. Upon Taylor expanding the last term, we find

$$f(t + P \cdot \Delta t, \mathbf{y}(t) + Q\Delta t \cdot f(t, \mathbf{y}(t))) = f(t, \mathbf{y}(t)) + P\Delta t \cdot f_t(t, \mathbf{y}(t))$$
$$+ Q\Delta t \cdot f_{\mathbf{y}}(t, \mathbf{y}(t)) \cdot f(t, \mathbf{y}(t)) + O(\Delta t^2) \tag{7.1.9}$$

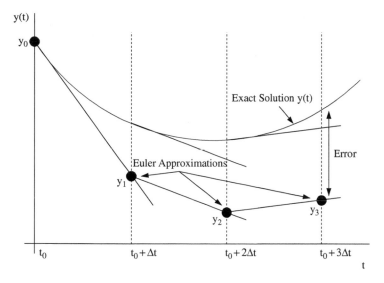

Figure 7.1: Graphical description of the iteration process used in the Euler method. Note that each subsequent approximation is generated from the slope of the previous point. This graphical illustration suggests that smaller steps Δt should be more accurate.

where f_t and $f_{\mathbf{y}}$ denote differentiation with respect to t and $\mathbf{y}$, respectively, use has been made of (7.1.1), and $O(\Delta t^2)$ denotes all terms that are of size Δt^2 and smaller. Plugging in this last result into the original iteration scheme (7.1.8) results in the following:

$$
\begin{aligned}
\mathbf{y}(t + \Delta t) = \mathbf{y}(t) &+ \Delta t(A + B)f(t, \mathbf{y}(t)) \\
&+ PB\Delta t^2 \cdot f_t(t, \mathbf{y}(t)) \\
&+ BQ\Delta t^2 \cdot f_{\mathbf{y}}(t, \mathbf{y}(t)) \cdot f(t, \mathbf{y}(t)) + O\left(\Delta t^3\right)
\end{aligned}
\tag{7.1.10}
$$

which is valid up to $O(\Delta t^2)$.

To proceed further, we simply note that the Taylor expansion for $\mathbf{y}(t + \Delta t)$ gives:

$$
\begin{aligned}
\mathbf{y}(t + \Delta t) = \mathbf{y}(t) &+ \Delta t \cdot f(t, \mathbf{y}(t)) + \frac{1}{2}\Delta t^2 \cdot f_t(t, \mathbf{y}(t)) \\
&+ \frac{1}{2}\Delta t^2 \cdot f_{\mathbf{y}}\left(t, \mathbf{y}(t)\right)f\left(t, \mathbf{y}(t)\right) + O\left(\Delta t^3\right).
\end{aligned}
\tag{7.1.11}
$$

Comparing this Taylor expansion with (7.1.10) gives the following relations:

$$
A + B = 1
\tag{7.1.12a}
$$

$$
PB = \frac{1}{2}
\tag{7.1.12b}
$$

$$
BQ = \frac{1}{2}
\tag{7.1.12c}
$$

which yields three equations for the four unknowns A, B, P and Q. Thus one degree of freedom is granted, and a wide variety of schemes can be implemented. Two of the more commonly used schemes are known as *Heun's method* and *modified Euler–Cauchy* (second-order Runge–Kutta). These schemes assume $A = 1/2$ and $A = 0$, respectively, and are given by:

$$\mathbf{y}(t + \Delta t) = \mathbf{y}(t) + \frac{\Delta t}{2} \left[f(t, \mathbf{y}(t)) + f(t + \Delta t, \mathbf{y}(t) + \Delta t \cdot f(t, \mathbf{y}(t))) \right] \qquad (7.1.13a)$$

$$\mathbf{y}(t + \Delta t) = \mathbf{y}(t) + \Delta t \cdot f \left(t + \frac{\Delta t}{2}, \mathbf{y}(t) + \frac{\Delta t}{2} \cdot f(t, \mathbf{y}(t)) \right). \qquad (7.1.13b)$$

Generally speaking, these methods for iterating forward in time given a single initial point are known as *Runge–Kutta methods*. By generalizing the assumption (7.1.8), we can construct stepping schemes which have arbitrary accuracy. Of course, the level of algebraic difficulty in deriving these higher accuracy schemes also increases significantly from Heun's method and modified Euler–Cauchy.

Fourth-order Runge–Kutta

Perhaps the most popular general stepping scheme used in practice is known as the *fourth-order Runge–Kutta method*. The term "fourth-order" refers to the fact that the Taylor series local truncation error is pushed to $O(\Delta t^5)$. The total cumulative (global) error is then $O(\Delta t^4)$ and is responsible for the scheme name of "fourth-order". The scheme is as follows:

$$\mathbf{y}_{n+1} = \mathbf{y}_n + \frac{\Delta t}{6} \left[f_1 + 2f_2 + 2f_3 + f_4 \right] \qquad (7.1.14)$$

where

$$f_1 = f(t_n, \mathbf{y}_n) \qquad (7.1.15a)$$

$$f_2 = f \left(t_n + \frac{\Delta t}{2}, \mathbf{y}_n + \frac{\Delta t}{2} f_1 \right) \qquad (7.1.15b)$$

$$f_3 = f \left(t_n + \frac{\Delta t}{2}, \mathbf{y}_n + \frac{\Delta t}{2} f_2 \right) \qquad (7.1.15c)$$

$$f_4 = f \left(t_n + \Delta t, \mathbf{y}_n + \Delta t \cdot f_3 \right). \qquad (7.1.15d)$$

This scheme gives a local truncation error which is $O(\Delta t^5)$. The cumulative (global) error in this case is fourth order so that for $t \sim O(1)$ the error is $O(\Delta t^4)$. The key to this method, as well as any of the other Runge–Kutta schemes, is the use of intermediate time-steps to improve accuracy. For the fourth-order scheme presented here, a graphical representation of this derivative sampling at intermediate time-steps is shown in Fig. 7.2.

Adams method: Multi-stepping techniques

The development of the Runge–Kutta schemes relies on the definition of the derivative and Taylor expansions. Another approach to solving (7.1.1) is to start with the fundamental theorem of calculus [8]. Thus the differential equation can be integrated over a time-step Δt to give

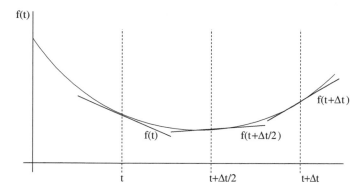

Figure 7.2: Graphical description of the initial, intermediate and final slopes used in the fourth-order Runge–Kutta iteration scheme over a time Δt.

$$\frac{d\mathbf{y}}{dt} = f(t, \mathbf{y}) \quad \Rightarrow \quad \mathbf{y}(t + \Delta t) - \mathbf{y}(t) = \int_t^{t+\Delta t} f(t, \mathbf{y}) dt. \tag{7.1.16}$$

And once again using our iteration notation we find

$$\mathbf{y}_{n+1} = \mathbf{y}_n + \int_{t_n}^{t_{n+1}} f(t, \mathbf{y}) dt. \tag{7.1.17}$$

This integral iteration relation is simply a restatement of (7.1.7) with $\Delta t \cdot \phi = \int_{t_n}^{t_{n+1}} f(t, \mathbf{y}) dt$. However, at this point, no approximations have been made and (7.1.17) is exact. The numerical solution will be found by approximating $f(t, \mathbf{y}) \approx p(t, \mathbf{y})$ where $p(t, \mathbf{y})$ is a polynomial. Thus the iteration scheme in this instance will be given by

$$\mathbf{y}_{n+1} \approx \mathbf{y}_n + \int_{t_n}^{t_{n+1}} p(t, \mathbf{y}) dt. \tag{7.1.18}$$

It only remains to determine the form of the polynomial to be used in the approximation.

The *Adams–Bashforth* suite of computational methods uses the current point and a determined number of past points to evaluate the future solution. As with the Runge–Kutta schemes, the order of accuracy is determined by the choice of ϕ. In the Adams–Bashforth case, this relates directly to the choice of the polynomial approximation $p(t, \mathbf{y})$. A first-order scheme can easily be constructed by allowing

$$p_1(t) = \text{constant} = f(t_n, \mathbf{y}_n), \tag{7.1.19}$$

where the present point and no past points are used to determine the value of the polynomial. Inserting this first-order approximation into (7.1.18) results in the previously found Euler scheme

$$\mathbf{y}_{n+1} = \mathbf{y}_n + \Delta t \cdot f(t_n, \mathbf{y}_n). \tag{7.1.20}$$

Alternatively, we could assume that the polynomial used both the current point and the previous point so that a second-order scheme resulted. The linear polynomial which passes through these two points is given by

$$p_2(t) = f_{n-1} + \frac{f_n - f_{n-1}}{\Delta t}(t - t_{n-1}). \tag{7.1.21}$$

When inserted into (7.1.18), this linear polynomial yields

$$\mathbf{y}_{n+1} = \mathbf{y}_n + \int_{t_n}^{t_{n+1}} \left(f_n + \frac{f_n - f_{n-1}}{\Delta t}(t - t_{n-1}) \right) dt. \tag{7.1.22}$$

Upon integration and evaluation at the upper and lower limits, we find the following second-order Adams–Bashforth scheme

$$\mathbf{y}_{n+1} = \mathbf{y}_n + \frac{\Delta t}{2} \left[3f\left(t_n, \mathbf{y}_n\right) - f\left(t_{n-1}, \mathbf{y}_{n-1}\right) \right]. \tag{7.1.23}$$

In contrast to the Runge–Kutta method, this is a *two-step algorithm* which requires two initial conditions. This technique can be easily generalized to include more past points and thus higher accuracy. However, as accuracy is increased, so are the number of initial conditions required to step forward one time-step Δt. Aside from the first-order accurate scheme, any implementation of Adams–Bashforth will require a *bootstrap* to generate a second "initial condition" for the solution iteration process.

The Adams–Bashforth scheme uses current and past points to approximate the polynomial $p(t, \mathbf{y})$ in (7.1.18). If instead a future point, the present, and the past is used, then the scheme is known as an *Adams–Moulton method*. As before, a first-order scheme can easily be constructed by allowing

$$p_1(t) = \text{constant} = f(t_{n+1}, \mathbf{y}_{n+1}), \tag{7.1.24}$$

where the future point and no past and present points are used to determine the value of the polynomial. Inserting this first-order approximation into (7.1.18) results in the *backward Euler scheme*

$$\mathbf{y}_{n+1} = \mathbf{y}_n + \Delta t \cdot f(t_{n+1}, \mathbf{y}_{n+1}). \tag{7.1.25}$$

Alternatively, we could assume that the polynomial used both the future point and the current point so that a second-order scheme resulted. The linear polynomial which passes through these two points is given by

$$p_2(t) = f_n + \frac{f_{n+1} - f_n}{\Delta t}(t - t_n). \tag{7.1.26}$$

Inserted into (7.1.18), this linear polynomial yields

$$\mathbf{y}_{n+1} = \mathbf{y}_n + \int_{t_n}^{t_{n+1}} \left(f_n + \frac{f_{n+1} - f_n}{\Delta t}(t - t_n) \right) dt. \tag{7.1.27}$$

Upon integration and evaluation at the upper and lower limits, we find the following second-order Adams–Moulton scheme

$$\mathbf{y}_{n+1} = \mathbf{y}_n + \frac{\Delta t}{2} \left[f(t_{n+1}, \mathbf{y}_{n+1}) + f(t_n, \mathbf{y}_n) \right]. \tag{7.1.28}$$

Once again this is a two-step algorithm. However, it is categorically different from the Adams–Bashforth methods since it results in an *implicit scheme*, i.e. the unknown value $\mathbf{y}_{n+1}$ is specified through a nonlinear equation (7.1.28). The solution of this nonlinear system can be very difficult, thus making *explicit schemes* such as Runge–Kutta and Adams–Bashforth, which are simple iterations, more easily handled. However, implicit schemes can have advantages when considering stability issues related to time-stepping.

One way to circumvent the difficulties of the implicit stepping method while still making use of its power is to use a *predictor–corrector method*. This scheme draws on the power of both the Adams–Bashforth and Adams–Moulton schemes. In particular, the second-order implicit scheme given by (7.1.28) requires the value of $f(t_{n+1}, \mathbf{y}_{n+1})$ in the right-hand side. If we can predict (approximate) this value, then we can use this predicted value to solve (7.1.28) explicitly. Thus we begin with a predictor step to estimate $\mathbf{y}_{n+1}$ so that $f(t_{n+1}, \mathbf{y}_{n+1})$ can be evaluated. We then insert this value into the right-hand side of (7.1.28) and explicitly find the corrected value of $\mathbf{y}_{n+1}$. The second-order predictor–corrector steps are then as follows:

$$\text{Predictor (Adams–Bashforth): } \mathbf{y}^P_{n+1} = \mathbf{y}_n + \frac{\Delta t}{2} \left[3 f_n - f_{n-1} \right] \tag{7.1.29a}$$

$$\text{Corrector (Adams–Moulton): } \mathbf{y}_{n+1} = \mathbf{y}_n + \frac{\Delta t}{2} \left[f\left(t_{n+1}, \mathbf{y}^P_{n+1}\right) + f(t_n, \mathbf{y}_n) \right]. \tag{7.1.29b}$$

Thus the scheme utilizes both explicit and implicit time-stepping schemes without having to solve a system of nonlinear equations.

Higher order differential equations

Thus far, we have considered systems of first-order equations. Higher order differential equations can be put into this form and the methods outlined here can be applied. For example, consider the third-order, nonhomogeneous, differential equation

$$\frac{d^3 u}{dt^3} + u^2 \frac{du}{dt} + \cos t \cdot u = g(t). \tag{7.1.30}$$

By defining

$$y_1 = u \tag{7.1.31a}$$

$$y_2 = \frac{du}{dt} \tag{7.1.31b}$$

$$y_3 = \frac{d^2 u}{dt^2}, \tag{7.1.31c}$$

we find that $dy_3/dt = d^3u/dt^3$. Using the original equation along with the definitions of y_i we find that

$$\frac{dy_1}{dt} = y_2 \tag{7.1.32a}$$

$$\frac{dy_2}{dt} = y_3 \tag{7.1.32b}$$

$$\frac{dy_3}{dt} = \frac{d^3u}{dt^3} = -u^2\frac{du}{dt} - \cos t \cdot u + g(t) = -y_1^2 y_2 - \cos t \cdot y_1 + g(t) \tag{7.1.32c}$$

which results in the original differential equation (7.1.1) considered previously

$$\frac{d\mathbf{y}}{dt} = \frac{d}{dt}\begin{pmatrix} y_1 \\ y_2 \\ y_3 \end{pmatrix} = \begin{pmatrix} y_2 \\ y_3 \\ -y_1^2 y_2 - \cos t \cdot y_1 + g(t) \end{pmatrix} = f(t, \mathbf{y}). \tag{7.1.33}$$

At this point, all the time-stepping techniques developed thus far can be applied to the problem. It is imperative to write any differential equation as a first-order system before solving it numerically with the time-stepping schemes developed here.

MATLAB commands

The time-stepping schemes considered here are all available in the MATLAB suite of differential equation solvers. The following are a few of the most common solvers:

- **ode23:** second-order Runge–Kutta routine;
- **ode45:** fourth-order Runge–Kutta routine;
- **ode113:** variable-order predictor–corrector routine;
- **ode15s:** variable-order Gear method for stiff problems [9, 10].

 7.2   Error Analysis for Time-Stepping Routines

Accuracy and *stability* are fundamental to numerical analysis and are the key factors in evaluating any numerical integration technique. Therefore, it is essential to evaluate the accuracy and stability of the time-stepping schemes developed. Rarely does it occur that both accuracy and stability work in concert. In fact, they are often offsetting and work directly against each other. Thus a highly accurate scheme may compromise stability, whereas a low-accuracy scheme may have excellent stability properties.

We begin by exploring accuracy. In the context of time-stepping schemes, the natural place to begin is with Taylor expansions. Thus we consider the expansion

$$\mathbf{y}(t + \Delta t) = \mathbf{y}(t) + \Delta t \cdot \frac{d\mathbf{y}(t)}{dt} + \frac{\Delta t^2}{2} \cdot \frac{d^2\mathbf{y}(c)}{dt^2} \tag{7.2.1}$$

where $c \in [t, t + \Delta t]$. Since we are considering $dy/dt = f(t, \mathbf{y})$, the above formula reduces to the Euler iteration scheme

$$\mathbf{y}_{n+1} = \mathbf{y}_n + \Delta t \cdot f\left(t_n, \mathbf{y}_n\right) + O\left(\Delta t^2\right). \tag{7.2.2}$$

It is clear from this that the truncation error is $O(\Delta t^2)$. Specifically, the truncation error is given by $\Delta t^2/2 \cdot d^2 \mathbf{y}(c)/dt^2$.

Of importance is how this truncation error contributes to the overall error in the numerical solution. Two types of error are important to identify: *local* and *global error*. Each is significant in its own right. However, in practice we are only concerned with the global (cumulative) error. The *global discretization error* is given by

$$E_k = \mathbf{y}(t_k) - \mathbf{y}_k \tag{7.2.3}$$

where $\mathbf{y}(t_k)$ is the exact solution and $\mathbf{y}_k$ is the numerical solution. The *local discretization error* is given by

$$\epsilon_{k+1} = \mathbf{y}\left(t_{k+1}\right) - \left(\mathbf{y}(t_k) + \Delta t \cdot \phi\right) \tag{7.2.4}$$

where $\mathbf{y}(t_{k+1})$ is the exact solution and $\mathbf{y}(t_k) + \Delta t \cdot \phi$ is a one-step approximation over the time interval $t \in [t_n, t_{n+1}]$.

For the Euler method, we can calculate both the local and global error. Given a time-step Δt and a specified time interval $t \in [a, b]$, we have after K steps that $\Delta t \cdot K = b - a$. Thus we find

$$\text{local:} \quad \epsilon_k = \frac{\Delta t^2}{2} \frac{d^2 \mathbf{y}(c_k)}{dt^2} \sim O(\Delta t^2) \tag{7.2.5a}$$

$$\text{global:} \quad E_k = \sum_{j=1}^{K} \frac{\Delta t^2}{2} \frac{d^2 \mathbf{y}(c_j)}{dt^2} \approx \frac{\Delta t^2}{2} \frac{d^2 \mathbf{y}(c)}{dt^2} \cdot K$$

$$= \frac{\Delta t^2}{2} \frac{d^2 \mathbf{y}(c)}{dt^2} \cdot \frac{b - a}{\Delta t} = \frac{b - a}{2} \Delta t \cdot \frac{d^2 \mathbf{y}(c)}{dt^2} \sim O(\Delta t) \tag{7.2.5b}$$

which gives a local error for the Euler scheme which is $O(\Delta t^2)$ and a global error which is $O(\Delta t)$. Thus the cumulative error is large for the Euler scheme, i.e. it is not very accurate.

A similar procedure can be carried out for all the schemes discussed thus far, including the multi-step Adams schemes. Table 7.1 illustrates various schemes and their associated local and global errors. The error analysis suggests that the error will always decrease in some power of

Table 7.1 Local and global discretization errors associated with various time-stepping schemes

Scheme	Local error ϵ_k	Global error E_k
Euler	$O(\Delta t^2)$	$O(\Delta t)$
Second-order Runge–Kutta	$O(\Delta t^3)$	$O(\Delta t^2)$
Fourth-order Runge–Kutta	$O(\Delta t^5)$	$O(\Delta t^4)$
Second-order Adams–Bashforth	$O(\Delta t^3)$	$O(\Delta t^2)$

Δt. Thus it is tempting to conclude that higher accuracy is easily achieved by taking smaller time-steps Δt. This would be true if not for round-off error in the computer.

Round-off and step-size

An unavoidable consequence of working with numerical computations is round-off error. When working with most computations, *double precision* numbers are used. This allows for 16-digit accuracy in the representation of a given number. This round-off has a significant impact upon numerical computations and the issue of time-stepping.

As an example of the impact of round-off, we consider the Euler approximation to the derivative

$$\frac{d\mathbf{y}}{dt} \approx \frac{\mathbf{y}_{n+1} - \mathbf{y}_n}{\Delta t} + \epsilon(\mathbf{y}_n, \Delta t) \tag{7.2.6}$$

where $\epsilon(\mathbf{y}_n, \Delta t)$ measures the truncation error. Upon evaluating this expression in the computer, round-off error occurs so that

$$\mathbf{y}_{n+1} = \mathbf{Y}_{n+1} + \mathbf{e}_{n+1}. \tag{7.2.7}$$

Thus the combined error between the round-off and truncation gives the following expression for the derivative:

$$\frac{d\mathbf{y}}{dt} = \frac{\mathbf{Y}_{n+1} - \mathbf{Y}_n}{\Delta t} + E_n(\mathbf{y}_n, \Delta t) \tag{7.2.8}$$

where the total error, E_n, is the combination of round-off and truncation such that

$$E_n = E_{\text{round}} + E_{\text{trunc}} = \frac{\mathbf{e}_{n+1} - \mathbf{e}_n}{\Delta t} - \frac{\Delta t}{2}\frac{d^2\mathbf{y}(c)}{dt^2}. \tag{7.2.9}$$

We now determine the maximum size of the error. In particular, we can bound the maximum value of round-off and the second derivative to be

$$|\mathbf{e}_{n+1}| \leq e_r \tag{7.2.10a}$$

$$|-\mathbf{e}_n| \leq e_r \tag{7.2.10b}$$

$$M = \max_{c \in [t_n, t_{n+1}]} \left\{ \left| \frac{d^2\mathbf{y}(c)}{dt^2} \right| \right\}. \tag{7.2.10c}$$

This then gives the maximum error to be

$$|E_n| \leq \frac{e_r + e_r}{\Delta t} + \frac{\Delta t}{2}M = \frac{2e_r}{\Delta t} + \frac{\Delta t M}{2}. \tag{7.2.11}$$

To minimize the error, we require that $\partial|E_n|/\partial(\Delta t) = 0$. Calculating this derivative gives

$$\frac{\partial|E_n|}{\partial(\Delta t)} = -\frac{2e_r}{\Delta t^2} + \frac{M}{2} = 0, \tag{7.2.12}$$

so that

$$\Delta t = \left(\frac{4e_r}{M}\right)^{1/2}. \tag{7.2.13}$$

This gives the step-size resulting in a minimum error. Thus the smallest step-size is not necessarily the most accurate. Rather, a balance between round-off error and truncation error is achieved to obtain the optimal step-size.

Stability

The accuracy of any scheme is certainly important. However, it is meaningless if the scheme is not stable numerically. The essense of a stable scheme: the numerical solutions do not blow up to infinity. As an example, consider the simple differential equation

$$\frac{dy}{dt} = \lambda y \tag{7.2.14}$$

with

$$y(0) = y_0. \tag{7.2.15}$$

The analytic solution is easily calculated to be $y(t) = y_0 \exp(\lambda t)$. However, if we solve this problem numerically with a forward Euler method we find

$$y_{n+1} = y_n + \Delta t \cdot \lambda y_n = (1 + \lambda \Delta t) y_n. \tag{7.2.16}$$

After N steps, we find this iteration scheme yields

$$y_N = (1 + \lambda \Delta t)^N y_0. \tag{7.2.17}$$

Given that we have a certain amount of round-off error, the numerical solution would then be given by

$$y_N = (1 + \lambda \Delta t)^N (y_0 + e). \tag{7.2.18}$$

The error then associated with this scheme is given by

$$E = (1 + \lambda \Delta t)^N e. \tag{7.2.19}$$

At this point, the following observations can be made. For $\lambda > 0$, the solution $y_N \to \infty$ in Eq. (7.2.18) as $N \to \infty$. So although the error also grows, it may not be significant in comparison to the size of the numerical solution.

In contrast, Eq. (7.2.18) for $\lambda < 0$ is markedly different. For this case, $y_N \to 0$ in Eq. (7.2.18) as $N \to \infty$. The error, however, can dominate in this case. In particular, we have the following two cases for the error given by (7.2.19):

$$\text{I:} \quad |1 + \lambda \Delta t| < 1 \quad \text{then} \quad E \to 0 \tag{7.2.20a}$$

$$\text{II:} \quad |1 + \lambda \Delta t| > 1 \quad \text{then} \quad E \to \infty. \tag{7.2.20b}$$

In case I, the scheme would be considered stable. However, case II holds and is unstable provided $\Delta t > -2/\lambda$.

A general theory of stability can be developed for any one-step time-stepping scheme. Consider the one-step recursion relation for an $M \times M$ system

$$\mathbf{y}_{n+1} = \mathbf{A}\mathbf{y}_n. \tag{7.2.21}$$

After N steps, the algorithm yields the solution

$$\mathbf{y}_N = \mathbf{A}^N \mathbf{y}_0, \tag{7.2.22}$$

where $\mathbf{y}_0$ is the initial vector. A well-known result from linear algebra is that

$$\mathbf{A}^N = \mathbf{S}\mathbf{\Lambda}^N\mathbf{S}^{-1} \tag{7.2.23}$$

where $\mathbf{S}$ is the matrix whose columns are the eigenvectors of $\mathbf{A}$, and

$$\mathbf{\Lambda} = \begin{pmatrix} \lambda_1 & 0 & \cdots & 0 \\ 0 & \lambda_2 & 0 & \cdots \\ \vdots & & \ddots & \vdots \\ 0 & \cdots & 0 & \lambda_M \end{pmatrix} \rightarrow \mathbf{\Lambda}^N = \begin{pmatrix} \lambda_1^N & 0 & \cdots & 0 \\ 0 & \lambda_2^N & 0 & \cdots \\ \vdots & & \ddots & \vdots \\ 0 & \cdots & 0 & \lambda_M^N \end{pmatrix} \tag{7.2.24}$$

is a diagonal matrix whose entries are the eigenvalues of $\mathbf{A}$. Thus upon calculating $\mathbf{\Lambda}^N$, we are only concerned with the eigenvalues. In particular, instability occurs if $|\lambda_i| > 1$ for $i = 1, 2, \ldots, M$. This method can be easily generalized to two-step schemes (Adams methods) by considering $\mathbf{y}_{n+1} = \mathbf{A}\mathbf{y}_n + \mathbf{B}\mathbf{y}_{n-1}$.

Lending further significance to this stability analysis is its connection with practical implementation. We contrast the difference in stability between the forward and backward Euler schemes. The forward Euler scheme has already been considered in (7.2.16)–(7.2.19). The backward Euler displays significant differences in stability. If we again consider (7.2.14) with (7.2.15), the backward Euler method gives the iteration scheme

$$y_{n+1} = y_n + \Delta t \cdot \lambda y_{n+1}, \tag{7.2.25}$$

which after N steps leads to

$$y_N = \left(\frac{1}{1 - \lambda \Delta t}\right)^N y_0. \tag{7.2.26}$$

The round-off error associated with this scheme is given by

$$E = \left(\frac{1}{1 - \lambda \Delta t}\right)^N e. \tag{7.2.27}$$

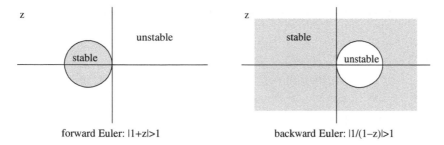

Figure 7.3: Regions for stable stepping (shaded) for the forward Euler and backward Euler schemes. The criteria for instability are also given for each stepping method.

By letting $z = \lambda \Delta t$ be a complex number, we find the following criteria to yield unstable behavior based upon (7.2.19) and (7.2.27):

$$\text{Forward Euler:} \quad |1 + z| > 1 \qquad (7.2.28a)$$

$$\text{Backward Euler:} \quad \left| \frac{1}{1 - z} \right| > 1. \qquad (7.2.28b)$$

Figure 7.3 shows the regions of stable and unstable behavior as a function of z. It is observed that the forward Euler scheme has a very small range of stability whereas the backward Euler scheme has a large range of stability. This large stability region is part of what makes implicit methods so attractive. Thus stability regions can be calculated. However, control of the accuracy is also essential.

7.3 Advanced Time-Stepping Algorithms

Before closing our analysis on time-stepping algorithms, a few alternative methods are considered in the interest of achieving better performance. Specifically, the commonly used adaptive time-stepping algorithm will be outlined. In addition, the exponential time-stepper will be developed for numerically stiff problems.

Adaptive time-stepping algorithm

Adaptive time-stepping algorithms are tremendously important in practice. The premise of the adaptive stepping technique is to take as large a time-step as possible while retaining a prescribed accuracy. All of the time-stepping algorithms used in MATLAB have a built-in adaptive stepper. Indeed, the fundamental premise of the standard differential equation solver in MATLAB is to guarantee a solution with a prescribed absolute and relative tolerance. The time-step Δt is chosen so that the tolerance constraints are met.

The following algorithm outlines the method employed in the adaptive stepping method routine.

1. Start with a default time-step Δt_0.

2. Use one of the iteration algorithms (such as fourth-order Runge–Kutta) to take a time-step Δt into the future. Represent the solution by $f_1(t + \Delta t)$. The initial $\Delta t = \Delta t_0$.

3. Now cut the time-step in half ($\Delta t/2$) and use the iteration algorithm to advance Δt into the future. This would require two iterative steps. Represent the solution by $f_2(t + \Delta t)$.

4. Compare the solutions using Δt and $\Delta t/2$. For instance, one may measure the L^2 norm between the two solutions: $E = \|f_1 - f_2\|$.

5. If the difference is above a prescribed tolerance, i.e. $E >$ tolerance, then cut the time-step in half again to $\Delta t/4$ and compare again. Continue cutting the time-step in half until the tolerance is achieved.

6. If the comparison is already below the prescribed tolerance, then the time-step can be doubled in size to $2\Delta t$. The comparison can be made again until the tolerance condition is violated.

This algorithm is a shell of what the algorithm might look like. Certainly a more sophisticated version can be constructed, but this illustrates the key concept of either making the time-step bigger or smaller as needed.

The advantages of such a scheme are enormous. It allows the iterative method for advancing the differential equation solution into the future to be maximally efficient. When the solution is changing slowly, the time-steps will be quite large, whereas when the solution changes rapidly, the time-step will automatically adjust and shorten in order to preserve the accuracy.

In MATLAB, the time-step can be adjusted by modifying the tolerance settings associated with the time-stepping algorithm. The following code adjusts the time-step so that a 10^{-4} accuracy is achieved both for relative tolerance and absolute tolerance.

```
TOL=1e-4; OPTIONS = odeset('RelTol',TOL,'AbsTol',TOL);
[t,y] = ode45('F',tspan,y0,OPTIONS);
```

The speed of the code is largely determined by the accuracy setting as determined from the **odeset** command. The default is a 10^{-6} tolerance for both relative and absolute error.

Exponential time-steppers

A nice example of a time-stepping technique that removes numerical stiffness generated from either large linear terms (or linear, high-order derivative terms) is the exponential time-stepping technique [13, 14]. This is the only stiff time-stepper that will be considered in detail in this book. As with any other stiff-stepping technique, advantage is taken of certain properties of the differential equation considered in order to make the algorithm faster, i.e. in order to maximize the step-size while keeping a fixed accuracy.

The prototype equation to be considered is the differential equation of the form [13]

$$\frac{d\mathbf{y}}{dt} = c\mathbf{y} + F(\mathbf{y}, t) \tag{7.3.1}$$

where $|c| \gg 1$. When c is large in magnitude, it dominates the selection of the time-step Δt. In fact, it forces the time-step Δt to be quite small in order to accurately resolve the future solution $\mathbf{y}(t + \Delta t)$. It should be noted that the large c term often arises from high-order and linear derivatives in problems involving partial differential equations. This will be considered in future sections.

One method of dealing with numerical stiffness induced by c is to attempt a solution via the integrating factor method. Multiplying Eq. (7.3.1) by the factor $\exp(-ct)$ gives the following set of algebraic reductions:

$$\frac{d\mathbf{y}}{dt} \exp(-ct) = c\mathbf{y} \exp(-ct) + F(\mathbf{y}, t) \exp(-ct)$$

$$\frac{d\mathbf{y}}{dt} \exp(-ct) - c\mathbf{y} \exp(-ct) = F(\mathbf{y}, t) \exp(-ct)$$

$$\frac{d}{dt} \left(\mathbf{y} \exp(-ct) \right) = F(\mathbf{y}, t) \exp(-ct). \tag{7.3.2}$$

Integrating both sides from time t to time $t + \Delta t$ yields

$$\mathbf{y}(t + \Delta t) \exp\left(-c(t + \Delta t) \right) - \mathbf{y}(t) \exp(-ct) = \int_t^{t+\Delta t} F(\mathbf{y}(\tau), \tau) \exp(-c\tau) d\tau. \tag{7.3.3}$$

Finally, by multiplying both sides by $\exp(c(t + \Delta t))$ and making a change of variables in the integral, the following formula is achieved:

$$\mathbf{y}(t + \Delta t) = \mathbf{y}(t) \exp(c\Delta t) + \exp(c\Delta t) \int_0^{\Delta t} F(\mathbf{y}(t + \tau), t + \tau) \exp(-c\tau) d\tau. \tag{7.3.4}$$

This formula is exact. Moreover, it has some of the basic characteristics of the Adams–Bashforth and Adams–Moulton schemes considered earlier where the integral approximation determines the accuracy and iteration of the scheme. But unlike the Adams methods, the linear term that is scaled with the parameter c is explicitly accounted for by the integrating factor.

The approximation of the integral yields the exponential steppers of interest. The simplest approximation is to assume that the function F takes on a constant value so that $F(\mathbf{y}(t + \tau), t + \tau) \approx F(\mathbf{y}(t), t)$. Integrating the integral now with respect to τ yields

$$\mathbf{y}_{n+1} = \mathbf{y}_n \exp(c\Delta t) + F_n \left(\exp(c\Delta t) - 1 \right)/c \tag{7.3.5}$$

where $\mathbf{y}_n = \mathbf{y}(t_n)$ and $F_n = F(\mathbf{y}_n, t_n)$. Note that in the limit as $c \ll 1$, this formula asymptotically approaches the Euler stepping formula.

As with the Adams–Bashforth method, a higher order approximation can be used for evaluating the integral. In particular, the following can be used

$$F = F_n + \tau \left(F_n - F_{n-1}\right) / \Delta t + O\left(\Delta t^2\right).$$ (7.3.6)

This approximation to the integrand gives an improved accuracy to the time-stepping method. When inserted into the formula (7.3.4), the following time-stepping algorithm is derived

$$\mathbf{y}_{n+1} = \mathbf{y}_n \exp(c\Delta t) + F_n \left[(1 + c\Delta t)\exp(c\Delta t) - 1 - 2c\Delta t\right] / \left(c^2 \Delta t\right)$$
$$+ F_{n-1} \left[1 + c\Delta t - \exp(c\Delta t)\right] / \left(c^2 \Delta t\right).$$ (7.3.7)

In the limit as $c \to 0$, this reduces to the second-order Adams–Bashforth scheme. But recall that the purpose of this scheme is to consider problems for which $c \gg 1$.

Cox and Matthews [13] continue with this idea in order to derive the equivalent of the fourth-order Runge–Kutta scheme with the exponential time-stepping explicitly accounted for. The claim is that this derivation is nontrivial and requires careful manipulation and the aid of symbolic computing. Regardless, the following exponential, fourth-order Runge–Kutta scheme is developed

$$\mathbf{y}_{n+1} = e^{c\Delta t}\mathbf{y}_n + \left[-4 - c\Delta t + e^{c\Delta t}\left(4 - 3c\Delta t + (c\Delta t)^2\right)\right] F\left(\mathbf{y}_n, t_n\right) / \left(c^3 \Delta t^2\right)$$
$$+ 2\left[2 + c\Delta t + e^{c\Delta t}(-2 + c\Delta t)\right] F(\mathbf{a}_n, t_n + \Delta t/2) + F(\mathbf{b}_n, t_n + \Delta t/2)$$
$$+ \left[-4 - 3c\Delta t - (c\Delta t)^2 + e^{c\Delta t}\left(4 - c\Delta t\right)\right] F(\mathbf{c}_n, t_n + \Delta t)$$ (7.3.8)

where

$$\mathbf{a}_n = \mathbf{y}_n e^{c\Delta t/2} + \left(e^{c\Delta t/2} - 1\right) F(\mathbf{y}_n, t_n)/c$$ (7.3.9a)

$$\mathbf{b}_n = \mathbf{y}_n e^{c\Delta t/2} + \left(e^{c\Delta t/2} - 1\right) F(\mathbf{a}_n, t_n + \Delta t/2)/c$$ (7.3.9b)

$$\mathbf{c}_n = \mathbf{a}_n e^{c\Delta t/2} + \left(e^{c\Delta t/2} - 1\right) \left[2F(\mathbf{b}_n, t_n + \Delta t/2) - F(\mathbf{y}_n, t_n)\right]/c.$$ (7.3.9c)

To implement this in practice, Kassam and Trefethen [14] show that special care must be taken in order to evaluate the coefficients $\mathbf{a}_n, \mathbf{b}_n$ and $\mathbf{c}_n$. This evaluation is intimately related to the well-known numerical difficulty in evaluating the function $(e^z - 1)/z$. However, using the method of Kassam and Trefethen [14], the exponential time-stepping algorithm becomes a tremendously efficient tool when $c \gg 1$. Indeed, Kassam and Trefethen [14] illustrate that an entire order or magnitude in step-size can be gained for higher order partial differential equations such as the Kuramoto–Sivashinky equation. In this case, the high c value is effectively created by linear, four-order diffusion. The order of magnitude in increased step-size Δt makes the implementation of the scheme above a must.

As a final comment, one can easily proceed with other schemes of this type. All that is needed is the starting point of the definition of the derivative or the fundamental theorem of calculus. Approximations are generated from there. The exponential scheme outlined above clearly takes advantage of the linear term by folding it into the integrating factor. All specialty stepping schemes typically do something of this form, or utilize a semi-implicit method, to maximize time-stepping performance and error reduction.

7.4   Boundary Value Problems: The Shooting Method

To this point, we have only considered the solutions of differential equations for which the initial conditions are known. However, many physical applications do not have specified initial conditions, but rather some given boundary (constraint) conditions. A simple example of such a problem is the second-order boundary value problem

$$\frac{d^2y}{dt^2} = f\left(t, y, \frac{dy}{dt}\right)$$ (7.4.1)

on $t \in [a, b]$ with the general boundary conditions

$$\alpha_1 y(a) + \beta_1 \frac{dy(a)}{dt} = \gamma_1$$ (7.4.2a)

$$\alpha_2 y(b) + \beta_2 \frac{dy(b)}{dt} = \gamma_2.$$ (7.4.2b)

Thus the solution is defined over a specific interval and must satisfy the relations (7.4.2) at the end points of the interval. Figure 7.4 gives a graphical representation of a generic boundary value problem solution. We discuss the algorithm necessary to make use of the time-stepping schemes in order to solve such a problem.

The shooting method

The boundary value problems constructed here require information at the present time ($t = a$) and a future time ($t = b$). However, the time-stepping schemes developed previously only require information about the starting time $t = a$. Some effort is then needed to reconcile the time-stepping schemes with the boundary value problems presented here.

We begin by reconsidering the generic boundary value problem

$$\frac{d^2y}{dt^2} = f\left(t, y, \frac{dy}{dt}\right)$$ (7.4.3)

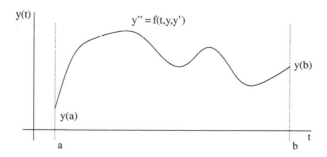

Figure 7.4: Graphical depiction of the structure of a typical solution to a boundary value problem with constraints at $t = a$ and $t = b$.

on $t \in [a, b]$ with the boundary conditions

$$y(a) = \alpha \tag{7.4.4a}$$

$$y(b) = \beta. \tag{7.4.4b}$$

The stepping schemes considered thus far for second-order differential equations involve a choice of the initial conditions $y(a)$ and $y'(a)$. We can still approach the boundary value problem from this framework by choosing the "initial" conditions

$$y(a) = \alpha \tag{7.4.5a}$$

$$\frac{dy(a)}{dt} = A, \tag{7.4.5b}$$

where the constant A is chosen so that as we advance the solution to $t = b$ we find $y(b) = \beta$. The shooting method gives an iterative procedure with which we can determine this constant A. Figure 7.5 illustrates the solution of the boundary value problem given two distinct values of A. In this case, the value of $A = A_1$ gives a value for the initial slope which is too low to satisfy the boundary conditions (7.4.4), whereas the value of $A = A_2$ is too large to satisfy (7.4.4).

Computational algorithm

The above example demonstrates that adjusting the value of A in (7.4.5b) can lead to a solution which satisfies (7.4.4b). We can solve this using a self-consistent algorithm to search for the appropriate value of A which satisfies the original problem. The basic algorithm is as follows:

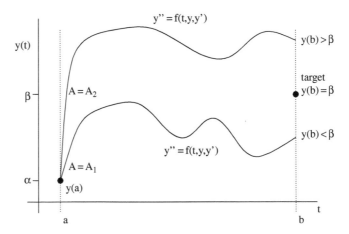

Figure 7.5: Solutions to the boundary value problem with $y(a) = \alpha$ and $y'(a) = A$. Here, two values of A are used to illustrate the solution behavior and its lack of matching the correct boundary value $y(b) = \beta$. However, the two solutions suggest that a bisection scheme could be used to find the correct solution and value of A.

1   Solve the differential equation using a time-stepping scheme with the initial conditions $y(a) = \alpha$ and $y'(a) = A$.

2   Evaluate the solution $y(b)$ at $t = b$ and compare this value with the target value of $y(b) = \beta$.

3   Adjust the value of A (either bigger or smaller) until a desired level of tolerance and accuracy is achieved. A bisection method for determining values of A, for instance, may be appropriate.

4   Once the specified accuracy has been achieved, the numerical solution is complete and is accurate to the level of the tolerance chosen and the discretization scheme used in the time-stepping.

We illustrate graphically a bisection process in Fig. 7.6 and show the convergence of the method to the numerical solution which satisfies the original boundary conditions $y(a) = \alpha$ and $y(b) = \beta$. This process can occur quickly so that convergence is achieved in a relatively low number of iterations provided the differential equation is well behaved.

Shooting example

To illustrate the implementation of the shooting method, consider the following boundary value problem

$$y'' + (x^2 - \sin x)y' - (\cos^2 x)y = 5 \qquad x \in [0, 1] \tag{7.4.6}$$

with the boundary conditions

$$y(0) = 3 \tag{7.4.7a}$$
$$y'(1) = 5. \tag{7.4.7b}$$

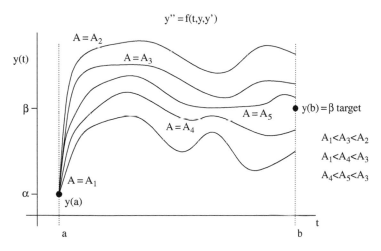

Figure 7.6: Graphical illustration of the shooting process which uses a bisection scheme to converge to the appropriate value of A for which $y(b) = \beta$.

The first step is to write the above boundary value problem as an equivalent system of equations by defining $y_1 = y$ and $y_2 = y'$. This yields

$$y_1' = y_2 \tag{7.4.8a}$$
$$y_2' = -(x^2 - \sin x)y_2 + (\cos^2 x)y_1 + 5 \tag{7.4.8b}$$
$$y_1(0) = 3 \quad y_2(1) = 5. \tag{7.4.8c}$$

The idea is to replace the second boundary condition above, $y_2(1) = 5$, with $y_1'(0) = A$ where A is to be determined in the shooting algorithm. From exploring the differential equations with different value of A, it is found that $A > -3$ in order for the derivate at $x = 1$ to go from below the value of 5 to above the value of 5. The following code finds the solution in a fairly straightforward manner.

```
xspan=[0 1];  % x range
A=-3; dA=0.5;  % initial derivative value

for j=1:100
    y=[3 A];  % initial condition
    [x,ysol]=ode45('bvpexam_rhs',xspan,y);

    if abs(ysol(end,2)-5)<10^(-6)  % check convergence
        break
    end

    if ysol(end,2)<5  % adjust launch angle
        A=A+dA;  % if below five, make A bigger
    else
        A=A-dA;  % if above five, make A smaller
        dA=dA/2;  % refine search now
    end
end
```

Here the right-hand side function is given by

bvpexam_rhs.m

```
function rhs=bvpexam_rhs(x,y)
rhs=[y(2);  -(x^2-sin(x))*y(2) + (cos(x)^2)*y(1) + 5];
```

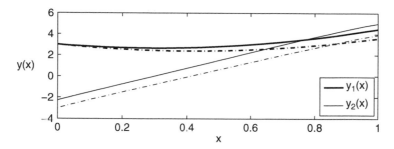

Figure 7.7: Solution of the boundary value problem depicting the solution $y(x) = y_1(x)$ (bolded lines) and its derivative $y'(x) = y_2(x)$. The dotted lines are the initial guess of the shooting algorithm for which $y'(1) = y_2(1) = A = -3$.

Figure 7.7 depicts the final solution $y(x) = y_1(x)$ along with the derivative $y'(x) = y_2(x)$. The dotted line is the initial guess ($A = -3$) for a solution and the starting point of the shooting algorithm. Note that the algorithm uses something like a bisection algorithm for refining the search for the appropriate value of A.

Eigenvalues and eigenfunctions: The infinite domain

Boundary value problems often arise as eigenvalue systems for which the eigenvalue and eigenfunction must both be determined. As an example of such a problem, we consider the second-order differential equation on the infinite line

$$\frac{d^2\psi_n}{dx^2} + \left[n(x) - \beta_n\right]\psi_n = 0 \tag{7.4.9}$$

with the boundary conditions $\psi_n(x) \to 0$ as $x \to \pm\infty$. For this example, we consider the spatial function $n(x)$ which is given by

$$n(x) = n_0 \begin{cases} 1 - |x|^2 & 0 \le |x| \le 1 \\ 0 & |x| > 1 \end{cases} \tag{7.4.10}$$

with n_0 being an arbitrary constant. Figure 7.8 shows the spatial dependence of $n(x)$. The parameter β_n in this problem is the eigenvalue. For each eigenvalue, we can calculate a normalized eigenfunction ψ_n. The standard normalization requires $\int_{-\infty}^{\infty} |\psi_n|^2 dx = 1$.

Although the boundary conditions are imposed as $x \to \pm\infty$, computationally we require a finite domain. We thus define our computational domain to be $x \in [-L, L]$ where $L \gg 1$. Since $n(x) = 0$ for $|x| > 1$, the governing equation reduces to

$$\frac{d^2\psi_n}{dx^2} - \beta_n\psi_n = 0 \quad |x| > 1 \tag{7.4.11}$$

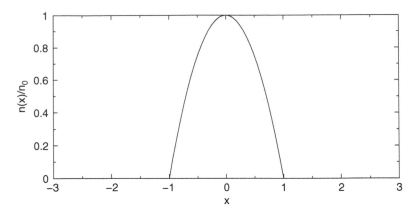

Figure 7.8: Plot of the spatial function $n(x)$.

which has the general solution

$$\psi_n = c_1 \exp\left(\sqrt{\beta_n}x\right) + c_2 \exp\left(-\sqrt{\beta_n}x\right) \tag{7.4.12}$$

for $\beta_n \geq 0$. Note that we can only consider values of $\beta_n \geq 0$ since for $\beta_n < 0$, the general solution becomes $\psi_n = c_1 \cos(\sqrt{|\beta_n|}x) + c_2 \sin(-\sqrt{|\beta_n|}x)$ which does not decay to zero as $x \to \pm\infty$. In order to ensure that the decay boundary conditions are satisfied, we must eliminate one of the two linearly independent solutions of the general solution. In particular, we must have

$$x \to \infty: \qquad \psi_n = c_2 \exp\left(-\sqrt{\beta_n}x\right) \tag{7.4.13a}$$

$$x \to -\infty: \qquad \psi_n = c_1 \exp\left(\sqrt{\beta_n}x\right). \tag{7.4.13b}$$

Thus the requirement that the solution decays at infinity eliminates one of the two linearly independent solutions. Alternatively, we could think of this situation as being a case where only one linearly independent solution is allowed as $x \to \pm\infty$. But a single linearly independent solution corresponds to a first-order differential equation. Therefore, the decay solutions (7.4.13) can equivalently be thought of as solutions to the following first-order equations:

$$x \to \infty: \qquad \frac{d\psi_n}{dx} + \sqrt{\beta_n}\psi_n = 0 \tag{7.4.14a}$$

$$x \to -\infty: \qquad \frac{d\psi_n}{dx} - \sqrt{\beta_n}\psi_n = 0. \tag{7.4.14b}$$

From a computational viewpoint then, the effective boundary conditions to be considered on the computational domain $x \in [-L, L]$ are the following

$$x = L: \qquad \frac{d\psi_n(L)}{dx} = -\sqrt{\beta_n}\psi_n(L) \tag{7.4.15a}$$

$$x = -L: \qquad \frac{d\psi_n(-L)}{dx} = \sqrt{\beta_n}\psi_n(-L). \tag{7.4.15b}$$

In order to solve the problem, we write the governing differential equation as a system of equations. Thus we let $x_1 = \psi_n$ and $x_2 = d\psi_n/dx$ which gives

$$x_1' = \psi_n' = x_2 \tag{7.4.16a}$$
$$x_2' = \psi_n'' = \left[\beta_n - n(x)\right]\psi_n = \left[\beta_n - n(x)\right]x_1. \tag{7.4.16b}$$

In matrix form, we can write the governing system as

$$\mathbf{x}' = \begin{pmatrix} 0 & 1 \\ \beta_n - n(x) & 0 \end{pmatrix}\mathbf{x} \tag{7.4.17}$$

where $\mathbf{x} = (x_1\ x_2)^T = (\psi_n\ d\psi_n/dx)^T$. The boundary conditions (7.4.15) are

$$x = L: \qquad x_2 = -\sqrt{\beta_n}x_1 \tag{7.4.18a}$$
$$x = -L: \qquad x_2 = \sqrt{\beta_n}x_1. \tag{7.4.18b}$$

The formulation of the boundary value problem is thus complete. It remains to develop an algorithm to find the eigenvalues β_n and corresponding eigenfunctions ψ_n. Figure 7.9 illustrates the first four eigenfunctions and their associated eigenvalues for $n_0 = 100$ and $L = 2$.

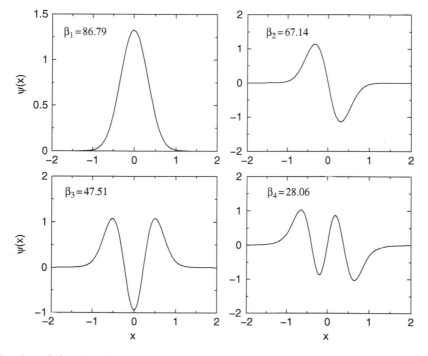

Figure 7.9: Plot of the first four eigenfunctions along with their eigenvalues β_n. For this example, $L = 2$ and $n_0 = 100$. These eigenmode structures are typical of those found in quantum mechanics and electromagnetic waveguides.

7.5 Implementation of Shooting and Convergence Studies

The implementation of the shooting scheme relies on the effective use of a time-stepping algorithm along with a root finding method for choosing the appropriate initial conditions which solve the boundary value problem. The specific system to be considered is similar to that developed in the last section. We consider

$$\mathbf{x}' = \begin{pmatrix} 0 & 1 \\ \beta_n - n(x) & 0 \end{pmatrix} \mathbf{x} \tag{7.5.1}$$

where $\mathbf{x} = (x_1 \ x_2)^T = (\psi_n \ d\psi_n/dx)^T$. The boundary conditions are simplified in this case to be

$$x = 1: \qquad \psi_n(1) = x_1(1) = 0 \tag{7.5.2a}$$

$$x = -1: \qquad \psi_n(-1) = x_1(-1) = 0. \tag{7.5.2b}$$

At this stage, we will also assume that $n(x) = n_0$ for simplicity.

With the problem thus defined, we turn our attention to the key aspects in the computational implementation of the boundary value problem solver. These are

- **FOR** loops
- **IF** statements
- time-stepping algorithms: **ode23, ode45, ode113, ode15s**
- step-size control
- code development and flow.

Every code will be controlled by a set of FOR loops and IF statements. It is imperative to have proper placement of these control statements in order for the code to operate successfully.

Convergence

In addition to developing a successful code, it is reasonable to ask whether your numerical solution is actually correct. Thus far, the premise has been that discretization should provide an accurate approximation to the true solution provided the time-step Δt is small enough. Although in general this philosophy is correct, every numerical algorithm should be carefully checked to determine if it indeed converges to the true solution. The time-stepping schemes considered previously already hint at how the solutions should converge: fourth-order Runge–Kutta converges like Δt^4, second-order Runge–Kutta converges like Δt^2, and second-order predictor–corrector schemes converge like Δt^2. Thus the algorithm for checking convergence is as follows:

1 Solve the differential equation using a time-step $\Delta t*$ which is very small. This solution will be considered the exact solution. Recall that we would in general like to take Δt as large as possible for efficiency purposes.

2   Using a much larger time-step Δt, solve the differential equation and compare the numerical solution with that generated from $\Delta t*$. Cut this time-step in half and compare once again. In fact, continue cutting the time-step in half: $\Delta t, \Delta t/2, \Delta t/4, \Delta t/8, \ldots$ in order to compare the difference in the exact solution to this hierarchy of solutions.

3   The difference between any run $\Delta t*$ and Δt is considered the error. Although there are many definitions of error, a practial error measurement is the root-mean square error $E = \left[(1/N) \sum_{i=1}^{N} |y_{\Delta t*} - y_{\Delta t}|^2 \right]^{1/2}$. Once calculated, it is possible to verify the convergence law of Δt^2, for instance, with a second-order Runge–Kutta.

Flow control

In order to begin coding, it is always prudent to construct the basic structure of the algorithm. In particular, it is good to determine the number of FOR loops and IF statements which may be required for the computations. What is especially important is determining the hierarchic structure for the loops. To solve the boundary value problem proposed here, we require two FOR loops and one IF statement block. The outermost FOR loop of the code should determine the number of eigenvalues and eigenmodes to be searched for. Within this FOR loop there exists a second FOR loop which iterates the shooting method so that the solution converges to the correct boundary value solution. This second FOR loop has a logical IF statement which needs to check whether the solution has indeed converged to the boundary value solution, or whether adjustment of the value of β_n is necessary and the iteration procedure needs to be continued. Figure 7.10 illustrates

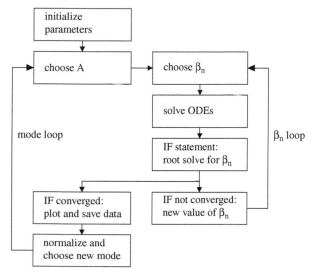

Figure 7.10: Basic algorithm structure for solving the boundary value problem. Two FOR loops are required to step through the values of β_n and A along with a single IF statement block to check for convergence of the solution.

the backbone of the numerical code for solving the boundary value problem. It includes the two FOR loops and logical IF statement block as the core of its algorithmic structure. For a nonlinear problem, a third FOR loop would be required for A in order to achieve the normalization of the eigenfunctions to unity.

The various pieces of the code are constructed here using the MATLAB programming language. We begin with the initialization of the parameters.

Initialization

```
clear all;  % clear all previously defined variables
close all;  % clear all previously defined figures

tol=10^(-4);  % define a tolerance level to be achieved
              % by the shooting algorithm
col=['r','b','g','c','m','k'];  % eigenfunction colors

n0=100;     % define the parameter n0
A=1;        % define the initial slope at x=-1
x0=[0 A];   % initial conditions: x1(-1)=0, x1'(-1)=A
xp=[-1 1];  % define the span of the computational domain
```

Upon completion of the initialization process for the parameters which are not involved in the main loop of the code, we move into the main FOR loop which searches out a specified number of eigenmodes. Embedded in this FOR loop is a second FOR loop which attempts different values of β_n until the correct eigenvalue is found. An IF statement is used to check the convergence of values of β_n to the appropriate value.

Main program

```
beta_start=n0;  % beginning value of beta
for modes=1:5   %  begin mode loop
  beta=beta_start;    % initial value of eigenvalue beta
  dbeta=n0/100;       % default step size in beta
  for j=1:1000  % begin convergence loop for beta
    [t,y]=ode45('shoot2',xp,x0,[],n0,beta); % solve ODEs
```

Continued

```
      if abs(y(end,1)-0) < tol   % check for convergence
         beta                     % write out eigenvalue
         break                    % get out of convergence loop
      end
      if (-1)^(modes+1)*y(end,1)>0 % this IF statement block
         beta=beta-dbeta;            % checks to see if beta
      else                           % needs to be higher or lower
         beta=beta+dbeta/2;          % and uses bisection to
         dbeta=dbeta/2;              % converge to the solution
      end                            %
   end   % end convergence loop
   beta_start=beta-0.1;   % after finding eigenvalue, pick
                          % new starting value for next mode
   norm=trapz(t,y(:,1).*y(:,1))   % calculate the normalization
   plot(t,y(:,1)/sqrt(norm),col(modes)); hold on   % plot modes
end   % end mode loop
```

The code uses *ode45*, which is a fourth-order Runge–Kutta method, to solve the differential equation and advance the solution. The function *shoot2.m* is called in this routine. For the differential equation considered here, the function *shoot2.m* would be the following:

shoot2.m

```
function rhs=shoot2(xspan,x,dummy,n0,beta)
rhs=[ x(2)
      (beta-n0)*x(1) ];
```

This code will find the first five eigenvalues and plot their corresponding normalized eigen-functions. The bisection method implemented to adjust the values of β_n to find the boundary value solution is based upon observations of the structure of the even and odd eigenmodes. In general, it is always a good idea to first explore the behavior of the solutions of the boundary value problem before writing the shooting routine. This will give important insights into the behavior of the solutions and will allow for a proper construction of an accurate and efficient bisection method. Figure 7.11 illustrates several characteristic features of this boundary value problem. In Figs. 7.11(a) and 7.11(b), the behavior of the solution near the first even and first odd solution is exhibited. From Fig. 7.11(a) it is seen that for the even modes increasing values of β bring the solution from $\psi_n(1) > 0$ to $\psi_n(1) < 0$. In contrast, odd modes go from $\psi_n(1) < 0$ to $\psi_n(1) > 0$ as β is increased. This observation forms the basis for the bisection method developed in the code. Figure 7.11(c) illustrates the first four normalized eigenmodes along with their corresponding eigenvalues.

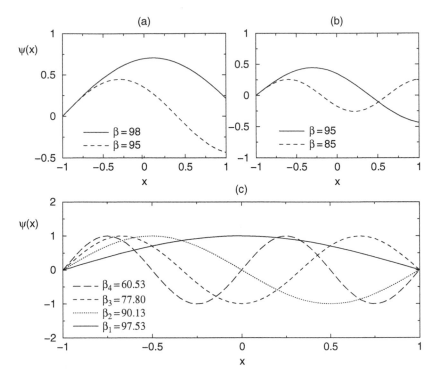

Figure 7.11 : In (a) and (b) the behavior of the solution near the first even and first odd solution is depicted. Note that for the even modes increasing values of β bring the solution from $\psi_n(1) > 0$ to $\psi_n(1) < 0$. In contrast, odd modes go from $\psi_n(1) < 0$ to $\psi_n(1) > 0$ as β is increased. In (c) the first four normalized eigenmodes along with their corresponding eigenvalues are illustrated for $n_0 = 100$.

7.6   Boundary Value Problems: Direct Solve and Relaxation

The shooting method is not the only method for solving boundary value problems. The direct method of solution relies on Taylor expanding the differential equation itself. For linear problems, this results in a matrix problem of the form $\mathbf{Ax} = \mathbf{b}$. For nonlinear problems, a nonlinear system of equations must be solved using a relaxation scheme, i.e. a Newton or secant method. The prototypical example of such a problem is the second-order boundary value problem

$$\frac{d^2 y}{dt^2} = f\left(t, y, \frac{dy}{dt}\right) \tag{7.6.1}$$

on $t \in [a, b]$ with the general boundary conditions

$$\alpha_1 y(a) + \beta_1 \frac{dy(a)}{dt} = \gamma_1 \tag{7.6.2a}$$

$$\alpha_2 y(b) + \beta_2 \frac{dy(b)}{dt} = \gamma_2. \tag{7.6.2b}$$

Thus the solution is defined over a specific interval and must satisfy the relations (7.6.2) at the end points of the interval.

Before considering the general case, we simplify the method by considering the linear boundary value problem

$$\frac{d^2y}{dt^2} = p(t)\frac{dy}{dt} + q(t)y + r(t) \tag{7.6.3}$$

on $t \in [a, b]$ with the simplified boundary conditions

$$y(a) = \alpha \tag{7.6.4a}$$
$$y(b) = \beta. \tag{7.6.4b}$$

Taylor expanding the differential equation and boundary conditions will generate the linear system of equations which solve the boundary value problem.

To see how the Taylor expansions are useful, consider the following two Taylor series:

$$f(t + \Delta t) = f(t) + \Delta t\frac{df(t)}{dt} + \frac{\Delta t^2}{2!}\frac{d^2f(t)}{dt^2} + \frac{\Delta t^3}{3!}\frac{d^3f(c_1)}{dt^3} \tag{7.6.5a}$$

$$f(t - \Delta t) = f(t) - \Delta t\frac{df(t)}{dt} + \frac{\Delta t^2}{2!}\frac{d^2f(t)}{dt^2} - \frac{\Delta t^3}{3!}\frac{d^3f(c_2)}{dt^3} \tag{7.6.5b}$$

where $c_1 \in [t, t+\Delta t]$ and $c_2 \in [t, t-\Delta t]$. Subtracting these two expressions gives

$$f(t + \Delta t) - f(t - \Delta t) = 2\Delta t\frac{df(t)}{dt} + \frac{\Delta t^3}{3!}\left(\frac{d^3f(c_1)}{dt^3} + \frac{d^3f(c_2)}{dt^3}\right). \tag{7.6.6}$$

By using the mean value theorem of calculus, we find $f'''(c) = (f'''(c_1) + f'''(c_2))/2$. Upon dividing the above expression by $2\Delta t$ and rearranging, we find the following expression for the first derivative:

$$\frac{df(t)}{dt} = \frac{f(t + \Delta t) - f(t - \Delta t)}{2\Delta t} - \frac{\Delta t^2}{6}\frac{d^3f(c)}{dt^3} \tag{7.6.7}$$

where the last term is the truncation error associated with the approximation of the first derivative using this particular Taylor series generated expression. Note that the truncation error in this case is $O(\Delta t^2)$. We could improve on this by continuing our Taylor expansion and truncating it at higher orders in Δt. This would lead to higher accuracy schemes. Further, we could also approximate the second, third, fourth and higher derivatives using this technique. It is also possible to generate backward and forward difference schemes by using points only behind or in front of the current point, respectively. Tables 4.1–4.3 summarize the second-order and fourth-order central difference schemes along with the forward- and backward-difference formulas which are accurate to second order.

To solve the simplified linear boundary value problem above, which is accurate to second order, we use Table 4.1 for the second and first derivatives. The boundary value problem then becomes

$$\frac{y(t+\Delta t)-2y(t)+y(t-\Delta t)}{\Delta t^2} = p(t)\frac{y(t+\Delta t)-y(t-\Delta t)}{2\Delta t}+q(t)y(t)+r(t) \tag{7.6.8}$$

with the boundary conditions $y(a) = \alpha$ and $y(b) = \beta$. We can rearrange this expression to read

$$\left[1 - \frac{\Delta t}{2}p(t)\right]y(t + \Delta t) - \left[2 + \Delta t^2 q(t)\right]y(t) + \left[1 + \frac{\Delta t}{2}p(t)\right]y(t - \Delta t) = \Delta t^2 r(t). \quad (7.6.9)$$

We discretize the computational domain and denote $t_0 = a$ to be the left boundary point and $t_N = b$ to be the right boundary point. This gives the boundary conditions

$$y(t_0) = y(a) = \alpha \quad (7.6.10a)$$
$$y(t_N) = y(b) = \beta. \quad (7.6.10b)$$

The remaining $N - 1$ points can be recast as a matrix problem $\mathbf{A}\mathbf{x} = \mathbf{b}$ where

$$\mathbf{A} = \begin{bmatrix} 2 + \Delta t^2 q(t_1) & -1 + \frac{\Delta t}{2}p(t_1) & 0 & \cdots & & & 0 \\ -1 - \frac{\Delta t}{2}p(t_2) & 2 + \Delta t^2 q(t_2) & -1 + \frac{\Delta t}{2}p(t_2) & 0 & \cdots & & \vdots \\ 0 & \ddots & \ddots & \ddots & & & \\ \vdots & & & & & & 0 \\ & & & & & & \\ \vdots & & & & \ddots & \ddots & -1 + \frac{\Delta t}{2}p(t_{N-2}) \\ 0 & \cdots & & & 0 & -1 - \frac{\Delta t}{2}p(t_{N-1}) & 2 + \Delta t^2 q(t_{N-1}) \end{bmatrix}$$
$$(7.6.11)$$

and

$$\mathbf{x} = \begin{bmatrix} y(t_1) \\ y(t_2) \\ \vdots \\ y(t_{N-2}) \\ y(t_{N-1}) \end{bmatrix} \quad \mathbf{b} = \begin{bmatrix} -\Delta t^2 r(t_1) + (1 + \Delta t p(t_1)/2)y(t_0) \\ -\Delta t^2 r(t_2) \\ \vdots \\ -\Delta t^2 r(t_{N-2}) \\ -\Delta t^2 r(t_{N-1}) + (1 - \Delta t p(t_{N-1})/2)y(t_N) \end{bmatrix}. \quad (7.6.12)$$

Thus the solution can be found by a direct solve of the linear system of equations.

Nonlinear systems

A similar solution procedure can be carried out for nonlinear systems. However, difficulties arise from solving the resulting set of nonlinear algebraic equations. We can once again consider the general differential equation and expand with second-order accurate schemes:

$$y'' = f(t, y, y') \rightarrow \frac{y(t+\Delta t) - 2y(t) + y(t-\Delta t)}{\Delta t^2} = f\left(t, y(t), \frac{y(t+\Delta t) - y(t-\Delta t)}{2\Delta t}\right). \qquad (7.6.13)$$

We discretize the computational domain and denote $t_0 = a$ to be the left boundary point and $t_N = b$ to be the right boundary point. Considering again the simplified boundary conditions $y(t_0) = y(a) = \alpha$ and $y(t_N) = y(b) = \beta$ gives the following nonlinear system for the remaining $N - 1$ points.

$$2y_1 - y_2 - \alpha + \Delta t^2 f(t_1, y_1, (y_2 - \alpha)/2\Delta t) = 0$$
$$-y_1 + 2y_2 - y_3 + \Delta t^2 f(t_2, y_2, (y_3 - y_1)/2\Delta t) = 0$$
$$\vdots$$
$$-y_{N-3} + 2y_{N-2} - y_{N-1} + \Delta t^2 f(t_{N-2}, y_{N-2}, (y_{N-1} - y_{N-3})/2\Delta t) = 0$$
$$-y_{N-2} + 2y_{N-1} - \beta + \Delta t^2 f(t_{N-1}, y_{N-1}, (\beta - y_{N-2})/2\Delta t) = 0.$$

This $(N - 1) \times (N - 1)$ nonlinear system of equations can be very difficult to solve and imposes a severe constraint on the usefulness of the scheme. However, there may be no other way of solving the problem and a solution to this system of equations must be computed. Further complicating the issue is the fact that for nonlinear systems such as these, there are no guarantees about the existence or uniqueness of solutions. The best approach is to use a *relaxation* scheme which is based upon Newton or secant method iterations.

7.7 Implementing MATLAB for Boundary Value Problems

Both a shooting technique and a direct discretization method have been developed here for solving boundary value problems. More generally, one would like to use a high-order method that is robust and capable of solving general, nonlinear boundary value problems. MATLAB provides a convenient and easy to use routine, known as **bvp4c**, that is capable of solving fairly sophisticated problems. The algorithm relies on an iteration structure for solving nonlinear systems of equations. In particular, **bvp4c** is a finite difference code that implements the three-stage Lobatto IIIa formula. This is a collocation formula and the collocation polynomial provides a C^1-continuous solution that is fourth-order accurate uniformly in $x \in [a, b]$. Mesh selection and error control are based on the residual of the continuous solution. Since it is an iteration scheme, its effectiveness will ultimately rely on your ability to provide the algorithm with an initial guess for the solution.

Two example codes will be demonstrated here. The first is a simple linear, constant coefficient second-order equation with fairly standard boundary conditions. The second example is nonlinear with an undetermined parameter (eigenvalue) that must also be determined. Both illustrate the power and ease of use of the built-in boundary value solver of MATLAB.

Linear boundary value problem

As a simple and particular example of a boundary value problem, consider the following:

$$y'' + 3y' + 6y = 5 \qquad (7.7.1)$$

on the domain $x \in [1, 3]$ and with boundary conditions

$$y(1) = 3 \qquad (7.7.2a)$$
$$y(3) + 2y'(3) = 5. \qquad (7.7.2b)$$

This problem can be solved in a fairly straightforward manner using analytic techniques. However, we will pursue here a numerical solution instead.

As with any differential equation solver, the equation must first be put into the form of a system of first-order equations. Thus by introducing the variables

$$y_1 = y(x) \qquad (7.7.3a)$$
$$y_2 = y'(x) \qquad (7.7.3b)$$

we can rewrite the governing equations as

$$y_1' = y_2 \qquad (7.7.4a)$$
$$y_2' = 5 - 3y_2 - 6y_1 \qquad (7.7.4b)$$

with the transformed boundary conditions

$$y_1(1) = 3 \qquad (7.7.5a)$$
$$y_1(3) + 2y_2(3) = 5. \qquad (7.7.5b)$$

In order to implement the boundary value problem in MATLAB, the boundary conditions need to be placed in the general form

$$f(y_1, y_2) = 0 \ \text{ at } \ x = x_L \qquad (7.7.6a)$$
$$g(y_1, y_2) = 0 \ \text{ at } \ x = x_R \qquad (7.7.6b)$$

where $f(y_1, y_2)$ and $g(y_1, y_2)$ are the boundary value functions at the left (x_L) and right (x_R) boundary points. This then allows us to rewrite the boundary conditions in (7.7.5) as the following:

$$f(y_1, y_2) = y_1 - 3 = 0 \qquad \text{ at } \ x_L = 1 \qquad (7.7.7a)$$
$$g(y_1, y_2) = y_1 + 2y_2 - 5 = 0 \quad \text{ at } \ x_R = 3. \qquad (7.7.7b)$$

The formulation of the boundary value problem is then completely specified by the differential equation (7.7.4) and its boundary conditions (7.7.7).

The boundary value solver **bvp4c** requires three pieces of information: the equation to be solved, its associated boundary conditions, and your initial guess for the solution. The first two lines of the following code performs all three of these functions:

```
init=bvpinit(linspace(1,3,10),[0 0]);
sol=bvp4c(@bvp_rhs,@bvp_bc,init);
x=linspace(1,3,100); BS=deval(sol,x);
plot(x,BS(1,:))
```

We dissect this code by first considering the first line of code for generating a MATLAB data structure for use as the initial data. In this initial line of code, the **linspace** command is used to define the initial mesh that goes from $x = 1$ to $x = 3$ with 10 equally spaced points. In this example, the initial values of $y_1(x) = y(x)$ and $y_2(x) = y'(x)$ are zero. If you have no guess, then this is probably your best guess to start with. Thus you not only guess the initial guess, but the initial grid on which to find the solution.

Once the guess and initial mesh is generated, two functions are called with the **@bvp_rhs** and **@bvp_bc** function calls. These are functions representing the differential equation and boundary conditions (7.7.4) and (7.7.7), respectively. These functions are easily constructed as the following:

bvp_rhs.m

```
function rhs=bvp_rhs(x,y)
rhs=[y(2);   5 - 3*y(2) - 6*y(1)];
```

Likewise, the boundary conditions are constructed from the code

bvp_rhs.m

```
function bc=bvp_bc(yL,yR)
bc=[yL(1)-3;   yR(1)+2*yR(2)-5];
```

Note that what is passed into the boundary condition function is the value of the vector **y** at the left (**yL**) and right (**yR**) of the computational domain, i.e. at the values of $x = 1$ and $x = 3$.

What is produced by **bvp4c** is a data structure **sol**. This data structure contains a variety of information about the problem, including the solution $y(x)$. To extract the solution, the final two lines of code in the main program are used. Specifically, the **deval** command evaluates the solution at the points specified by the vector **x**, which in this case is a linear space of 100 points between one and three. The solution can then simply be plotted once the values of $y(x)$ have been extracted. Figure 7.12 shows the solution and its derivative that are produced from the boundary value solver.

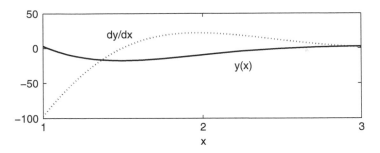

Figure 7.12: The solution of the boundary value problem (7.7.4) with (7.7.7). The solid line is $y(x)$ while the dotted line is its derivative $dy(x)/dx$.

Nonlinear eigenvalue problem

As a more sophisticated example of a boundary value problem, we will consider a fully nonlinear eigenvalue problem. Thus consider the following:

$$y'' + (100 - \beta)y + \gamma y^3 = 0 \tag{7.7.8}$$

on the domain $x \in [-1, 1]$ and with boundary conditions

$$y(-1) = 0 \tag{7.7.9a}$$
$$y(1) = 0. \tag{7.7.9b}$$

This problem cannot be solved using analytic techniques due to the complexity introduced by the nonlinearity. But a numerical solution can be fairly easily constructed. Note the similarity between this problem and that considered in the shooting section.

As before, the equation must first be put into the form of a system of first-order equations. Thus by introducing the variables

$$y_1 = y(x) \tag{7.7.10a}$$
$$y_2 = y'(x) \tag{7.7.10b}$$

we can rewrite the governing equations as

$$y_1' = y_2 \tag{7.7.11a}$$
$$y_2' = (\beta - 100)y_1 - \gamma y_1^3 \tag{7.7.11b}$$

with the transformed boundary conditions

$$y_1(-1) = 0 \tag{7.7.12a}$$
$$y_1(1) = 0. \tag{7.7.12b}$$

The formulation of the boundary value problem is then completely specified by the differential equation (7.7.11) and its boundary conditions (7.7.12). Note that unlike before, we do not know the parameter β. Thus it must be determined along with the solution. As with the initial conditions, we will also guess an initial value of β and let **bvp4c** converge to the appropriate value of β. Note that for such nonlinear problems, the effectiveness of **bvp4c** relies almost exclusively on providing a good initial guess.

As before, the boundary value solver **bvp4c** requires three pieces of information: the equation to be solved, its associated boundary conditions, and your initial guess for the solution and the parameter β. The first line of the code gives the initial guess for β while the next two lines of the following code perform the three remaining functions:

```
beta=99;
init2=bvpinit(linspace(-1,1,50),@mat4init,beta);
sol2=(bvp4c(@bvp_rhs2,@bvp_bc2,init2));
x2=linspace(-1,1,100);  BS2=deval(sol2,x2);
figure(2), plot(x2,BS2(1,:))
```

In this case, there are now three function calls: **@bvp_rhs2**, **@bvp_bc2** and **@mat4init**. This calls the differential equation, its boundary values and its initial guess, respectively.

We begin with the initial guess. Our implementation example of the shooting method showed that the first linear solution behaved like a cosine function. Thus we can guess that $y_1(x) = y(x) = \cos[(\pi/2)x]$ where the factor of $\pi/2$ is chosen to make the solution zero at $x = \pm 1$. Note that initially 50 points are chosen between $x = -1$ and $x = 1$. The following function generates the initial data to begin the iteration process.

mat4init.m

```
function yinit = mat4init(x)
yinit = [cos((pi/2)*x); -(pi/2)*sin((pi/2)*x)];
```

Again, it must be emphasized that the success of **bvp4c** relies almost exclusively on the above subroutine and guess. Without a good guess, especially for nonlinear problems, you may find a solution, but just not the one you want.

The equation itself is handled in the subroutine **bvp_rhs2.m**. The following is the construction of the right-hand side function with $\gamma = 1$.

bvp_rhs2.m

```
function rhs=bvp_rhs2(x,y,beta)
rhs=[y(2); ((beta-100)*y(1)-y(1)^3)];
```

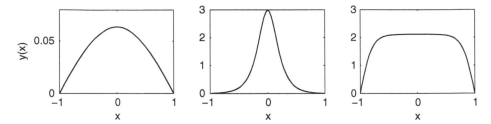

Figure 7.13: Three different nonlinear eigenvalue solutions to (7.7.11) and its boundary conditions (7.7.12). The left panel has the constraint condition $y'(-1) = 0.1$ with $\gamma = 1$, the middle panel has $y'(-1) = 0.1$ with $\gamma = 10$ and the right panel has $y'(-1) = 10$ with $\gamma = -10$. Such solutions are called *nonlinear ground states* of the eigenvalue problem.

Finally, the implementation of the boundary conditions, or constraints, must be imposed. There is something very important to note here: we started with two constraints since we have a second-order differential equation, i.e. $y(\pm 1) = 0$. However, since we do not know the value of β, the system is currently an underdetermined system of equations. In order to remedy this, we need to impose one more constraint on the system. This is somewhat arbitrary unless there is a natural constraint for you to choose in the system. Here, we will simply choose $dy(-1)/dx = 0.1$, i.e. we will impose the launch angle at the left. This results in the boundary value imposition routine:

bvp_bc2.m

```
function bc=bvp_bc2(yl,yr,beta)
bc=[yl(1); yl(2)-0.1; yr(1)];
```

With these three constraints, the **bvp4c** iteration routine not only finds the solution, but also the appropriate value of β that satisfies the above constraints.

Executing the routine and subroutines generates the solution to the boundary value problem. One can also experiment with different guesses and values of β to generate new and interesting solutions. Three such solutions are demonstrated in Fig. 7.13 as a function of the β value and the guess angle $dy(-1)/dx$.

 ## 7.8   Linear Operators and Computing Spectra

As a final application of boundary value problems, we will consider the ability to accurately compute the spectrum of linear operators. Linear operators often arise in the context of evaluating the stability of solutions to partial differential equations. Indeed, stability plays a key role in many branches of science and engineering, including aspects of fluid mechanics, pattern formation with reaction–diffusion models, high-speed transmission of optical information, and the feasibility of MHD fusion devices, to name a few. If one can find solutions for a given particular differential equation, then the stability of that solution becomes critical in determining the ultimate behavior of the system. Specifically, if a physical phenomenon is observable and persists, then the corresponding solution to a valid mathematical model should be stable. If, however, instability is established, the nature of the unstable modes suggest what patterns may develop from the unstable solutions. Finally, for many problems of physical interest, fundamental mathematical models are well established. However, in many cases these fundamental models are too complicated to allow for detailed analysis, thus leading to the study of simpler approximate (linear) models using reductive perturbation methods.

To further illustrate these concepts, consider a generic partial differential equation of the form

$$\frac{\partial u}{\partial t} = N\left(x, u, \frac{\partial u}{\partial x}, \frac{\partial^2 u}{\partial x^2}, \cdots\right) \tag{7.8.1}$$

where the function N is generically nonconstant coefficient in x and a nonlinear function of $u(x, t)$ and its derivatives. Here, we will assume that this function is well behaved so that solutions can exist by the basic Cauchy–Kovaleskaya theorem. In addition, there are boundary conditions associated with Eq. (7.8.1). Such boundary constraints may be imposed on either a finite or infinite domain, and an example of each case will be considered in what follows.

In what follows, only the stability of equilibrium solutions will be considered. Equilibrium solutions are found when $\partial u/\partial t = 0$. Thus, there would exist a time-independent solution $U(x)$ such that

$$N\left(x, U, \frac{\partial U}{\partial x}, \frac{\partial^2 U}{\partial x^2}, \cdots\right) = 0. \tag{7.8.2}$$

If one could find such a solution $U(x)$, then its stability could be computed. Again, if the solution is stable, then it may be observable in the physical system. If it is unstable, then it will not be observed in practice. However, the unstable modes of such a system may potentially give clues to the resulting behavior of the system.

To generate a linear stability problem for the equilibrium solution $U(x)$ of (7.8.1), one would *linearize* about the solution so that

$$u(x, t) = U(x) + \epsilon v(x, t) \tag{7.8.3}$$

where the parameter $\epsilon \ll 1$ so that the function $v(x, t)$ is only a small perturbation from the equilibrium solution. If the function $v(x, t) \to \infty$ as $t \to \infty$, the system is said to be unstable. If $v(x, t) \to 0$ as $t \to \infty$, the system is said to be asymptotically stable. If $v(x, t)$ remains $O(1)$ as $t \to \infty$, the system is said to simply be stable. Thus a determination of what happens to $v(x, t)$ is required. Plugging in (7.8.3) into (7.8.1) and Taylor expanding using the fact that $\epsilon \ll 1$ gives the equation

$$\frac{\partial v}{\partial t} = \mathcal{L}\left[U(x)\right] v + O(\epsilon) \tag{7.8.4}$$

where the $O(\epsilon)$ represents all the terms from the Taylor expansion that are $O(\epsilon)$ or smaller. Finally, by assuming that the function $v(x, t)$ takes the following form:

$$v(x, t) = w(x) \exp(\lambda t) \tag{7.8.5}$$

the following *linear* eigenvalue problem (spectral problem) results

$$\mathcal{L}\big[U(x)\big]v = \lambda v \tag{7.8.6}$$

where $\mathcal{L}$ is the linear operator associated with the stability of the equilibrium solution $U(x)$. Note that by our definitions of stability above and the solution form (7.8.5), the equilibrium solution is unstable if any real part of the eigenvalue is positive: $\Re\{\lambda\} > 0$. Stability is established if $\Re\{\lambda\} \le 0$, with asymptotic stability occurring if $\Re\{\lambda\} < 0$. Thus computing the eigenvalues of the linear operator is the critical step in evaluating stability.

Sturm–Liouville theory

Of course, anybody familiar with Sturm–Liouville theory [7] will recognize the importance of computing the spectra of the linearized operators. For Sturm–Liouville problems, the linear operator and its associated eigenvalues take on the form

$$\mathcal{L}v = -\frac{d}{dx}\left[p(x)\frac{dv}{dx}\right] + q(x)v = \lambda w(x)v \tag{7.8.7}$$

where $p(x) > 0$, $q(x)$ and $w(x) > 0$ are specified on the interval $x \in [a, b]$. In the simplest case, these are continuous on the entire interval. The associated boundary conditions are given by

$$\alpha_1 v(a) + \alpha_2 \frac{dv(a)}{dx} = 0 \quad \text{and} \quad \beta_1 v(b) + \beta_2 \frac{dv(b)}{dx} = 0. \tag{7.8.8}$$

It is known that the Sturm–Liouville problem has some very nice properties, including the fact that the eigenvalues are all real and distinct. Many classic physical systems of interest naturally fall within Sturm–Liouville theory, thus propagating a variety of special functions such as Bessel functions, Legendre functions, Laguerre polynomials, etc. Indeed, much of special-function theory involves Sturm–Liouville theory at its core.

Derivate operators

Although much is known in the literature about Sturm–Liouville theory, the theory often relates properties about the eigenfunctions and eigenvectors in the abstract. To compute the actual spectra of (7.8.7), we will advocate a numerical procedure based upon finite difference discretization. Shooting methods can also be applied, but they will not be considered in what follows. What is immediately apparent from (7.8.7), and more generally (7.8.6), is the presence of derivative operators. From the finite difference perspective, these are no more than matrix operations on the discretized (vector) solution. Following the ideas of Section 4.1, we can construct a number of first and second derivate matrices that are continually used in practice. The domain $x \in [a, b]$ is discretized into $N + 1$ intervals so that $x_0 = a$ and $x_{N+1} = b$.

To begin, we will consider constructing first-derivative matrices that are $O(\Delta x^2)$ accurate. In this case, Table 4.1 applies. The simplest case to consider is when *Dirichlet boundary conditions* are applied, i.e. $v(a) = v(b) = 0$. The Dirichlet boundaries imply that only the interior points $v(x_j)$ where $j = 1, 2, \cdots, N$ must be solved for. The resulting derivative matrix is given by

$$\frac{\partial}{\partial x} \text{ with } u(a) - u(b) = 0 \rightarrow \mathbf{A}_1 = \frac{1}{2\Delta x} \begin{bmatrix} 0 & 1 & 0 & 0 & \cdots & 0 \\ -1 & 0 & 1 & 0 & \cdots & \\ 0 & -1 & 0 & 1 & \cdots & \\ & & \vdots & & & \\ & & \cdots & -1 & 0 & 1 \\ 0 & \cdots & & 0 & -1 & 0 \end{bmatrix}. \tag{7.8.9}$$

If the boundary conditions are changed to *Neumann boundary conditions* with $dv(a)/dx = dv(b)/dx = 0$, then the derivative matrix changes. Specifically, if use is made of the forward- and backward- difference formulas of Table 4.3, then it is easy to show that

$$\frac{dv_0}{dx} = \frac{-3v_0 + 4v_1 - v_2}{2\Delta x} = 0 \rightarrow v_0 = \frac{4}{3}v_1 - \frac{1}{3}v_2. \tag{7.8.10}$$

Similarly at the right boundary, one can find

$$\frac{dv_{N+1}}{dx} = \frac{-3v_{N+1} + 4v_N - v_{N-1}}{2\Delta x} = 0 \rightarrow v_{N+1} = \frac{4}{3}v_N - \frac{1}{3}v_{N-1}. \tag{7.8.11}$$

Thus both boundary values, v_0 and v_{N+1}, are determined from the interior points alone. The final piece of information necessary is the following:

$$\frac{dv_1}{dx} = \frac{v_2 - v_0}{2\Delta x} = \frac{v_2 - [(4/3)v_1 - (1/3)v_2]}{2\Delta x} = \frac{(4/3)(v_2 - v_1)}{2\Delta x}. \tag{7.8.12}$$

A similar calculation can be done for dv_N/dx to yield the derivative matrix

$$\frac{\partial}{\partial x} \text{ with } \frac{du(a)}{dx} = \frac{du(b)}{dx} = 0 \rightarrow \mathbf{A}_2 = \frac{1}{2\Delta x} \begin{bmatrix} -4/3 & 4/3 & 0 & 0 & \cdots & 0 \\ -1 & 0 & 1 & 0 & \cdots & \\ 0 & -1 & 0 & 1 & \cdots & \\ & & \vdots & & & \\ & & \cdots & -1 & 0 & 1 \\ 0 & \cdots & & 0 & -4/3 & 4/3 \end{bmatrix}. \tag{7.8.13}$$

Finally, we consider the case of *periodic boundary conditions* for which $v(a) = v(b)$. This implies that $v_0 = v_{N+1}$. Thus when considering this case, the unknowns include v_0 through v_N, i.e. there are $N + 1$ unknowns as opposed to N unknowns. The derivative matrix in this case is

$$\frac{\partial}{\partial x} \text{ with } u(a) = u(b) \quad \rightarrow \quad \mathbf{A}_3 = \frac{1}{2\Delta x} \begin{bmatrix} 0 & 1 & 0 & 0 & \cdots & -1 \\ -1 & 0 & 1 & 0 & \cdots & \\ 0 & -1 & 0 & 1 & \cdots & \\ & & & \vdots & & \\ & \cdots & & -1 & 0 & 1 \\ 1 & \cdots & & 0 & -1 & 0 \end{bmatrix} \tag{7.8.14}$$

where the top right and bottom left corners come from the fact that, for instance, $dv_N/dx = (v_{N+1} - v_{N-1})/(2\Delta x) = (v_0 - v_{N-1})/(2\Delta x)$.

Second-derivative analogs can also be created for these matrices using the same basic ideas as for first derivatives. In this case, we find for Dirichlet boundary conditions

$$\frac{\partial^2}{\partial x^2} \text{ with } u(a) = u(b) = 0 \quad \rightarrow \quad \mathbf{B}_1 = \frac{1}{\Delta x^2} \begin{bmatrix} -2 & 1 & 0 & 0 & \cdots & 0 \\ 1 & -2 & 1 & 0 & \cdots & \\ 0 & 1 & -2 & 1 & \cdots & \\ & & & \vdots & & \\ & \cdots & & 1 & -2 & 1 \\ 0 & \cdots & & 0 & 1 & -2 \end{bmatrix}. \tag{7.8.15}$$

For Neumann boundary conditions, the following applies

$$\frac{\partial^2}{\partial x^2} \text{ with } \frac{du(a)}{dx} = \frac{du(b)}{dx} = 0 \quad \rightarrow \quad \mathbf{B}_2 = \frac{1}{\Delta x^2} \begin{bmatrix} -2/3 & 2/3 & 0 & 0 & \cdots & 0 \\ 1 & -2 & 1 & 0 & \cdots & \\ 0 & 1 & -2 & 1 & \cdots & \\ & & & \vdots & & \\ & \cdots & & 1 & -2 & 1 \\ 0 & \cdots & & 0 & 2/3 & -2/3 \end{bmatrix}. \tag{7.8.16}$$

where the boundaries are again evaluated from the interior points through the relations (7.8.10) and (7.8.11). Finally, for periodic boundaries the matrix is

$$\frac{\partial^2}{\partial x^2} \text{ with } u(a) = u(b) \quad \rightarrow \quad \mathbf{B}_3 = \frac{1}{\Delta x^2} \begin{bmatrix} -2 & 1 & 0 & 0 & \cdots & 1 \\ 1 & -2 & 1 & 0 & \cdots & \\ 0 & 1 & -2 & 1 & \cdots & \\ & & & \vdots & & \\ & \cdots & & 1 & -2 & 1 \\ 1 & \cdots & & 0 & 1 & -2 \end{bmatrix}. \tag{7.8.17}$$

The derivative operators shown here are fairly standard and can easily be implemented in MATLAB. The following code constructs matrices $\mathbf{A}_1$–$\mathbf{A}_3$ and $\mathbf{B}_1$ $\mathbf{B}_3$. First, there is the construction of the first-derivative matrices:

```
% assume N and dx given
A=zeros(N,N);
for j=1:N-1;
   A(j,j+1)=1;
   A(j+1,j)=-1;
end
A1=A/(2*dx);   % Dirichlet matrix

A2=A;
A2(1,1)=-4/3;  A2(1,2)=4/3;
A2(N,N)=4/3;   A2(N,N-1)=-4/3;
A2=A2/(2*dx);  % Neumann matrix

A3=A;
A3(N,1)=1;  A3(1,N)=-1;
A3=A3/(2*dx);   % periodic BC matrix
```

The second-derivative matrices can be constructed similarly with the following MATLAB code

```
% assume N and dx given
B=zeros(N,N);
for j=1:N
   B(j,j)=-2;  % diagonal term
end
for j=1:N-1;
   B(j,j+1)=1;  % off-diagonals
   B(j+1,j)=1;
end
B1=B/(dx^2);  % Dirichlet matrix

B2=B;
B2(1,1)=-2/3;  B2(1,2)=2/3;
B2(N,N)=-2/3;  B2(N,N-1)=2/3;
A2=A2/(dx^2); % Neumann matrix

B3=B;
B3(N,1)=1;  B3(1,N)=-1;
B3=B3/(dx^2);   % periodic BC matrix
```

Given how often such derivative matrices are used, the above code could be used as part of function call to simply pull out the derivative matrix required. In addition, we will later learn how to encode such matrices in a sparse manner since they are mostly zeros.

Example: The harmonic oscillator

To see an example of how to compute the spectra of a linear operator, consider the example of the harmonic oscillator which has the well-known Hermite function solutions. This is a Sturm–Liouville problem with $p(x) = w(x) = 1$ and $q(x) = x^2$. The $q(x)$ plays the role of a trapping potential in quantum mechanics. Typically, the harmonic oscillator is specified on the domain $x \in [-\infty, \infty]$ with solutions $v(\pm\infty) \to 0$. Instead of specifying that the domain is infinite, we will consider a finite domain of size $x \in [-a, a]$ with the Dirichlet boundary conditions $v(\pm L) = 0$. Thus we have the eigenvalue problem

$$\mathcal{L}v = -\frac{d^2v}{dx^2} + x^2v = \lambda v \quad \text{with} \quad v(\pm L) = 0. \tag{7.8.18}$$

Our objective is to construct the spectra and eigenfunctions of this linear operator. We can even compare them directly to the known solutions to see how accurate the reconstruction is. The following code discretizes the domain and computes the spectra.

```
L=4;  % domain size
N=200; % discretization of interior
x=linspace(-L,L,N+2); % add boundary points
dx=x(2)-x(1); % compute dx

% NOTE: compute B1 from previous code

P=zeros(N,N);
for j=1:N
  P(j,j)=x(j+1)^2; % potenial x^2
end

linL=-B1+ P; % linear operator
[V,D]=eigs(linL,5,'sm')
```

This code computes the second-order accurate spectra of the harmonic oscillator. Here, $\Delta x \approx 0.04$ so that an accuracy of 10^{-3} is expected. Only the five smallest eigenvalues are returned in the code. These five eigenvalues are theoretically known to be given by

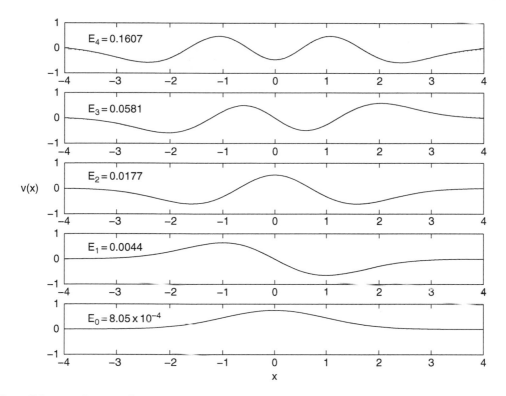

Figure 7.14: First five eigenfunctions of the harmonic oscillator computed using finite difference discretization. Shown are the computed eigenfunctions (solid line) versus the exact solution (dotted line). The two lines are almost indistinguishable from each other. Thus the norm of the difference of the solutions is given by E_j for the different eigenfunctions.

$\lambda_n = (2n + 1)$ with $n = 0, 1, 2, \cdots$. In our calculation we find the first five $\lambda_n = 1, 3, 5, 7, 9 \approx$ 0.9999, 2.9995, 4.9991, 7.0009, 9.0152. Thus the accuracy is as prescribed. The first five eigenfunctions are shown in Fig. 7.14. As can be seen, the finite difference method is fairly easy to implement and yields fast and accurate results.

8 Finite Difference Methods

Finite difference methods are based exclusively on Taylor expansions. They are one of the most powerful methods available since they are relatively easy to implement, can handle fairly complicated boundary conditions, and allow for explicit calculations of the computational error. The result of discretizing any given problem is the need to solve a large linear system of equations or perhaps manipulate large, sparse matrices. All this will be dealt with in the following sections.

8.1 Finite Difference Discretization

To discuss the solution of a given problem with the finite difference method, we consider a specific example from atmospheric sciences. The quasi-two-dimensional motion of the atmosphere can be modeled by the advection–diffusion behavior for the vorticity $\omega(x, y, t)$ which is coupled to the streamfunction $\psi(x, y, t)$:

$$\frac{\partial \omega}{\partial t} + [\psi, \omega] = \nu \nabla^2 \omega \tag{8.1.1a}$$

$$\nabla^2 \psi = \omega \tag{8.1.1b}$$

where

$$[\psi, \omega] = \frac{\partial \psi}{\partial x}\frac{\partial \omega}{\partial y} - \frac{\partial \psi}{\partial y}\frac{\partial \omega}{\partial x} \tag{8.1.2}$$

and $\nabla^2 = \partial_x^2 + \partial_y^2$ is the two-dimensional Laplacian. Note that this equation has both an advection component (hyperbolic) from $[\psi, \omega]$ and a diffusion component (parabolic) from $\nu \nabla^2 \omega$. We will assume that we are given the initial value of the vorticity

$$\omega(x, y, t = 0) = \omega_0(x, y). \tag{8.1.3}$$

Additionally, we will proceed to solve this problem with periodic boundary conditions. This gives the following set of boundary conditions

$$\omega(-L, y, t) = \omega(L, y, t) \tag{8.1.4a}$$

$$\omega(x, -L, t) = \omega(x, L, t) \tag{8.1.4b}$$

$$\psi(-L, y, t) = \psi(L, y, t) \tag{8.1.4c}$$

$$\psi(x, -L, t) = \psi(x, L, t) \tag{8.1.4d}$$

where we are solving on the computational domain $x \in [-L, L]$ and $y \in [-L, L]$.

Basic algorithm structure

Before discretizing the governing partial differential equation, it is important to clarify what the basic solution procedure will be. Two physical quantities need to be solved as functions of time:

$$\psi(x, y, t) \quad \text{streamfunction} \tag{8.1.5a}$$

$$\omega(x, y, t) \quad \text{vorticity.} \tag{8.1.5b}$$

We are given the initial vorticity $\omega_0(x, y)$ and periodic boundary conditions. The solution procedure is as follows:

1 **Elliptic solve:** Solve the elliptic problem $\nabla^2 \psi = \omega_0$ to find the streamfunction at time zero $\psi(x, y, t = 0) = \psi_0$.

2 **Time-stepping:** Given initial ω_0 and ψ_0, solve the advection–diffusion problem by time-stepping with a given method. The Euler method is illustrated below

$$\omega(x, y, t + \Delta t) = \omega(x, y, t) + \Delta t \left(\nu \nabla^2 \omega(x, y, t) - [\psi(x, y, t), \omega(x, y, t)] \right).$$

This advances the solution Δt into the future.

3 **Loop:** With the updated value of $\omega(x, y, \Delta t)$, we can repeat the process by again solving for $\psi(x, y, \Delta t)$ and updating the vorticity once again.

This gives the basic algorithmic structure which must be implemented in order to generate the solution for the vorticity and streamfunction as functions of time. It only remains to discretize the problem and solve.

Step 1: Elliptic solve

We begin by discretizing the elliptic solve problem for the streamfunction $\psi(x, y, t)$. The governing equation in this case is

$$\nabla^2 \psi = \frac{\partial^2 \psi}{\partial x^2} + \frac{\partial^2 \psi}{\partial y^2} = \omega. \tag{8.1.6}$$

Using the central difference formulas of Section 7.6 reduces the governing equation to a set of linearly coupled equations. In particular, we find for a second-order accurate central difference scheme that the elliptic equation reduces to

$$\frac{\psi(x + \Delta x, y, t) - 2\psi(x, y, t) + \psi(x - \Delta x, y, t)}{\Delta x^2} \qquad (8.1.7)$$

$$+ \frac{\psi(x, y + \Delta y, t) - 2\psi(x, y, t) + \psi(x, y - \Delta y, t)}{\Delta y^2} = \omega(x, y, t).$$

Thus the solution at each point depends upon itself and four neighboring points. This creates a five-point stencil for solving this equation. Figure 8.1 illustrates the stencil which arises from discretization. For convenience we denote

$$\psi_{mn} = \psi(x_m, y_n, t). \qquad (8.1.8)$$

By letting $\Delta x^2 = \Delta y^2 = \delta^2$, the discretized equations reduce to

$$-4\psi_{mn} + \psi_{(m-1)n} + \psi_{(m+1)n} + \psi_{m(n-1)} + \psi_{m(n+1)} = \delta^2 \omega_{mn} \qquad (8.1.9)$$

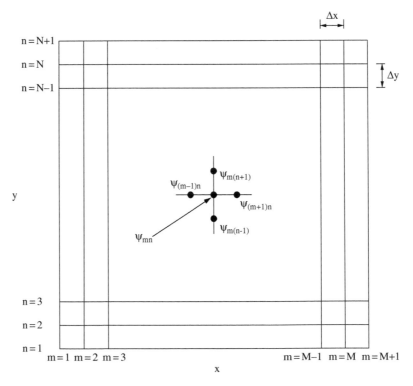

Figure 8.1: Discretization stencil for solving for the streamfunction with second-order accurate central difference schemes. Note that $\psi_{mn} = \psi(x_m, y_n)$.

with periodic boundary conditions imposing the following constraints

$$\psi_{1n} = \psi_{(N+1)n} \tag{8.1.10a}$$

$$\psi_{m1} = \psi_{m(N+1)} \tag{8.1.10b}$$

where $N + 1$ is the total number of discretization points in the computational domain in both the x- and y-directions.

As a simple example, consider the four-point system for which $N = 4$. For this case, we have the following sets of equations

$$-4\psi_{11} + \psi_{41} + \psi_{21} + \psi_{14} + \psi_{12} = \delta^2 \omega_{11}$$
$$-4\psi_{12} + \psi_{42} + \psi_{22} + \psi_{11} + \psi_{13} = \delta^2 \omega_{12}$$
$$\vdots \tag{8.1.11}$$
$$-4\psi_{21} + \psi_{11} + \psi_{31} + \psi_{24} + \psi_{22} = \delta^2 \omega_{21}$$
$$\vdots$$

which results in the sparse matrix (banded matrix) system

$$\mathbf{A}\psi = \delta^2 \omega \tag{8.1.12}$$

where

$$\mathbf{A} = \begin{bmatrix}
-4 & 1 & 0 & 1 & 1 & 0 & 0 & 0 & 0 & 0 & 0 & 0 & 1 & 0 & 0 & 0 \\
1 & -4 & 1 & 0 & 0 & 1 & 0 & 0 & 0 & 0 & 0 & 0 & 0 & 1 & 0 & 0 \\
0 & 1 & -4 & 1 & 0 & 0 & 1 & 0 & 0 & 0 & 0 & 0 & 0 & 0 & 1 & 0 \\
1 & 0 & 1 & -4 & 0 & 0 & 0 & 1 & 0 & 0 & 0 & 0 & 0 & 0 & 0 & 1 \\
1 & 0 & 0 & 0 & -4 & 1 & 0 & 1 & 1 & 0 & 0 & 0 & 0 & 0 & 0 & 0 \\
0 & 1 & 0 & 0 & 1 & -4 & 1 & 0 & 0 & 1 & 0 & 0 & 0 & 0 & 0 & 0 \\
0 & 0 & 1 & 0 & 0 & 1 & -4 & 1 & 0 & 0 & 1 & 0 & 0 & 0 & 0 & 0 \\
0 & 0 & 0 & 1 & 1 & 0 & 1 & -4 & 0 & 0 & 0 & 1 & 0 & 0 & 0 & 0 \\
0 & 0 & 0 & 0 & 1 & 0 & 0 & 0 & -4 & 1 & 0 & 1 & 1 & 0 & 0 & 0 \\
0 & 0 & 0 & 0 & 0 & 1 & 0 & 0 & 1 & -4 & 1 & 0 & 0 & 1 & 0 & 0 \\
0 & 0 & 0 & 0 & 0 & 0 & 1 & 0 & 0 & 1 & -4 & 1 & 0 & 0 & 1 & 0 \\
0 & 0 & 0 & 0 & 0 & 0 & 0 & 1 & 1 & 0 & 1 & -4 & 0 & 0 & 0 & 1 \\
1 & 0 & 0 & 0 & 0 & 0 & 0 & 0 & 1 & 0 & 0 & 0 & -4 & 1 & 0 & 1 \\
0 & 1 & 0 & 0 & 0 & 0 & 0 & 0 & 0 & 1 & 0 & 0 & 1 & -4 & 1 & 0 \\
0 & 0 & 1 & 0 & 0 & 0 & 0 & 0 & 0 & 0 & 1 & 0 & 0 & 1 & -4 & 1 \\
0 & 0 & 0 & 1 & 0 & 0 & 0 & 0 & 0 & 0 & 0 & 1 & 1 & 0 & 1 & -4
\end{bmatrix} \tag{8.1.13}$$

and

$$\psi = (\psi_{11}\ \psi_{12}\ \psi_{13}\ \psi_{14}\ \psi_{21}\ \psi_{22}\ \psi_{23}\ \psi_{24}\ \psi_{31}\ \psi_{32}\ \psi_{33}\ \psi_{34}\ \psi_{41}\ \psi_{42}\ \psi_{43}\ \psi_{44})^{\mathrm{T}} \tag{8.1.14a}$$

$$\omega = (\omega_{11}\ \omega_{12}\ \omega_{13}\ \omega_{14}\ \omega_{21}\ \omega_{22}\ \omega_{23}\ \omega_{24}\ \omega_{31}\ \omega_{32}\ \omega_{33}\ \omega_{34}\ \omega_{41}\ \omega_{42}\ \omega_{43}\ \omega_{44})^{\mathrm{T}}. \tag{8.1.14b}$$

Any matrix solver can then be used to generate the values of the two-dimensional streamfunction which are contained completely in the vector ψ.

Step 2: Time-stepping

After generating the matrix $\mathbf{A}$ and the value of the streamfunction $\psi(x, y, t)$, we use this updated value along with the current value of the vorticity to take a time-step Δt into the future. The appropriate equation is the advection–diffusion evolution equation:

$$\frac{\partial \omega}{\partial t} + [\psi, \omega] = \nu \nabla^2 \omega. \tag{8.1.15}$$

Using the definition of the bracketed term and the Laplacian, this equation is

$$\frac{\partial \omega}{\partial t} = \frac{\partial \psi}{\partial y} \frac{\partial \omega}{\partial x} - \frac{\partial \psi}{\partial x} \frac{\partial \omega}{\partial y} + \nu \left(\frac{\partial^2 \omega}{\partial x^2} + \frac{\partial^2 \omega}{\partial y^2} \right). \tag{8.1.16}$$

Second-order central-differencing discretization then yields

$$
\begin{aligned}
\frac{\partial \omega}{\partial t} = & \left(\frac{\psi(x, y+\Delta y, t) - \psi(x, y-\Delta y, t)}{2\Delta y} \right) \left(\frac{\omega(x+\Delta x, y, t) - \omega(x-\Delta x, y, t)}{2\Delta x} \right) \\
& - \left(\frac{\psi(x+\Delta x, y, t) - \psi(x-\Delta x, y, t)}{2\Delta x} \right) \left(\frac{\omega(x, y+\Delta y, t) - \omega(x, y-\Delta y, t)}{2\Delta y} \right) \\
& + \nu \left\{ \frac{\omega(x+\Delta x, y, t) - 2\omega(x, y, t) + \omega(x-\Delta x, y, t)}{\Delta x^2} \right. \\
& \left. + \frac{\omega(x, y+\Delta y, t) - 2\omega(x, y, t) + \omega(x, y-\Delta y, t)}{\Delta y^2} \right\}. \tag{8.1.17}
\end{aligned}
$$

This is simply a large system of differential equations which can be stepped forward in time with any convenient time-stepping algorithm such as fourth-order Runge–Kutta. In particular, given that there are $N + 1$ points and periodic boundary conditions, this reduces the system of differential equations to an $N \times N$ coupled system. Once we have updated the value of the vorticity, we must again update the value of the streamfunction to once again update the vorticity. This loop continues until the solution at the desired future time is achieved. Figure 8.2 illustrates how the five-point, two-dimensional stencil advances the solution.

The behavior of the vorticity is illustrated in Fig. 8.3 where the solution is advanced for eight time units. The initial condition used in this simulation is

$$\omega_0 = \omega(x, y, t = 0) = \exp\left(-2x^2 - \frac{y^2}{20} \right). \tag{8.1.18}$$

This stretched Gaussian is seen to rotate while advecting and diffusing vorticity. Multiple vortex solutions can also be considered along with oppositely signed vortices.

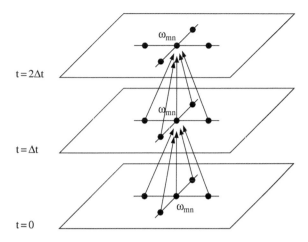

Figure 8.2: Discretization stencil resulting from center-differencing of the advection–diffusion equations. Note that for explicit stepping schemes the future solution only depends upon the present. Thus we are not required to solve a large linear system of equations.

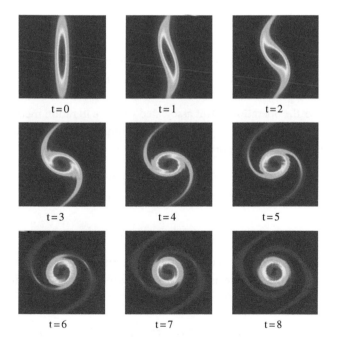

Figure 8.3: Time evolution of the vorticity $\omega(x, y, t)$ over eight time units with $\nu = 0.001$ and a spatial domain $x \in [-10, 10]$ and $y \in [-10, 10]$. The initial condition was a stretched Gaussian of the form $\omega(x, y, 0) = \exp(-2x^2 - y^2/20)$.

8.2   Advanced Iterative Solution Methods for Ax = b

In addition to the standard techniques of Gaussian elimination or LU decomposition for solving $\mathbf{Ax} = \mathbf{b}$, a wide range of iterative techniques are available.

Application to advection–diffusion

When discretizing many systems of interest, such as the advection–diffusion problem, we are left with a system of equations that is naturally geared toward iterative methods. Discretization of the stream/function previously yielded the system

$$-4\psi_{mn} + \psi_{(m+1)n} + \psi_{(m-1)n} + \psi_{m(n+1)} + \psi_{m(n-1)} = \delta^2 \omega_{mn}. \tag{8.2.1}$$

The matrix $\mathbf{A}$ in this case is represented by the left-hand side of the equation. Letting ψ_{mn} be the diagonal term, the iteration procedure yields

$$\psi_{mn}^{k+1} = \frac{\psi_{(m+1)n}^{k} + \psi_{(m-1)n}^{k} + \psi_{m(n+1)}^{k} + \psi_{m(n-1)}^{k} - \delta^2 \omega_{mn}}{4}. \tag{8.2.2}$$

Note that the diagonal term has a coefficient of $|-4| = 4$ and the sum of the off-diagonal elements is $|1| + |1| + |1| + |1| = 4$. Thus the system is at the borderline of being diagonally dominant. So although convergence is not guaranteed, it is highly likely that we could get the Jacobi scheme to converge.

Finally, we consider the operation count associated with the iteration methods. This will allow us to compare this solution technique with Gaussian elimination and LU decomposition. The following basic algorithmic steps are involved:

1   Update each ψ_{mn} which costs N operations times the number of nonzero diagonals D.

2   For each ψ_{mn}, perform the appropriate additions and subtractions. In this case there are five operations.

3   Iterate until the desired convergence which costs K operations.

Thus the total number of operations is $O(N \cdot D \cdot 5 \cdot K)$. If the number of iterations K can be kept small, then iteration provides a viable alternative to the direct solution techniques.

8.3   Fast Poisson Solvers: The Fourier Transform

Other techniques exist for solving many computational problems which are not based upon the standard Taylor series discretization. For instance, we have considered solving the streamfunction equation

$$\nabla^2 \psi = \omega \tag{8.3.1}$$

by discretizing in both the x- and y-directions and solving the associated linear problem $\mathbf{Ax} = \mathbf{b}$. At best, we can use a factorization scheme to solve this problem in $O(N^2)$ operations. Although iteration schemes have the possibility of outperforming this, it is not guaranteed.

Another alternative is to use the fast Fourier transform (FFT). The FFT is an integral transform defined over the entire line $x \in [-\infty, \infty]$. Given computational practicalities, however, we transform over a finite domain $x \in [-L, L]$ and assume periodic boundary conditions due to the oscillatory behavior of the kernel of the Fourier transform. The Fourier transform and its inverse can be defined as

$$F(k) = \frac{1}{\sqrt{2\pi}} \int_{-\infty}^{\infty} e^{-ikx} f(x) dx \tag{8.3.2a}$$

$$f(x) = \frac{1}{\sqrt{2\pi}} \int_{-\infty}^{\infty} e^{ikx} F(k) dk. \tag{8.3.2b}$$

There are other equivalent definitions. However, this definition will serve to illustrate the power and functionality of the Fourier transform method. We again note that formally, the transform is over the entire real line $x \in [-\infty, \infty]$ whereas our computational domain is only over a finite domain $x \in [-L, L]$. Further, the kernel of the transform, $\exp(\pm ikx)$, describes oscillatory behavior. Thus the Fourier transform is essentially an eigenfunction expansion over all continuous wavenumbers k. And once we are on a finite domain $x \in [-L, L]$, the continuous eigenfunction expansion becomes a discrete sum of eigenfunctions and associated wavenumbers (eigenvalues).

Derivative relations

The critical property in the usage of Fourier transforms concerns derivative relations. To see how these properties are generated, we begin by considering the Fourier transform of $f'(x)$. We denote the Fourier transform of $f(x)$ as $\widehat{f(x)}$. Thus we find

$$\widehat{f'(x)} = \frac{1}{\sqrt{2\pi}} \int_{-\infty}^{\infty} e^{-ikx} f'(x) dx = f(x) e^{-ikx} |_{-\infty}^{\infty} + \frac{ik}{\sqrt{2\pi}} \int_{-\infty}^{\infty} e^{-ikx} f(x) dx. \tag{8.3.3}$$

Assuming that $f(x) \to 0$ as $x \to \pm\infty$ results in

$$\widehat{f'(x)} = \frac{ik}{\sqrt{2\pi}} \int_{-\infty}^{\infty} e^{-ikx} f(x) dx = ik\widehat{f(x)}. \tag{8.3.4}$$

Thus the basic relation $\widehat{f'} = ik\widehat{f}$ is established. It is easy to generalize this argument to an arbitrary number of derivatives. The final result is the following relation between Fourier transforms of the derivative and the Fourier transform itself

$$\widehat{f^{(n)}} = (ik)^n \widehat{f}. \tag{8.3.5}$$

This property is what makes Fourier transforms so useful and practical.

As an example of the Fourier transform, consider the following differential equation

$$y'' - \omega^2 y = -f(x) \quad x \in [-\infty, \infty].$$ (8.3.6)

We can solve this by applying the Fourier transform to both sides. This gives the following reduction

$$\widehat{y''} - \omega^2 \widehat{y} = -\widehat{f}$$

$$-k^2 \widehat{y} - \omega^2 \widehat{y} = -\widehat{f}$$

$$(k^2 + \omega^2)\widehat{y} = \widehat{f}$$

$$\widehat{y} = \frac{\widehat{f}}{k^2 + \omega^2}.$$ (8.3.7)

To find the solution $y(x)$, we invert the last expression above to yield

$$y(x) = \frac{1}{\sqrt{2\pi}} \int_{-\infty}^{\infty} e^{ikx} \frac{\widehat{f}}{k^2 + \omega^2} dk.$$ (8.3.8)

This gives the solution in terms of an integral which can be evaluated analytically or numerically.

The fast Fourier transform

The fast Fourier transform routine was developed specifically to perform the forward and backward Fourier transforms. In the mid-1960s, Cooley and Tukey developed what is now commonly known as the FFT algorithm [15]. Their algorithm was named one of the top ten algorithms of the twentieth century for one reason: the operation count for solving a system dropped to $O(N \log N)$. For N large, this operation count grows almost linearly like N. Thus it represents a great leap forward from Gaussian elimination and LU decomposition. The key features of the FFT routine are as follows:

1   It has a low operation count: $O(N \log N)$.

2   It finds the transform on an interval $x \in [-L, L]$. Since the integration kernel $\exp(\pm ikx)$ is oscillatory, it implies that the solutions on this finite interval have periodic boundary conditions.

3   The key to lowering the operation count to $O(N \log N)$ is in discretizing the range $x \in [-L, L]$ into 2^n points, i.e. the number of points should be 2, 4, 8, 16, 32, 64, 128, 256, $\cdots$.

4   The FFT has excellent accuracy properties, typically well beyond that of standard discretization schemes.

We will consider the underlying FFT algorithm in detail at a later time. For more information at the present, see [15] for a broader overview.

The streamfunction

The FFT algorithm provides a fast and efficient method for solving the streamfunction equation

$$\nabla^2 \psi = \omega \tag{8.3.9}$$

given the vorticity ω. In two dimensions, this equation is equivalent to

$$\frac{\partial^2 \psi}{\partial x^2} + \frac{\partial^2 \psi}{\partial y^2} = \omega. \tag{8.3.10}$$

Denoting the Fourier transform in x as $\widehat{f(x)}$ and the Fourier transform in y as $\widetilde{g(y)}$, we transform the equation. We begin by transforming in x:

$$\widehat{\frac{\partial^2 \psi}{\partial x^2}} + \widehat{\frac{\partial^2 \psi}{\partial y^2}} = \widehat{\omega} \quad \rightarrow \quad -k_x^2 \widehat{\psi} + \frac{\partial^2 \widehat{\psi}}{\partial y^2} = \widehat{\omega}, \tag{8.3.11}$$

where k_x are the wavenumbers in the x-direction. Transforming now in the y-direction gives

$$-k_x^2 \widetilde{\widehat{\psi}} + \widetilde{\frac{\partial^2 \widehat{\psi}}{\partial y^2}} = \widetilde{\widehat{\omega}} \quad \rightarrow \quad -k_x^2 \widetilde{\widehat{\psi}} + -k_y^2 \widetilde{\widehat{\psi}} = \widetilde{\widehat{\omega}}. \tag{8.3.12}$$

This can be rearranged to obtain the final result

$$\widetilde{\widehat{\psi}} = -\frac{\widetilde{\widehat{\omega}}}{k_x^2 + k_y^2}. \tag{8.3.13}$$

The remaining step is to inverse-transform in x and y to get back to the solution $\psi(x, y)$.

There is one mathematical difficulty which must be addressed. The streamfunction equation with periodic boundary conditions does not have a unique solution. Thus if $\psi_0(x, y, t)$ is a solution, so is $\psi_0(x, y, t) + c$ where c is an arbitrary constant. When solving this problem with FFTs, the FFT will arbitrarily add a constant to the solution. Fundamentally, we are only interested in derivatives of the streamfunction. Therefore, this constant is inconsequential. When solving with direct methods for $\mathbf{Ax} = \mathbf{b}$, the nonuniqueness gives a singular matrix $\mathbf{A}$. Thus solving with Gaussian elimination, LU decomposition or iterative methods is problematic. But since the arbitray constant does not matter, we can simply pin the streamfunction to some prescribed value on our computational domain. This will fix the constant c and give a unique solution to $\mathbf{Ax} = \mathbf{b}$. For instance, we could impose the following constraint condition $\psi(-L, -L, t) = 0$. Such a condition pins the value of $\psi(x, y, t)$ at the left-hand corner of the computational domain and fixes c.

MATLAB commands

The commands for executing the fast Fourier transform and its inverse are as follows

- $fft(x)$: Forward Fourier transform a vector $\mathbf{x}$.
- $ifft(x)$: Inverse Fourier transform a vector $\mathbf{x}$.

8.4   Comparison of Solution Techniques for Ax = b: Rules of Thumb

The practical implementation of the mathematical tools available in MATLAB is crucial. This section will focus on the use of some of the more sophisticated routines in MATLAB which are cornerstones to scientific computing. Included in this section will be a discussion of the fast Fourier transform routines (*fft, ifft, fftshift, ifftshift, fft2, ifft2*), sparse matrix construction (*spdiag, spy*), and high-end iterative techniques for solving **Ax = b** (*bicgstab, gmres*). These routines should be studied carefully since they are the building blocks of any serious scientific computing code.

Fast Fourier transform: FFT, IFFT, FFTSHIFT, IFFTSHIFT

The fast Fourier transform will be the first subject discussed. Its implementation is straightforward. Given a function which has been discretized with 2^n points and represented by a vector **x**, the FFT is found with the command *fft(x)*. Aside from transforming the function, the algorithm associated with the FFT does three major things: it shifts the data so that $x \in [0, L] \rightarrow [-L, 0]$ and $x \in [-L, 0] \rightarrow [0, L]$, additionally it multiplies every other mode by -1, and it assumes you are working on a 2π periodic domain. These properties are a consequence of the FFT algorithm discussed in detail at a later time.

To see the practical implications of the FFT, we consider the transform of a Gaussian function. The transform can be calculated analytically so that we have the exact relations:

$$f(x) = \exp(-\alpha x^2) \quad \rightarrow \quad \widehat{f}(k) = \frac{1}{\sqrt{2\alpha}} \exp\left(-\frac{k^2}{4\alpha}\right). \tag{8.4.1}$$

A simple MATLAB code to verify this with $\alpha = 1$ is as follows

```
clear all; close all;  % clear all variables and figures

L=20;    % define the computational domain [-L/2,L/2]
n=128;   % define the number of Fourier modes 2^n

x2=linspace(-L/2,L/2,n+1);  % define the domain discretization
x=x2(1:n);    % consider only the first n points: periodicity

u=exp(-x.*x);          % function to take a derivative of
ut=fft(u);             % FFT the function
utshift=fftshift(ut);  % shift FFT
```

Continued

```
figure(1), plot(x,u)         % plot initial gaussian
figure(2), plot(abs(ut))       % plot unshifted transform
figure(3), plot(abs(utshift)) % plot shifted transform
```

The second figure generated by this script shows how the pulse is shifted. By using the command *fftshift*, we can shift the transformed function back to its mathematically correct positions as shown in the third figure generated. However, before inverting the transformation, it is crucial that the transform is shifted back to the form of the second figure. The command *ifftshift* does this. In general, unless you need to plot the spectrum, it is better not to deal with the *fftshift* and *ifftshift* commands. A graphical representation of the *fft* procedure and its shifting properties is illustrated in Fig. 8.4 where a Gaussian is transformed and shifted by the *fft* routine.

To take a derivative, we need to calculate the k values associated with the transformation. The following example does this. Recall that the FFT assumes a 2π periodic domain which gets shifted.

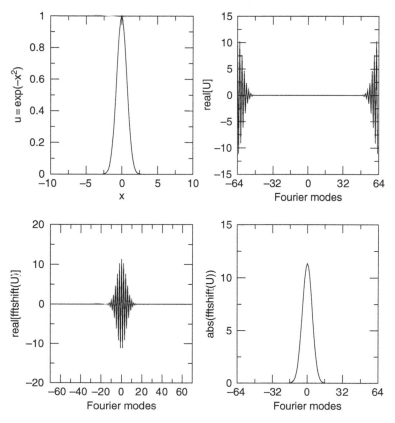

Figure 8.4: Fast Fourier transform of Gaussian data illustrating the shifting properties of the FFT routine. Note that the *fftshift* command restores the transform to its mathematically correct, unshifted state.

Thus the calculation of the k values needs to shift and rescale to the 2π domain. The following example differentiates the function $f(x) = \text{sech}(x)$ three times. The first derivative is compared with the analytic value of $f'(x) = -\text{sech}(x)\tanh(x)$.

```
clear all; close all;  % clear all variables and figures

L=20;    % define the computational domain [-L/2,L/2]
n=128;   % define the number of Fourier modes 2^n

x2=linspace(-L/2,L/2,n+1);  % define the domain discretization
x=x2(1:n);   % consider only the first n points:  periodicity
u=sech(x);     % function to take a derivative of
ut=fft(u);     % FFT the function
k=(2*pi/L)*[0:(n/2-1) (-n/2):-1]; % k rescaled to 2pi domain

ut1=i*k.*ut;        % first derivative
ut2=-k.*k.*ut;      % second derivative
ut3=-i*k.*k.*k.*ut; % third derivative

u1=ifft(ut1); u2=ifft(ut2); u3=ifft(ut3); % inverse transform
u1exact=-sech(x).*tanh(x);   % analytic first derivative

figure(1)
plot(x,u,'r',x,u1,'g',x,u1exact,'go',x,u2,'b',x,u3,'c') % plot
```

The routine accounts for the periodic boundaries, the correct k values, and differentiation. Note that no shifting was necessary since we constructed the k values in the shifted space.

For transforming in higher dimensions, a couple of choices in MATLAB are possible. For 2D transformations, it is recommended to use the commands *fft2* and *ifft2*. These will transform a matrix **A**, which represents data in the x- and y-directions, respectively, along the rows and then columns. For higher dimensions, the *fft* command can be modified to *fft(x,[],N)* where N is the number of dimensions.

Sparse matrices: SPDIAGS, SPY

Under discretization, most physical problems yield sparse matrices, i.e. matrices which are largely composed of zeros. For instance, the matrices (7.6.11) and (8.1.13) are sparse matrices generated under the discretization of a boundary value problem and Poisson equation, respectively. The *spdiag* command allows for the construction of sparse matrices in a relatively simple fashion. The sparse matrix is then saved using a minimal amount of memory and all matrix operations

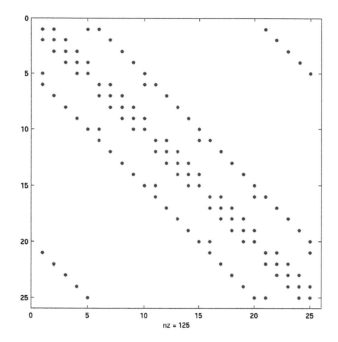

Figure 8.5: Sparse matrix structure for the Laplacian operator using second-order discretization. The output is generated via the *spy* command in MATLAB.

are conducted as usual. The *spy* command allows you to look at the nonzero components of the matrix structure (see Fig. 8.5). As an example, we construct the matrix given by (8.1.13) for the case of $N = 5$ in both the *x*- and *y*-directions.

```
clear all; close all;  % clear all variables and figures

m=5;     % N value in x and y directions
n=m*m;   % total size of matrix

e0=zeros(n,1);   % vector of zeros
e1=ones(n,1);    % vector of ones

e2=e1;     % copy the one vector
e4=e0;     % copy the zero vector
for j=1:m
 e2(m*j)=0;  % overwrite every m^th value with zero
 e4(m*j)=1;  % overwirte every m^th value with one
end
```

Continued

```
e3(2:n,1)=e2(1:n-1,1); e3(1,1)=e2(n,1); % shift to correct
e5(2:n,1)=e4(1:n-1,1); e5(1,1)=e4(n,1); % positions

% place diagonal elements
matA=spdiags([e1 e1 e5 e2 -4*e1 e3 e4 e1 e1], ...
     [-(n-m) -m -m+1 -1 0 1 m-1 m (n-m)],n,n);
spy(matA) % view the matrix structure
```

The appropriate construction of the sparse matrix is critical to solving any problem. It is essential to carefully check the matrix for accuracy before using it in an application. As can be seen from this example, it takes a bit of work to get the matrix correct. However, once constructed properly, it can be used with confidence in all other matrix applications and operations.

Iterative methods: BICGSTAB, GMRES

Iterative techniques need also to be considered. There are a wide variety of built-in iterative techniques in MATLAB. Two of the more promising methods are discussed here: the bi-conjugate stabilized gradient method (*bicgstab*) and the generalized minimum residual method (*gmres*). Both are easily implemented in MATLAB.

Recall that these iteration methods are for solving the linear system $\mathbf{Ax} = \mathbf{b}$. The basic call to the generalized minimum residual method is

```
>>x=gmres(A,b);
```

Likewise, the bi-conjugate stabilized gradient method is called by

```
>>x=bicgstab(A,b);
```

It is rare that you would use these commands without options. The iteration scheme stops upon convergence, failure to converge, or when the maximum number of iterations has been achieved. By default, the initial guess for the iteration procedure is the zero vector. The default number of maximum iterations is 10, which is rarely enough iterations for the purposes of scientific computing.

Thus it is important to understand the different options available with these iteration techniques. The most general user-specified command line for either *gmres* or *bicgstab* is as follows

```
>>[x,flag,relres,iter]=bicgstab(A,b,tol,maxit,M1,M2,x0);
>>[x,flag,relres,iter]=gmres(A,b,restart,tol,maxit,M1,M2,x0);
```

We already know that the matrix **A** and vector **b** come from the original problem. The remaining parameters are as follows:

$$tol = \text{specified tolerance for convergence}$$
$$maxit = \text{maximum number of iterations}$$
$$M1, M2 = \text{preconditioning matrices}$$
$$x0 = \text{initial guess vector}$$
$$restart = \text{restart of iterations (gmres only).}$$

In addition to these input parameters, *gmres* or *bicgstab* will give the relative residual, *relres*, and the number of iterations performed, *iter*. The *flag* variable gives information on whether the scheme converged in the maximum allowable iterations or not.

8.5   Overcoming Computational Difficulties

In developing algorithms for any given problem, you should always try to maximize the use of information available to you about the specific problem. Specifically, a well-developed numerical code will make extensive use of analytic information concerning the problem. Judicious use of properties of the specific problem can lead to significant reduction in the amount of computational time and effort needed to perform a given calculation.

Here, various practical issues which may arise in the computational implementation of a given method are considered. Analysis provides a strong foundation for understanding the issues which can be problematic on a computational level. Thus the focus here will be on details which need to be addressed before straightforward computing is performed.

Streamfunction equations: Nonuniqueness

We return to the consideration of the streamfunction equation

$$\nabla^2 \psi = \omega \tag{8.5.1}$$

with the periodic boundary conditions

$$\psi(-L, y, t) = \psi(L, y, t) \tag{8.5.2a}$$
$$\psi(x, -L, t) = \psi(x, L, t). \tag{8.5.2b}$$

The mathematical problem which arises in solving this Poisson equation has been considered previously. Namely, the solution can only be determined to an arbitrary constant. Thus if ψ_0 is a solution to the streamfunction equation, so is

$$\psi = \psi_0 + c, \tag{8.5.3}$$

where c is an arbitrary constant. This gives an infinite number of solutions to the problem. Thus upon discretizing the equation in x and y and formulating the matrix formulation of the problem $\mathbf{Ax} = \mathbf{b}$, we find that the matrix $\mathbf{A}$, which is given by (8.1.13), is singular, i.e. det $\mathbf{A} = 0$.

Obviously, the fact that the matrix $\mathbf{A}$ is singular will create computational problems. However, does this nonuniqueness jeopardize the validity of the physical model? Recall that the streamfunction is a fictitious quantity which captures the fluid velocity through the quantities $\partial\psi/\partial x$ and $\partial\psi/\partial y$. In particular, we have the x and y velocity components

$$u = -\frac{\partial\psi}{\partial y} \quad (x\text{-component of velocity}) \tag{8.5.4a}$$

$$v = \frac{\partial\psi}{\partial x} \quad (y\text{-component of velocity}). \tag{8.5.4b}$$

Thus the arbitrary constant c drops out when calculating physically meaningful quantities. Further, when considering the advection–diffusion equation

$$\frac{\partial\omega}{\partial t} + [\psi, \omega] = \nu\nabla^2\omega \tag{8.5.5}$$

where

$$[\psi, \omega] = \frac{\partial\psi}{\partial x}\frac{\partial\omega}{\partial y} - \frac{\partial\psi}{\partial y}\frac{\partial\omega}{\partial x}, \tag{8.5.6}$$

only the derivative of the streamfunction is important, which again removes the constant c from the problem formulation.

We can thus conclude that the nonuniqueness from the constant c does not generate any problems for the physical model considered. However, we still have the mathematical problem of dealing with a singular matrix. It should be noted that this problem also occurs when all the boundary conditions are of the Neuman type, i.e. $\partial\psi/\partial n = 0$ where n denotes the outward normal direction.

To overcome this problem numerically, we simply observe that we can arbitrarily add a constant to the solution. Or alternatively, we can pin down the value of the streamfunction at a single location in the computational domain. This constraint fixes the arbitrary constant problem and removes the singularity from the matrix $\mathbf{A}$. Thus to fix the problem, we can simply pick an arbitrary point in our computational domain ψ_{mn} and fix its value. Essentially, this will alter a single component of the sparse matrix (8.1.13). And in fact, this is the simplest thing to do. For instance, given the construction of the matrix (8.1.13), we could simply add the following line of MATLAB code:

```
A(1,1)=0;
```

Then det $\mathbf{A} \neq 0$ and the matrix can be used in any of the linear solution methods. Note that the choice of the matrix component and its value are completely arbitrary. However, in this example, if you choose to alter this matrix component, to overcome the matrix singularity you must have $A(1, 1) \neq -4$.

Fast Fourier transforms: Divide by zero

In addition to solving the streamfunction equation by standard discretization and $\mathbf{Ax} = \mathbf{b}$, we could use the Fourier transform method. In particular, Fourier transforming in both x and y reduces the streamfunction equation

$$\nabla^2 \psi = \omega \tag{8.5.7}$$

to Eq. (8.3.13)

$$\widetilde{\widehat{\psi}} = -\frac{\widetilde{\widehat{\omega}}}{k_x^2 + k_y^2}. \tag{8.5.8}$$

Here we have denoted the Fourier transform in x as $\widehat{f(x)}$ and the Fourier transform in y as $\widetilde{g(y)}$. The final step is to inverse-transform in x and y to get back to the solution $\psi(x, y)$. It is recommended that the routines *fft2* and *ifft2* be used to perform the transformations. However, *fft* and *ifft* may also be used in loops or with the dimension option set to 2.

An observation concerning (8.5.8) is that there will be a divide by zero when $k_x = k_y = 0$ at the zero mode. Two options are commonly used to overcome this problem. The first is to modify (8.5.8) so that

$$\widetilde{\widehat{\psi}} = -\frac{\widetilde{\widehat{\omega}}}{k_x^2 + k_y^2 + eps} \tag{8.5.9}$$

where *eps* is the command for generating a machine precision number which is on the order of $O(10^{-15})$. It essentially adds a round-off to the denominator which removes the divide by zero problem. A second option, which is more highly recommended, is to redefine the $\mathbf{k}_x$ and $\mathbf{k}_y$ vectors associated with the wavenumbers in the x- and y-directions. Specifically, after defining the $\mathbf{k}_x$ and $\mathbf{k}_y$, we could simply add the command line

```
kx(1)=10^(-6);
ky(1)=10^(-6);
```

The values of $kx(1) = ky(1) = 0$ by default. This would make the values small but finite so that the divide by zero problem is effectively removed with only a small amount of error added to the problem.

Sparse derivative matrices: Advection terms

The sparse matrix (8.1.13) represents the discretization of the Laplacian which is accurate to second order. However, when calculating the advection terms given by

$$[\psi, \omega] = \frac{\partial \psi}{\partial x}\frac{\partial \omega}{\partial y} - \frac{\partial \psi}{\partial y}\frac{\partial \omega}{\partial x}, \tag{8.5.10}$$

only the first derivative is required in both x and y. Associated with each of these derivative terms is a sparse matrix which must be calculated in order to evaluate the advection term.

The calculation of the x derivative will be considered here. The y derivative can be calculated in a similar fashion. Obviously, it is crucial that the sparse matrices representing these operations be correct. Consider then the second-order discretization of

$$\frac{\partial \omega}{\partial x} = \frac{\omega(x + \Delta x, y) - \omega(x - \Delta x, y)}{2\Delta x}. \tag{8.5.11}$$

Using the notation developed previously, we define $\omega(x_m, y_n) = \omega_{mn}$. Thus the discretization yields

$$\frac{\partial \omega_{mn}}{\partial x} = \frac{\omega_{(m+1)n} - \omega_{(m-1)n}}{2\Delta x}, \tag{8.5.12}$$

with periodic boundary conditions. The first few terms of this linear system are as follows:

$$\frac{\partial \omega_{11}}{\partial x} = \frac{\omega_{21} - \omega_{n1}}{2\Delta x}$$

$$\frac{\partial \omega_{12}}{\partial x} = \frac{\omega_{22} - \omega_{n2}}{2\Delta x}$$

$$\vdots \tag{8.5.13}$$

$$\frac{\partial \omega_{21}}{\partial x} = \frac{\omega_{31} - \omega_{11}}{2\Delta x}$$

$$\vdots$$

From these individual terms, the linear system $\partial \omega / \partial x = \mathbf{B}\omega$ can be constructed. The critical component is the sparse matrix $\mathbf{B}$. Note that we have used the usual definition of the vector ω

$$\omega = \begin{pmatrix} \omega_{11} \\ \omega_{12} \\ \vdots \\ \omega_{1n} \\ \omega_{21} \\ \omega_{22} \\ \vdots \\ \omega_{n(n-1)} \\ \omega_{nn} \end{pmatrix}. \tag{8.5.14}$$

It can be verified then that the sparse matrix $\mathbf{B}$ is given by the matrix

$$\mathbf{B} = \frac{1}{2\Delta x} \begin{bmatrix} \mathbf{0} & \mathbf{I} & \mathbf{0} & \cdots & \mathbf{0} & -\mathbf{I} \\ -\mathbf{I} & \mathbf{0} & \mathbf{I} & \mathbf{0} & \cdots & \mathbf{0} \\ \mathbf{0} & \ddots & \ddots & \ddots & & \\ \vdots & & & & & \vdots \\ & & & & & \mathbf{0} \\ \vdots & \cdots & \mathbf{0} & -\mathbf{I} & \mathbf{0} & \mathbf{I} \\ \mathbf{I} & \mathbf{0} & \cdots & \mathbf{0} & -\mathbf{I} & \mathbf{0} \end{bmatrix}, \tag{8.5.15}$$

where $\mathbf{0}$ is an $n \times n$ zero matrix and $\mathbf{I}$ is an $n \times n$ identity matrix. Recall that the matrix $\mathbf{B}$ is an $n^2 \times n^2$ matrix. The sparse matrix associated with this operation can be constructed with the *spdiag* command. Following the same analysis, the y derivative matrix can also be constructed. If we call this matrix $\mathbf{C}$, then the advection operator can simply be written as

$$[\psi, \omega] = \frac{\partial \psi}{\partial x} \frac{\partial \omega}{\partial y} - \frac{\partial \psi}{\partial y} \frac{\partial \omega}{\partial x} = (\mathbf{B}\psi)(\mathbf{C}\omega) - (\mathbf{C}\psi)(\mathbf{B}\omega). \tag{8.5.16}$$

This forms part of the right-hand side of the system of differential equations which is then solved using a standard time-stepping algorithm.

Time and Space Stepping Schemes: Method of Lines

With the Fourier transform and discretization in hand, we can turn towards the solution of partial differential equations whose solutions need to be advanced forward in time. Thus, in addition to spatial discretization, we will need to discretize in time in a self-consistent way so as to advance the solution to a desired future time. Issues of stability and accuracy are, of course, at the heart of a discussion on time- and space-stepping schemes.

Basic Time-Stepping Schemes

Basic time-stepping schemes involve discretization in both space and time. Issues of numerical stability as the solution is propagated in time and accuracy from the space–time discretization are of greatest concern for any implementation. To begin this study, we will consider very simple and well-known examples from partial differential equations. The first equation we consider is the heat (diffusion) equation

$$\frac{\partial u}{\partial t} = \kappa \frac{\partial^2 u}{\partial x^2} \tag{9.1.1}$$

with the periodic boundary conditions $u(-L, t) = u(L, t)$. We discretize the spatial derivative with a second-order scheme (see Table 4.1) so that

$$\frac{\partial u}{\partial t} = \frac{\kappa}{\Delta x^2} \left[u(x + \Delta x, t) - 2u(x, t) + u(x - \Delta x, t) \right]. \tag{9.1.2}$$

This approximation reduces the partial differential equation to a system of ordinary differential equations. We have already considered a variety of time-stepping schemes for differential equations and we can apply them directly to this resulting system.

The ODE system

To define the system of ODEs, we discretize and define the values of the vector **u** in the following way.

$$u(-L, t) = u_1$$
$$u(-L + \Delta x, t) = u_2$$
$$\vdots$$
$$u(L - 2\Delta x, t) = u_{n-1}$$
$$u(L - \Delta x, t) = u_n$$
$$u(L, t) = u_{n+1}.$$

Recall from periodicity that $u_1 = u_{n+1}$. Thus our system of differential equations solves for the vector

$$\mathbf{u} = \begin{pmatrix} u_1 \\ u_2 \\ \vdots \\ u_n \end{pmatrix}. \tag{9.1.3}$$

The governing equation (9.1.2) is then reformulated as the differential equation system

$$\frac{d\mathbf{u}}{dt} = \frac{\kappa}{\Delta x^2} \mathbf{A}\mathbf{u}, \tag{9.1.4}$$

where **A** is given by the sparse matrix

$$\mathbf{A} = \begin{bmatrix} -2 & 1 & 0 & \cdots & & 0 & 1 \\ 1 & -2 & 1 & 0 & \cdots & & 0 \\ 0 & \ddots & \ddots & \ddots & & & \\ \vdots & & & & & & \vdots \\ & & & & & & 0 \\ \vdots & \cdots & 0 & 1 & -2 & 1 \\ 1 & 0 & \cdots & & 0 & 1 & -2 \end{bmatrix}, \tag{9.1.5}$$

and the values of 1 on the upper right and lower left of the matrix result from the periodic boundary conditions.

MATLAB implementation

The system of differential equations can now be easily solved with a standard time-stepping algorithm such as *ode23* or *ode45*. The basic algorithm would be as follows

1   Build the sparse matrix **A**.

```
e1=ones(n,1);   % build a vector of ones
A=spdiags([e1 -2*e1 e1],[-1 0 1],n,n);   % diagonals
A(1,n)=1;   A(n,1)=1;   % periodic boundaries
```

2   Generate the desired initial condition vector $\mathbf{u} = \mathbf{u}_0$.

3   Call an ODE solver from the MATLAB suite. The matrix **A**, the diffusion constant κ and spatial step Δx need to be passed into this routine.

```
[t,y]=ode45('rhs',tspan,u0,[],k,dx,A);
```

The function *rhs.m* should be of the following form

```
function rhs=rhs(tspan,u,dummy,k,dx,A)
rhs=(k/dx^2)*A*u;
```

4   Plot the results as a function of time and space.

The algorithm is thus fairly routine and requires very little effort in programming since we can make use of the standard time-stepping algorithms already available in MATLAB.

2D MATLAB implementation

In the case of two dimensions, the calculation becomes slightly more difficult since the 2D data are represented by a matrix and the ODE solvers require a vector input for the initial data. For this case, the governing equation is

$$\frac{\partial u}{\partial t} = \kappa \left(\frac{\partial^2 u}{\partial x^2} + \frac{\partial^2 u}{\partial y^2} \right).$$

(9.1.6)

Discretizing in x and y gives a right-hand side which takes on the form of (8.1.7). Provided $\Delta x = \Delta y = \delta$ are the same, the system can be reduced to the linear system

$$\frac{d\mathbf{u}}{dt} = \frac{\kappa}{\delta^2} \mathbf{A} \mathbf{u},$$

(9.1.7)

where we have arranged the vector **u** in a similar fashion to (8.1.14) so that

$$
\mathbf{u} = \begin{pmatrix}
u_{11} \\
u_{12} \\
\vdots \\
u_{1n} \\
u_{21} \\
u_{22} \\
\vdots \\
u_{n(n-1)} \\
u_{nn}
\end{pmatrix},
\tag{9.1.8}
$$

where we have defined $u_{jk} = u(x_j, y_k)$. The matrix **A** is a nine diagonal matrix given by (8.1.13). The sparse implementation of this matrix is also given previously.

Again, the system of differential equations can now be easily solved with a standard time-stepping algorithm such as *ode23* or *ode45*. The basic algorithm follows the same course as the 1D case, but extra care is taken in arranging the 2D data into a vector.

1   Build the sparse matrix **A** (8.1.13).

2   Generate the desired initial condition matrix $\mathbf{U} = \mathbf{U}_0$ and reshape it to a vector $\mathbf{u} = \mathbf{u}_0$. This example considers the case of a simple Gaussian as the initial condition. The *reshape* and *meshgrid* commands are important for computational implementation.

```
Lx=20;   % spatial domain of x
Ly=20;   % spatial domain of y
nx=100;  % number of discretization points in x
ny=100;  % number of discretization points in y
N=nx*ny; % elements in reshaped initial condition

x2=linspace(-Lx/2,Lx/2,nx+1); % account for periodicity
x=x2(1:nx);        % x-vector
y2=linspace(-Ly/2,Ly/2,ny+1); % account for periodicity
y=y2(1:ny);        % y-vector

[X,Y]=meshgrid(x,y);  % set up for 2D initial conditions
U=exp(-X.^2-Y.^2);    % generate a Gaussian matrix
u=reshape(U,N,1);     % reshape into a vector
```

3   Call an ODE solver from the MATLAB suite. The matrix **B**, the diffusion constant κ and spatial step $\Delta x = \Delta y = dx$ need to be passed into this routine.

```
[t,y]=ode45('rhs',tspan,u,[],k,dx,B);
```

The function *rhs.m* should be of the following form

```
function rhs=rhs(tspan,u,dummy,k,dx,B)
rhs=(k/dx^2)*B*u;
```

4   Reshape and plot the results as a function of time and space.

The algorithm is again fairly routine and requires very little effort in programming since we can make use of the standard time-stepping algorithms.

Method of lines

Fundamentally, these methods use the data at a single slice of time to generate a solution Δt in the future. This is then used to generate a solution $2\Delta t$ into the future. The process continues

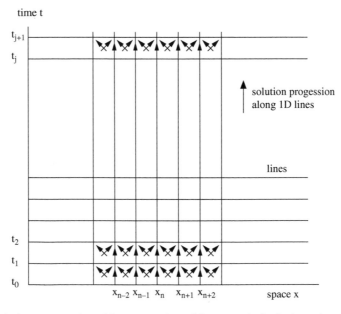

Figure 9.1: Graphical representation of the progression of the numerical solution using the method of lines. Here a second-order discretization scheme is considered which couples each spatial point with its nearest neighbor.

until the desired future time is achieved. This process of solving a partial differential equation is known as the *method of lines*. Each *line* is the value of the solution at a given time slice. The lines are used to update the solution to a new timeline and progressively generate future solutions. Figure 9.1 depicts the process involved in the method of lines for the 1D case. The 2D case was illustrated previously in Fig. 8.2.

9.2 Time-Stepping Schemes: Explicit and Implicit Methods

Now that several technical and computational details have been addressed, we continue to develop methods for time-stepping the solution into the future. Some of the more common schemes will be considered along with a graphical representation of the scheme. Every scheme eventually leads to an iteration procedure which the computer can use to advance the solution in time.

We will begin by considering the simplest partial differential equations. Often, it is difficult to do much analysis with complicated equations. Therefore, considering simple equations is not merely an exercise, but rather they are typically the only equations we can make analytical progress with.

As an example, we consider the one-way wave equation

$$\frac{\partial u}{\partial t} = c \frac{\partial u}{\partial x}.$$
(9.2.1)

The simplest discretization of this equation is to first central-difference in the x-direction. This yields

$$\frac{du_n}{dt} = \frac{c}{2\Delta x} (u_{n+1} - u_{n-1}),$$
(9.2.2)

where $u_n = u(x_n, t)$. We can then step forward with an Euler time-stepping method. Denoting $u_n^{(m)} = u(x_n, t_m)$, and applying the method of lines iteration scheme gives

$$u_n^{(m+1)} = u_n^{(m)} + \frac{c\Delta t}{2\Delta x} \left(u_{n+1}^{(m)} - u_{n-1}^{(m)} \right).$$
(9.2.3)

This simple scheme has the four-point stencil shown in Fig. 9.2. To illustrate more clearly the iteration procedure, we rewrite the discretized equation in the form

$$u_n^{(m+1)} = u_n^{(m)} + \frac{\lambda}{2} \left(u_{n+1}^{(m)} - u_{n-1}^{(m)} \right)$$
(9.2.4)

where

$$\lambda = \frac{c\Delta t}{\Delta x}$$
(9.2.5)

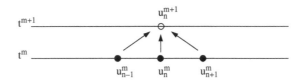

Figure 9.2: Four-point stencil for second-order spatial discretization and Euler time-stepping of the one-way wave equation.

is known as the CFL (Courant, Friedrichs and Lewy) number [16]. The iteration procedure assumes that the solution does not change significantly from one time-step to the next, i.e. $u_n^{(m)} \approx u_n^{(m+1)}$. The accuracy and stability of this scheme is controlled almost exclusively by the CFL number λ. This parameter relates the spatial and time discretization schemes in (9.2.5). Note that decreasing Δx without decreasing Δt leads to an increase in λ which can result in instabilities. Smaller values of Δt suggest smaller values of λ and improved numerical stability properties. In practice, you want to take Δx and Δt as large as possible for computational speed and efficiency without generating instabilities.

There are practical considerations to keep in mind relating to the CFL number. First, given a spatial discretization step-size Δx, you should choose the time discretization so that the CFL number is kept in check. Often a given scheme will only work for CFL conditions below a certain value, thus the importance of choosing a small enough time-step. Second, if indeed you choose to work with very small Δt, then although stability properties are improved with a lower CFL number, the code will also slow down accordingly. Thus achieving good stability results is often counter-productive to fast numerical solutions.

Central differencing in time

We can discretize the time-step in a similar fashion to the spatial discretization. Instead of the Euler time-stepping scheme used above, we could central-difference in time using Table 4.1. Thus after spatial discretization we have

$$\frac{du_n}{dt} = \frac{c}{2\Delta x} \left(u_{n+1} - u_{n-1} \right), \tag{9.2.6}$$

as before. And using a central difference scheme in time now yields

$$\frac{u_n^{(m+1)} - u_n^{(m-1)}}{2\Delta t} = \frac{c}{2\Delta x} \left(u_{n+1}^{(m)} - u_{n-1}^{(m)} \right). \tag{9.2.7}$$

This last expression can be rearranged to give

$$u_n^{(m+1)} = u_n^{(m-1)} + \lambda \left(u_{n+1}^{(m)} - u_{n-1}^{(m)} \right). \tag{9.2.8}$$

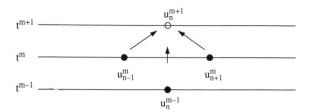

Figure 9.3: Four-point stencil for second-order spatial discretization and central-difference time-stepping of the one-way wave equation.

This iterative scheme is called *leap-frog (2,2)* since it is $O(\Delta t^2)$ accurate in time and $O(\Delta x^2)$ accurate in space. It uses a four-point stencil as shown in Fig. 9.3. Note that the solution utilizes two time slices to leap-frog to the next time slice. Thus the scheme is not self-starting since only one time slice (initial condition) is given.

Improved accuracy

We can improve the accuracy of any of the above schemes by using higher order central differencing methods. The fourth-order accurate scheme from Table 4.2 gives

$$\frac{\partial u}{\partial x} = \frac{-u(x + 2\Delta x) + 8u(x + \Delta x) - 8u(x - \Delta x) + u(x - 2\Delta x)}{12\Delta x}. \tag{9.2.9}$$

Combining this fourth-order spatial scheme with second-order central differencing in time gives the iterative scheme

$$u_n^{(m+1)} = u_n^{(m-1)} + \lambda \left[\frac{4}{3} \left(u_{n+1}^{(m)} - u_{n-1}^{(m)} \right) - \frac{1}{6} \left(u_{n+2}^{(m)} - u_{n-2}^{(m)} \right) \right]. \tag{9.2.10}$$

This scheme, which is based upon a six-point stencil, is called *leap-frog (2,4)*. It is typical that for *(2,4)* schemes, the maximum CFL number for stable computations is reduced from the basic *(2,2)* scheme.

Lax–Wendroff

Another alternative to discretizing in time and space involves a clever use of the Taylor expansion

$$u(x, t + \Delta t) = u(x, t) + \Delta t \frac{\partial u(x, t)}{\partial t} + \frac{\Delta t^2}{2!} \frac{\partial^2 u(x, t)}{\partial t^2} + O(\Delta t^3). \tag{9.2.11}$$

But we note from the governing one-way wave equation

$$\frac{\partial u}{\partial t} = c \frac{\partial u}{\partial x} \approx \frac{c}{2\Delta x} (u_{n+1} - u_{n-1}). \tag{9.2.12a}$$

Taking the derivative of the equation results in the relation

$$\frac{\partial^2 u}{\partial t^2} = c^2 \frac{\partial^2 u}{\partial x^2} \approx \frac{c^2}{\Delta x^2} \left(u_{n+1} - 2u_n + u_{n-1} \right). \tag{9.2.13a}$$

These two expressions for $\partial u / \partial t$ and $\partial^2 u / \partial t^2$ can be substituted into the Taylor series expression to yield the iterative scheme

$$u_n^{(m+1)} = u_n^{(m)} + \frac{\lambda}{2} \left(u_{n+1}^{(m)} - u_{n-1}^{(m)} \right) + \frac{\lambda^2}{2} \left(u_{n+1}^{(m)} - 2u_n^{(m)} + u_{n-1}^{(m)} \right). \tag{9.2.14}$$

This iterative scheme is similar to the Euler method. However, it introduces an important *stabilizing* diffusion term which is proportional to λ^2. This is known as the *Lax–Wendroff* scheme. Although useful for this example, it is difficult to implement in practice for variable coefficient problems. It illustrates, however, the variety and creativity in developing iteration schemes for advancing the solution in time and space.

Backward Euler: Implicit scheme

The backward Euler method uses the future time for discretizing the spatial domain. Thus upon discretizing in space and time we arrive at the iteration scheme

$$u_n^{(m+1)} = u_n^{(m)} + \frac{\lambda}{2} \left(u_{n+1}^{(m+1)} - u_{n-1}^{(m+1)} \right). \tag{9.2.15}$$

This gives the tridiagonal system

$$u_n^{(m)} = -\frac{\lambda}{2} u_{n+1}^{(m+1)} + u_n^{(m+1)} + \frac{\lambda}{2} u_{n-1}^{(m+1)}, \tag{9.2.16}$$

which can be written in matrix form as

$$\mathbf{A} \mathbf{u}^{(m+1)} = \mathbf{u}^{(m)} \tag{9.2.17}$$

where

$$\mathbf{A} = \frac{1}{2} \begin{pmatrix} 2 & -\lambda & \cdots & 0 \\ \lambda & 2 & -\lambda & \cdots \\ \vdots & & \ddots & \vdots \\ 0 & \cdots & \lambda & 2 \end{pmatrix}. \tag{9.2.18}$$

Thus before stepping forward in time, we must solve a matrix problem. This can severely affect the computational time of a given scheme. The only thing which may make this method viable is if the CFL condition is such that much larger time-steps are allowed, thus overcoming the limitations imposed by the matrix solve.

Table 9.1   Stability of time-stepping schemes as a function of the CFL number

Scheme	Stability
Forward Euler	unstable for all λ
Backward Euler	stable for all λ
Leap-frog (2,2)	stable for $\lambda \leq 1$
Leap-frog (2,4)	stable for $\lambda \leq 0.707$

MacCormack scheme

In the MacCormack scheme, the variable coefficient problem of the Lax–Wendroff scheme and the matrix solve associated with the backward Euler are circumvented by using a predictor–corrector method. The computation thus occurs in two pieces:

$$u_n^{(P)} = u_n^{(m)} + \lambda \left(u_{n+1}^{(m)} - u_{n-1}^{(m)} \right) \tag{9.2.19a}$$

$$u_n^{(m+1)} = \frac{1}{2} \left[u_n^{(m)} + u_n^{(P)} + \lambda \left(u_{n+1}^{(P)} - u_{n-1}^{(P)} \right) \right]. \tag{9.2.19b}$$

This method essentially combines forward and backward Euler schemes so that we avoid the matrix solve and the variable coefficient problem.

The CFL condition will be discussed in detail in the next section. For now, the basic stability properties of many of the schemes considered here are given in Table 9.1. The stability of the various schemes holds only for the one-way wave equation considered here as a prototypical example. Each partial differential equation needs to be considered and classified individually with regards to stability.

9.3   Stability Analysis

In the preceeding section, we considered a variety of schemes which can solve the time–space problem posed by partial differential equations. However, it remains undetermined which scheme is best for implementation purposes. Two criteria provide a basis for judgement: accuracy and stability. For each scheme, we already know the accuracy due to the discretization error. However, a determination of stability is still required.

To start to understand stability, we once again consider the one-way wave equation

$$\frac{\partial u}{\partial t} = c \frac{\partial u}{\partial x}. \tag{9.3.1}$$

Using Euler time-stepping and central differencing in space gives the iteration procedure

$$u_n^{(m+1)} = u_n^{(m)} + \frac{\lambda}{2} \left(u_{n+1}^{(m)} - u_{n-1}^{(m)} \right) \tag{9.3.2}$$

where λ is the CFL number.

von Neumann analysis

To determine the stability of a given scheme, we perform a *von Neumann analysis* [17]. This assumes the solution is of the form

$$u_n^{(m)} = g^m \exp(i\xi_n h) \quad \xi \in [-\pi/h, \pi/h] \tag{9.3.3}$$

where $h = \Delta x$ is the spatial discretization parameter. Essentially this assumes the solution can be constructed of Fourier modes. The key then is to determine what happens to g^m as $m \to \infty$. Two possibilities exist

$$\lim_{m \to \infty} |g|^m \to \infty \quad \text{unstable scheme} \tag{9.3.4a}$$

$$\lim_{m \to \infty} |g|^m \le 1 \ (< \infty) \quad \text{stable scheme.} \tag{9.3.4b}$$

Thus this stability check is very much like that performed for the time-stepping schemes developed for ordinary differential equations.

Forward Euler for one-way wave equation

The Euler discretization of the one-way wave equation produces the iterative scheme (9.3.2). Plugging in the ansatz (9.3.3) gives the following relations:

$$g^{m+1} \exp(i\xi_n h) = g^m \exp(i\xi_n h) + \frac{\lambda}{2} \left(g^m \exp(i\xi_{n+1}h) - g^m \exp(i\xi_{n-1}h) \right)$$

$$g^{m+1} \exp(i\xi_n h) = g^m \left(\exp(i\xi_n h) + \frac{\lambda}{2} \left(\exp(i\xi_{n+1}h) - \exp(i\xi_{n-1}h) \right) \right)$$

$$g = 1 + \frac{\lambda}{2} \left[\exp(ih(\xi_{n+1} - \xi_n)) - \exp(ih(\xi_{n-1} - \xi_n)) \right]. \tag{9.3.5}$$

Letting $\xi_n = n\zeta$ reduces the equations further to

$$g(\zeta) = 1 + \frac{\lambda}{2}[\exp(i\zeta h) - \exp(-i\zeta h)]$$

$$g(\zeta) = 1 + i\lambda \sin(\zeta h). \tag{9.3.6}$$

From this we can deduce that

$$|g(\zeta)| = \sqrt{g^*(\zeta)g(\zeta)} = \sqrt{1 + \lambda^2 \sin^2 \zeta h}. \tag{9.3.7}$$

Thus

$$|g(\zeta)| \ge 1 \quad \to \quad \lim_{m \to \infty} |g|^m \to \infty \quad \text{unstable scheme.} \tag{9.3.8}$$

Thus the forward Euler time-stepping scheme is unstable for any value of λ. The implementation of this scheme will force the solution to infinity due to numerical instability.

Backward Euler for one-way wave equation

The Euler discretization of the one-way wave equation produces the iterative scheme

$$u_n^{(m+1)} = u_n^{(m)} + \frac{\lambda}{2}\left(u_{n+1}^{(m+1)} - u_{n-1}^{(m+1)}\right). \tag{9.3.9}$$

Plugging in the ansatz (9.3.3) gives the following relations:

$$g^{m+1}\exp(i\xi_n h) = g^m \exp(i\xi_n h) + \frac{\lambda}{2}\left(g^{m+1}\exp(i\xi_{n+1}h) - g^{m+1}\exp(i\xi_{n-1}h)\right)$$

$$g = 1 + \frac{\lambda}{2}g\left[\exp(ih(\xi_{n+1} - \xi_n)) - \exp(ih(\xi_{n-1} - \xi_n))\right]. \tag{9.3.10}$$

Letting $\xi_n = n\zeta$ reduces the equations further to

$$g(\zeta) = 1 + \frac{\lambda}{2}g[\exp(i\zeta h) - \exp(-i\zeta h)]$$
$$g(\zeta) = 1 + i\lambda g \sin(\zeta h)$$
$$g[1 - i\lambda \sin(\zeta h)] = 1$$
$$g(\zeta) = \frac{1}{1 - i\lambda \sin \zeta h}. \tag{9.3.11}$$

From this we can deduce that

$$|g(\zeta)| = \sqrt{g^*(\zeta)g(\zeta)} = \sqrt{\frac{1}{1 + \lambda^2 \sin^2 \zeta h}}. \tag{9.3.12}$$

Thus

$$|g(\zeta)| \leq 1 \quad \rightarrow \quad \lim_{m\to\infty} |g|^m \leq 1 \quad \text{unconditionally stable.} \tag{9.3.13}$$

Thus the backward Euler time-stepping scheme is stable for any value of λ. The implementation of this scheme will not force the solution to infinity.

Lax–Wendroff for one-way wave equation

The Lax–Wendroff scheme is not as transparent as the forward and backward Euler schemes.

$$u_n^{(m+1)} = u_n^{(m)} + \frac{\lambda}{2}\left(u_{n+1}^{(m)} - u_{n-1}^{(m)}\right) + \frac{\lambda^2}{2}\left(u_{n+1}^{(m)} - 2u_n^{(m)} + u_{n-1}^{(m)}\right). \tag{9.3.14}$$

Plugging in the standard ansatz (9.3.3) gives

$$g^{m+1} \exp(i\xi_n h) = g^m \exp(i\xi_n h) + \frac{\lambda}{2} g^m \left(\exp(i\xi_{n+1} h) - \exp(i\xi_{n-1} h) \right)$$

$$+ \frac{\lambda^2}{2} g^m \left(\exp(i\xi_{n+1} h) - 2 \exp(i\xi_n h) + \exp(i\xi_{n-1} h) \right)$$

$$g = 1 + \frac{\lambda}{2} \left[\exp(ih(\xi_{n+1} - \xi_n)) - \exp(ih(\xi_{n-1} - \xi_n)) \right]$$

$$+ \frac{\lambda^2}{2} \left[\exp(ih(\xi_{n+1} - \xi_n)) + \exp(ih(\xi_{n-1} - \xi_n)) - 2 \right]. \tag{9.3.15}$$

Letting $\xi_n = n\zeta$ reduces the equations further to

$$g(\zeta) = 1 + \frac{\lambda}{2} [\exp(i\zeta h) - \exp(-i\zeta h)] + \frac{\lambda^2}{2} [\exp(i\zeta h) + \exp(-i\zeta h) - 2]$$

$$g(\zeta) = 1 + i\lambda \sin \zeta h + \lambda^2 (\cos \zeta h - 1)$$

$$g(\zeta) = 1 + i\lambda \sin \zeta h - 2\lambda^2 \sin^2(\zeta h/2). \tag{9.3.16}$$

This results in the relation

$$|g(\zeta)|^2 = \lambda^4 \left[4 \sin^4(\zeta h/2) \right] + \lambda^2 \left[\sin^2 \zeta h - 4 \sin^2(\zeta h/2) \right] + 1. \tag{9.3.17}$$

This expression determines the range of values for the CFL number λ for which $|g| \leq 1$. Ultimately, the stability of the Lax–Wendroff scheme for the one-way wave equation is determined.

Leap-frog (2,2) for one-way wave equation

The leap-frog discretization for the one-way wave equation yields the iteration scheme

$$u_n^{(m+1)} = u_n^{(m-1)} + \lambda \left(u_{n+1}^{(m)} - u_{n-1}^{(m)} \right). \tag{9.3.18}$$

Plugging in the ansatz (9.3.3) gives the following relations:

$$g^{m+1} \exp(i\xi_n h) = g^{m-1} \exp(i\xi_n h) + \lambda g^m \left(\exp(i\xi_{n+1} h) - \exp(i\xi_{n-1} h) \right)$$

$$g^2 = 1 + \lambda g \left[\exp(ih(\xi_{n+1} - \xi_n)) - \exp(ih(\xi_{n-1} - \xi_n)) \right]. \tag{9.3.19}$$

Letting $\xi_n = n\zeta$ reduces the equations further to

$$g^2 = 1 + \lambda g [\exp(i\zeta h) - \exp(-i\zeta h)]$$

$$g - \frac{1}{g} = 2i\lambda \sin(\zeta h). \tag{9.3.20}$$

To assure stability, it can be shown that we require

$$\lambda \leq 1. \tag{9.3.21}$$

Thus the leap-frog (2,2) time-stepping scheme is stable for values of $\lambda \leq 1$. The implementation of this scheme with this restriction will not force the solution to infinity.

A couple of general remarks should be made concerning the von Neumann analysis.

- It is a general result that a scheme which is forward in time and centered in space is unstable for the one-way wave equation. This assumes a standard forward discretization, not something like Runge–Kutta.
- von Neumann analysis is rarely enough to guarantee stability, i.e. it is necessary but not sufficient.
- Many other mechanisms for unstable growth are not captured by von Neumann analysis.
- Nonlinearity usually kills the von Neumann analysis immediately. Thus a large variety of nonlinear partial differential equations are beyond the scope of a von Neumann analysis.
- Accuracy versus stability: it is always better to worry about accuracy. An unstable scheme will quickly become apparent by causing the solution to blow-up to infinity, whereas an inaccurate scheme will simply give you a wrong result without indicating a problem.

Comparison of Time-Stepping Schemes

A few open questions remain concerning the stability and accuracy issues of the time- and space-stepping schemes developed. In particular, it is not clear how the stability results derived in the previous section apply to other partial differential equations.

Consider, for instance, the leap-frog (2,2) method applied to the one-way wave equation

$$\frac{\partial u}{\partial t} = c\frac{\partial u}{\partial x}.$$ (9.4.1)

The leap-frog discretization for the one-way wave equation yielded the iteration scheme

$$u_n^{(m+1)} = u_n^{(m-1)} + \lambda \left(u_{n+1}^{(m)} - u_{n-1}^{(m)} \right),$$ (9.4.2)

where $\lambda = c\Delta t/\Delta x$ is the CFL number. The von Neumann stability analysis based upon the ansatz $u_n^{(m)} = g^m \exp(i\xi_n h)$ results in the expression

$$g - \frac{1}{g} = 2i\lambda \sin(\zeta h),$$ (9.4.3)

where $\xi_n = n\zeta$. Thus the scheme was stable provided $\lambda \leq 1$. Note that to double the accuracy, both the time and space discretizations Δt and Δx need to be simultaneously halved.

Diffusion equation

We will again consider the leap-frog (2,2) discretization applied to the diffusion equation. The stability properties will be found to be very different from those of the one-way wave equation.

The difference in the one-way wave equation and diffusion equation is an extra x derivative so that

$$\frac{\partial u}{\partial t} = c\frac{\partial^2 u}{\partial x^2}.$$

(9.4.4)

Discretizing the spatial second derivative yields

$$\frac{\partial u_n^{(m)}}{\partial t} = \frac{c}{\Delta x^2}\left(u_{n+1}^{(m)} - 2u_n^{(m)} + u_{n-1}^{(m)}\right).$$

(9.4.5)

Second-order center-differencing in time then yields

$$u_n^{(m+1)} = u_n^{(m-1)} + 2\lambda\left(u_{n+1}^{(m)} - 2u_n^{(m)} + u_{n-1}^{(m)}\right),$$

(9.4.6)

where now the CFL number is given by

$$\lambda = \frac{c\Delta t}{\Delta x^2}.$$

(9.4.7)

In this case, to double the accuracy and hold the CFL number constant requires cutting the time-step by a factor of 4. This is generally true of any partial differential equation with the highest derivative term having two derivatives. The von Neumann stability analysis based upon the ansatz $u_n^{(m)} = g^m \exp(i\xi_n h)$ results in the expression

$$g^{m+1}\exp(i\xi_n h) = g^{m-1}\exp(i\xi_n h) + 2\lambda g^m\left(\exp(i\xi_{n+1}h) - 2\exp(i\xi_n h) + \exp(i\xi_{n-1}h)\right)$$

$$g - \frac{1}{g} = 2\lambda(\exp(i\zeta h) + \exp(-i\zeta h) - 2)$$

$$g - \frac{1}{g} = 2\lambda(i\sin(\zeta h) - 1),$$

(9.4.8)

where we let $\xi_n = n\zeta$. Unlike the one-way wave equation, the addition of the term $-2u_n^{(m)}$ in the discretization makes $|g(\zeta)| \geq 1$ for all values of λ. Thus the leap-frog (2,2) is unstable for all CFL numbers for the diffusion equation.

Interestingly enough, the forward Euler method which was unstable when applied to the one-way wave equation can be stable for the diffusion equation. Using an Euler discretization instead of a center-difference scheme in time yields

$$u_n^{(m+1)} = u_n^{(m)} + \lambda\left(u_{n+1}^{(m)} - 2u_n^{(m)} + u_{n-1}^{(m)}\right).$$

(9.4.9)

The von Neumann stability analysis based upon the ansatz $u_n^{(m)} = g^m \exp(i\xi_n h)$ results in the expression

$$g = 1 + \lambda(i\sin(\zeta h) - 1)$$

(9.4.10)

This gives

$$|g| = 1 - 2\lambda + \lambda^2(1 + \sin^2 \zeta h),$$

(9.4.11)

so that

$$|g| \leq 1 \quad \text{for} \quad \lambda \leq \frac{1}{2} . \tag{9.4.12}$$

Thus the maximum step-size in time is given by $\Delta t = \Delta x^2/2c$. Again, it is clear that to double accuracy, the time-step must be reduced by a factor of 4.

Hyper-diffusion

Higher derivatives in x require central differencing schemes which successively add powers of Δx to the denominator. This makes for severe time-stepping restrictions in the CFL number. As a simple example, consider the fourth-order diffusion equation

$$\frac{\partial u}{\partial t} = -c \frac{\partial^4 u}{\partial x^4} . \tag{9.4.13}$$

Using a forward Euler method in time and a central difference scheme in space gives the iteration scheme

$$u_n^{(m+1)} = u_n^{(m)} - \lambda \left(u_{n+2}^{(m)} - 4u_{n+1}^{(m)} + 6u_n^{(m)} - 4u_{n-1}^{(m)} + u_{n-2}^{(m)} \right) , \tag{9.4.14}$$

where the CFL number λ is now given by

$$\lambda = \frac{c\Delta t}{\Delta x^4} . \tag{9.4.15}$$

Thus doubling the accuracy requires a drop in the step-size to $\Delta t/16$, i.e. the run time takes 32 times as long since there are twice as many spatial points and 16 times as many time-steps. This kind of behavior is often referred to as numerical stiffness.

Numerical stiffness of this sort is not a result of the central differencing scheme. Rather, it is an inherent problem with hyper-diffusion. For instance, we could consider solving the fourth-order diffusion equation (9.4.13) with fast Fourier transforms. Transforming the equation in the x-direction gives

$$\frac{\partial \widehat{u}}{\partial t} = -c(ik)^4 \widehat{u} = -ck^4 \widehat{u}. \tag{9.4.16}$$

This can then be solved with a differential equation time-stepping routine. The time-step of any of the standard time-stepping routines is based upon the size of the right-hand side of the equation. If there are $n = 128$ Fourier modes, then $k_{\max} = 64$. But note that

$$(k_{\max})^4 = (64)^4 = 16.8 \times 10^6 , \tag{9.4.17}$$

which is a very large value. The time-stepping algorithm will have to adjust to these large values which are generated strictly from the physical effect of the higher order diffusion.

There are a few key issues in dealing with numerical stiffness.

- Use a variable time-stepping routine which uses smaller steps when necessary but large steps if possible. Every built-in MATLAB differential equation solver uses an adaptive stepping routine to advance the solution.

- If numerical stiffness is a problem, then it is often useful to use an implicit scheme. This will generally allow for larger time-steps. The time-stepping algorithm *ode113* uses a predictor–corrector method which partially utilizes an implicit scheme.

- If stiffness comes from the behavior of the solution itself, i.e. you are considering a singular problem, then it is advantageous to use a solver specifically built for this stiffness. The time-stepping algorithm *ode15s* relies on Gear methods and is well suited to this type of problem.

- From a practical viewpoint, beware of the accuracy of *ode23* or *ode45* when strong nonlinearity or singular behavior is expected.

9.5 Operator Splitting Techniques

With either the finite difference or spectral methods, the governing partial differential equations are transformed into a system of differential equations which are advanced in time with any of the standard time-stepping schemes. A von Neumann analysis can often suggest the appropriateness of a scheme. For instance, we have the following:

A. **Wave behavior:** $\partial u/\partial t = \partial u/\partial x$
   - forward Euler: unstable for all λ
   - leap-frog (2,2): stable $\lambda \leq 1$

B. **Diffusion behavior:** $\partial u/\partial t = \partial^2 u/\partial x^2$
   - forward Euler: stable $\lambda \leq 1/2$
   - leap-frog (2,2): unstable for all λ

Thus the physical behavior of the governing equation dictates the kind of scheme which must be implemented. But what if we wanted to consider the equation

$$\frac{\partial u}{\partial t} = \frac{\partial u}{\partial x} + \frac{\partial^2 u}{\partial x^2}?$$

(9.5.1)

This has elements of both diffusion and wave propagation. What time-stepping scheme is appropriate for such an equation? Certainly the Euler method seems to work well for diffusion, but destabilizes wave propagation. In contrast, leap-frog (2,2) works well for wave behavior and destabilizes diffusion.

Operator splitting

The key idea behind operator splitting is to decouple the various physical effects of the problem from each other. Over very small time-steps, for instance, one can imagine that diffusion would essentially act independently of the wave propagation and vice versa in (9.5.1). Just as in (9.5.1), the advection–diffusion can also be thought of as decoupling. So we can consider

$$\frac{\partial \omega}{\partial t} + [\psi, \omega] = \nu \nabla^2 \omega \tag{9.5.2}$$

where the bracketed terms represent the advection (wave propagation) and the right-hand side represents the diffusion. Again there is a combination of wave dynamics and diffusive spreading.

Over a very small time interval Δt, it is reasonable to conjecture that the diffusion process is independent of the advection. Thus we could split the calculation into the following two pieces:

$$\Delta t: \quad \frac{\partial \omega}{\partial t} + [\psi, \omega] = 0 \quad \text{advection only} \tag{9.5.3a}$$

$$\Delta t: \quad \frac{\partial \omega}{\partial t} = \nu \nabla^2 \omega \quad \text{diffusion only.} \tag{9.5.3b}$$

This then allows us to time-step each physical effect independently over Δt. Advantage can then be taken of an appropriate time-stepping scheme which is stable for those particular terms. For instance, in this decoupling we could solve the advection (wave propagation) terms with a leap-frog (2,2) scheme and the diffusion terms with a forward Euler method.

Additional advantages of splitting

There can be additional advantages to the splitting scheme. To see this, we consider the nonlinear Schrödinger equation

$$i\frac{\partial u}{\partial t} + \frac{1}{2}\frac{\partial^2 u}{\partial x^2} + |u|^2 u = 0 \tag{9.5.4}$$

where we split the operations so that

$$\textbf{I.} \quad \Delta t: \quad i\frac{\partial u}{\partial t} + \frac{1}{2}\frac{\partial^2 u}{\partial x^2} = 0 \tag{9.5.5a}$$

$$\textbf{II.} \quad \Delta t: \quad i\frac{\partial u}{\partial t} + |u|^2 u = 0. \tag{9.5.5b}$$

Thus the solution will be decomposed into a linear and nonlinear part. We begin by solving the linear part **I**. Fourier transforming yields

$$i\frac{d\widehat{u}}{dt} - \frac{k^2}{2}\widehat{u} = 0 \tag{9.5.6}$$

which has the solution

$$\widehat{u} = \widehat{u}_0 \exp\left(-i\frac{k^2}{2}t\right). \tag{9.5.7}$$

So at each step, we simply need to calculate the transform of the initial condition $\widehat{u}_0$, multiply by $\exp(-ik^2t/2)$, and then invert.

The next step is to solve the nonlinear evolution **II** over a time-step Δt. This is easily done since the nonlinear evolution admits the exact solution

$$u = u_0 \exp\left(i|u_0|^2 t\right) \tag{9.5.8}$$

where u_0 is the initial condition. This part of the calculation then requires no computation, i.e. we have an exact, analytic solution. The fact that we can take advantage of this property is why the splitting algorithm is so powerful.

To summarize then, the split-step method for the nonlinear Schrödinger equation yields the following algorithm.

1   Dispersion: $u_1 = \text{FFT}^{-1}\left[\widehat{u_0} \exp(-ik^2 \Delta t/2)\right]$

2   Nonlinearity: $u_2 = u_1 \exp(i|u_1|^2 \Delta t)$

3   Solution: $u(t + \Delta t) = u_2$.

Note the advantage that is taken of the analytic properties of the solution of each aspect of the split operators.

Symmetrized splitting

Although we will not perform an error analysis on this scheme, it is not hard to see that the error will depend heavily on the time-step Δt. Thus our scheme for solving

$$\frac{\partial u}{\partial t} = Lu + N(u) \tag{9.5.9}$$

involves the splitting

$$\textbf{I.} \ \ \Delta t: \quad \frac{\partial u}{\partial t} + Lu = 0 \tag{9.5.10a}$$

$$\textbf{II.} \ \ \Delta t: \quad \frac{\partial u}{\partial t} + N(u) = 0. \tag{9.5.10b}$$

To drop the error down by another $O(\Delta t)$, we use what is called a Strang splitting technique [19] which is based upon the Trotter product formula. This essentially involves symmetrizing the splitting algorithm so that

$$\textbf{I.} \ \ \frac{\Delta t}{2}: \quad \frac{\partial u}{\partial t} + Lu = 0 \tag{9.5.11a}$$

$$\textbf{II.} \ \ \Delta t: \quad \frac{\partial u}{\partial t} + N(u) = 0 \tag{9.5.11b}$$

$$\textbf{III.} \ \ \frac{\Delta t}{2}: \quad \frac{\partial u}{\partial t} + Lu = 0. \tag{9.5.11c}$$

This essentially cuts the time-step in half and allows the error to drop down an order of magnitude. It should always be used so as to keep the error in check.

9.6   Optimizing Computational Performance: Rules of Thumb

Computational performance is always crucial in choosing a numerical scheme. Speed, accuracy and stability all play key roles in determining the appropriate choice of method for solving. We will consider three prototypical equations:

$$\frac{\partial u}{\partial t} = \frac{\partial u}{\partial x} \qquad \text{one-way wave equation} \qquad (9.6.1a)$$

$$\frac{\partial u}{\partial t} = \frac{\partial^2 u}{\partial x^2} \qquad \text{diffusion equation} \qquad (9.6.1b)$$

$$i\frac{\partial u}{\partial t} = \frac{1}{2}\frac{\partial^2 u}{\partial x^2} + |u|^2 u \qquad \text{nonlinear Schrödinger equation.} \qquad (9.6.1c)$$

Periodic boundary conditions will be assumed in each case. The purpose of this section is to build a numerical routine which will solve these problems using the iteration procedures outlined in the previous two sections. Of specific interest will be the setting of the CFL number and the consequences of violating the stability criteria associated with it.

The equations are considered in one dimension such that the first and second derivative are given by

$$\frac{\partial u}{\partial x} \rightarrow \frac{1}{2\Delta x}
\begin{bmatrix}
0 & 1 & 0 & \cdots & & 0 & -1 \\
-1 & 0 & 1 & 0 & \cdots & & 0 \\
0 & \ddots & \ddots & \ddots & & & \\
\vdots & & & & & & 0 \\
& & & & & & \vdots \\
\vdots & \cdots & 0 & -1 & 0 & 1 \\
1 & 0 & \cdots & & 0 & -1 & 0
\end{bmatrix}
\begin{pmatrix} u_1 \\ u_2 \\ \vdots \\ u_n \end{pmatrix}
\qquad (9.6.2)$$

and

$$\frac{\partial^2 u}{\partial x^2} \rightarrow \frac{1}{\Delta x^2}
\begin{bmatrix}
-2 & 1 & 0 & \cdots & & 0 & 1 \\
1 & -2 & 1 & 0 & \cdots & & 0 \\
0 & \ddots & \ddots & \ddots & & & \\
\vdots & & & & & & 0 \\
& & & & & & \vdots \\
\vdots & \cdots & 0 & 1 & -2 & 1 \\
1 & 0 & \cdots & & 0 & 1 & -2
\end{bmatrix}
\begin{pmatrix} u_1 \\ u_2 \\ \vdots \\ u_n \end{pmatrix}.
\qquad (9.6.3)$$

From the previous sections, we have the following discretization schemes for the one-way wave equation (9.6.1):

Euler (unstable): $\quad u_n^{(m+1)} = u_n^{(m)} + \dfrac{\lambda}{2}\left(u_{n+1}^{(m)} - u_{n-1}^{(m)}\right)$ (9.6.4a)

leap-frog (2,2) (stable for $\lambda \leq 1$): $\quad u_n^{(m+1)} = u_n^{(m-1)} + \lambda\left(u_{n+1}^{(m)} - u_{n-1}^{(m)}\right)$ (9.6.4b)

where the CFL number is given by $\lambda = \Delta t/\Delta x$. Similarly for the diffusion equation (9.6.1)

Euler (stable for $\lambda \leq 1/2$): $\quad u_n^{(m+1)} = u_n^{(m)} + \lambda\left(u_{n+1}^{(m)} - 2u_n^{(m)} + u_{n-1}^{(m)}\right)$ (9.6.5a)

leap-frog (2,2) (unstable): $\quad u_n^{(m+1)} = u_n^{(m-1)} + 2\lambda\left(u_{n+1}^{(m)} - 2u_n^{(m)} + u_{n-1}^{(m)}\right)$ (9.6.5b)

where now the CFL number is given by $\lambda = \Delta t/\Delta x^2$. The nonlinear Schrödinger equation discretizes to the following form:

$$\frac{\partial u_n^{(m)}}{\partial t} = -\frac{i}{2\Delta x^2}\left(u_{n+1}^{(m)} - 2u_n^{(m)} + u_{n-1}^{(m)}\right) - i|u_n^{(m)}|^2 u_n^{(m)}.$$ (9.6.6)

We will explore Euler and leap-frog (2,2) time-stepping with this equation.

One-way wave equation

We first consider the leap-frog (2,2) scheme applied to the one-way wave equation. Figure 9.4 depicts the evolution of an initial Gaussian pulse. For this case, the CFL is 0.5 so that stable evolution is analytically predicted. The solution propagates to the left as expected from the exact solution. The leap-frog (2,2) scheme becomes unstable for $\lambda \geq 1$ and the system is always unstable for the Euler time-stepping scheme. Figure 9.5 depicts the unstable evolution of the leap-frog (2,2) scheme with CFL $= 2$ and the Euler time-stepping scheme. The initial conditions

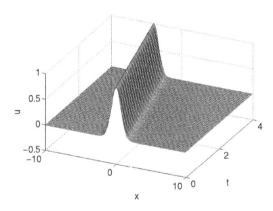

Figure 9.4: Evolution of the one-way wave equation with the leap-frog (2,2) scheme and with CFL $= 0.5$. The stable traveling wave solution is propagated in this case to the left.

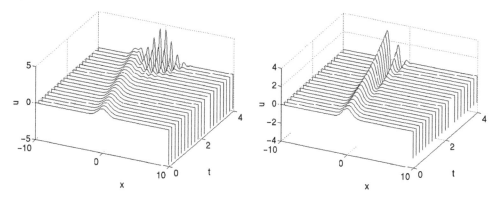

Figure 9.5: Evolution of the one-way wave equation using the leap-frog (2,2) scheme with CFL = 2 (left) along with the Euler time-stepping scheme (right). The analysis predicts stable evolution of leap-frog provided the CFL ≤ 1. Thus the onset of numerical instability near $t \approx 3$ for the CFL = 2 case is not surprising. Likewise, the Euler scheme is expected to be unstable for all CFL.

used are identical to that in Fig. 9.4. Since we have predicted that the leap-frog numerical scheme is only stable provided $\lambda < 1$, it is not surprising that the figure on the left goes unstable. Likewise, the figure on the right shows the numerical instability generated in the Euler scheme. Note that both of these unstable evolutions develop high-frequency oscillations which eventually blow up. The MATLAB code used to generate the leap-frog and Euler iterative solutions is given by

```
clear all; close all;  % clear previous figures and values

% initialize grid size, time, and CFL number
Time=4;
L=20;
n=200;
x2=linspace(-L/2,L/2,n+1);
x=x2(1:n);
dx=x(2)-x(1);
dt=0.2;
CFL=dt/dx
time_steps=Time/dt;
t=0:dt:Time;

% initial conditions
u0=exp(-x.^2)';
u1=exp(-(x+dt).^2)';
usol(:,1)=u0;
usol(:,2)=u1;

% sparse matrix for derivative term
e1=ones(n,1);
A=spdiags([-e1  e1],[-1 1],n,n);
```

Continued

```
A(1,n)=-1;  A(n,1)=1;
% leap frog (2,2) or euler iteration scheme
for j=1:time_steps-1
%   u2 = u0 + CFL*A*u1;  % leap frog (2,2)
%   u0 = u1; u1 = u2;    % leap frog (2,2)
    u2 = u1 + 0.5*CFL*A*u1;  % euler
    u1 = u2;                 % euler
    usol(:,j+2)=u2;
end

% plot the data
waterfall(x,t,usol');
map=[0 0 0];
colormap(map);

% set x and y limits and fontsize
set(gca,'Xlim',[-L/2 L/2],'Xtick',[-L/2 0 L/2],'FontSize',[20]);
set(gca,'Ylim',[0 Time],'ytick',[0 Time/2 Time],'FontSize',[20]);
view(25,40)

% set axis labels and fonts
xl=xlabel('x'); yl=ylabel('t'); zl=zlabel('u');
set(xl,'FontSize',[20]);set(yl,'FontSize',[20]);set(zl,'FontSize',[20]);

print -djpeg -r0 fig.jpg % print jpeg at screen resolution
```

Heat equation

In a similar fashion, we investigate the evolution of the diffusion equation when the space–time discretization is given by the leap-frog (2,2) scheme or Euler stepping. Figure 9.6 shows the

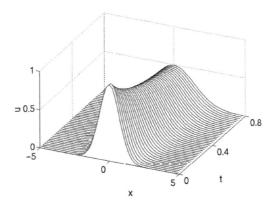

Figure 9.6: Stable evolution of the heat equation with the Euler scheme with CFL = 0.5. The initial Gaussian is diffused in this case.

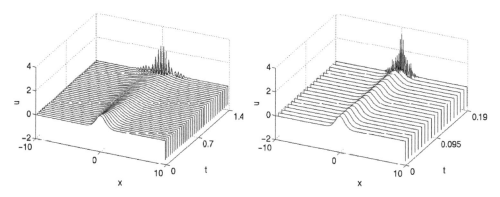

Figure 9.7: Evolution of the heat equation with the Euler time-stepping scheme (left) and leap-frog (2,2) scheme (right) with CFL $= 1$. The analysis predicts that both these schemes are unstable. Thus the onset of numerical instability is observed.

expected diffusion behavior for the stable Euler scheme ($\lambda \leq 0.5$). In contrast, Fig. 9.7 shows the numerical instabilities which are generated from violating the CFL constraint for the Euler scheme or using the always unstable leap-frog (2,2) scheme for the diffusion equation. The numerical code used to generate these solutions follows that given previously for the one-way wave equation. However, the sparse matrix is now given by

```
% sparse matrix for second derivative term
e1=ones(n,1);
A=spdiags([e1 -2*e1 e1],[-1 0 1],n,n);
A(1,n)=1;   A(n,1)=1;
```

Further, the iterative process is now

```
% leap frog (2,2) or euler iteration scheme
for j=1:time_steps-1
   u2 = u0 + 2*CFL*A*u1;  % leap frog (2,2)
   u0 = u1; u1 = u2;      % leap frog (2,2)
%  u2 = u1 + CFL*A*u1;    % euler
%  u1 = u2;               % euler
   usol(:,j+2)=u2;
end
```

where we recall that the CFL condition is now given by $\lambda = \Delta t / \Delta x^2$, i.e.

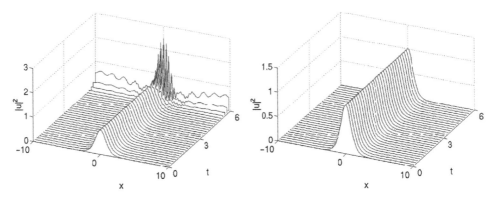

Figure 9.8: Evolution of the nonlinear Schrödinger equation with the Euler time-stepping scheme (left) and leap-frog (2,2) scheme (right) with CFL = 0.05.

```
CFL=dt/dx/dx
```

This solves the one-dimensional heat equation with periodic boundary conditions.

Nonlinear Schrödinger equation

The nonlinear Schrödinger equation can easily be discretized by the above techniques. However, as with most nonlinear equations, it is a bit more difficult to perform a von Neumann analysis. Therefore, we explore the behavior for this system for two different discretization schemes: Euler and leap-frog (2,2). The CFL number will be the same with both schemes ($\lambda = 0.05$) and the stability will be investigated through numerical computations. Figure 9.8 shows the evolution of the exact one-soliton solution of the nonlinear Schrödinger equation ($u(x, 0) = \text{sech}(x)$) over six units of time. The Euler scheme is observed to lead to numerical instability whereas the leap-frog (2,2) scheme is stable. In general, the leap-frog schemes work well for wave propagation problems while Euler methods are better for problems of a diffusive nature.

The MATLAB code modifications necessary to solve the nonlinear Schrödinger equation are trivial. Specifically, the iteration scheme requires change. For the stable leap-frog scheme, the following command structure is required

```
u2 = u0 + -i*CFL*A*u1- i*2*dt*(conj(u1).*u1).*u1;
u0 = u1; u1 = u2;
```

Note that i is automatically defined in MATLAB as $i = \sqrt{-1}$. Thus it is imperative that you do not use the variable i as a counter in your FOR loops. You will solve a very different equation if you are not careful with this definition.

Spectral Methods

10

Spectral methods are one of the most powerful solution techniques for ordinary and partial differential equations. The best-known example of a spectral method is the Fourier transform. We have already made use of the Fourier transform using FFT routines. Other spectral techniques exist which render a variety of problems easily tractable and often at significant computational savings.

10.1 Fast Fourier Transforms and Cosine/Sine Transform

From our previous chapters, we are already familiar with the Fourier transform and some of its properties. At the time, the distinct advantage to using the FFT was its computational efficiency of solving a problem in $O(N \log N)$. This section explores the underlying mathematical reasons for such performance.

One of the abstract definitions of a Fourier transform pair is given by

$$F(k) = \int_{-\infty}^{\infty} e^{-ikx} f(x) dx \tag{10.1.1a}$$

$$f(x) = \frac{1}{2\pi} \int_{-\infty}^{\infty} e^{ikx} F(k) dk. \tag{10.1.1b}$$

On a practical level, the value of the Fourier transform revolves squarely around the derivative relationship

$$\widehat{f^{(n)}} = (ik)^n \widehat{f} \tag{10.1.2}$$

which results from the definition of the Fourier transform and integration by parts. Recall that we denote the Fourier transform of $f(x)$ as $\widehat{f(x)}$.

When considering a computational domain, the solution can only be found on a finite length domain. Thus the definition of the Fourier transform needs to be modified in order to account for the finite sized computational domain. Instead of expanding in terms of a continuous integral for values of wavenumber k and cosines and sines ($\exp(ikx)$), we expand in a Fourier series

$$F(k) = \sum_{n=1}^{N} f(n) \exp\left[-i\frac{2\pi(k-1)}{N}(n-1)\right] \quad 1 \leq k \leq N \tag{10.1.3a}$$

$$f(n) = \frac{1}{N}\sum_{k=1}^{N} F(k) \exp\left[i\frac{2\pi(k-1)}{N}(n-1)\right] \quad 1 \leq n \leq N. \tag{10.1.3b}$$

Thus the Fourier transform is nothing but an expansion in a basis of cosine and sine functions. If we define the fundamental oscillatory piece as

$$w^{nk} = \exp\left(\frac{2i\pi(k-1)(n-1)}{N}\right) = \cos\left(\frac{2\pi(k-1)(n-1)}{N}\right) + i\sin\left(\frac{2\pi(k-1)(n-1)}{N}\right), \tag{10.1.4}$$

then the Fourier transform results in the expression

$$F_n = \sum_{k=0}^{N-1} w^{nk} f_k, \quad 0 \leq n \leq N-1. \tag{10.1.5}$$

Thus the calculation of the Fourier transform involves a double sum and an $O(N^2)$ operation. Thus, at first, it would appear that the Fourier transform method is the same operation count as LU decomposition. The basis functions used for the Fourier transform, sine transform and cosine transform are depicted in Fig. 10.1. The process of solving a differential or partial differential equation involves evaluating the coefficient of each of the modes. Note that this expansion, unlike the finite difference method, is a *global* expansion in that every basis function is evaluated on the entire domain.

The Fourier, sine and cosine transforms behave very differently at the boundaries. Specifically, the Fourier transform assumes periodic boundary conditions whereas the sine and cosine transforms assume pinned and no-flux boundaries, respectively. The cosine and sine transform are often chosen for their boundary properties. Thus for a given problem, an evaluation must be made of the type of transform to be used based upon the boundary conditions needing to be satisfied. Table 10.1 illustrates the three different expansions and their associated boundary conditions. The appropriate MATLAB command is also given.

Fast Fourier transforms: Cooley–Tukey algorithm

To see how the FFT gets around the computational restriction of an $O(N^2)$ scheme, we consider the Cooley–Tukey algorithm which drops the computations to $O(N \log N)$. To consider the algorithm, we begin with the $N \times N$ matrix $\mathbf{F}_N$ whose components are given by

$$(F_N)_{jk} = w_n^{jk} = \exp(i2\pi jk/N). \tag{10.1.6}$$

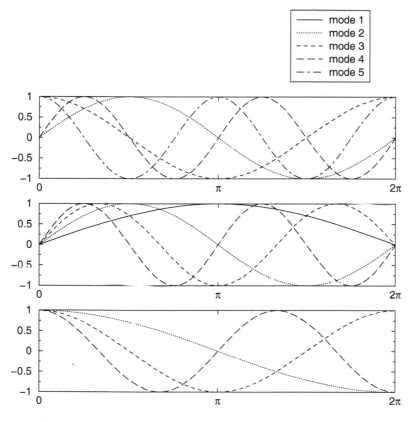

Figure 10.1: Basis functions used for a Fourier mode expansion (top), a sine expansion (middle) and a cosine expansion (bottom).

Table 10.1 MATLAB functions for Fourier, sine and cosine transforms and their associated boundary conditions. To invert the expansions, the MATLAB commands are *ifft*, *idst* and *idct*, repectively

Command	Expansion	Boundary conditions
fft	$F_k = \sum_{j=0}^{2N-1} f_j \exp(i\pi jk/N)$	periodic: $f(0) = f(L)$
dst	$F_k = \sum_{j=1}^{N-1} f_j \sin(\pi jk/N)$	pinned: $f(0) = f(L) = 0$
dct	$F_k = \sum_{j=0}^{N-2} f_j \cos(\pi jk/2N)$	no-flux: $f'(0) = f'(L) = 0$

The coefficients of the matrix are points on the unit circle since $|w_n^{jk}| = 1$. They are also the basis functions for the Fourier transform.

The FFT starts with the following trivial observation

$$w_{2n}^2 = w_n, \tag{10.1.7}$$

which is easy to show since $w_n = \exp(i2\pi/n)$ and

$$w_{2n}^2 = \exp(i2\pi/(2n))\exp(i2\pi/(2n)) = \exp(i2\pi/n) = w_n. \tag{10.1.8}$$

The consequences of this simple relationship are enormous and are at the core of the success of the FFT algorithm. Essentially, the FFT is a matrix operation

$$\mathbf{y} = \mathbf{F}_N\mathbf{x} \tag{10.1.9}$$

which can now be split into two separate operations. Thus defining

$$\mathbf{x}^e = \begin{pmatrix} x_0 \\ x_2 \\ x_4 \\ \vdots \\ x_{N-2} \end{pmatrix} \quad \text{and} \quad \mathbf{x}^o = \begin{pmatrix} x_1 \\ x_3 \\ x_5 \\ \vdots \\ x_{N-1} \end{pmatrix} \tag{10.1.10}$$

which are both vectors of length $M = N/2$, we can form the two $M \times M$ systems

$$\mathbf{y}^e = \mathbf{F}_M\mathbf{x}^e \quad \text{and} \quad \mathbf{y}^o = \mathbf{F}_M\mathbf{x}^o \tag{10.1.11}$$

for the even coefficient terms $\mathbf{x}^e, \mathbf{y}^e$ and odd coefficient terms $\mathbf{x}^o, \mathbf{y}^o$. Thus the computation size goes from $O(N^2)$ to $O(2M^2) = O(N^2/2)$. However, we must be able to reconstruct the original $\mathbf{y}$

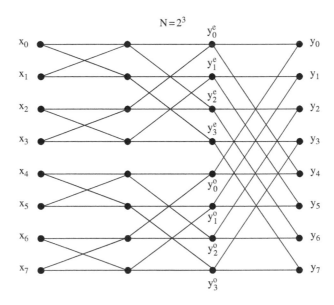

Figure 10.2: Graphical description of the fast Fourier transform process which systematically continues to factor the problem in two. This process allows the FFT routine to drop to $O(N\log N)$ operations.

from the smaller system of $\mathbf{y}^e$ and $\mathbf{y}^o$. And indeed we can reconstruct the original $\mathbf{y}$. In particular, it can be shown that component by component

$$y_n = y_n^e + w_N^n y_n^o \quad n = 0, 1, 2, \ldots, M - 1 \tag{10.1.12a}$$

$$y_{n+M} = y_n^e - w_N^n y_n^o \quad n = 0, 1, 2, \ldots, M - 1. \tag{10.1.12b}$$

This is where the shift occurs in the FFT routine which maps the domain $x \in [0, L]$ to $[-L, 0]$ and $x \in [-L, 0]$ to $[0, L]$. The command *fftshift* undoes this shift. Details of this construction can be found elsewhere [15, 19].

There is no reason to stop the splitting process at this point. In fact, provided we choose the size of our domain and matrix $\mathbf{F}_N$ so that N is a power of 2, then we can continue to split the system until we have a simple algebraic, i.e. a 1×1 system, solution to perform. The process is illustrated graphically in Fig. 10.2 where the switching and factorization are illustrated. Once the final level is reached, the algebraic expression is solved and the process is reversed. This factorization process renders the FFT scheme $O(N \log N)$.

 ## 10.2   Chebychev Polynomials and Transform

The fast Fourier transform is only one of many possible expansion bases, i.e. there is nothing special about expanding in cosines and sines. Of course, the FFT expansion does have the unusual property of factorization which drops it to an $O(N \log N)$ scheme. Regardless, there are a myriad of other expansion bases which can be considered. The primary motivation for considering other expansions is based upon the specifics of the given governing equations and its physical boundaries and constraints. Special functions are often prime candidates for use as expansion bases. The following are some important examples

- Bessel functions: radial, 2D problems
- Legendre polynomials: 3D Laplaces equation
- Hermite–Gauss polynomials: Schrödinger with harmonic potential
- Spherical harmonics: radial, 3D problems
- Chebychev polynomials: bounded 1D domains.

The two key properties required to use any expansion basis successfully are its orthogonality properties and the calculation of the norm. Regardless, all the above expansions appear to require $O(N^2)$ calculations.

Chebychev polynomials

Although not often encountered in mathematics courses, the Chebychev polynomial is an important special function from a computational viewpoint. The reason for its prominence in the computational setting will become apparent momentarily. For the present, we note that the Chebychev polynomials are solutions to the differential equation

$$\sqrt{1-x^2}\frac{d}{dx}\left(\sqrt{1-x^2}\frac{dT_n}{dx}\right)+n^2T_n=0 \quad x\in[-1,1].$$

(10.2.1)

This is a self-adjoint Sturm–Lioville problem. Thus the following properties are known:

1   Eigenvalues are real: $\lambda_n = n^2$

2   Eigenfunctions are real: $T_n(x)$

3   Eigenfunctions are orthogonal:

$$\int_{-1}^{1}(1-x^2)^{-1/2}T_n(x)T_m(x)dx = \frac{\pi}{2}c_n\delta_{nm}$$

(10.2.2)

where $c_0 = 2, c_n = 1(n > 0)$ and δ_{nm} is the delta function

4   Eigenfunctions form a complete basis.

Each Chebychev polynomial (of degree n) is defined by

$$T_n(\cos\theta) = \cos n\theta.$$

(10.2.3)

Thus we find

$$T_0(x) = 1$$ (10.2.4a)

$$T_1(x) = x$$ (10.2.4b)

$$T_2(x) = 2x^2 - 1$$ (10.2.4c)

$$T_3(x) = 4x^3 - 3x$$ (10.2.4d)

$$T_4(x) = 8x^4 - 8x^2 + 1.$$ (10.2.4e)

The behavior of the first five Chebychev polynomials is illustrated in Fig. 10.3.

It is appropriate to ask why the Chebychev polynomials, of all the special functions listed, are of such computational interest. Especially given that the equation which the $T_n(x)$ satisfy, and their functional form shown in Fig. 10.3, appear to be no better than Bessel, Hermite–Gauss, or any other special function. The distinction with the Chebychev polynomials is that you can transform them so that use can be made of the $O(N\log N)$ discrete cosine transform. This effectively renders the Chebychev expansion scheme an $O(N\log N)$ transformation. Specifically, we transform from the interval $x\in[-1,1]$ by letting

$$x = \cos\theta \quad \theta\in[0,\pi].$$

(10.2.5)

Thus when considering a function $f(x)$, we have $f(\cos\theta) = g(\theta)$. Under differentiation we find

$$\frac{dg}{d\theta} = -f'\cdot\sin\theta.$$

(10.2.6)

Thus $dg/d\theta = 0$ at $\theta = 0, \pi$, i.e. no-flux boundary conditions are satisfied. This allows us to use the *dct* (discrete cosine transform) to solve a given problem in the new transformed variables.

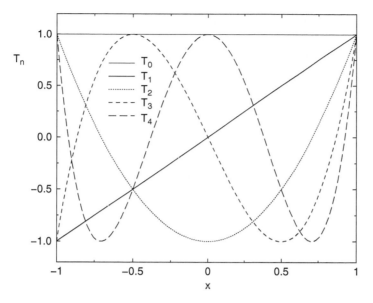

Figure 10.3: The first five Chebychev polynomials over the interval of definition $x \in [-1, 1]$.

The Chebychev expansion is thus given by

$$f(x) = \sum_{k=0}^{\infty} a_k T_k(x) \tag{10.2.7}$$

where the coefficients a_k are determined from orthogonality and inner products to be

$$a_k = \int_{-1}^{1} \frac{1}{\sqrt{1 - x^2}} f(x) T_k(x) dx. \tag{10.2.8}$$

It is these coefficients which are calculated in $O(N \log N)$ time. Some of the properties of the Chebychev polynomials are as follows:

- $T_{n+1} = 2x T_n(x) - T_{n-1}(x)$
- $|T_n(x)| \leq 1, \ |T_n'(x)| \leq n^2$
- $T_n(\pm 1) = (\pm 1)^n$
- $d^p/dx^p(T_n(\pm 1)) = (\pm 1)^{n+p} \prod_{k=0}^{p-1}(n^2 - k^2)/(2k + 1)$
- If n is even (odd), $T_n(x)$ is even (odd).

There are a couple of critical practical issues which must be considered when using the Chebychev scheme. Specifically the grid generation and spatial resolution are a little more difficult to handle. In using the discrete cosine transform on the variable $\theta \in [0, \pi]$, we recall that our original variable is actually $x = \cos \theta$ where $x \in [-1, 1]$. Thus the discretization of the θ variable leads to

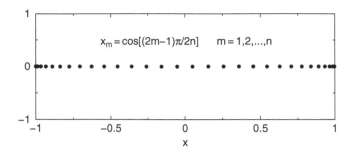

Figure 10.4: Clustered grid generation for $n = 30$ points using the Chebychev polynomials. Note that although the points are uniformly spaced in θ, they are clustered due to the fact that $x_m = \cos[(2m - 1)\pi/2n]$ where $m = 1, 2, \ldots, n$.

$$x_m = \cos\left(\frac{(2m - 1)\pi}{2n}\right) \quad m = 1, 2, \ldots, n. \tag{10.2.9}$$

Thus although the grid points are uniformly spaced in θ, the grid points are clustered in the original x variable. Specifically, there is a clustering of grid points at the boundaries. The Chebychev scheme then automatically has higher resolution at the boundaries of the computational domain. The clustering of the grid points at the boundary is illustrated in Fig. 10.4. So, as the resolution is increased, it is important to be aware that the resolution increase is not uniform across the computational domain.

Solving differential equations

As with any other solution method, a solution scheme must have an efficient way of relating derivatives to the function itself. For the FFT method, there was a very convenient relationship between the transform of a function and the transform of its derivatives. Although not as transparent as the FFT method, we can also relate the Chebychev transform derivatives to the Chebychev transform itself.

Defining L to be a linear operator so that

$$Lf(x) = \sum_{n=0}^{\infty} b_n T_n(x), \tag{10.2.10}$$

then with $f(x) = \sum_{n=0}^{\infty} a_n T_n(x)$ we find

- $Lf = f'(x) : c_n b_n = 2 \sum_{p=n+1(p+n \text{ odd})}^{\infty} p a_p$
- $Lf = xf(x) : b_n = (c_{n-1} a_{n-1} + a_{n+1})/2$
- $Lf = x^2 f(x) : b_n = (c_{n-2} a_{n-2} + (c_n + c_{n-1})a_n + a_{n+2})/4$

where $c_0 = 2, c_n = 0(n < 0), c_n = 1(n > 0), d_n = 1(n \geq 0)$, and $d_n = 0(n < 0)$.

10.3 **Spectral Method Implementation**

In this section, we develop an algorithm which implements a spectral method solution technique. We begin by considering the general partial differential equation

$$\frac{\partial u}{\partial t} = Lu + N(u) \tag{10.3.1}$$

where L is a linear, constant coefficient operator, i.e. it can take the form $L = ad^2/dx^2 + bd/dx + c$ where a, b, and c are constants. The second term $N(u)$ includes the nonlinear and nonconstant coefficient terms. An example of this would be $N(u) = u^3 + f(x)u + g(x)d^2u/dx^2$.

By applying a Fourier transform, the equations reduce to the system of differential equations

$$\frac{d\widehat{u}}{dt} = \alpha(k)\widehat{u} + \widehat{N(u)}. \tag{10.3.2}$$

This system can be stepped forward in time with any of the standard time-stepping techniques. Typically *ode45* or *ode23* is a good first attempt.

The parameter $\alpha(k)$ arises from the linear operator Lu and is easily determined from Fourier transforming. Specifically, if we consider the linear operator

$$Lu = a\frac{d^2u}{dx^2} + b\frac{du}{dx} + cu \tag{10.3.3}$$

then upon transforming this becomes

$$(ik)^2 a\widehat{u} + b(ik)\widehat{u} + c\widehat{u}$$
$$= (-k^2 a + ibk + c)\widehat{u}$$
$$= \alpha(k)\widehat{u}. \tag{10.3.4}$$

The parameter $\alpha(k)$ therefore takes into account all constant coefficient, linear differentiation terms.

The nonlinear terms are a bit more difficult to handle, but they are still relatively easy. Consider the following examples

1 $f(x)du/dx$
   - determine $du/dx \rightarrow \widehat{du/dx} = ik\widehat{u}, du/dx = FFT^{-1}(ik\widehat{u})$
   - multiply by $f(x) \rightarrow f(x)du/dx$
   - Fourier transform $FFT(f(x)du/dx)$.

2 u^3
   - Fourier transform $FFT(u^3)$.

3 $u^3 d^2u/dx^2$
   - determine $d^2u/dx^2 \rightarrow \widehat{d^2u/dx^2} = (ik)^2\widehat{u}, d^2u/dx^2 = FFT^{-1}(-k^2\widehat{u})$
   - multiply by $u^3 \rightarrow u^3 d^2u/dx^2$
   - Fourier transform $FFT(u^3 d^2u/dx^2)$.

These examples give an outline of how the nonlinear, nonconstant coefficient schemes would work in practice.

To illustrate the implementation of these ideas, we solve the advection–diffusion equations spectrally. Thus far the equations have been considered largely with finite difference techniques. However, the MATLAB codes presented here solve the equations using the FFT for both the streamfunction and vorticity evolution. We begin by initializing the appropriate numerical parameters

```
clear all; close all;

tspan=0:2:10;
nu=0.001;
Lx=20; Ly=20; nx=64; ny=64; N=nx*ny;

x2=linspace(-Lx/2,Lx/2,nx+1); x=x2(1:nx);
y2=linspace(-Ly/2,Ly/2,ny+1); y=y2(1:ny);
```

Thus the computational domain is $x \in [-10, 10]$ and $y \in [-10, 10]$. The diffusion parameter is chosen to be $\nu = 0.001$. The initial conditions are then defined as a stretched Gaussian

```
% INITIAL CONDITIONS
[X,Y]=meshgrid(x,y);
w=1*exp(-0.25*X.^2-Y.^2);
figure(1), pcolor(abs(w)); shading interp; colorbar; drawnow
```

The next step is to define the spectral k values in both the x- and y-directions. This allows us to solve for the streamfunction (8.5.8) spectrally. Once the streamfunction is determined, the vorticity can be stepped forward in time.

```
% SPECTRAL K VALUES
kx=(2*pi/Lx)*[0:(nx/2-1)  (-nx/2):-1]'; kx(1)=10^(-6);
ky=(2*pi/Ly)*[0:(ny/2-1)  (-ny/2):-1]'; ky(1)=10^(-6);
[KX,KY]=meshgrid(kx,ky);
K=KX.^2+KY.^2;

%  Solve and plot
wt=fft2(w)
wt2=reshape(wt,N,1);
[t,wtsol]=ode45('spc_rhs',tspan,wt2,[],nx,ny,N,KX,KY,K,nu);
```

Continued

```
for j=1:length(tspan)
    w=ifft2(reshape(wtsol(j,:),nx,ny));
    subplot(2,3,j), pcolor(x,y,real(w)); shading interp
end
```

The right-hand side of the system of differential equations which results from Fourier transforming is contained within the function *spc_rhs.m*. The outline of this routine is provided below. Note that the matrix components are reshaped into a vector and a large system of differential equations is solved.

```
function rhs=spc_rhs(tspan,wt2,dummy,nx,ny,N,KX,KY,K,nu);
wt=reshape(wt2,nx,ny);
psit=-wt./K;

psix=real(ifft2(i*KX.*psit));
psiy=real(ifft2(i*KY.*psit));

wx=real(ifft2(i*KX.*wt));
wy=real(ifft2(i*KY.*wt));

rhs=reshape(-nu*K.*wt+fft2(wx.*psiy-wy.*psix),N,1);
```

The code will quickly and efficiently solve the advection–diffusion equations in two dimensions. Figure 8.3 demonstrates the evolution of the initial stretched Gaussian vortex over $t \in [0,8]$.

10.4 Pseudo-Spectral Techniques with Filtering

The decomposition of the solution into Fourier mode components does not always lead to high-performance computing. Specifically when some form of numerical stiffness is present, computational performance can suffer dramatically. However, there are methods available which can effectively eliminate some of the numerical stiffness by making use of analytic insight into the problem.

We consider again the example of hyper-diffusion for which the governing equations are

$$\frac{\partial u}{\partial t} = \frac{\partial^4 u}{\partial x^4}. \tag{10.4.1}$$

In the Fourier domain, this becomes

$$\frac{d\widehat{u}}{dt} = -(ik)^4\widehat{u} = -k^4\widehat{u} \tag{10.4.2}$$

which has the solution

$$\widehat{u} = \widehat{u}_0 \exp(-k^4 t). \tag{10.4.3}$$

However, a time-stepping scheme would obviously solve (10.4.2) directly without making use of the analytic solution.

For $n = 128$ Fourier modes, the wavenumber k ranges in values from $k \in [-64, 64]$. Thus the largest value of k for the hyper-diffusion equation is

$$k_{\text{max}}^4 = (64)^4 = 16,777,216, \tag{10.4.4}$$

or roughly $k_{\text{max}}^4 = 1.7 \times 10^7$. For $n = 1024$ Fourier modes, this is

$$k_{\text{max}}^4 = 6.8 \times 10^{10}. \tag{10.4.5}$$

Thus even if our solution is small at the high wavenumbers, this can create problems. For instance, if the solution at high wavenumbers is $O(10^{-6})$, then

$$\frac{d\widehat{u}}{dt} = -(10^{10})(10^{-6}) = -10^4. \tag{10.4.6}$$

Such large numbers on the right-hand side of the equations force time-stepping schemes like *ode23* and *ode45* to take much smaller time-steps in order to maintain tolerance constraints. This is a form of numerical stiffness which can be circumvented.

Filtered pseudo-spectral

There are a couple of ways to help get around the above mentioned numerical stiffness. We again consider the very general partial differential equation

$$\frac{\partial u}{\partial t} = Lu + N(u) \tag{10.4.7}$$

where as before L is a linear, constant coefficient operator, i.e. it can take the form $L = ad^2/dx^2 + bd/dx + c$ where $a, b,$ and c are constants. The second term $N(u)$ includes the nonlinear and nonconstant coefficient terms. An example of this would be $N(u) = u^3 + f(x)u + g(x)d^2u/dx^2$.

Previously, we transformed the equation by Fourier transforming and constructing a system of differential equations. However, there is a better way to handle this general equation and remove some numerical stiffness at the same time. To introduce the technique, we consider the first-order differential equation

$$\frac{dy}{dt} + p(t)y = g(t). \tag{10.4.8}$$

We can multiply by the integrating factor $\mu(t)$ so that

$$\mu\frac{dy}{dt} + \mu p(t)y = \mu g(t). \tag{10.4.9}$$

We note from the chain rule that $(\mu y)' = \mu'y + \mu y'$ where the prime denotes differentiation with respect to t. Thus the differential equation becomes

$$\frac{d}{dt}(\mu y) = \mu g(t), \tag{10.4.10}$$

provided $d\mu/dt = \mu p(t)$. The solution then becomes

$$y = \frac{1}{\mu}\left[\int \mu(t)g(t)dt + c\right] \quad \mu(t) = \exp\left(\int p(t)dt\right) \tag{10.4.11}$$

which is the standard integrating factor method of a first course on differential equations.

   We use the key ideas of the integrating factor method to help solve the general partial differential equation and remove stiffness. Fourier transforming (10.4.7) results in the spectral system of differential equations

$$\frac{d\widehat{u}}{dt} = \alpha(k)\widehat{u} + \widehat{N(u)}. \tag{10.4.12}$$

This can be rewritten

$$\frac{d\widehat{u}}{dt} - \alpha(k)\widehat{u} = \widehat{N(u)}. \tag{10.4.13}$$

Multiplying by $\exp(-\alpha(k)t)$ gives

$$\frac{d\widehat{u}}{dt}\exp(-\alpha(k)t) - \alpha(k)\widehat{u}\exp(-\alpha(k)t) = \exp(-\alpha(k)t)\widehat{N(u)}$$

$$\frac{d}{dt}[\widehat{u}\exp(-\alpha(k)t)] = \exp(-\alpha(k)t)\widehat{N(u)}.$$

By defining $\widehat{v} = \widehat{u}\exp(-\alpha(k)t)$, the system of equations reduces to

$$\frac{d\widehat{v}}{dt} = \exp(-\alpha(k)t)\widehat{N(u)} \tag{10.4.14a}$$

$$\widehat{u} = \widehat{v}\exp(\alpha(k)t). \tag{10.4.14b}$$

Thus the linear, constant coefficient terms are solved for explicitly, and the numerical stiffness associated with the Lu term is effectively eliminated.

Example: Fisher–Kolmogorov equation

As an example of the implementation of the filtered pseudo-spectral scheme, we consider the Fisher–Kolmogorov equation

$$\frac{\partial u}{\partial t} = \frac{\partial^2 u}{\partial x^2} + u^3 + cu. \tag{10.4.15}$$

Fourier transforming the equation yields

$$\frac{d\widehat{u}}{dt} = (ik)^2\widehat{u} + \widehat{u^3} + \widehat{cu} \tag{10.4.16}$$

which can be rewritten

$$\frac{d\widehat{u}}{dt} + (k^2 - c)\widehat{u} = \widehat{u^3}. \tag{10.4.17}$$

Thus $\alpha(k) = c - k^2$ and

$$\frac{d\widehat{v}}{dt} = \exp[-(c - k^2)t]\widehat{u^3} \tag{10.4.18a}$$

$$\widehat{u} = \widehat{v}\exp[(c - k^2)t]. \tag{10.4.18b}$$

It should be noted that the solutions u must continually be updating the value of u^3 which is being transformed in the right-hand side of the equations. Thus after every time-step Δt, the new u should be used to evaluate $\widehat{u^3}$.

There are a few practical issues that should be considered when implementing this technique:

- When solving the general equation

$$\frac{\partial u}{\partial t} = Lu + N(u) \tag{10.4.19}$$

  it is important to only step forward Δt in time with

$$\frac{d\widehat{v}}{dt} = \exp(-\alpha(k)t)\widehat{N(u)}$$
$$\widehat{u} = \widehat{v}\exp(\alpha(k)t)$$

  before the nonlinear term $\widehat{N(u)}$ is updated.

- The computational saving for this method generally does not manifest itself unless there are more than two spatial derivatives in the highest derivative of Lu.

- Care must be taken in handling your time-step Δt in MATLAB since it uses adaptive time-stepping.

Comparison of Spectral and Finite Difference Methods

Before closing the discussion on the spectral method, we investigate the advantages and disadvantages associated with the spectral and finite difference schemes. Of particular interest are the issues of accuracy, implementation, computational efficiency and boundary conditions. The strengths and weaknesses of the schemes will be discussed.

A. **Accuracy**
   - **Finite differences:** Accuracy is determined by the Δx and Δy chosen in the discretization. Accuracy is fairly easy to compute and is generally much worse than spectral methods.

   - **Spectral method:** Spectral methods rely on a global expansion and are often called *spectrally accurate*. In particular, spectral methods have *infinite order accuracy*. Although the details of what this means will not be discussed here, it will suffice to say that they are generally of much higher accuracy than finite differences.

B. **Implementation**
   - **Finite differences:** The greatest difficulty in implementing the finite difference schemes is generating the correct sparse matrices. Many of these matrices are very complicated with higher order schemes and in higher dimensions. Further, when solving the resulting system $\mathbf{A}\mathbf{x} = \mathbf{b}$, it should always be checked whether $\det \mathbf{A} = 0$. The MATLAB command *cond(A)* checks the condition number of the matrix. If $cond(A) > 10^{15}$, then $\det \mathbf{A} \approx 0$ and steps must be taken in order to solve the problem correctly.

   - **Spectral method:** The difficulty with using FFTs is the continual switching between the time or space domain and the spectral domain. Thus it is imperative to know exactly when and where in the algorithm this switching must take place.

C. **Computational efficiency**
   - **Finite differences:** The computational time for finite differences is determined by the size of the matrices and vectors in solving $\mathbf{A}\mathbf{x} = \mathbf{b}$. Generally speaking, you can guarantee $O(N^2)$ efficiency by using LU decomposition. At times, iterative schemes can lower the operation count, but there are no guarantees about this.

   - **Spectral method:** The FFT algorithm is an $O(N \log N)$ operation. Thus it is almost always guaranteed to be faster than the finite difference solution method which is $O(N^2)$. Recall that this efficiency improvement comes with an increased accuracy as well, thus making the spectral method highly advantageous when implemented.

D. **Boundary conditions**
   - **Finite differences:** Of the above categories, spectral methods are generally better in every regard. However, finite differences are clearly superior when considering boundary conditions. Implementing the generic boundary conditions

$$\alpha u(L) + \beta \frac{du(L)}{dx} = \gamma \qquad (10.4.20)$$

is easily done in the finite difference framework. Also, more complicated computational domains may be considered. Generally any computational domain which can be constructed of rectangles is easily handled by finite difference methods.

- **Spectral method:** Boundary conditions are the critical limitation on using the FFT method. Specifically, only periodic boundary conditions can be considered. The use of the discrete sine or cosine transform allows for the consideration of pinned or no-flux boundary conditions, but only odd or even solutions are admitted, respectively.

10.5   Boundary Conditions and the Chebychev Transform

Thus far, we have focused on the use of the FFT as the primary tool for spectral methods and their implementation. However, many problems do not in fact have periodic boundary conditions and the accuracy and speed of the FFT is rendered useless. The Chebychev polynomials are a set of mathematical functions which still allow for the construction of a spectral method which is both fast and accurate. The underlying concept is that Chebychev polynomials can be related to sines and cosines, and therefore they can be connected to the FFT routine.

Before constructing the details of the Chebychev method, we begin by considering three methods for handling nonperiodic boundary conditions.

Method 1: Periodic extension with FFTs

Since the FFT only handles periodic boundary conditions, we can periodically extend a general function $f(x)$ in order to make the function itself now periodic. The FFT routine can now be used. However, the periodic extension will in general generate discontinuities in the periodically extended function. The discontinuities give rise to Gibb's phenomenon: strong oscillations and errors are accumulated at the jump locations. This will greatly affect the accuracy of the scheme. So although spectral accuracy and speed is retained away from discontinuities, the errors and the jumps will begin to propagate out to the rest of the computational domain.

To see this phenomenon, Fig. 10.5(a) considers a function which is periodically extended as shown in Fig. 10.5(b). The FFT approximation to this function is shown in Fig. 10.5(c) where the Gibb's oscillations are clearly seen at the jump locations. The oscillations clearly impact the usefulness of this periodic extension technique.

Method 2: Polynomial approximation with equi-spaced points

In moving away from an FFT basis expansion method, we can consider the most straightforward method available: polynomial approximation. Thus we simply discretize the given function $f(x)$ with $N + 1$ equally spaced points and fit an Nth degree polynomial through it. This amounts to letting

$$f(x) \approx a_0 + a_1 x + a_2 x^2 + a_3 x^3 + \cdots + a_n x^N \qquad (10.5.1)$$

where the coefficients a_n are determined by an $(N + 1) \times (N + 1)$ system of equations. This method easily satisfies the prescribed nonperiodic boundary conditions. Further, differentiation

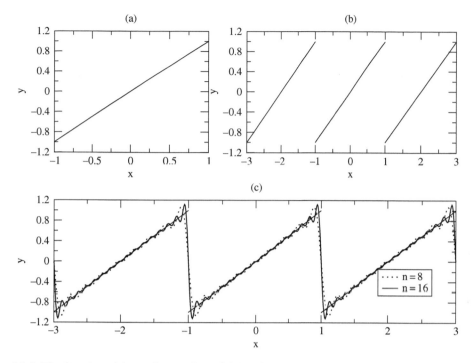

Figure 10.5: The function $y(x) = x$ for $x \in [-1, 1]$ (a) and its periodic extension (b). The FFT approximation is shown in (c) for $n = 8$ Fourier modes and $n = 16$ Fourier modes. Note the Gibb's oscillations.

of such an approximation is trivial. In this case, however, Runge phenomena (polynomial oscillations) generally occur. This is because a polynomial of degree n generally has $N - 1$ combined maxima and minima.

The Runge phenomena can easily be illustrated with the simple example function $f(x) = (1 + 16x^2)^{-1}$. Figure 10.6(a)–(b) illustrates the large oscillations which develop near the boundaries due to Runge phenomena for $N = 12$ and $N = 24$. As with the Gibb's oscillations generated by FFTs, the Runge oscillations render a simple polynomial approximation based upon equally spaced points useless.

Method 3: Polynomial approximation with clustered points

There exists a modification to the straightforward polynomial approximation given above. Specifically, this modification constructs a polynomial approximation on a clustered grid as opposed to the equal spacing which led to Runge phenomena. This polynomial approximation involves a clustered grid which is transformed to fit onto the unit circle, i.e. the Chebychev points. Thus we have the transformation

$$x_n = \cos(n\pi/N) \tag{10.5.2}$$

where $n = 0, 1, 2, \ldots, N$. This helps to greatly reduce the effects of the Runge phenomena.

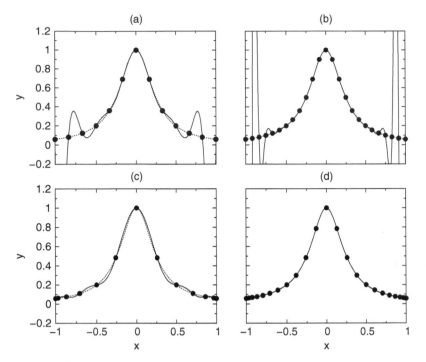

Figure 10.6: The function $y(x) = (1 + 16x^2)^{-1}$ for $x \in [-1, 1]$ (dotted line) with the bold points indicating the grid points for equi-spaced points (a) $n = 12$ and (b) $n = 24$ and Chebychev clustered points (c) $n = 12$ and (d) $n = 24$. The solid lines are the polynomial approximations generated using the equi-spaced points (a)–(b) and clustered points (c)–(d), respectively.

The clustered grid approximation results in a polynomial approximation shown in Fig. 10.6(c)–(d). There are reasons for this great improvement using a clustered grid [18]. However, we will not discuss them since they are beyond the scope of this book. This clustered grid suggests an accurate and easy way to represent a function which does not have periodic boundaries.

Clustered points and Chebychev differentiation

The clustered grid given in method 3 above is on the Chebychev points. The resulting algorithm for constructing the polynomial approximation and differentiating it is as follows.

1   Let p be a unique polynomial of degree $\leq N$ with $p(x_n) = V_n$, $0 \leq n \leq N$, where $V(x)$ is the function we are approximating.

2   Differentiate by setting $w_n = p'(x_n)$.

The second step in the process of calculating the derivative is essentially the matrix multiplication

$$\mathbf{w} = \mathbf{D}_N \mathbf{v} \tag{10.5.3}$$

where $\mathbf{D}_N$ represents the action of differentiation. By using interpolation of the Lagrange form [11], the matrix elements of $p(x)$ can be constructed along with the $(N+1) \times (N+1)$ matrix $\mathbf{D}_N$. This results in each matrix element $(D_N)_{ij}$ being given by

$$(D_N)_{00} = \frac{2N^2 + 1}{6} \tag{10.5.4a}$$

$$(D_N)_{NN} = -\frac{2N^2 + 1}{6} \tag{10.5.4b}$$

$$(D_N)_{jj} = -\frac{x_j}{2(1 - x_j^2)} \quad j = 1, 2, \ldots, N-1 \tag{10.5.4c}$$

$$(D_N)_{ij} = \frac{c_i(-1)^{i+j}}{c_j(x_i - x_j)} \quad i, j = 0, 1, \ldots, N \; (i \neq j) \tag{10.5.4d}$$

where the parameter $c_j = 2$ for $j = 0$ or N or $c_j = 1$ otherwise.

Calculating the individual matrix elements results in the matrix $\mathbf{D}_N$

$$\mathbf{D}_N = \begin{pmatrix} \frac{2N^2+1}{6} & \frac{2(-1)^j}{1-x_j} & \frac{(-1)^N}{2} \\ \frac{-(-1)^i}{2(1-x_i)} & \ddots & \frac{(-1)^{i+j}}{x_i-x_j} & \frac{(-1)^{N+i}}{2(1+x_j)} \\ & \frac{-x_j}{2(1-x_j^2)} & \\ \frac{(-1)^{i+j}}{x_i-x_j} & \ddots \\ \frac{-(-1)^N}{2} & \frac{-2(-1)^{N+j}}{1+x_j} & -\frac{2N^2+1}{6} \end{pmatrix} \tag{10.5.5}$$

which is a full matrix, i.e. it is not sparse. To calculate second, third, fourth and higher derivatives, simply raise the matrix to the appropriate power:

- $\mathbf{D}_N^2$ – second derivative
- $\mathbf{D}_N^3$ – third derivative
- $\mathbf{D}_N^4$ – fourth derivative
- $\mathbf{D}_N^m$ – mth derivative.

Boundaries

The construction of the differentiation matrix $\mathbf{D}_N$ does not explicitly include the boundary conditions. Thus the general differentiation given by (10.5.3) must be modified to include the given boundary conditions. Consider, for example, the simple boundary conditions

$$v(-1) = v(1) = 0. \tag{10.5.6}$$

The given differentiation matrix is then written as

$$
\begin{pmatrix} w_0 \\ w_1 \\ \vdots \\ w_{N-1} \\ w_N \end{pmatrix} = \begin{pmatrix} & \text{top row} & \\ \hline \text{first} & & \text{last} \\ \text{column} & (D_N) & \text{column} \\ \hline & \text{bottom row} & \end{pmatrix} \begin{pmatrix} v_0 \\ v_1 \\ \vdots \\ v_{N-1} \\ v_N \end{pmatrix}.
\tag{10.5.7}
$$

In order to satisfy the boundary conditions, we must manually set $v_0 = v_N = 0$. Thus only the interior points in the differentiation matrix are relevant. Note that for more general boundary conditions $v(-1) = \alpha$ and $v(1) = \beta$, we would simply set $v_0 = \alpha$ and $v_N = \beta$. The remaining $(N-1) \times (N-1)$ system is given by

$$
\tilde{\mathbf{w}} = \tilde{\mathbf{D}}_N \tilde{\mathbf{v}}
\tag{10.5.8}
$$

where $\tilde{\mathbf{w}} = (w_1 w_2 \cdots w_{N-1})^T$, $\tilde{\mathbf{v}} = (v_1 v_2 \cdots v_{N-1})^T$ and we construct the matrix $\tilde{\mathbf{D}}_N$ with the simple MATLAB command:

```
tildeD=D(2:N,2:N)
```

Note that the new matrix created is the old $\mathbf{D}_N$ matrix with the top and bottom rows and the first and last columns removed.

Connecting to the FFT

We have already discussed the connection of the Chebychev polynomial with the FFT algorithm. Thus we can connect the differentiation matrix with the FFT routine. After transforming via (10.5.2), then for real data the discrete Fourier transform can be used. For complex data, the regular FFT is used. Note that for the Chebychev polynomials

$$
\frac{\partial T_n(\pm 1)}{\partial x} = 0
\tag{10.5.9}
$$

so that no-flux boundaries are already satisfied. To impose pinned boundary conditions $v(\pm 1) = 0$, then the differentiation matrix must be imposed as shown above.

 ## 10.6   Implementing the Chebychev Transform

In order to make use of the Chebychev polynomials, we must generate our given function on the clustered grid given by

$$
x_j = \cos(j\pi/N) \quad j = 0, 1, 2, \ldots, N.
\tag{10.6.1}
$$

This clustering will give higher resolution near the boundaries than the interior of the computational domain. The Chebychev differentiation matrix

$$\mathbf{D}_N \tag{10.6.2}$$

can then be constructed. Recall that the elements of this matrix are given by

$$(D_N)_{00} = \frac{2N^2 + 1}{6} \tag{10.6.3a}$$

$$(D_N)_{NN} = -\frac{2N^2 + 1}{6} \tag{10.6.3b}$$

$$(D_N)_{jj} = -\frac{x_j}{2(1 - x_j^2)} \quad j = 1, 2, \ldots, N - 1 \tag{10.6.3c}$$

$$(D_N)_{ij} = \frac{c_i(-1)^{i+j}}{c_j(x_i - x_j)} \quad i, j = 0, 1, \ldots, N \ (i \neq j) \tag{10.6.3d}$$

where the parameter $c_j = 2$ for $j = 0$ or N or $c_j = 1$ otherwise. Thus given the number of discretization points N, we can build the matrix $\mathbf{D}_N$ and also the associated clustered grid. The following MATLAB code simply required the number of points N to generate both of these fundamental quantities. Recall that it is assumed that the computational domain has been scaled to $x \in [-1, 1]$.

```
% from Trefethen, Spectral methods in MATLAB (SIAM, 2000)
% cheb.m - compute the matrix D_N
function [D,x]=cheb(N)
if N==0, D=0; x=1; return; end
x=cos(pi*(0:N)/N)';
c=[2; ones(N-1,1); 2].*(-1).^(0:N)';
X=repmat(x,1,N+1);
dX=X-X';
D=(c*(1./c)')./(dX+eye(N+1));   % off diagonals
D=D-diag(sum(D'));              % diagonals
```

To test the differentiation, we consider two functions for which we know the exact values for the derivative and second derivative. Consider then

```
x=[-1:0.01:1]
u=exp(x).*sin(5*x);
v=sech(x);
```

The first and second derivative of each of these functions is given by

```
ux=exp(x).*sin(5*x) + 5*exp(x).*cos(5*x);
uxx=-24*exp(x).*sin(5*x) + 10*exp(x).*cos(5*x);

vx=-sech(x).*tanh(x);
vxx=sech(x)-2*sech(x).^3;
```

We can also use the Chebychev differentiation matrix to numerically calculate the values of the first and second derivatives. All that is required is the number of discretation points N and the routine *cheb.m*.

```
N=20
[D,x2]=cheb(N) % x2-clustered grid, D-differentiation matrix
D2=D^2;        % D2 - second derivative matrix
```

Given the differentiation matrix $\mathbf{D}_N$ and clustered grid, the given function and its derivatives can be constructed numerically.

```
u2=exp(x2).*sin(5*x2);
v2=sech(x2);

u2x=D*u2;    % first derivatives
v2x=D*v2;

u2xx=D2*u2;  % second derivatives
v2xx=D2*v2;
```

A comparison between the exact values for the differentiated functions and their approximations is shown in Fig. 10.7. As is expected the agreement is best for higher values of N. The MATLAB code for generating these graphical figures is given by

```
figure 1; plot(x,u,'r',x2,u2,'.',x,v,'g',x2,v2,'.');
figure 2; plot(x,ux,'r',x2,u2x,'.',x,vx,'g',x2,v2x,'.'):
figure 3; plot(x,uxx,'r',x2,u2xx,'.',x,vxx,'g',x2,v2xx,'.');
```

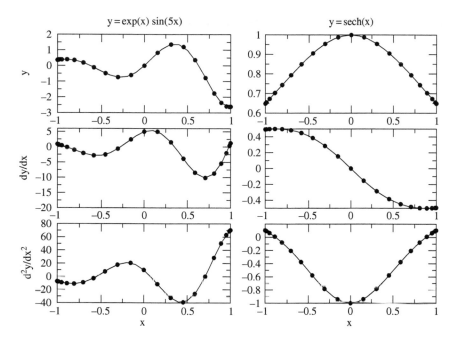

Figure 10.7: The function $y(x) = \exp(x) \sin 5x$ (left) and $y(x) = \operatorname{sech} x$ (right) for $x \in [-1, 1]$ and their first and second derivatives. The dots indicate the numerical values while the solid line is the exact solution. For these calculations, $N = 20$ in the differentiation matrix $\mathbf{D}_N$.

Differentiation matrix in 2D

Unlike finite difference schemes which result in sparse differentiation matrices, the Chebychev differentiation matrix is full. The construction of 2D differentiation matrices thus would seem to be a complicated matter. However, the use of the *kron* command in MATLAB makes this calculation trivial. In particular, the 2D Laplacian operator L given by

$$Lu = \frac{\partial^2 u}{\partial x^2} + \frac{\partial^2 u}{\partial y^2} \tag{10.6.4}$$

can be constructed with

```
L = kron(I,D2) + kron(D2,I)
```

where $D2 = D^2$ and I is the identity matrix of size $N \times N$. The $D2$ in each slot takes the x and y derivatives, respectively.

Solving PDEs with the Chebychev differentiation matrix

To illustrate the use of the Chebychev differentiation matrix, the two-dimensional heat equation is considered:

$$\frac{\partial u}{\partial t} = \nabla^2 u = \frac{\partial^2 u}{\partial x^2} + \frac{\partial^2 u}{\partial y^2}$$

(10.6.5)

on the domain $x, y \in [-1, 1]$ with the Dirichlet boundary conditions $u = 0$.

The number of Chebychev points is first chosen and the differentiation matrix constructed in one dimension:

```
clear all; close all;
N=30;
[D,x]=cheb(N);
D(N+1,:)=zeros(1,N+1);
D(1,:)=zeros(1,N+1);
D2=D^2; y=x;
```

The Neumann boundary conditions were imposed by modifying the first and last rows of the differentiation matrix D. Specifically, those rows were set to zero.

The two-dimensional Laplacian is then constructed from the one-dimensional differentiation matrix by using the *kron* command

```
I=eye(length(D2));
L=kron(I,D2)+kron(D2,I);   % 2D Laplacian
```

The initial conditions for this simulation will be a Gaussian centered at the origin.

```
[X,Y]=meshgrid(x,y);
U=exp(-(X.^2+Y.^2)/0.1);
surfl(x,y,U), shading interp, colormap(gray), drawnow
```

Note that the vectors x and y in the *meshgrid* command correspond to the Chebychev points for a given N. The final step is to build a loop which will advance and plot the solution in time. This requires the use of the *reshape* command, since we are in two dimensions, and an ODE solver such as *ode23*.

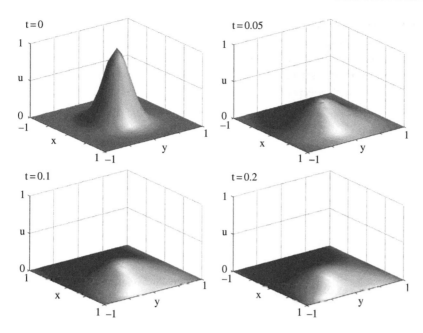

Figure 10.8: Evolution of an initial two-dimensional Gaussian governed by the heat equation on the domain $x, y \in [-1, 1]$ with Dirichlet boundary conditions $u = 0$. The Chebychev differentiation matrix is used to calculate the Laplacian with $N = 30$.

```
u=reshape(U,(N+1)^2,1);
for j=1:4
  [t,ysol]=ode23('heatrhs2D',[0 0.05],u,[],L);
  u=ysol(end,:);
  U=reshape(u,N+1,N+1);
  surfl(x,y,U), shading interp, colormap(gray), drawnow
end
```

This code will advance the solution $\Delta t = 0.05$ in the future and plot the solution. Figure 10.8 depicts the two-dimensional diffusive behavior given the Dirichlet boundary conditions using the Chebychev differentiation matrix.

 10.7 **Computing Spectra: The Floquet–Fourier–Hill Method**

Spectral transforms, such as Fourier and Chebychev, have many advantages including their performance speed of $O(N \log N)$ and their spectral accuracy properties. Given these advantages, it is desirable to see where else such methods can play a role in practice. One area that we now return to is the idea of computing spectra of linear operators. This was introduced via finite difference discretization in Section 7.8. But just as finite difference discretization of PDEs can be replaced in certain cases by spectral discretization of PDEs, so can the computation of spectra of linearized operators be replaced with spectral based methods [20].

The methodology developed here will be applied to finding spectra (all eigenvalues) of the eigenvalue problem

$$\mathcal{L}v = \lambda v \tag{10.7.1}$$

where the linear operator is assumed to be of the form

$$\mathcal{L} = \sum_{k=0}^{M} f_k(x)\partial_x^k = f_0(x) + f_1(x)\partial_x + \cdots + f_M(x)\partial_x^M \tag{10.7.2}$$

with the periodic constraint $f_k(x + L) = f_k(x)$. Thus there are restrictions on how broadly the technique can be applied. Generically, it is for periodic problems only. However, for problems of infinite extent where the solutions $v(\pm\infty) \to 0$, the method is often successful when considering a large, but finite sized domain, i.e. the periodic boundary conditions suffice to approximate the decaying to zero solutions at the boundary. The Floquet–Fourier–Hill method can also be generalized to a vector version [20].

The numerical technique developed here is based upon ideas of Fourier analysis and Floquet theory. George Hill first used variations of this technique in 1886 [21] on what is now called Hill's equation. One of the great open questions of his day concerned the stability of orbits in planetary and lunar motion. As such, the method will be called the Floquet–Fourier–Hill (FFH) method [20]. The method allows for the efficient computation of the spectra of a linear operator by exploiting numerous advantages over finite difference schemes, including accuracy and speed.

The coefficients of the operator $\mathcal{L}$ in (10.7.2) are periodic with period L, thus they can be represented by a Fourier series expansion

$$f_k(x) = \sum_{j=-\infty}^{\infty} \hat{f}_{k,j} \exp(i2\pi jx/L), \qquad k = 0, \cdots, M \tag{10.7.3}$$

where the Fourier coefficients are determined from the inner product integral

$$\hat{f}_{k,j} = \frac{1}{L} \int_{-L/2}^{L/2} f_k(x) \exp(-i2\pi jx/L)dx, \qquad k = 0, \cdots, M. \tag{10.7.4}$$

Recall that the variable i again represents the imaginary unit. Thus far, this is standard Fourier theory. In fact, the FFT algorithm computes such Fourier coefficients in $O(N \log N)$ time.

Floquet theory dictates that every bounded solution of the fundamental equation (10.7.1) is of the form [22, 23]

$$w(x) = \exp(i\mu x)\phi(x) \tag{10.7.5}$$

with $\phi(x + L) = \phi(x)$ for any fixed λ and $\mu \in [0, 2\pi/L]$. The factor $\exp(i\mu x)$ is referred to as a Floquet multiplier and $i\mu$ is the Floquet exponent [22, 23]. In different areas of science, Floquet theory may be known as monodromy theory (Hamiltonian systems, etc.) or Bloch theory (solid state theory, etc). In the context of Bloch theory the Floquet exponent is often referred to as the quasi-momentum.

Since the function $\phi(x)$ is also periodic with a period L, it can also be expanded in a Fourier series so that

$$w(x) = \exp(i\mu x) \sum_{j=-\infty}^{\infty} \hat{\phi}_j \exp(i2\pi jx/L) = \sum_{j=-\infty}^{\infty} \hat{\phi}_j \exp(ix[\mu + 2\pi j/L]) \tag{10.7.6}$$

where

$$\hat{\phi}_j = \frac{1}{L} \int_{-L/2}^{L/2} \phi(x) \exp(-i2\pi jx/L)dx \tag{10.7.7}$$

is the jth Fourier coefficient of $\phi(x)$. Upon multiplying (10.7.1) by $\exp(-i\mu x)$ and noting that any term of the resulting equation is periodic, the nth Fourier coefficient of the eigenvalue equation becomes, after some manipulation and using orthogonality [20],

$$\sum_{m=-\infty}^{\infty} \left(\sum_{k=0}^{M} \hat{f}_{k,n-m} \left[i \left(\mu + \frac{2\pi m}{L} \right) \right]^k \right) \hat{\phi}_m = \lambda \hat{\phi}_m. \tag{10.7.8}$$

Thus the original eigenvalue problem has been transformed into the Fourier domain and the eigenvalue problem

$$\hat{\mathcal{L}}(\mu)\hat{\phi} = \lambda \hat{\phi} \tag{10.7.9}$$

with $\hat{\phi} = (\cdots, \hat{\phi}_{-2}, \hat{\phi}_{-1}, \hat{\phi}_0, \hat{\phi}_1, \hat{\phi}_2, \cdots)^T$ and where the μ-dependence of $\hat{\mathcal{L}}$ is explicitly indicated. This is a bi-infinite matrix with the linear operator in the spectral domain defined by

$$\hat{\mathcal{L}}(\mu)_{nm} = \sum_{k=0}^{M} \hat{f}_{k,n-m} \left[i \left(\mu + \frac{2\pi m}{L} \right) \right]^k. \tag{10.7.10}$$

Note that to this point, *no approximations were used*. The problem was simply transformed to the Fourier domain. To compute the eigenvalue spectra, the bi-infinite matrix (10.7.9) is approximated by limiting the number of Fourier modes. Specifically, a cut-off wavenumber, N, is chosen which results in a matrix system of size $2N + 1$:

$$\hat{\mathcal{L}}_N(\mu)\hat{\phi}_N = \lambda_N \hat{\phi}_N \tag{10.7.11}$$

where the λ_N are now numerical approximations to the true eigenvalues λ. Convergence of this scheme is considered further in Ref. [20]. Note that the **eigs** command still plays the underlying and fundamental role, but now for a transformed matrix and vector space.

Example: Schrödinger operators

To implement this scheme, an example is considered arising from Schrödinger operators. Specifically, the following two operators are considered:

$$\mathcal{L}_-v = -\frac{d^2v}{dx^2} + (1 - 2\operatorname{sech}^2 x)v = \lambda v \tag{10.7.12a}$$

$$\mathcal{L}_+v = -\frac{d^2v}{dx^2} + (1 - 6\operatorname{sech}^2 x)v = \lambda v. \tag{10.7.12b}$$

These operators are chosen since their spectrum is known completely. In particular, the operator $\mathcal{L}_-$ has a discrete eigenvalue at zero with eigenfunction $v_{\lambda=0} = \operatorname{sech} x$. It also has a continuum of eigenvalues for $\lambda \in [1, \infty)$. The operator $\mathcal{L}_+$ has two discrete eigenvalues at $\lambda = -3$ and $\lambda = 0$ with eigenfunctions $v_{\lambda=-3} = \operatorname{sech}^2 x$ and $v_{\lambda=0} = \operatorname{sech} x \tanh x$. It also has a continuum of eigenvalues for $\lambda \in [1, \infty)$.

For these two examples, we have the following relations, respectively, back to (10.7.2)

$$f_0(x) = 1 - 2\operatorname{sech}^2 x, \quad f_1(x) = 0, \quad f_2(x) = -1 \tag{10.7.13a}$$

$$f_0(x) = 1 - 6\operatorname{sech}^2 x, \quad f_1(x) = 0, \quad f_2(x) = -1. \tag{10.7.13b}$$

The only Fourier expansion that needs to be done then is for the term $\operatorname{sech}^2 x$. The following MATLAB code constructs the FFH eigenvalue problem.

```
L=60;  % box size  -L,L
n=200;  % modes
TOL=10^(-8);  % For integration

for j=1:n+1  % compute Fourier coefficients
  a(j)=quad(inline('(1/2/L)*cos((j-1)*pi*x/L).*sech(x).^2', ...
      'x','j','L'),-L,L,TOL,[],j,L);
end


A=zeros(n+1);  % Make matrix of coefficients
for j=1:n+1
   A(j,j)=a(1);
end
for jj=2:n+1
   for j=jj:n+1
      A(j+1-jj,j)=a(jj);
      A(j,j+1-jj)=a(jj);
   end
end
```

Continued

```
D2=zeros(n+1);   % include derivative and constant
for j=1:n+1            %   -1 + d^2/dx^2
   D2(j,j)=-1-((pi/L)^2)*(n/2+1-j)^2;
end


Lplus=-(D2+6*A);   % L+
[V,W]=eig(Lplus);    % eigenvalues/eigenvectors

Lminus=-(D2+2*A);   % L-
[V2,W2]=eig(Lminus); % eigenvalues/eigenvectors
```

In this example where the operator is of the Sturm–Liouville form, the inner product only needs to be computed with respect to $\cos(n\pi x/L)$ versus $\exp(in\pi x/L)$. Figure 10.9 shows the computed spectrum of both the $\mathcal{L}_+$ and $\mathcal{L}_-$ operators. The eigenfunctions are also shown for the three discrete eigenvalues, two for $\mathcal{L}_+$ and one for $\mathcal{L}_-$, respectively. The computation uses $\Delta x = 0.6$

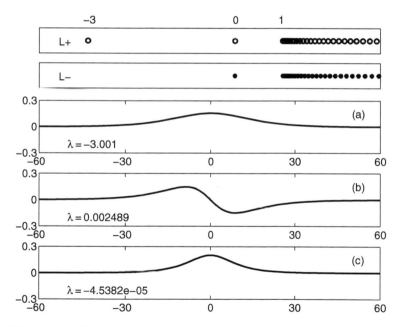

Figure 10.9: The top panels are the spectra of the operators $\mathcal{L}_+$ and $\mathcal{L}_-$. The operator $\mathcal{L}_+$ is known to have two discrete eigenvalues at $\lambda = -3$ and $\lambda = 0$ with eigenfunctions $v_{\lambda=-3} = \text{sech}^2 x$ and $v_{\lambda=0} = \text{sech}\, x \tanh x$. It also has a continuum of eigenvalues for $\lambda \in [1, \infty)$. Panels (a) and (b) demonstrate these two eigenfunctions computed numerically along with the approximate evaluation of the eigenvalue. The operator $\mathcal{L}_-$ has a discrete eigenvalue at zero with eigenfunction $v_{\lambda=0} = \text{sech}\, x$. It also has a continuum of eigenvalues for $\lambda \in [1, \infty)$. Its eigenfunction and computed eigenvalue is demonstrated in panel (c).

which would only give an accuracy of 10^{-1}. However, the computation with FFH is accurate to 10^{-3} or more. Note that using the **quad** command is a highly inefficient way to compute the Fourier coefficients. Using the FFT algorithm is much faster and ultimately the way to do the computation. However, the above does illustrate the FFH process in its entirety.

Example: Mathieu's equation

As a second example of computing spectra in a highly efficient way, consider the Mathieu equation of mathematical physics

$$\frac{d^2 v}{dx^2} + \left[\lambda - 2q\cos(\omega x)\right] v = 0. \tag{10.7.14}$$

This gives the linear operator $\mathcal{L} = -d^2/dx^2 + 2q\cos(\omega x)$. The equation originates from the Helmholtz equation through the use of separation of variables. Here we are interested in determining all λ values for which bounded solutions exist. Many texts on perturbation methods use this equation as one of their prototypical examples. One of their goals is to determine the edges of the spectrum for varying q.

In this case, we have the following relations back to (10.7.2)

$$f_0(x) = 2q\cos(\omega x), \quad f_1(x) = 0, \quad f_2(x) = -1. \tag{10.7.15}$$

Thus a fairly simple structure exists with $f_0(x)$ already being represented as Fourier modes if we choose ω to be an integer. In what follows, ω is chosen to be an integer and the eigenvalues are found as a function of the variable q.

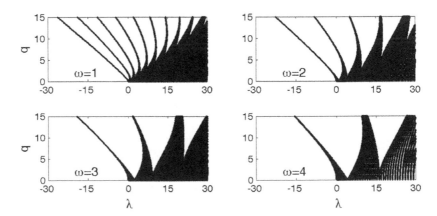

Figure 10.10: Spectrum of Mathieu's equation for four different values of ω. Every black point is an approximate point of the spectrum and no filling-in of spectral regions was done. Standard perturbation theory is typically only able to approximate the spectrum near $q = 0$.

```
for q=0:0.1:15  %  q values to consider

n=10;  %  number of Fourier modes
cuts=40; % number of mu slices
freq=2;  % frequency omega
lam=[];
for jcut=-freq/2:freq/cuts:freq/2
    A=zeros(2*n+1);
    for j=1:2*n+1
        A(j,j)=(n+1-j+jcut)^2;  % derivative elements
    end
    for j=1:2*n+1-freq
        A(j+freq,j)=q;          %  off-diagonals with strength q
        A(j,j+freq)=q;          %
    end
    [V,W]=eig(A);               %  compute eigenvalues
    lam=[lam; diag(W)];  %  track all eigenvalues
  end
  plot(real(lam),imag(lam)+q,'k.'), hold on, axis([-30 30 0 15])

end
```

Unlike the previous example, no Fourier mode projection needs to be done since $f_0(x) = 2q \cos(\omega x)$ is already in Fourier mode form. Thus only the coefficient $2q$ is needed at frequency ω. Figure 10.10 shows the resulting band-gap structure of eigenvalues for four different values of ω. All points in the plot are computed points and no filling-in of the spectrum was done.

To see how to generalize the FFH method to a vector model and/or higher dimensions, see Ref. [20]. In general, one should think about this spectral based FFH method as simply taking advantage of spectral accuracy and speed. Just as PDE based solvers aim to take advantage of these features for accurately and quickly solving a given problem, the FFH should always be used when either an infinite domain (with decaying boundary conditions) or periodic solutions are of interest.

Finite Element Methods

The numerical methods considered thus far, i.e. finite difference and spectral, are powerful tools for solving a wide variety of physical problems. However, neither method is well suited for complicated boundary configurations or complicated boundary conditions associated with such domains. The finite element method is ideally suited for complex boundaries and very general boundary conditions. This method, although perhaps not as fast as finite difference and spectral, is instrumental in solving a wide range of problems which is beyond the grasp of other techniques.

11.1 Finite Element Basis

The finite difference method approximates a solution by discretizing and solving the matrix problem $\mathbf{Ax} = \mathbf{b}$. Spectral methods use the FFT to expand the solution in global sine and consine basis functions. Finite elements are another form of finite difference in which the approximation to the solution is made by interpolating over patches of our solution region. Ultimately, to find the solution we will once again solve a matrix problem $\mathbf{Ax} = \mathbf{b}$. Five essential steps are required in the finite element method:

1 Discretization of the computational domain
2 Selection of the interpolating functions
3 Derivation of characteristic matrices and vectors
4 Assembly of characteristic matrices and vectors
5 Solution of the matrix problem $\mathbf{Ax} = \mathbf{b}$.

As will be seen, each step in the finite element method is mathematically significant and relatively challenging. However, there are a large number of commercially available packages that take care of implementing the above ideas. Keep in mind that the primary reason to develop this technique is boundary conditions: both complicated domains and general boundary conditions.

Domain discretization

Finite element domain discretization usually involves the use of a commercial package. Two commonly used packages are the MATLAB PDE Toolbox and FEMLAB (which is built on the MATLAB platform). Writing your own code to generate an unstructured computational grid is a difficult task and well beyond the scope of this book. Thus we will use commercial packages to generate the computational mesh necessary for a given problem. The key idea is to discretize the domain with triangular elements (or another appropriate element function). Figure 11.1 shows the discretization of a domain with complicated boundaries inside the computationally relevant region. Note that the triangular discretization is such that it adaptively sizes the triangles so that higher resolution is automatically in place where needed. The key features of the discretization are as follows:

- The width and height of all discretization triangles should be similar.
- All shapes used to span the computational domain should be approximated by polygons.
- A commercial package is almost always used to generate the grid unless you are doing research in this area.

Interpolating functions

Each element in the finite element basis relies on a piecewise approximation. Thus the solution to the problem is approximated by a simple function in each triangular (or polygonal) element.

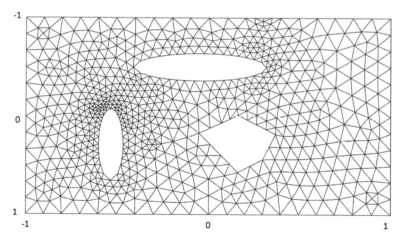

Figure 11.1: Finite element triangular discretization provided by the MATLAB PDE Toolbox. Note the increased resolution near corners and edges.

As might be expected, the accuracy of the scheme depends upon the choice of the approximating function chosen.

Polynomials are the most commonly used interpolating functions. The simpler the function, the less computationally intensive. However, accuracy is usually increased with the use of higher order polynomials. Three groups of elements are usually considered in the basic finite element scheme:

- Simplex elements: linear polynomials are used
- Complex elements: higher order polynomials are used
- Multiplex elements: rectangles are used instead of triangles.

An example of each of the three finite elements is given in Figs. 11.2 and 11.3 for one-dimensional and two-dimensional finite elements.

1D simplex

To consider the finite element implementation, we begin by considering the one-dimensional problem. Figure 11.4 shows the linear polynomial used to approximate the solution between points x_i and x_j. The approximation of the function is thus

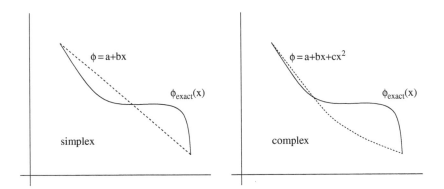

Figure 11.2: Finite element discretization for simplex and complex finite elements. Essentially the finite elements are approximating polynomials.

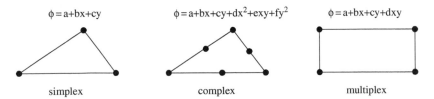

Figure 11.3: Finite element discretization for simplex, complex and multiplex finite elements in two-dimensions. The finite elements are approximating surfaces.

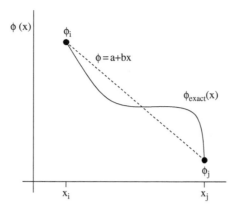

Figure 11.4: One-dimensional approximation using finite elements.

$$\psi(x) - a + bx = (1 \; x)\begin{pmatrix} a \\ b \end{pmatrix}.$$ (11.1.1)

The coefficients a and b are determined by enforcing that the function goes through the end points. This gives

$$\phi_i = a + bx_i$$ (11.1.2a)

$$\phi_j = a + bx_j$$ (11.1.2b)

which is a 2×2 system for the unknown coefficients. In matrix form this can be written

$$\phi = \mathbf{A}\mathbf{a} \;\rightarrow\; \phi = \begin{pmatrix} \phi_i \\ \phi_j \end{pmatrix}, \; \mathbf{A} = \begin{pmatrix} 1 & x_i \\ 1 & x_j \end{pmatrix}, \; \mathbf{a} = \begin{pmatrix} a \\ b \end{pmatrix}.$$ (11.1.3)

Solving for $\mathbf{a}$ gives

$$\mathbf{a} = \mathbf{A}^{-1}\phi = \frac{1}{x_j - x_i}\begin{pmatrix} x_j & -x_i \\ -1 & 1 \end{pmatrix}\begin{pmatrix} \phi_i \\ \phi_j \end{pmatrix} = \frac{1}{l}\begin{pmatrix} x_j & -x_i \\ -1 & 1 \end{pmatrix}\begin{pmatrix} \phi_i \\ \phi_j \end{pmatrix}$$ (11.1.4)

where $l = x_j - x_i$. Recalling that $\phi = (1 \; x)\mathbf{a}$ then gives

$$\phi = \frac{1}{l}(1 \; x)\begin{pmatrix} x_j & -x_i \\ -1 & 1 \end{pmatrix}\begin{pmatrix} \phi_i \\ \phi_j \end{pmatrix}.$$ (11.1.5)

Multiplying this expression out yields the approximate solution

$$\phi(x) = N_i(x)\phi_i + N_j(x)\phi_j$$ (11.1.6)

where $N_i(x)$ and $N_j(x)$ are the Lagrange polynomial coefficients [11] (shape functions)

$$N_i(x) = \frac{1}{l}(x_j - x)$$ (11.1.7a)

$$N_j(x) = \frac{1}{l}(x - x_i).$$ (11.1.7b)

Note the following properties: $N_i(x_i) = 1, N_i(x_j) = 0$ and $N_j(x_i) = 0, N_j(x_j) = 1$. This completes the generic construction of the polynomial which approximates the solution in one dimension.

2D simplex

In two dimensions, the construction becomes a bit more difficult. However, the same approach is taken. In this case, we approximate the solution over a region with a plane (see Fig. 11.5)

$$\phi(x) = a + bx + cy. \tag{11.1.8}$$

The coefficients a, b and c are determined by enforcing that the function goes through the end points. This gives

$$\phi_i = a + bx_i + cy_i \tag{11.1.9a}$$

$$\phi_j = a + bx_j + cy_j \tag{11.1.9b}$$

$$\phi_k = a + bx_k + cy_k \tag{11.1.9c}$$

which is now a 3×3 system for the unknown coefficients. In matrix form this can be written

$$\phi = \mathbf{Aa} \;\rightarrow\; \phi = \begin{pmatrix} \phi_i \\ \phi_j \\ \phi_k \end{pmatrix}, \; \mathbf{A} = \begin{pmatrix} 1 & x_i & y_i \\ 1 & x_j & y_j \\ 1 & x_k & y_k \end{pmatrix}, \; \mathbf{a} = \begin{pmatrix} a \\ b \\ c \end{pmatrix}. \tag{11.1.10}$$

The geometry of this problem is reflected in Fig. 11.5 where each of the points of the discretization triangle is illustrated.

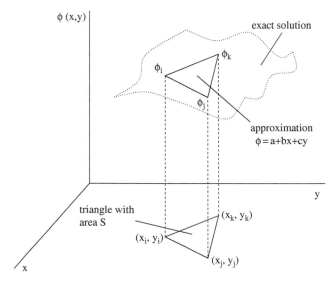

Figure 11.5: Two-dimensional triangular approximation of the solution using the finite element method.

Solving for **a** gives

$$\mathbf{a} = \mathbf{A}^{-1}\phi = \frac{1}{2S} \begin{pmatrix} x_j y_k - x_k y_j & x_k y_i - x_i y_k & x_i y_j - x_j y_i \\ y_j - y_k & y_k - y_i & y_i - y_j \\ x_k - x_j & x_i - x_k & x_j - x_i \end{pmatrix} \begin{pmatrix} \phi_i \\ \phi_j \\ \phi_k \end{pmatrix} \qquad (11.1.11)$$

where S is the area of the triangle projected onto the $x-y$ plane (see Fig. 11.5). Recalling that $\phi = \mathbf{A}\mathbf{a}$ gives

$$\phi(x) = N_i(x, y)\phi_i + N_j(x, y)\phi_j + N_k(x, y)\phi_k \qquad (11.1.12)$$

where $N_i(x, y)$, $N_j(x, y)$, $N_k(x, y)$ are the Lagrange polynomial coefficients [11] (shape functions)

$$N_i(x, y) = \frac{1}{2S}\left[(x_j y_k - x_k y_j) + (y_j - y_k)x + (x_k - x_j)y\right] \qquad (11.1.13a)$$

$$N_j(x, y) = \frac{1}{2S}\left[(x_k y_i - x_i y_k) + (y_k - y_i)x + (x_i - x_k)y\right] \qquad (11.1.13b)$$

$$N_k(x, y) = \frac{1}{2S}\left[(x_i y_j - x_j y_i) + (y_i - y_j)x + (x_j - x_i)y\right]. \qquad (11.1.13c)$$

Note the following end point properties: $N_i(x_i, y_i) = 1$, $N_j(x_j, y_j) = N_k(x_k, y_k) = 0$ and $N_j(x_j, y_j) = 1$, $N_j(x_i, y_i) = N_j(x_k, y_k) = 0$ and $N_k(x_k, y_k) = 1$, $N_k(x_i, y_i) = N_k(x_j, y_j) = 0$. This completes the generic construction of the polynomial which approximates the solution in one dimension, and the generic construction of the surface which approximates the solution in two dimensions.

11.2 Discretizing with Finite Elements and Boundaries

In the previous section, an outline was given for the construction of the polynomials which approximate the solution over a finite element. Once constructed, however, they must be used to solve the physical problem of interest. The finite element solution method relies on solving the governing set of partial differential equations in the *weak formulation* of the problem, i.e. the integral formulation of the problem. This is because we will be using linear interpolation pieces whose derivatives don't necessarily match accross elements. In the integral formulation, this does not pose a problem.

We begin by considering the elliptic partial differential equation

$$\frac{\partial}{\partial x}\left(p(x, y)\frac{\partial u}{\partial x}\right) + \frac{\partial}{\partial y}\left(q(x, y)\frac{\partial u}{\partial y}\right) + r(x, y)u = f(x, y) \qquad (11.2.1)$$

where, over part of the boundary,

$$u(x, y) = g(x, y) \quad \text{on } S_1 \qquad (11.2.2)$$

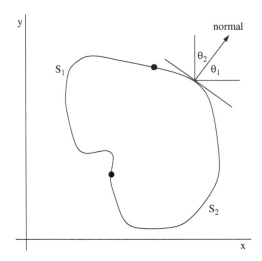

Figure 11.6: Computational domain of interest which has the two domain regions S_1 and S_2.

and

$$p(x,y)\frac{\partial u}{\partial x}\cos\theta_1 + q(x,y)\frac{\partial u}{\partial y}\cos\theta_2 + g_1(x,y)u = g_2(x,y) \quad \text{on } S_2. \tag{11.2.3}$$

The boundaries and domain are illustrated in Fig. 11.6. Note that the normal derivative determines the angles θ_1 and θ_2. A domain such as this would be very difficult to handle with finite difference techniques. And further, because of the boundary conditions, spectral methods cannot be implemented. Only the finite element method renders the problem tractable.

The variational principle

To formulate the problem correctly for the finite element method, the governing equations and their associated boundary conditions are recast in an integral form. This recasting of the problem involves a variational principle. Although we will not discuss the calculus of variations here [24], the highlights of this method will be presented.

The variational method expresses the governing partial differential as an integral which is to be minimized. In particular, the functional to be minimized is given by

$$I(\phi) = \iiint_V F\left(\phi, \frac{\partial\phi}{\partial x}, \frac{\partial\phi}{\partial y}, \frac{\partial\phi}{\partial z}\right) dV + \iint_S g\left(\phi, \frac{\partial\phi}{\partial x}, \frac{\partial\phi}{\partial y}, \frac{\partial\phi}{\partial z}\right) dS, \tag{11.2.4}$$

where a three-dimensional problem is being considered. The derivation of the functions F, which captures the governing equation, and g, which captures the boundary conditions, follow the Euler–Lagrange equations for minimizing variations:

$$\frac{\delta F}{\delta\phi} = \frac{\partial}{\partial x}\left(\frac{\partial F}{\partial(\phi_x)}\right) + \frac{\partial}{\partial y}\left(\frac{\partial F}{\partial(\phi_y)}\right) + \frac{\partial}{\partial z}\left(\frac{\partial F}{\partial(\phi_z)}\right) - \frac{\partial F}{\partial\phi} = 0. \tag{11.2.5}$$

This is essentially a generalization of the concept of the zero derivative which minimizes a function. Here we are minimizing a functional.

For our governing elliptic problem (11.2.1) with boundary conditions given by those of S_2 (11.2.3), we find the functional $I(u)$ to be given by

$$I(u) = \frac{1}{2} \iint_D dxdy \left[p(x,y) \left(\frac{\partial u}{\partial x}\right)^2 + q(x,y) \left(\frac{\partial u}{\partial y}\right)^2 - r(x,y)u^2 + 2f(x,y)u \right]$$
$$+ \int_{S_2} dS \left[-g_2(x,y)u + \frac{1}{2}g_1(x,y)u^2 \right]. \tag{11.2.6}$$

Thus the interior domain over which we integrate D has the integrand

$$I_D = \frac{1}{2} \left[p(x,y) \left(\frac{\partial u}{\partial x}\right)^2 + q(x,y) \left(\frac{\partial u}{\partial y}\right)^2 - r(x,y)u^2 + 2f(x,y)u \right]. \tag{11.2.7}$$

This integrand was derived via variational calculus as the minimization over the functional space of interest. To confirm this, we apply the variational derivative as given by (11.2.5) and find

$$\frac{\delta I_D}{\delta u} = \frac{\partial}{\partial x}\left(\frac{\partial I_D}{\partial(u_x)}\right) + \frac{\partial}{\partial y}\left(\frac{\partial I_D}{\partial(u_y)}\right) - \frac{\partial I_D}{\partial u} = 0$$
$$= \frac{\partial}{\partial x}\left(\frac{1}{2}p(x,y) \cdot 2\frac{\partial u}{\partial x}\right) + \frac{\partial}{\partial y}\left(\frac{1}{2}q(x,y) \cdot 2\frac{\partial u}{\partial y}\right) - \left(-\frac{1}{2}r(x,y) \cdot 2u + f(x,y)\right)$$
$$= \frac{\partial}{\partial x}\left(p(x,y)\frac{\partial u}{\partial x}\right) + \frac{\partial}{\partial y}\left(q(x,y)\frac{\partial u}{\partial y}\right) + r(x,y)u - f(x,y) = 0. \tag{11.2.8}$$

Upon rearranging, this gives back the governing equation (11.2.1)

$$\frac{\partial}{\partial x}\left(p(x,y)\frac{\partial u}{\partial x}\right) + \frac{\partial}{\partial y}\left(q(x,y)\frac{\partial u}{\partial y}\right) + r(x,y)u = f(x,y). \tag{11.2.9}$$

A similar procedure can be followed on the boundary terms to derive the appropriate boundary conditions (11.2.2) or (11.2.3). Once the integral formulation has been achieved, the finite element method can be implemented with the following key ideas:

1. The solution is expressed in the *weak* (integral) form since the linear (simplex) interpolating functions give that the second derivates (e.g. ∂_x^2 and ∂_y^2) are zero. Thus the elliptic operator (11.2.1) would have no contribution from a finite element of the form $\phi = a_1 + a_2 x + a_3 y$. However, in the integral formulation, the second derivative terms are proportional to $(\partial u/\partial x)^2 + (\partial u/\partial y)^2$ which gives $\phi = a_2^2 + a_3^2$.

2. A solution is sought which takes the form

$$u(x,y) = \sum_{i=1}^{m} \gamma_i \phi_i(x,y) \tag{11.2.10}$$

where $\phi_i(x,y)$ are the linearly independent piecewise-linear polynomials and $\gamma_1, \gamma_2, \ldots, \gamma_m$ are constants.

- $\gamma_{n+1}, \gamma_{n+2}, \ldots, \gamma_m$ ensure that the boundary conditions are satisfied, i.e. those elements which touch the boundary must meet certain restrictions.

- $\gamma_1, \gamma_2, \ldots, \gamma_n$ ensure that the integrand $I(u)$ in the interior of the computational domain is minimized, i.e. $\partial I/\partial \gamma_i = 0$ for $i = 1, 2, 3, \ldots, n$.

Solution method

We begin with the governing equation in integral form (11.2.6) and assume an expansion of the form (11.2.10). This gives

$$
\begin{aligned}
I(u) = I&\left(\sum_{i=1}^{m} \gamma_i \phi_i(x,y)\right) \\
= \frac{1}{2} &\iint_D dxdy \left[p(x,y)\left(\sum_{i=1}^{m}\gamma_i\frac{\partial\phi_i}{\partial x}\right)^2 + q(x,y)\left(\sum_{i=1}^{m}\gamma_i\frac{\partial\phi_i}{\partial y}\right)^2 \right. \\
&\left. - r(x,y)\left(\sum_{i=1}^{m}\gamma_i\phi_i(x,y)\right)^2 + 2f(x,y)\left(\sum_{i=1}^{m}\gamma_i\phi_i(x,y)\right) \right] \\
+ &\int_{S_2} dS \left[-g_2(x,y)\sum_{i=1}^{m}\gamma_i\phi_i(x,y) + \frac{1}{2}g_1(x,y)\left(\sum_{i=1}^{m}\gamma_i\phi_i(x,y)\right)^2 \right].
\end{aligned}
\tag{11.2.11}
$$

This completes the expansion portion.

The functional $I(u)$ must now be differentiated with respect to all the interior points and minimized. This involves differentiating the above with respect to γ_i where $i = 1, 2, 3, \ldots, n$. This results in the rather complicated expression

$$
\begin{aligned}
\frac{\partial I}{\partial \gamma_j} = &\iint_D dxdy \left[p(x,y)\sum_{i=1}^{m}\gamma_i\frac{\partial\phi_i}{\partial x}\frac{\partial\phi_j}{\partial x} + q(x,y)\sum_{i=1}^{m}\gamma_i\frac{\partial\phi_i}{\partial y}\frac{\partial\phi_j}{\partial y} \right. \\
&\left. - r(x,y)\sum_{i=1}^{m}\gamma_i\phi_i\phi_j + f(x,y)\phi_j \right] \\
+ &\int_{S_2} dS \left[-g_2(x,y)\phi_j + g_1(x,y)\sum_{i=1}^{m}\gamma_i\phi_i\phi_j \right] = 0.
\end{aligned}
\tag{11.2.12}
$$

Term by term, this results in an expression for the γ_i given by

$$
\begin{aligned}
\sum_{i=1}^{m} &\left\{ \iint_D dxdy \left[p\frac{\partial\phi_i}{\partial x}\frac{\partial\phi_j}{\partial x} + q\frac{\partial\phi_i}{\partial y}\frac{\partial\phi_j}{\partial y} - r\phi_i\phi_j \right] + \int_{S_2} dSg_1\phi_i\phi_j \right\} \gamma_i \\
&+ \iint_D dxdy f\phi_j - \int_{S_2} dSg_2(x,y)\phi_j = 0.
\end{aligned}
\tag{11.2.13}
$$

But this expression is simply a matrix solve $\mathbf{Ax} = \mathbf{b}$ for the unknowns γ_i. In particular, we have

$$
\mathbf{x} = \begin{pmatrix} \gamma_1 \\ \gamma_2 \\ \vdots \\ \gamma_n \end{pmatrix} \qquad \mathbf{b} = \begin{pmatrix} \beta_1 \\ \beta_2 \\ \vdots \\ \beta_n \end{pmatrix} \qquad \mathbf{A} = (\alpha_{ij}) \tag{11.2.14}
$$

where

$$
\beta_i = -\iint_D dx\,dy\, f\,\phi_i + \int_{S_2} dS g_2 \phi_i - \sum_{k=n+1}^{m} \alpha_{ik} \gamma_k \tag{11.2.15a}
$$

$$
\alpha_{ij} = \iint_D dx\,dy \left[p \frac{\partial \phi_i}{\partial x} \frac{\partial \phi_j}{\partial x} + q \frac{\partial \phi_i}{\partial y} \frac{\partial \phi_j}{\partial y} - r\phi_i\phi_j \right] + \int_{S_2} dS g_1 \phi_i \phi_j. \tag{11.2.15b}
$$

Recall that each element ϕ_i is given by the simplex approximation

$$
\phi_i = \sum_{i=1}^{3} N_j^{(i)}(x,y)\phi_j^{(i)} = \sum_{i=1}^{3} \left(a_j^{(i)} + b_j^{(i)}x + c_j^{(i)}y \right) \phi_j^{(i)}. \tag{11.2.16}
$$

This concludes the construction of the approximate solution in the finite element basis. From an algorithmic point of view, the following procedure would be carried out to generate the solution.

1 Discretize the computational domain into triangles. $T_1, T_2, \ldots, T_k$ are the interior triangles and $T_{k+1}, T_{k+2}, \ldots, T_m$ are triangles that have at least one edge which touches the boundary.

2 For $l = k+1, k+2, \ldots, m$, determine the values of the vertices on the triangles which touch the boundary.

3 Generate the shape functions

$$
N_j^{(i)} = a_j^{(i)} + b_j^{(i)}x + c_j^{(i)}y \tag{11.2.17}
$$

where $i = 1, 2, 3, \ldots, m$ and $j = 1, 2, 3, \ldots, m$.

4 Compute the integrals for matrix elements α_{ij} and vector elements β_j in the interior and boundary.

5 Construct the matrix $\mathbf{A}$ and vector $\mathbf{b}$.

6 Solve $\mathbf{Ax} = \mathbf{b}$.

7 Plot the solution $u(x,y) = \sum_{i=1}^{m} \gamma_i \phi_i(x,y)$.

For a wide variety of problems, the above procedures can simply be automated. That is exactly what commercial packages do. Thus once the coefficients and boundary conditions associated with (11.2.1), (11.2.2) and (11.2.3) are known, the solution procedure is straightforward.

11.3   MATLAB for Partial Differential Equations

Given the ubiquity of partial differential equations, it is not surprising that MATLAB has a built-in PDE solver: **pdepe**. Thus the time and space discretization, as well as time-stepping within the CFL tolerances, are handled directly as a subroutine call to MATLAB. This is similar to using a differential equation solver such as **ode45**. The following specific PDE can be solved with **pdepe**:

$$c\left(x, t, u, \frac{\partial u}{\partial x}\right) \frac{\partial u}{\partial t} = x^{-m} \frac{\partial}{\partial x}\left[x^m f\left(x, t, u, \frac{\partial u}{\partial x}\right)\right] + s\left(x, t, u, \frac{\partial u}{\partial x}\right) \tag{11.3.1}$$

where $u(x, t)$ generically is a vector field, $t \in [t_-, t_+]$ is the finite time range to be solved for, $x \in [a, b]$ is the spatial domain of interest, and $m = 0, 1$ or 2. The integer m arises from considering the Laplacian operator in cylindrical and spherical coordinates for which $m = 1$ and $m = 2$, respectively. Note that $c(x, t, u, u_x)$ is a diagonal matrix with identically zero or positive coefficients. The functions $f(x, t, u, u_x)$ and $s(x, t, u, u_x)$ correspond to a flux and source term, respectively. All three functions allow for time and spatially dependent behavior that can be nonlinear.

In addition to the PDE, boundary conditions must also be specified. The specific form required for **pdepe** can be nonlinear and time dependent so that:

$$p(x, t, u) + q(x, t)g(x, t, u, u_x) = 0 \text{ at } x = a, b. \tag{11.3.2}$$

As before $p(x, t, u)$ and $g(x, t, u, u_x)$ is time and spatially dependent behavior that can be nonlinear. The function $q(x, t)$ is simply time and space dependent.

Heat equation

To start, consider the simplest PDE: the heat equation:

$$\pi^2 \frac{\partial u}{\partial t} = \frac{\partial^2 u}{\partial x^2} \tag{11.3.3}$$

where the solution is defined on the domain $x \in [0, 1]$ with the boundary conditions

$$u(0, t) = 0 \tag{11.3.4a}$$

$$\pi \exp(-t) + \frac{\partial u(1, t)}{\partial t} = 0 \tag{11.3.4b}$$

and initial conditions

$$u(x, 0) = \sin(\pi x). \tag{11.3.5}$$

This specifies the PDE completely and allows for the construction of a unique solution. In MATLAB, the **pdepe** function call relies on three subroutines that specify the PDE, initial conditions and boundary conditions. Before calling these, the time and space grid must be defined. The following code defines the time domain, spatial discretization and plots the solution

```
m = 0;
x = linspace(0,1,20);
t = linspace(0,2,5);
u = pdepe(m,'pdex1pde','pdex1ic','pdex1bc',x,t);
surf(x,t,u)
```

It only remains to specify the three subroutines. The PDE subroutine is constructed as follows.

pdex1pde.m

```
function [c,f,s] = pdex1pde(x,t,u,DuDx)
c = pi^2;
f = DuDx;
s = 0;
```

In this case, the implementation is fairly straightfoward since $s(x, t, u, u_x) = 0$, $m = 0$, $c(x, t, u, u_x) = \pi^2$ and $f(x, t, u, u_x) = u_x$. The initial condition is quite easy and can be done in one line.

pdex1ic.m

```
function u0 = pdex1ic(x)
u0 = sin(pi*x);
```

Finally, the left and right boundaries can be implemented by specifying the function q, q and g at the right and left.

pdex1bc.m

```
function [pl,ql,pr,qr] = pdex1bc(xl,ul,xr,ur,t)
pl = ul;
ql = 0;
pr = pi * exp(-t);
qr = 1;
```

Combined, the subroutines quickly and efficiently solve the heat equation with a time-dependent boundary condition.

FitzHugh–Nagumo equation

Overall, the combination of (11.3.1) with boundary conditions (11.3.2) allows for a fairly broad range of problems to solve. As a more sophisticated example, the FitzHugh–Nagumo equation is considered which models the voltage dynamics in neurons. The FitzHugh–Nagumo model supports the propagation of voltage spikes that are held together in a coherent structure through the interaction of diffusion, nonlinearity and coupling to a recovery field. Thus we will consider the following system of PDEs:

$$\frac{\partial V}{\partial t} = D\frac{\partial^2 V}{\partial x^2} + V(a - V)(V - 1) - W \tag{11.3.6a}$$

$$\frac{\partial W}{\partial t} = bV - cW \tag{11.3.6b}$$

where $V(x, t)$ measures the voltage in the neuron. The following initial conditions will be applied that generate a voltage spike moving from left to right in time

$$V(x, 0) = \exp(-x^2) \tag{11.3.7a}$$

$$W(x, 0) = 0.2\exp(-(x + 2)^2). \tag{11.3.7b}$$

Finally, boundary conditions must be imposed on the PDE system. For convenience, no-flux boundary conditions will be applied at both ends of the computational domain so that

$$\frac{\partial V}{\partial x} = 0 \text{ and } \frac{\partial W}{\partial x} = 0 \text{ at } x = a, b. \tag{11.3.8}$$

The partial differential equation along with the boundary conditions and initial conditions completely specify the system.

The FitzHugh–Nagumo PDE system is a vector sytem for the field variables $V(x, t)$ and $W(x, t)$. Before calling on **pdepe**, the time and space domain discretization is defined along with the parameter m:

```
m = 0;
x = linspace(-10,30,400);   % space
t = linspace(0,400,20);     % time
```

In this example, the time and space domains are discretized with equally spaced Δt and Δx. However, an arbitrary spacing can be specified. This is especially important in the spatial domain as there can be, for instance, boundary layer effects near the boundary where a higher density of grid points should be imposed.

The next step is to call upon **pdepe** in order to generate the solution. The following code picks the parameters a, b and c in the FitzHugh–Nagumo equation and passes them into the PDE solver. The final result is the generation of a matrix **sol** that contains the solution matrices for both $V(x, t)$ and $W(x, t)$.

```
A=0.1;   B=0.01; C=0.01;
sol = pdepe(m,'pde_fn_pde','pde_fn_ic','pde_fn_bc',x,t,[],A,B,C);

u1 = sol(:,:,1);
u2 = sol(:,:,2);

waterfall(x,t,u1), map=[0 0 0]; colormap(map), view(15,60)
```

Upon solving, the matrix **sol** is generated which is $20 \times 400 \times 2$. The first layer of this matrix cube is the voltage $V(x, t)$ in time and space ($\mathbf{u_1}$) and the second layer is the recovery field $W(x, t)$ in time and space ($\mathbf{u_2}$). The voltage, and the propagating wave, is plotted using **waterfall**.

What is left is to define the PDE itself (**pdepe_fn_pde.m**), its boundary conditions (**pdepe_fn_bc.m**) and its initial conditions (**pdepe_fn_ic.m**). We begin by developing the function call associated with the PDE itself.

pdepe_fn_pde.m

```
function [c,f,s] = pde_fn_pde(x,t,u,DuDx,A,B,C)
c = [1; 1];                  % c diagonal terms
f = [0.01; 0] .* DuDx;    % diffusion term

rhsV = u(1)*(A-u(1))*(u(1)-1) - u(2);
rhsW = B*u(1)-C*u(2);
s = [rhsV; rhsW];
```

Note that the specification of $c(x, t, u, u_x)$ in the PDE must be a diagonal matrix. Thus the PDE call only allow for the placement of these diagonal elements into a vector that must have positive coefficients. The diffusion operator and the derivative are specified by **f** and **DuDx**, respectively. Once the appropriate terms are constructed, the right-hand side is fully formed in the matrix **s**. Note the similarity between this and the construction involved in **ode45**. We next implement the initial conditions.

pdepe_fn_ic.m

```
function u0 = pde_fn_ic(x,A,B,C)
u0 = [1*exp(-((x)/1).^2)
      0.2*exp(-((x+2)/1).^2)];
```

Arbitrary initial conditions can be applied, but preferably, one would implement initial conditions consistent with the boundary conditions. The boundary conditions are implemented in the following code.

pdepe_fn_bc.m

```
function [pl,ql,pr,qr] = pde_fn_bc(xl,ul,xr,ur,t,A,B,C)
pl = [0; 0];
ql = [1; 1];
pr = [0; 0];
qr = [1; 1];
```

Here **pl** and **ql** are the functions $p(x, t, u)$ and $q(x, t)$, respectively, at the left boundary point while **pr** and **qr** are the functions $p(x, t, u)$ and $q(x, t)$ at the right boundary point.

Figure 11.7 demonstrates the formation of a spike and its propagation using the **pdepe** methodology. Note that the time-step used is not determined by the initial time vector used for output. Rather, just like the adaptive differential equation steppers, a step-size Δt is chosen to ensure a default accuracy of 10^{-6}. The spatial discretization, however, is absolutely critical as the method uses the user-specified mesh in space. Thus it is imperative to choose a fine enough mesh in order to avoid numerical inaccuracies and instabilities.

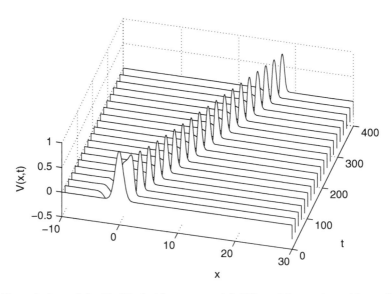

Figure 11.7: The solution of the FitzHugh–Nagumo partial differential equation with no-flux boundary conditions.

11.4 MATLAB Partial Differential Equations Toolbox

In addition to the **pdepe** function call, MATLAB has a finite element based PDE solver. Unlike **pdepe**, which provides solutions to one-dimensional parabolic and elliptic type PDEs, the PDE Toolbox allows for the solution of linear, two-dimensional PDE systems of parabolic, elliptic and hyperbolic type along with eigenvalue problems. In particular, the toolbox allows for the solution of

$$elliptic: \quad Lu = f(x, y, t) \tag{11.4.1a}$$

$$eigenvalue: \quad Lu = \lambda d(x, y, t)u \tag{11.4.1b}$$

$$parabolic: \quad d(x, y, t)\frac{\partial u}{\partial t} + Lu = f(x, y, t) \tag{11.4.1c}$$

$$hyperbolic: \quad d(x, y, t)\frac{\partial^2 u}{\partial t^2} + Lu = f(x, y, t) \tag{11.4.1d}$$

where L denotes the spatial linear operator

$$Lu = -\nabla \cdot \big(c(x, y, t)\nabla u\big) + a(x, y, t)u \tag{11.4.2}$$

where x and y are on a given domain Ω. The functions $d(x, y, t)$, $c(x, y, t)$, $f(x, y, t)$ and $a(x, y, t)$ can be fairly arbitrary, but should be, generically, well behaved on the domain Ω.

The boundary of the domain, as will be shown, can be quite complicated. On each portion of the domain, boundary conditions must be specified which are of the Dirichlet or Neumann type. In particular, each portion of the boundary must be specified as one of the following:

$$Dirichlet: \quad h(x, y, t)u = r(x, y, t) \tag{11.4.3a}$$

$$Neumann: \quad c(x, y, t)(\mathbf{n} \cdot \nabla u) + q(x, y, t)u = g(x, y, t) \tag{11.4.3b}$$

where $h(x, y, t)$, $r(x, y, t)$, $c(x, y, t)$, $q(x, y, t)$ and $g(x, y, t)$ can be fairly arbitrary and $\mathbf{n}$ is the unit normal vector on the boundary. Thus the flux across the boundary is specified with the Neumann boundary conditions.

In addition to the PDE itself and its boundary conditions, the initial condition must be specified for the parabolic and hyperbolic manifestation of the equation. The initial condition is a function specified over the entire domain Ω.

Although the PDE Toolbox is fairly general in terms of its nonconstant coefficient specifications, it is a fairly limited solver given that it can only solve linear problems in two dimensions with up to two space and time derivatives. If considering a problem that can be specified by the conditions laid out above, then it can provide a fast and efficient solution method with very little development costs and overhead. Parenthetically, it should be noted that a wide variety of other professional grade tools exist that also require specifying the above constraints. And in general, every one of these PDE toolboxes is limited by a constrained form of PDE and boundary conditions. To access the toolbox in MATLAB, simply type

```
pdetool
```

in MATLAB and a graphical user interface (GUI) will appear guiding one through the setup of the PDE solver.

Figure 11.8 shows the basic user interface associated with the MATLAB PDE Toolbox GUI. An initial domain and mesh are displayed for demonstration purposes. In what follows, a step-by-step outline of the use of the GUI will be presented. Indeed, the GUI allows for a natural progression that specifies (i) the domain on which the PDE is to be considered, (ii) the specific PDE to be solved, (iii) the boundary conditions on each portion of the domain, and (iv) the resolution and grid for which to generate the solution. Generically, each of these steps is performed by following the graphical icons aligned on the top of the GUI (see Fig. 11.9).

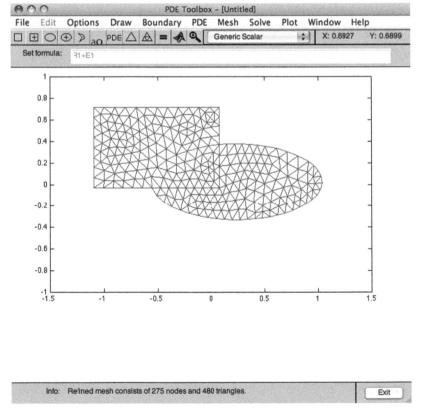

Figure 11.8: Graphical user interface of the partial differential equations GUI. The interface is activated by the command **pdetool**.

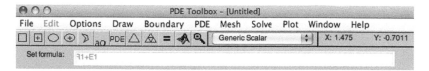

Figure 11.9: User interface for PDE specification along with boundary conditions and domain. The icons can be activated from left to right in order to (i) specify the domain, (ii) specify the boundary conditions, (iii) specify the PDE, (iv) implement a mesh, and (v) solve the PDE.

Domain specification

To begin, the domain is specified using a variety of drawing Tools in the PDE Toolbox. In particular, one can draw rectangles (boxes), ellipses (circles) and/or polygons. The first five icons of the top bar of the GUI, demonstrated in Fig. 11.9, show the various options available. In particular, the first icon allows one to draw a rectangle (box) where two corners of the rectangle are specified (first icon) or the center and outside of the rectangle are specified (second icon). Similarly, the third and fourth icons specify the outsides of an ellipse (circle) or its center and semi-major/minor axes. The fifth icon allows one to specify a closed polygon for a boundary. One can also import a boundary shape and/or rotate a given object of interest. The **set formula** Fig. 11.9 allows you to specify wether the domains drawn are added together, or in case of generating holes in the domain, subtracted out. The rectangles, ellipses and polygons are denoted by *Rk*, *Ek* and *Pk* where *k* is the numbering of the object. Figure 11.8, for instance, shows the formula $R1 + E1$ since the domain is comprised of a rectangle and an ellipse.

Boundary condition specification

The sixth icon is given by $\partial\Omega$ and represents the specification of the boundary conditions. Either Dirichlet or Neumann are specified in the form of (11.4.3). Figure 11.10 shows the GUI that appears for specifying both types of boundary conditions. The default settings for Dirichlet boundary conditions are $h(x,y,t) = 1$ and $r(x,y,t) = 0$. The default settings for Neumann boundaries are $q(x,y,t) = g(x,y,t) = 0$.

PDE specification

The PDE to be solved is specified by the seventh icon labeled PDE. This will allow the user to specify one of the PDEs given in (11.4.1). Figure 11.11 demonstrates the PDE specification GUI and its four options. A parabolic PDE specification is demonstrated in the figure. The default settings are $c(x,y,t) = d(x,y,t) = 1$, $a(x,y,t) = 0$ and $f(x,y,t) = 10$.

Grid generation and refinement

A finite element grid can be generated by applying the eighth icon (a triangle). This will automatically generate a set of triangles that cover the domain of interest where a higher density of

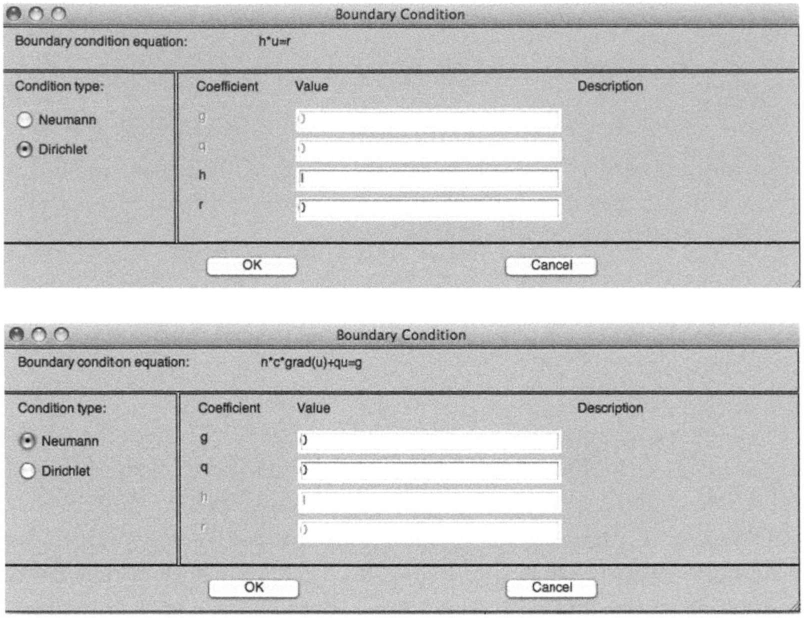

Figure 11.10: User interface for PDE boundary condition specification. Simply double click on a given boundary and specify Dirichlet (top figure) or Neumann (bottom figure) on the boundary selected. Note that the boundary will appear red if it is Dirichlet and blue if it is Neumann.

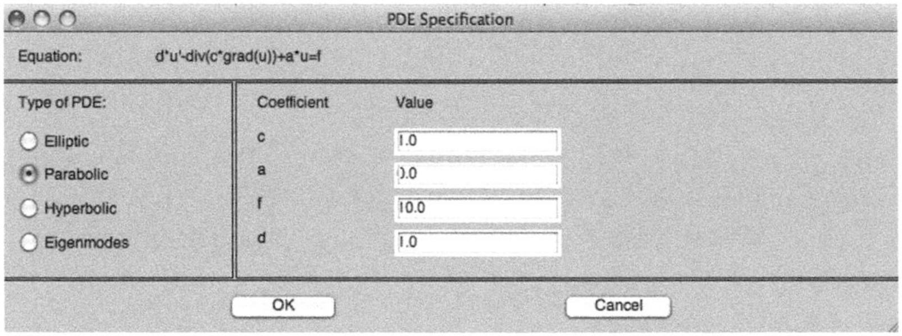

Figure 11.11: User interface for PDE specification. Four choices are given representing the equations given in (11.4.1).

triangular (discretization) points are automatically placed at locations where geometry changes occur. To refine the mesh, the ninth icon (a triangle within a triangle) can be executed. This refinement can be applied a number of times in order to achieve a desired grid resolution.

Solving the PDE and plotting results

The PDE can then be solved by executing the tenth icon which is an equals sign. This will solve the specified PDE with prescribed boundary conditions and domain. In order to specify initial

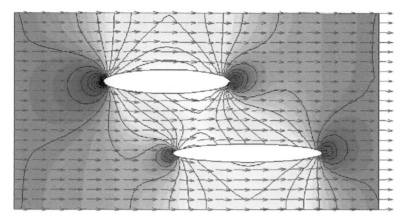

Figure 11.12: Plot of the solution of Laplace's equation with flux conditions on the left and right boundaries and no-flux on the top and bottom and internal boundaries. This represents a potential solution to the streamfunction equation where the gradient (velocity) field is plotted representing the flow lines.

conditions and/or the time domain, the solve button must be pushed and the **parameters** button from the drag down menu executed. For a parabolic PDE, the initial condition is $u(x, y, 0) = 0$, while for a hyperbolic PDE, the initial conditions are $u(x, y, 0) = u_x(x, y, 0) = 0$. The tolerance and accuracy for the time-stepping routine can also be adjusted from this drag down menu. The solution will automatically appear once the computation is completed. The MATLAB icon, which is the eleventh icon, allows for full control of the plotting routines.

As an example, Laplace's equation is solved on a rectangular domain with elliptical internal boundary constraints (see Fig. 11.12). This can potentially model the flow field, in terms of the streamfunction, in a complicated geometry. The boundaries are all subject to Neumann boundary conditions with the top and bottom and internal boundaries specifying no-flux conditions. The left and right boundary impose a flow of the field from left-to-right. The gradient of the solution field, i.e. the velocity of the flow field, is plotted so that the steady state flow can be constructed over a complicated domain.

PART III

Computational Methods for Data Analysis

Statistical Methods and Their Applications

12

Our ultimate goal is to analyze highly generic data arising from applications as diverse as imaging, biological sciences, atmospheric sciences, or finance, to name a few specific examples. In all these application areas, there is a fundamental reliance on extracting meaningful trends and information from large data sets. Primarily, this is motivated by the fact that in many of these systems, the degree of complexity, or the governing equations, is unknown or impossible to extract. Thus one must rely on data, its statistical properties, and its analysis in the context of spectral methods and linear algebra.

12.1 Basic Probability Concepts

To understand the methods required for data analysis, a review of probability theory and statistical concepts is necessary. Many of the ideas presented in this section are intuitively understood by most students in the mathematical, biological, physical and engineering sciences. Regardless, a review will serve to refresh one with the concepts and will further help understand their implementation in MATLAB.

Sample space and events

Often one performs an experiment whose outcome is not known in advance. However, while the outcome is not known, suppose that the set of all possible outcomes is known. The set of all possible outcomes is the *sample space* of the experiment and is denoted by S. Some specific examples of sample spaces are the following:

1   If the experiment consists of flipping a coin, then

$$S = \{H, T\}$$

where H denotes an outcome of heads and T denotes an outcome of tails.

2   If the experiment consists of flipping two coins, then

$$S = \{(H, H), (H, T), (T, H), (T, T)\}$$

where the outcome (H, H) denotes both coins being heads, (H, T) denotes the first coin being heads and the second tails, (T, H) denotes the first coin being tails and the second being heads, and (T, T) denotes both coins being tails.

3   If the experiment consists of tossing a die, then

$$S = \{1, 2, 3, 4, 5, 6\}$$

where the outcome i means that the number i appeared on the die, $i = 1, 2, 3, 4, 5, 6$.

Any subset of the sample space is referred to as an event, and it is denoted by E. For the sample spaces given above, we can also define an event.

1   If the experiment consists of flipping a single coin, then

$$E = \{H\}$$

denotes the event that a head appears on the coin.

2   If the experiment consists of flipping two coins, then

$$E = \{(H, H), (H, T)\}$$

denotes the event where a head appears on the first coin.

3   If the experiment consists of tossing a die, then

$$E = \{2, 4, 6\}$$

would be the event that an even number appears on the toss.

For any two events E_1 and E_2 of a sample space S, the *union* of these events $E_1 \cup E_2$ consists of all points which are either in E_1 or in E_2 or both in E_1 and E_2. Thus the event $E_1 \cup E_2$ will occur if either E_1 or E_2 occurs. For these same two events E_1 and E_2 we can also define their *intersection* as follows: $E_1 E_2$ consists of all points which are in both E_1 and E_2, thus requiring that both E_1 and E_2 occur simultaneously. Figure 12.1 gives a simple graphical interpretation of the probability space S along with the events E_1 and E_2 and their union $E_1 \cup E_2$ and intersection $E_1 E_2$.

To again make connection to the sample spaces and event examples given previously, consider the following unions and intersections:

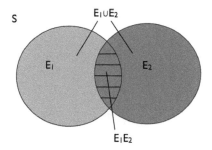

Figure 12.1: Venn diagram showing the sample space S along with the events E_1 and E_2. The union of the two events $E_1 \cup E_2$ is denoted by the gray regions while the intersection E_1E_2 is depicted in the cross-hatched region where the two events overlap.

1   If the experiment consists of flipping a single coin and $E_1 = \{H\}$ and $E_2 = \{T\}$ then

$$E_1 \cup E_2 = \{(H, T)\} = S$$

so that $E_1 \cup E_2$ is the entire sample space S and would occur if the coin flip produces either heads or tails.

2   If the experiment consists of flipping a single coin and $E_1 = \{H\}$ and $E_2 = \{T\}$ then

$$E_1E_2 = \emptyset$$

where the null event $\emptyset$ cannot occur since the coin cannot be both heads and tails simultaneously. Indeed if two events are such that $E_1E_2 = \emptyset$, then they are said to be mutually exclusive.

3   If the experiment consists of tossing a die and if $E_1 = \{1, 3, 5\}$ and $E_2 = \{1, 2, 3\}$ then

$$E_1E_2 = \{1, 3\}$$

so that the intersection of these two events would be a roll of the die of either 1 or 3.

The union and intersection of more than two events can be easily generalized from these ideas. Consider a set of events $E_1, E_2, \cdots, E_N$. Then the union is defined as $\cup_{n=1}^{N} E_n = E_1 \cup E_2 \cup \cdots \cup E_N$ and the intersection is defined as $\prod_{n=1}^{N} E_n = E_1E_2 \cdots E_N$. Thus an event which is in any of the E_n belongs to the union whereas an event needs to be in all E_n in order to be part of the intersection ($n = 1, 2, \cdots, N$). The complement of an event E, denoted E^c, consists of all points in the sample space S not in E. Thus $E \cup E^c = S$ and $EE^c = \emptyset$.

Probabilities defined on events

With the concept of sample spaces S and events E in hand, the probability of an event $P(E)$ can be calculated. Thus the following conditions are necessary in assigning a number to $P(E)$:

- $0 \leq P(E) \leq 1$
- $P(S) = 1$
- For any sequence of events $E_1, E_2, \cdots, E_N$ that are mutually exclusive so that $E_i E_j = \emptyset$ when $i \neq j$ then

$$P\left(\cup_{n=1}^{N} E_n\right) = \sum_{n=1}^{N} P(E_n).$$

The notation $P(E)$ is the probability of the event E occurring. The concept of probability is quite intuitive and it can be quite easily applied to our previous examples.

1   If the experiment consists of flipping a fair coin, then

$$P(\{H\}) = P(\{T\}) = 1/2$$

where H denotes an outcome of heads and T denotes an outcome of tails. On the other hand, a biased coin that was twice as likely to produce heads over tails would result in

$$P(\{(H\}) = 2/3, \;\; P(\{T\}) = 1/3$$

2   If the experiment consists of flipping two fair coins, then

$$P(\{(H, H)\} = 1/4$$

where the outcome (H, H) denotes both coins being heads.

3   If the experiment consists of tossing a fair die, then

$$P(\{1\}) = P(\{2\}) = P(\{3\}) = P(\{4\}) = P(\{5\}) = P(\{6\}) = 1/6$$

and the probability of producing an even number is

$$P(\{2, 4, 6\}) = P(\{2\}) + P(\{4\}) + P(\{6\}) = 1/2.$$

This is a formal definition of the probabilities being functions of events on sample spaces. On the other hand, our intuitive concept of the probability is that if an event is repeated over and over again, then with probability one, the proportion of time an event E occurs is just $P(E)$. This is the frequentist's viewpoint.

A few more facts should be placed here. First, since E and E^c are mutually exclusive and $E \cup E^c = S$, then $P(S) = P(E \cup E^c) = P(E) + P(E^c) = 1$. Additionally, we can compute the formula for $P(E_1 \cup E_2)$ which is the probability that either E_1 or E_2 occurs. From Fig. 12.1 it can be seen that the probability $P(E_1) + P(E_2)$, which represents all points in E_1 and all points in E_2, is given by the gray regions. Notice, in this calculation, that the overlap region is counted twice. Thus the intersection must be subtracted so that we find

$$P(E_1 \cup E_2) = P(E_1) + P(E_2) - P(E_1 E_2). \tag{12.1.1}$$

If the two events E_1 and E_2 are mutually exclusive then $E_1 E_2 = \emptyset$ and $P(E_1 \cup E_2) = P(E_1) + P(E_2)$ (note that $P(\emptyset) = 0$).

As an example, consider tossing two fair coins so that each of the points in the sample space $S = \{(H, H), (H, T), (T, H), (T, T)\}$ is equally likely to occur, or said another way, each has probability of 1/4 to occur. Then let $E_1 = \{(H, H), (H, T)\}$ be the event where the first coin is a head, and let $E_2 = \{(H, H), (T, H)\}$ be the event that the second coin is a head. Then

$$P(E_1 \cup E_2) = P(E_1) + P(E_2) - P(E_1 E_2)$$
$$= \frac{1}{2} + \frac{1}{2} - P\{(H, H)\} = \frac{1}{2} + \frac{1}{2} - \frac{1}{4} = \frac{3}{4}.$$

Of course, this could have also easily been computed directly from the union $P(E_1 \cup E_2) = P(\{(H, H), (H, T), (T, H)\}) = 3/4$.

Conditional probabilities

Conditional probabilities concern, in some sense, the interdependency of events. Specifically, given that the event E_1 has occurred, what is the probability of E_2 occurring? This conditional probability is denoted as $P(E_2|E_1)$. A general formula for the conditional probability can be derived from the following argument: If the event E_1 occurs, then in order for E_2 to occur, it is necessary that the actual occurrence be in the intersection $E_1 E_2$. Thus in the conditional probability way of thinking, E_1 becomes our new sample space, and the event that $E_1 E_2$ occurs will equal the probability of $E_1 E_2$ relative to the probability of E_1. Thus we have the conditional probability definition

$$P(E_2|E_1) = \frac{P(E_2 E_1)}{P(E_1)}. \tag{12.1.2}$$

The following are examples of some basic conditional probability events.

1   If the experiment consists of flipping two fair coins, what is the probability that both are heads given that at least one of them is heads? With the sample space $S = \{(H, H), (H, T), (T, H), (T, T)\}$ and each outcome as likely to occur as the other, let E_2 denote the event that both coins are heads, and E_1 denote the event that at least one of them is heads. Then

$$P(E_2|E_1) = \frac{P(E_2 E_1)}{P(E_1)} = \frac{P(\{(H, H)\})}{P(\{(H, H), (H, T), (T, H)\})} = \frac{1/4}{3/4} = \frac{1}{3}.$$

2   Suppose cards numbered one through ten are placed in a hat, mixed and drawn. If we are told that the number on the drawn card is at least five, then what is the probability that it is ten? In this case, E_2 is the probability ($= 1/10$) that the card is a ten, while E_1 is the probability ($= 6/10$) that it is at least a five. Then

$$P(E_2|E_1) = \frac{1/10}{6/10} = \frac{1}{6}.$$

Independent events

Independent events will be of significant interest in this book. More specifically, the definition of independent events determines whether or not an event E_1 has any impact, or is correlated, with a second event E_2. By definition, two events are said to be *independent* if

$$P(E_1E_2) = P(E_1)P(E_2).$$

(12.1.3)

By Eq. (12.1.2), this implies that E_1 and E_2 are independent if

$$P(E_2|E_1) = P(E_2).$$

(12.1.4)

In other words, the probability of obtaining E_2 does not depend upon whether E_1 occurred. Two events that are not independent are said to be *dependent*.

As an example, consider tossing two fair die where each possible outcome is an equal 1/36. Then let E_1 denote the event that the sum of the dice is six, E_2 denote the event that the first die equals four and E_3 denote the event that the sum of the dice is seven. Note that

$$P(E_1E_2) = P(\{(4,2)\}) = 1/36$$

$$P(E_1)P(E_2) = 5/36 \times 1/6 = 5/216$$

$$P(E_3E_2) = P(\{(4,3)\}) = 1/36$$

$$P(E_3)P(E_2) = 1/6 \times 1/6 = 1/36.$$

Thus we can find that E_2 and E_1 cannot be independent while E_3 and E_2 are, in fact, independent. Why is this true? In the first case where a total of six (E_2) on the die is sought, then the roll of the first die is important as it must be below six in order to satisfy this condition. On the other hand, a total of seven can be accomplished (E_3) regardless of the first die roll.

Bayes's formula

Consider two events E_1 and E_2. Then the event E_1 can be expressed as follows

$$E_1 = E_1E_2 \cup E_1E_2^c.$$

(12.1.5)

This is true since in order for a point to be in E_1, it must either be in E_1 and E_2, or it must be in E_1 and not in E_2. Since the intersections E_1E_2 and $E_1E_2^c$ are obviously mutually exclusive, then

$$P(E_1) = P(E_1E_2) + P(E_1E_2^c)$$

$$= P(E_1|E_2)P(E_2) + P(E_1|E_2^c)P(E_2^c)$$

$$= P(E_1|E_2)P(E_2) + P(E_1|E_2^c)(1 - P(E_2)).$$

(12.1.6)

This states that the probability of the event E_1 is a weighted average of the conditional probability of E_1 given that E_2 has occurred and the conditional probability of E_1 given that E_2 has not occurred, with each conditional probability being given as much weight as the event it is conditioned on has of occurring. To illustrate the application of Bayes's formula, consider the following two examples:

1   Consider two boxes: the first containing two white and seven black balls, and the second containing five white and six black balls. Now flip a fair coin and draw a ball from the first or second box depending on whether you get heads or tails. What is the conditional probability that the outcome of the toss was heads given that a white ball was selected? For this problem, let W be the event that a white ball was drawn and H be the event that the coin came up heads. The solution to this problem involves the calculation of $P(H|W)$. This can be calculated as follows:

$$P(H|W) = \frac{P(HW)}{P(W)} = \frac{P(W|H)P(H)}{P(W)}$$

$$= \frac{P(W|H)P(H)}{P(W|H)P(H) + P(W|H^c)P(H^c)}$$

$$= \frac{2/9 \times 1/2}{2/9 \times 1/2 + 5/11 \times 1/2} = \frac{22}{67}. \qquad (12.1.7)$$

2   A laboratory blood test is 95% effective in determining a certain disease when it is present. However, the test also yields a false positive result for 1% of the healthy persons tested. If 0.5% of the population actually has the disease, what is the probability a person has the disease given that their test result is positive? For this problem, let D be the event that the person tested has the disease, and E the event that their test is positive. The problem involves calculation of $P(D|E)$ which can be obtained by

$$P(D|E) = \frac{P(DE)}{P(E)}$$

$$= \frac{P(E|D)P(D)}{P(E|D)P(D) + P(E|D^c)P(D^c)}$$

$$= \frac{(0.95)(0.005)}{(0.95)(0.005) + (0.01)(0.995)}$$

$$= \frac{95}{294} \approx 0.323 \rightarrow 32\%.$$

In this example, it is clear that if the disease is very rare so that $P(D) \rightarrow 0$, it becomes very difficult to actually test for it since even with a 95% accuracy, the chance of the false positives means that the chance of a person actually having the disease is quite small, making the test not so worthwhile. Indeed, if this calculation is redone with the probability that one in a million people of the population have the disease ($P(D) = 10^{-6}$), then $P(D|E) \approx 1 \times 10^{-4} \rightarrow 0.01\%$.

Equation (12.1.6) can be generalized in the following manner: suppose $E_1, E_2, \cdots, E_N$ are mutually exclusive events such that $\cup_{n=1}^{N} E_n = S$. Thus exactly one of the events E_n occurs. Further, consider an event H so that

$$P(H) = \sum_{n=1}^{N} P(HE_n) = \sum_{n=1}^{N} P(H|E_n)P(E_n).$$

(12.1.8)

Thus for the given events E_n, one of which must occur, $P(H)$ can be computed by conditioning upon which of the E_n actually occurs. Said another way, $P(H)$ is equal to the weighted average of $P(H|E_n)$, each term being weighted by the probability of the event on which it is conditioned. Finally, using this result we can compute

$$P(E_n|H) = \frac{P(HE_n)}{P(H)} = \frac{P(H|E_n)P(E_n)}{\sum_{n=1}^{N} P(H|E_n)P(E_n)}$$

(12.1.9)

which is known as Bayes's formula.

 ## 12.2 Random Variables and Statistical Concepts

Ultimately in probability theory we are interested in the idea of a *random variable*. This is typically the outcome of some experiment that has an undetermined outcome, but whose sample space S and potential events E can be characterized. In the examples already presented, the toss of a die and flip of a coin represent two experiments whose outcome is not known until the experiment is performed. A random variable is typically defined as some real-valued function on the sample space. The following are some examples.

1 Let X denote the random variable which is defined as the sum of two fair dice. Then the random variable is assigned the values

$$P\{X = 2\} = P\{(1, 1)\} = 1/36$$
$$P\{X = 3\} = P\{(1, 2), (2, 1)\} = 2/36$$
$$P\{X = 4\} = P\{(1, 3), (3, 1), (2, 2,)\} = 3/36$$
$$P\{X = 5\} = P\{(1, 4), (4, 1), (2, 3), (3, 2)\} = 4/36$$
$$P\{X = 6\} = P\{(1, 5), (5, 1), (2, 4), (4, 2), (3, 3)\} = 5/36$$
$$P\{X = 7\} = P\{(1, 6), (6, 1), (2, 5), (5, 2), (3, 4), (4, 3)\} = 6/36$$
$$P\{X = 8\} = P\{(2, 6), (6, 2), (3, 5), (5, 3), (4, 4)\} = 5/36$$
$$P\{X = 9\} = P\{(3, 6), (6, 3), (4, 5), (5, 4)\} = 4/36$$
$$P\{X = 10\} = P\{(4, 6), (6, 4), (5, 5)\} = 3/36$$
$$P\{X = 11\} = P\{(5, 6), (6, 5)\} = 2/36$$
$$P\{X = 12\} = P\{(6, 6)\} = 1/36.$$

Thus the random variable X can take on integer values between 2 and 12 with the probability for each assigned above. Since the above events are all possible outcomes which are mutually exclusive, then

$$P\left(\cup_{n=2}^{12}\{X=n\}\right)=\sum_{n=2}^{12}P\{X=n\}=1.$$

Figure 12.2 illustrates the probability as a function of X as well as its distribution function.

2   Suppose a coin is tossed having probability p of coming up heads, until the first head appears. Then define the random variable N which denotes the number of flips required to generate the first head. Thus

$$P\{N=1\}=P\{H\}=p$$
$$P\{N=2\}=P\{(T,H)\}=(1-p)p$$
$$P\{N=3\}=P\{(T,T,H)\}=(1-p)^2p$$
$$\vdots$$
$$P\{N=n\}=P\{(T,T,\cdots,T,H)\}=(1-p)^{n-1}p$$

and again the random variable takes on discrete values.

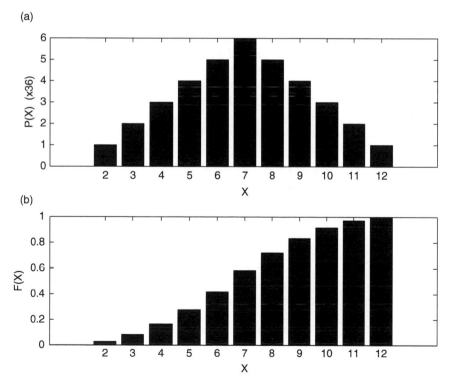

Figure 12.2: (a) Random variable probability assignments $P(X)$ for the sum of two fair dice, and (b) the distribution function $F(X)$ associated with the toss of two fair dice.

In the above examples, the random variables took on a finite, or countable, number of possible values. Such random variables are *discrete* random variables. The cumulative distribution function, or more simply the distribution function $F(\cdot)$ of the random variable X (see Fig. 12.2) is defined for any real number b $(-\infty < b < \infty)$ by

$$F(b) = P\{X \leq b\}. \tag{12.2.1}$$

Thus the $F(b)$ denotes the probability of a random variable X taking on a value which is less than or equal to b. Some properties of the distribution function are as follows:

- $F(b)$ is a nondecreasing function of b;
- $\lim_{b \to \infty} F(b) = F(\infty) = 1$;
- $\lim_{b \to -\infty} F(b) = F(-\infty) = 0$.

Figure 12.2(b) illustrates these properties quite nicely. Any probability question about X can be answered in terms of the distribution function. The most common and useful example of this is to consider

$$P\{a < X \leq b\} = F(b) - F(a) \tag{12.2.2}$$

for all $a < b$. This result, which gives the probability of $X \in (a, b]$, is fairly intuitive as it involves first computing the probability that $X \leq b$ $(F(b))$ and then subtracting from this the probability that $X \leq a$ $(F(a))$.

For a discrete random variable, the overall probability function can be defined using the *probability mass function* of X as

$$p(a) = P\{X = a\}. \tag{12.2.3}$$

The probability mass function $p(a)$ is positive for at most a countable number of values a so that

$$p(x) = \begin{cases} > 0 & x = x_n \ (n = 1, 2, \cdots, N) \\ = 0 & x \neq x_n. \end{cases} \tag{12.2.4}$$

Further, since X must take on one of the values of x_n, then

$$\sum_{n=1}^{N} p(x_n) = 1. \tag{12.2.5}$$

The cumulative distribution is defined as follows:

$$F(a) = \sum_{x_n < a} p(x_n). \tag{12.2.6}$$

In contrast to discrete random variables, *continuous random variables* have an uncountable set of possible values. Many of the definitions and ideas presented thus far for discrete random variables are then easily transferred to the continuous context. Letting X be a continuous random variable, then there exists a nonnegative function $f(x)$ defined for all real $x \in (-\infty, \infty)$ having the property that for any set of real numbers

$$P\{X \in B\} = \int_B f(x)dx.$$ (12.2.7)

The function $f(x)$ is called the *probability density function* of the random variable X. Thus the probability that X will be in B is determined by integrating over the set B. Since X must take on some value in $x \in (-\infty, \infty)$, then

$$P\{X \in (-\infty, \infty)\} = \int_{-\infty}^{\infty} f(x)dx = 1.$$ (12.2.8)

Any probability statement about X can be completely determined in terms of $f(x)$. If we determine the probability that the event $B \in [a, b]$ then

$$P\{a \le X \le b\} = \int_a^b f(x)dx.$$ (12.2.9)

If in this calculation we let $a = b$, then

$$P\{X = a\} = \int_a^a f(x)dx = 0.$$ (12.2.10)

Thus the probability that a continuous random variable assumes a *particular* value is zero.

The cumulative distribution function $F(\cdot)$ can also be defined for a continuous random variable. Indeed, it has a simple relationship to the probability density function $f(x)$. Specifically,

$$F(a) = P\{X \in (-\infty, a]\} = \int_{-\infty}^a f(x)dx.$$ (12.2.11)

Thus the density is the derivative of the cumulative distribution function. An important interpretation of the density function is obtained by considering the following

$$P\left\{a - \frac{\epsilon}{2} \le X \le a + \frac{\epsilon}{2}\right\} = \int_{a-\epsilon/2}^{a+\epsilon 2} f(x)dx \approx \epsilon f(a).$$ (12.2.12)

The *sifting* property of this integral illustrates that the density function $f(a)$ measures how likely it is that the random variable will be near a.

Having established key properties and definitions concerning discrete and random variables, it is illustrative to consider some of the most common random variables used in practice. Thus consider first the following discrete random variables:

1 **Bernoulli random variable** Consider a trial whose outcome can either be termed a success or failure. Let the random variable $X = 1$ if there is a success and $X = 0$ if there is failure. Then the probability mass function of X is given by

$$p(0) = P\{X = 0\} = 1 - p$$
$$p(1) = P\{X = 1\} = p$$

where p $(0 \le p \le 1)$ is the probability that the trial is a success.

2 **Binomial random variable** Suppose n independent trials are performed, each of which results in a success with probability p and failure with probability $1 - p$. Denote X as the number of successes that occur in n such trials. Then X is a binomial random variable with parameters (n, p) and the probability mass function is given by

$$p(j) = \binom{n}{j} p^j (1 - p)^{n-j}$$

where $\binom{n}{j} = n!/((n - j)!j!)$ is the number of different groups of j objects that can be chosen from a set of n objects. For instance, if $n = 3$ and $j = 2$, then this binomial coefficient is $3!/(1!2!)=3$ representing the fact that there are three ways to have two successes in three trials: $(s, s, f), (s, f, s), (f, s, s)$.

3 **Poisson random variable** A Poisson random variable with parameter λ is defined such that

$$p(n) = P\{X = n\} = e^{-\lambda} \frac{\lambda_n}{n!} \quad n = 0, 1, 2, \cdots$$

for some $\lambda > 0$. The Poisson random variable has an enormous range of applications across the sciences as it represents, to some extent, the exponential distribution of many systems.

In a similar fashion, a few of the more common continuous random variables can be highlighted here:

1 **Uniform random variable** A random variable is said to be uniformly distributed over the interval $(0, 1)$ if its probability density function is given by

$$f(x) = \begin{cases} 1, & 0 < x < 1 \\ 0, & \text{otherwise.} \end{cases}$$

More generally, the definition can be extended to any interval (α, β) with the density function being given by

$$f(x) = \begin{cases} \dfrac{1}{\beta - \alpha}, & \alpha < x < \beta \\ 0, & \text{otherwise.} \end{cases}$$

The associated distribution function is then given by

$$F(a) = \begin{cases} 0, & a < \alpha \\ \dfrac{a - \alpha}{\beta - \alpha}, & \alpha < a < \beta \\ 1, & a \geq \beta. \end{cases}$$

2 **Exponential random variable** Given a positive constant $\lambda > 0$, the density function for this random variable is given by

$$f(x) = \begin{cases} \lambda e^{-\lambda x}, & x \geq 0 \\ 0, & x \leq 0. \end{cases}$$

This is the continuous version of the Poisson distribution considered for the discrete case. Its distribution function is given by

$$F(a) = \int_0^a \lambda e^{-\lambda x} dx = 1 - e^{-\lambda a}, \quad a > 0.$$

3 **Normal random variable** Perhaps the most important of them all is the normal random variable with parameters μ and σ^2 defined by

$$f(x) = \frac{1}{\sqrt{2\pi\sigma^2}} e^{-(x-\mu)^2/2\sigma^2}, \quad -\infty < x < \infty.$$

Although we have not defined them yet, the parameters μ and σ refer to the mean and variance of this distribution, respectively. Further, the Gaussian shape produces the ubiquitous bell-shaped curve that is always talked about in the context of large samples, or perhaps grading. Its distribution is given by the error function $erf(a)$, one of the most common functions for look-up tables in mathematics.

Expectation, moments and variance of random variables

The expected value of a random variable is defined as follows:

$$E[X] = \sum_{x:p(x)>0} xp(x)$$

for a discrete random variable, and

$$E[X] = \int_{-\infty}^{\infty} xf(x)dx$$

for a continous random variable. Both of these definitions illustrate that the expectation is a weighted average whose weight is the probability that X assumes that value. Typically, the expectation is interpreted as the average value of the random variable. Some examples will serve to illustrate this point:

1 The expected value $E[X]$ of the roll of a fair die is given by

$$E[X] = 1(1/6) + 2(1/6) + 3(1/6) + 4(1/6) + 5(1/6) + 6(1/6) = 7/2.$$

2 The expectation $E[X]$ of a Bernoulli variable is

$$E[X] = 0(1 - p) + 1(p) = p$$

showing that the expected number of successes in a single trial is just the probability that the trial will be a success.

3 The expectation $E[X]$ of a binomial or Poisson random variable is

$$E[X] = np \text{ binomial}$$
$$E[X] = \lambda \text{ Poisson.}$$

4   The expectation of a continuous uniform random variable is

$$E[X] = \int_{\alpha}^{\beta} \frac{x}{\beta - \alpha} dx = \frac{\alpha + \beta}{2}.$$

5   The expectation $E[X]$ of an exponential or normal random variable is

$$E[X] = 1/\lambda \text{ exponential}$$
$$E[X] = \mu \quad \text{normal}.$$

The idea of expectation can be generalized beyond the standard expectation $E[X]$. Indeed, we can consider the expectation of any function of the random variable X. Thus to compute the expectation of some function $g(X)$, the discrete and continuous expectations become:

$$E[g(X)] = \sum_{x:p(x)>0} g(x)p(x) \tag{12.2.13a}$$

$$E[g(X)] = \int_{-\infty}^{\infty} g(x)f(x)dx. \tag{12.2.13b}$$

This also then leads us to the idea of higher moments of a random variable, i.e. $g(X) = X^n$ so that

$$E[X^n] = \sum_{x:p(x)>0} x^n p(x) \tag{12.2.14a}$$

$$E[X^n] = \int_{-\infty}^{\infty} x^n f(x)dx. \tag{12.2.14b}$$

The first moment is essentially the mean, the second moment will be related to the variance, and the third and fourth moments measure the *skewness* (asymmetry of the probability distribution) and the *kurtosis* (the flatness or sharpness) of the distribution.

Now that these probability ideas are in place, the quantities of greatest and most practical interest can be considered, namely the ideas of mean and variance (and standard deviation, which is defined as the square root of the variance). The mean has already been defined, while the variance of a random variable is given by

$$Var[X] = E[(X - E[X])^2] = E[X^2] - E[X]^2. \tag{12.2.15}$$

The second form can be easily manipulated from the first. Note that depending upon your scientific community of origin $E[X] = \bar{x} = \langle x \rangle$. It should be noted that the normal distribution has $Var(X) = \sigma^2$, standard deviation σ and skewness of zero.

Joint probability distribution and covariance

The mathematical framework is now in place to discuss what is perhaps, for this book, the most important use of our probability and statistical thinking, namely the relation between two or more random variables. Ultimately, this assesses whether or not two variables depend upon each other. A joint cumulative probability distribution function of two random variables X and Y is given by

$$F(a, b) = P\{X \leq a, Y \leq b\}, \quad -\infty < a, b < \infty. \tag{12.2.16}$$

The distribution of X or Y can be obtained by considering

$$F_X(a) = P\{X \leq a\} = P\{X \leq a, Y \leq \infty\} = F(a, \infty) \tag{12.2.17a}$$

$$F_Y(b) = P\{Y \leq b\} = P\{Y \leq b, X \leq \infty\} = F(\infty, b). \tag{12.2.17b}$$

The random variables X and Y are jointly continuous if there exists a function $f(x, y)$ defined for real x and y so that for any set of real numbers A and B

$$P\{X \in A, Y \in B\} = \int_A \int_B f(x, y) dx dy. \tag{12.2.18}$$

The function $f(x, y)$ is called the joint probability density function of X and Y.

The probability density of X or Y can be extracted from $f(x, y)$ as follows:

$$P\{X \in A\} = P\{X \in A, Y \in (-\infty, \infty)\} = \int_{-\infty}^{\infty} \int_A f(x, y) dx dy = \int_A f_X(x) dx \tag{12.2.19}$$

where $f_X(x) = \int_{-\infty}^{\infty} f(x, y) dy$. Similarly, the probability density function of Y is $f_Y(y) = \int_{-\infty}^{\infty} f(x, y) dx$. As previously, to determine a function of the random variables X and Y, then

$$E[g(X, Y)] = \begin{cases} \sum_y \sum_x g(x, y) p(x, y) & \text{discrete} \\ \int_{-\infty}^{\infty} \int_{-\infty}^{\infty} g(x, y) f(x, y) dx dy & \text{continuous.} \end{cases} \tag{12.2.20}$$

Jointly distributed variables lead naturally to a discussion of whether the variables are independent. Two random variables are said to be independent if

$$P\{X \leq a, Y \leq b\} = P\{X \leq a\} P\{Y \leq b\}. \tag{12.2.21}$$

In terms of the joint distribution function F of X and Y, independence is established if $F(a, b) = F_X(a) F_Y(b)$ for all a, b. The independence assumption essentially reduces to the following:

$$f(x, y) = f_X(x) f_Y(y). \tag{12.2.22}$$

This further implies

$$E[g(X)h(Y)] = E[g(X)]E[h(Y)]. \tag{12.2.23}$$

Mathematically, the case of independence allows for essentially a separation of the random variables X and Y.

What is, to our current purposes, more interesting is the idea of dependent random variables. For such variables, the concept of covariance can be introduced. This is defined as

$$\mathrm{Cov}(X, Y) = E[(X - E[X])(Y - E[Y])] = E[XY] - E[X]E[Y]. \qquad (12.2.24)$$

If two variables are independent then $\mathrm{Cov}(X,Y)=0$. Alternatively, as $\mathrm{Cov}(X,Y)$ approaches unity, it implies that X and Y are essentially directly, or strongly, correlated random variables. This will serve as a basic measure of the statistical dependence of our data sets. It should be noted that although two independent variables implies $\mathrm{Cov}(X,Y)=0$, the converse is not necessarily true, i.e. $\mathrm{Cov}(X,Y)=0$ does not imply independence of X and Y.

MATLAB commands

The following are some of the basic commands in MATLAB that we will use for computing the above mentioned probability and statistical concepts.

1 **rand(m,n)**: This command generates an $m \times n$ matrix of random numbers from the uniform distribution on the unit interval.

2 **randn(m,n)**: This command generates an $m \times n$ matrix of normally distributed random numbers with zero mean and unit variance.

3 **mean(X)**: This command returns the mean or average value of each column of the matrix **X**.

4 **var(X)**: This computes the variance of each row of the matrix **X**. To produce an unbiased estimator, the variance of the population is normalized by $N - 1$ where N is the sample size.

5 **std(X)**: This computes the standard deviation of each row of the matrix **X**. To produce an unbiased estimator, the standard deviation of the population is normalized by $N - 1$ where N is the sample size.

6 **cov(X)**: If **X** is a vector, it returns the variance. For matrices, where each row is an observation, and each column a variable, the command returns a covariance matrix. Along the diagonal of this matrix is the variance of each column.

 ## 12.3 Hypothesis Testing and Statistical Significance

With the key concepts of probability theory in hand, statistical testing of data can now be pursued. But before doing so, a few key results concerning a large number of random variables or large number of trials should be explicitly stated.

Limit theorems

Two theorems of probability theory are important to establish before discussing testing of hypothesis via statistics: they are the *strong law of large numbers* and the *central limit theorem*. They are, perhaps, the most well-known and most important practical results in probability theory.

Theorem (Strong law of large numbers): *Let $X_1, X_2, \cdots$ be a sequence of independent random variables having a common distribution, and let $E[X_j] = \mu$. Then, with probability one,*

$$\frac{X_1 + X_2 + \cdots + X_n}{n} \to \mu \ \text{ as } \ n \to \infty. \tag{12.3.1}$$

As an example of the strong law of large numbers, suppose that a sequence of independent trials is performed. Let E be a fixed event and denote by $P(E)$ the probability that E occurs on any particular trial. Let

$$X_j = \begin{cases} 1, & \text{if } E \text{ occurs on the } j\text{th trial} \\ 0, & \text{if } E \text{ does not occur on the } j\text{th trial.} \end{cases} \tag{12.3.2}$$

The theorem states that with probability one

$$\frac{X_1 + X_2 + \cdots + X_n}{n} \to E[X] = P(E). \tag{12.3.3}$$

Since $X_1 + X_2 + \cdots + X_n$ represents the number of times that the event E occurs in n trials, this result can be interpreted as the fact that, with probability one, the limiting proportion of time that the event E occurs is just $P(E)$.

Theorem (Central limit theorem): *Let $X_1, X_2, \cdots$ be a sequence of independent random variables each with mean μ and variance σ^2. Then the distribution of the quantity*

$$\frac{X_1 + X_2 + \cdots + X_n - n\mu}{\sigma\sqrt{n}} \tag{12.3.4}$$

tends to the standard normal distribution as $n \to \infty$ so that

$$P\left\{\frac{X_1 + X_2 + \cdots + X_n - n\mu}{\sigma\sqrt{n}} \leq a\right\} \to \frac{1}{\sqrt{2\pi}} \int_{-\infty}^{a} e^{-x^2/2} dx \tag{12.3.5}$$

as $n \to \infty$.

The most important part of this result: it holds for *any* distribution of X_j. Thus for a large enough sampling of data, i.e. a sequence of independent random variables that goes to infinity, one should observe the ubiquitous bell-shaped probability curve of the normal distribution. This is a powerful result indeed, provided you have a large enough sampling. And herein lies both its power, and susceptibility to overstating results. The fact is, the normal distribution has some very beautiful mathematical properties, leading us to consider this central limit almost exclusively when testing for statistical significance.

As an example, consider a binomially distributed random variable X with parameters n (n is the number of events) and p (p is the probability of a success of an event). The distribution of

$$\frac{X - E[X]}{\sqrt{Var[X]}} = \frac{X - np}{\sqrt{np(1 - p)}} \tag{12.3.6}$$

approaches the standard normal distribution as $n \to \infty$. Empirically, it has been found that the normal approximation will, in general, be quite good for values of n satisfying $np(1 - p) \geq 10$.

Statistical decisions

In practice, the power of probability and statistics comes into play in making decisions, or drawing conclusions, based upon a limited sampling of information of an entire *population* (of data). Indeed, it is often impossible or impractical to examine the entire population. Thus a sample of the population is considered instead. This is called the *sample*. The goal is to infer facts or trends about the entire population with this limited sample size. The process of obtaining samples is called *sampling*, while the process of drawing conclusions about this sample is called *statistical inference*. Using these ideas to make decisions about a population based upon the sample information is termed *statistical decisions*.

In general, these definitions are quite intuitive and also quite well known to us. Statistical inference, for instance, is used extensively in polling data for elections. In such a scenario, county, state, and national election projections are made well before the election based upon sampling a small portion of the population. Statistical decision making on the other hand is used extensively to determine whether a new medical treatment or drug is, in fact, effective in curing a disease or ailment.

In what follows, the focus will be upon statistical decision making. In attempting to reach a decision, it is useful to make assumptions or guesses about the population involved. Much like an analysis proof, you can reach a conclusion by assuming something is true and proving it to be so, or assuming the opposite is true and proving that it is false, for instance, by a counter-example. There is no difference with statistical decision making. In this case, all the statistical tests are hypothesis driven. Thus a *statistical hypothesis* is proposed and its validity investigated. As with analysis, some hypotheses are explicitly constructed for the purpose of rejection, or nullifying the hypothesis. For example, to decide whether a coin is fair or loaded, the hypothesis is formulated that the coin is fair with $p = 0.5$ as the probability of heads. To decide if one procedure is better or worse than another, the hypothesis formulated is that there is no difference between the procedures. Such hypotheses are called *null hypotheses* and denoted by H_0. Any hypothesis which differs from a given hypothesis is called an *alternative hypothesis* and is denoted by H_1.

Tests for statistical significance

The ultimate goal in statistical decision making is to develop tests that are capable of allowing us to accept or reject a hypothesis with a certain, quantifiable confidence level. Procedures that allow us to do this are called *tests of hypothesis, tests of significance* or *rules of decision*. Overall, they fall within the aegis of the ideas of statistical significance.

In making such statistical decisions, there is some probability that an erroneous conclusion will be reached. Such errors are classified as follows:

- **Type I error**: If a hypothesis is rejected when it should be accepted.
- **Type II error**: If a hypothesis is accepted when it should be rejected.

As an example, suppose a fair coin ($p = 0.5$ for achieving heads or tails) is tossed 100 times and unbelievably, 100 heads in a row results. Any statistical test based upon assuming that the coin is fair is going to be rejected and a Type I error will occur. If, on the other hand, a loaded coin twice as likely to produce a heads than a tails ($p = 2/3$ for achieving a heads and $1 - p = 1/3$ for producing a tails) is tossed 100 times and heads and tails both appear 50 times, then a hypothesis about a biased coin will be rejected and a Type II error will be made.

What is important in making the statistical decision is the *level of significance* of the test. This probability is often denoted by α, and in practice it is often taken to be 0.05 or 0.01. These correspond to a 5% or 1% level of significance, or alternatively, we are 95% or 99% confident, respectively, that our decision is correct. Basically, in these tests, we will be wrong 5% or 1% of the time.

Hypothesis testing with the normal distribution

For large samples, the central limit theorem asserts that the sample statistics have a normal distribution (or nearly so) with mean μ_s and standard deviation σ_s. Thus many statistical decision tests are specifically geared to the normal distribution. Consider the normal distribution with mean μ and variance σ^2 such that the density is given by

$$f(x) = \frac{1}{\sqrt{2\pi\sigma^2}} e^{-(x-\mu)^2/2\sigma^2} \qquad -\infty < x < \infty \tag{12.3.7}$$

and where the cumulative distribution is given by

$$F(x) = P(X \le x) = \frac{1}{\sqrt{2\pi\sigma^2}} \int_{-\infty}^{x} e^{-(x-\mu)^2/2\sigma^2} dx. \tag{12.3.8}$$

With the change of variable

$$Z = \frac{X - \mu}{\sigma} \tag{12.3.9}$$

the *standard normal density function* is given by

$$f(z) = \frac{1}{\sqrt{2\pi}} e^{-x^2/2} \tag{12.3.10}$$

with the corresponding distribution function (related to the error function)

$$F(z) = \frac{1}{\sqrt{2\pi}} \int_{-\infty}^{z} e^{-u^2/2} du = \frac{1}{2}\left[1 + \text{erf}\left(\frac{z}{\sqrt{2}}\right)\right]. \tag{12.3.11}$$

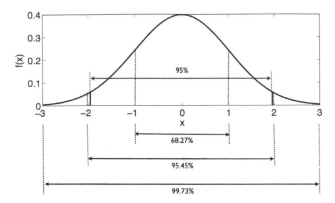

Figure 12.3: Standard normal distribution curve $f(x)$ with the first, second and third standard deviation markers below the graph along with the corresponding percentage of the population contained within each. Above the graph, markers are placed at $x = -1.96$ and $x = 1.96$ within which 95% of the population is contained. Outside of this range, a statistical test with 95% confidence would result in a conclusion that a given hypothesis is false.

The standard normal density function has mean zero ($\mu = 0$) and unit variance ($\sigma = 1$). The standard normal curve is shown in Fig. 12.3. In this graph, several key features are demonstrated. On the bottom, the first, second, and third standard deviation markers indicate that 68.72%, 95.45% and 99.73% percent of the population are contained within the respective standard deviation markers. Above the graph, a line is drawn at $x = -1.96$ and $x = 1.96$ indicated by the 95% confidence markers. Indeed, the graph indicates two key regions: the *region of acceptance of the hypothesis* for $x \in [-1.96, 1.96]$ and the *region of nonsignificance* for $|x| > 1.96$.

In the above example, interest was given to the extreme values of the statistics, or the corresponding z-value on both sides of the mean. Thus both tails of the distribution mattered. This is then called a *two-tailed test*. However, some hypothesis tests are only concerned with one side of the mean. Such a test, for example, may be evaluating whether one process is better than another versus whether it is better or worse. This is called a *one-tailed test*. Some examples will serve to illustrate the concepts. It should also be noted that a variety of tests can be formulated for determining confidence intervals, including the Student's t-test (involving measurements on the mean), the chi-square test (involving tests on the variance), and the F-test (involving ratios of variances). These will not be discussed in detail here.

Example: A fair coin

Design a decision rule to test the null hypothesis H_0 that a coin is fair if a sample of 64 tosses of the coin is taken and if the level of significance is 0.05 or 0.01.

The coin toss is an example of a binomial distribution as a success (heads) and failure (tails). Under the null hypothesis that the coin is fair ($p = 0.5$ for achieving either heads or tails) the binomial mean and standard deviation can be easily calculated for n events:

$$\mu = np = 64 \times 0.5 = 32$$

$$\sigma = \sqrt{np(1-p)} = \sqrt{64 \times 0.5 \times 0.5} = 4.$$

The statistical decision must now be formulated in terms of the standard normal variable Z where

$$Z = \frac{X - \mu}{\sigma} = \frac{X - 32}{4}.$$

For the hypothesis to hold with 95% confidence (or 99% confidence), then $z \in [-1.96, 1.96]$. Thus the following must be satisfied:

$$-1.96 \leq \frac{X - 32}{4} \leq 1.96 \quad \rightarrow \quad 24.16 \leq X \leq 39.84.$$

Thus if in a trial of 64 flips heads appears between 25 and 39 times inclusive, the hypothesis is true with 95% confidence. For 99% confidence, heads must appear between 22 and 42 times inclusive.

Example: Class scores

An examination was given to two classes of 40 and 50 students, respectively. In the first class, the mean grade was 74 with a standard deviation of 8, while in the second class the mean was 78 with standard deviation 7. Is there a statistically significant difference between the performance of the two classes at a level of significance of 0.05?

Consider that the two classes come from two populations having respective means μ_1 and μ_2. Then the following null hypothesis and alternative hypothesis can be formulated:

$H_0 : \mu_1 = \mu_2$, and the difference is merely due to chance

$H_1 : \mu_1 \neq \mu_2$, and there is a significant difference between classes.

Under the null hypothesis H_0, both classes come from the same population. The mean and standard deviation of the differences of these is given by

$$\mu_{\bar{X}_1 - \bar{X}_2} = 0$$

$$\sigma_{\bar{X}_1 - \bar{X}_2} = \sqrt{\frac{\sigma_1^2}{n_1} + \frac{\sigma_2^2}{n_2}} = \sqrt{\frac{8^2}{40} + \frac{7^2}{50}} = 1.606$$

where the standard deviations have been used as estimates for σ_1 and σ_2. The problem is thus formulated in terms of a new variable representing the difference of the means

$$Z = \frac{\bar{X}_1 - \bar{X}_2}{\sigma_{\bar{X}_1 - \bar{X}_2}} = \frac{74 - 78}{1.606} = -2.49.$$

(a) For a two-tailed test, the results are significant if $|Z| > 1.96$. Thus it must be concluded that there is a significant difference in performance between the two classes with the second class probably being better

(b) For the same test with 0.01 significance, then the results are significant for $|Z| > 2.58$ so that at a 0.01 significance level there is no statistical difference between the classes.

Example: Rope breaking (Student's t-distribution)

A test of strength of six ropes manufactured by a company showed a breaking strength of 7750 lb and a standard deviation of 145 lb, whereas the manufacturer claimed a mean breaking strength of 8000 lb. Can the manufacturers claim be supported at a significance level of 0.05 or 0.01?

The decision is between the hypotheses

$$H_0 : \mu = 8000 \text{ lb, the manufacturer is justified}$$

$$H_1 : \mu < 8000 \text{ lb, the manufacturer's claim is unjustified}$$

where a one-tailed test is to be considered. Under the null hypothesis H_0 and the small sample size, the Student's t-test can be used so that

$$T = \frac{\bar{X} - \mu}{S}\sqrt{n-1} = \frac{7750 - 8000}{145}\sqrt{6-1} = -3.86.$$

As $n \to \infty$, the Student's t-distribution goes to the normal distribution. But in this case of a small sample, the degree of freedom of this distribution is $6 - 1 = 5$ and a 0.05 (0.01) significance is achieved with $T > -2.01$ (or $T > -3.36$). In either case, with $T = -3.86$ it is extremely unlikely that the manufacture's claim is justified.

Time–Frequency Analysis: Fourier Transforms and Wavelets

13

Fourier transforms, and more generally, spectral transforms, are one of the most powerful and efficient techniques for solving a wide variety of problems arising in the physical, biological and engineering sciences. The key idea of the Fourier transform, for instance, is to represent functions and their derivatives as sums of cosines and sines. This operation can be done with the fast Fourier transform (FFT) which is an $O(N \log N)$ operation. Thus the FFT is faster than most linear solvers of $O(N^2)$. The basic properties and implementation of the FFT will be considered here along with other time–frequency analysis techniques, including wavelets.

13.1 Basics of Fourier Series and the Fourier Transform

Fourier introduced the concept of representing a given function by a trigonometric series of sines and cosines:

$$f(x) = \frac{a_0}{2} + \sum_{n=1}^{\infty} \left(a_n \cos nx + b_n \sin nx \right) \quad x \in (-\pi, \pi].$$ (13.1.1)

At the time, this was a rather startling assertion, especially as the claim held even if the function $f(x)$, for instance, was discontinuous. Indeed, when working with such series solutions, many questions arise concerning the convergence of the series itself. Another, of course, concerns the determination of the expansion coefficients a_n and b_n. Assume for the moment that Eq. (13.1.1) is a uniformly convergent series. Then multiplying by $\cos mx$ gives the following uniformly convergent series

$$f(x) \cos mx = \frac{a_0}{2} \cos mx + \sum_{n=1}^{\infty} \left(a_n \cos nx \cos mx + b_n \sin nx \cos mx \right).$$ (13.1.2)

Integrating both sides from $x \in [-\pi, \pi]$ yields the following

$$\int_{-\pi}^{\pi} f(x) \cos mx dx = \frac{a_0}{2} \int_{-\pi}^{\pi} \cos mx dx$$

$$+ \sum_{n=1}^{\infty} \left(a_n \int_{-\pi}^{\pi} \cos nx \cos mx dx + b_n \int_{-\pi}^{\pi} \sin nx \cos mx dx \right). \tag{13.1.3}$$

But the sine and cosine expressions above are subject to the following *orthogonality* properties

$$\int_{-\pi}^{\pi} \sin nx \cos mx dx = 0 \quad \forall \; n, m \tag{13.1.4a}$$

$$\int_{-\pi}^{\pi} \cos nx \cos mx dx = \begin{cases} 0 & n \neq m \\ \pi & n = m \end{cases} \tag{13.1.4b}$$

$$\int_{-\pi}^{\pi} \sin nx \sin mx dx = \begin{cases} 0 & n \neq m \\ \pi & n = m. \end{cases} \tag{13.1.4c}$$

Thus we find the following formulas for the coefficients a_n and b_n

$$a_n = \frac{1}{\pi} \int_{-\pi}^{\pi} f(x) \cos nx dx \quad n \geq 0 \tag{13.1.5a}$$

$$b_n = \frac{1}{\pi} \int_{-\pi}^{\pi} f(x) \sin nx dx \quad n > 0 \tag{13.1.5b}$$

which are called the Fourier coefficients.

The expansion (13.1.1) must, by construction, produce 2π-periodic functions since they are constructed from sines and cosines on the interval $x \in (-\pi, \pi]$. In addition to these 2π-periodic solutions, one can consider expanding a function $f(x)$ defined on $x \in (0, \pi]$ as an *even periodic function* on $x \in (-\pi, \pi]$ by employing only the cosine portion of the expansion (13.1.1) or as an *odd periodic function* on $x \in (-\pi, \pi]$ by employing only the sine portion of the expansion (13.1.1).

Using these basic ideas, the complex version of the expansion produces the Fourier series on the domain $x \in [-L, L]$ which is given by

$$f(x) = \sum_{-\infty}^{\infty} c_n e^{in\pi x/L} \quad x \in [-L, L] \tag{13.1.6}$$

with the corresponding Fourier coefficients

$$c_n = \frac{1}{2L} \int_{-L}^{L} f(x) e^{-in\pi x/L} dx. \tag{13.1.7}$$

Although the Fourier series is now complex, the function $f(x)$ is still assumed to be real. Thus the following observations are of note: (i) c_0 is real and $c_{-n} = c_n*$, (ii) if $f(x)$ is even, all c_n are real, and (iii) if $f(x)$ is odd, $c_0 = 0$ and all c_n are purely imaginary. These results follow from Euler's identity: $\exp(\pm ix) = \cos x \pm i \sin x$. The convergence properties of this representation will not be considered here except that it will be noted that at a discontinuity of $f(x)$, the Fourier series converges to the mean of the left and right values of the discontinuity.

The Fourier transform

The *Fourier transform* is an integral transform defined over the entire line $x \in [-\infty, \infty]$. The Fourier transform and its inverse are defined as

$$F(k) = \frac{1}{\sqrt{2\pi}} \int_{-\infty}^{\infty} e^{-ikx} f(x) dx \qquad (13.1.8a)$$

$$f(x) = \frac{1}{\sqrt{2\pi}} \int_{-\infty}^{\infty} e^{ikx} F(k) dk. \qquad (13.1.8b)$$

There are other equivalent definitions. However, this definition will serve to illustrate the power and functionality of the Fourier transform method. We again note that formally, the transform is over the entire real line $x \in [-\infty, \infty]$ whereas our computational domain is only over a finite domain $x \in [-L, L]$. Further, the kernel of the transform, $\exp(\pm ikx)$, describes oscillatory behavior. Thus the Fourier transform is essentially an eigenfunction expansion over all continuous wavenumbers k. And once we are on a finite domain $x \in [-L, L]$, the continuous eigenfunction expansion becomes a discrete sum of eigenfunctions and associated wavenumbers (eigenvalues).

Derivative relations

The critical property in the usage of Fourier transforms concerns derivative relations. To see how these properties are generated, we begin by considering the Fourier transform of $f'(x)$. We denote the Fourier transform of $f(x)$ as $\widehat{f(x)}$. Thus we find

$$\widehat{f'(x)} = \frac{1}{\sqrt{2\pi}} \int_{-\infty}^{\infty} e^{-ikx} f'(x) dx = f(x) e^{-ikx}|_{-\infty}^{\infty} + \frac{ik}{\sqrt{2\pi}} \int_{-\infty}^{\infty} e^{-ikx} f(x) dx. \qquad (13.1.9)$$

Assuming that $f(x) \to 0$ as $x \to \pm\infty$ results in

$$\widehat{f'(x)} = \frac{ik}{\sqrt{2\pi}} \int_{-\infty}^{\infty} e^{-ikx} f(x) dx = ik\widehat{f(x)}. \qquad (13.1.10)$$

Thus the basic relation $\widehat{f'} = ik\widehat{f}$ is established. It is easy to generalize this argument to an arbitrary number of derivatives. The final result is the following relation between Fourier transforms of the derivative and the Fourier transform itself

$$\widehat{f^{(n)}} = (ik)^n \widehat{f}. \qquad (13.1.11)$$

This property is what makes Fourier transforms so useful and practical.

As an example of the Fourier transform, consider the following differential equation

$$y'' - \omega^2 y = -f(x) \quad x \in [-\infty, \infty]. \qquad (13.1.12)$$

We can solve this by applying the Fourier transform to both sides. This gives the following reduction

$$\widehat{y''} - \omega^2 \widehat{y} = -\widehat{f}$$
$$-k^2 \widehat{y} - \omega^2 \widehat{y} = -\widehat{f}$$
$$(k^2 + \omega^2)\widehat{y} = \widehat{f}$$
$$\widehat{y} = \frac{\widehat{f}}{k^2 + \omega^2}. \tag{13.1.13}$$

To find the solution $y(x)$, we invert the last expression above to yield

$$y(x) = \frac{1}{\sqrt{2\pi}} \int_{-\infty}^{\infty} e^{ikx} \frac{\widehat{f}}{k^2 + \omega^2} dk. \tag{13.1.14}$$

This gives the solution in terms of an integral which can be evaluated analytically or numerically.

The fast Fourier transform

The fast Fourier transform routine was developed specifically to perform the forward and backward Fourier transforms. In the mid-1960s, Cooley and Tukey developed what is now commonly known as the FFT algorithm [15]. Their algorithm was named one of the top ten algorithms of the twentieth century for one reason: the operation count for solving a system dropped to $O(N \log N)$. For N large, this operation count grows almost linearly like N. Thus it represents a great leap forward from Gaussian elimination and LU decomposition ($O(N^3)$). The key features of the FFT routine are as follows:

1   It has a low operation count: $O(N \log N)$.

2   It finds the transform on an interval $x \in [-L, L]$. Since the integration kernel $\exp(ikx)$ is oscillatory, it implies that the solutions on this finite interval have periodic boundary conditions.

3   The key to lowering the operation count to $O(N \log N)$ is in discretizing the range $x \in [-L, L]$ into 2^n points, i.e. the number of points should be $2, 4, 8, 16, 32, 64, 128, 256, \cdots$.

4   The FFT has excellent accuracy properties, typically well beyond that of standard discretization schemes.

We will consider the underlying FFT algorithm in detail at a later time. For more information at the present, see [15] for a broader overview.

The practical implementation of the mathematical tools available in MATLAB is crucial. This section will focus on the use of some of the more sophisticated routines in MATLAB which are cornerstones to scientific computing. Included in this section will be a discussion of the fast Fourier transform routines (*fft, ifft, fftshift, ifftshift, fft2, ifft2*), sparse matrix construction (*spdiag, spy*), and high-end iterative techniques for solving $\mathbf{Ax} = \mathbf{b}$ (*bicgstab, gmres*). These routines should be studied carefully since they are the building blocks of any serious scientific computing code.

Fast Fourier transform: FFT, IFFT, FFTSHIFT, IFFTSHIFT

The fast Fourier transform will be the first subject discussed. Its implementation is straightforward. Given a function which has been discretized with 2^n points and represented by a vector **x**, the FFT is found with the command *fft(x)*. Aside from transforming the function, the algorithm associated with the FFT does three major things: it shifts the data so that $x \in [0, L] \to [-L, 0]$ and $x \in [-L, 0] \to [0, L]$, additionally it multiplies every other mode by -1, and it assumes you are working on a 2π-periodic domain. These properties are a consequence of the FFT algorithm discussed in detail at a later time.

To see the practical implications of the FFT, we consider the transform of a Gaussian function. The transform can be calculated analytically so that we have the exact relations:

$$f(x) = \exp(-\alpha x^2) \quad \to \quad \widehat{f}(k) = \frac{1}{\sqrt{2\alpha}} \exp\left(-\frac{k^2}{4\alpha}\right). \tag{13.1.15}$$

A simple MATLAB code to verify this with $\alpha = 1$ is as follows

```
clear all; close all;  % clear all variables and figures

L=20;   % define the computational domain [-L/2,L/2]
n=128;  % define the number of Fourier modes 2^n

x2=linspace(-L/2,L/2,n+1);  % define the domain discretization
x=x2(1:n);   % consider only the first n points:  periodicity

u=exp(-x.*x);           % function to take a derivative of
ut=fft(u);              % FFT the function
utshift=fftshift(ut);   % shift FFT

figure(1), plot(x,u)        % plot initial gaussian
figure(2), plot(abs(ut))        % plot unshifted transform
figure(3), plot(abs(utshift))  % plot shifted transform
```

The second figure generated by this script shows how the pulse is shifted. By using the command *fftshift*, we can shift the transformed function back to its mathematically correct positions as shown in the third figure generated. However, before inverting the transformation, it is crucial that the transform is shifted back to the form of the second figure. The command *ifftshift* does this. In general, unless you need to plot the spectrum, it is better not to deal with the *fftshift* and *ifftshift* commands. A graphical representation of the *fft* procedure and its shifting properties is illustrated in Fig. 8.4 where a Gaussian is transformed and shifted by the *fft* routine.

FFT versus finite difference differentiation

Taylor series methods for generating approximations to derivatives of a given function can also be used. In the Taylor series method, the approximations are *local* since they are given by nearest neighbor coupling formulas of finite differences. Using the FFT, the approximation to the derivative is *global* since it uses an expansion basis of cosines and sines which extend over the entire domain of the given function.

The calculation of the derivative using the FFT is performed trivially from the derivative relation formula (13.1.11). As an example of how to implement this differentiation formula, consider a specific function:

$$u(x) = \text{sech}(x) \tag{13.1.16}$$

which has the following derivative relations

$$\frac{du}{dx} = -\text{sech}(x)\tanh(x) \tag{13.1.17a}$$

$$\frac{d^2u}{dx^2} = \text{sech}(x) - 2\,\text{sech}^3(x). \tag{13.1.17b}$$

A comparison can be made between the differentiation accuracy of finite difference formulas of Tables 4.1–4.2 and the spectral method using FFTs (see Fig. 13.1). The following code generates the derivative using the FFT method along with the finite difference $O(\Delta x^2)$ and $O(\Delta x^4)$ approximations considered with finite differences.

```
clear all; close all;   % clear all variables and figures


L=20;    % define the computational domain [-L/2,L/2]
n=128;   % define the number of Fourier modes 2^n


x2=linspace(-L/2,L/2,n+1);   % define the domain discretization
x=x2(1:n);    % consider only the first n points:  periodicity
dx=x(2)-x(1);   % dx value needed for finite difference
u=sech(x);    % function to take a derivative of
ut=fft(u);    % FFT the function
k=(2*pi/L)*[0:(n/2-1) (-n/2):-1]; % k rescaled to 2pi domain


% FFT calculation of derivatives


ut1=i*k.*ut;        % first derivative
ut2=-k.*k.*ut;        % second derivative
u1=real(ifft(ut1)); u2=real(ifft(ut2)); % inverse transform
u1exact=-sech(x).*tanh(x);   % analytic first derivative
u2exact=sech(x)-2*sech(x).^3;   % analytic second derivative
```

Continued

```
% Finite difference calculation of first derivative

% 2nd-order accurate
ux(1)=(-3*u(1)+4*u(2)-u(3))/(2*dx);
for j=2:n-1
    ux(j)=(u(j+1)-u(j-1))/(2*dx);
end
ux(n)=(3*u(n)-4*u(n-1)+u(n-2))/(2*dx);

% 4th-order accurate
ux2(1)=(-3*u(1)+4*u(2)-u(3))/(2*dx);
ux2(2)=(-3*u(2)+4*u(3)-u(4))/(2*dx);
for j=3:n-2
   ux2(j)=(-u(j+2)+8*u(j+1)-8*u(j-1)+u(j-2))/(12*dx);
end
ux2(n-1)=(3*u(n-1)-4*u(n-2)+u(n-3))/(2*dx);
ux2(n)=(3*u(n)-4*u(n-1)+u(n-2))/(2*dx);

figure(1)
plot(x,u,'r',x,u1,'g',x,u1exact,'go',x,u2,'b',x,u2exact,'bo')
figure(2)
subplot(3,1,1), plot(x,u1exact,'ks-',x,u1,'k',x,ux,'ko',x,ux2,'k*')
axis([-1.15 -0.75 0.47 0.5])
subplot(3,1,2), plot(x,u1exact,'ks-',x,u1,'kv',x,ux,'ko',x,ux2,'k*')
axis([-0.9376 -0.9374 0.49848 0.49850])
subplot(3,1,3), plot(x,u1exact,'ks-',x,u1,'kv',x,ux,'ko',x,ux2,'k*')
axis([-0.9376 -0.9374 0.498487 0.498488])
```

Note that the real part is taken after inverse Fourier transforming due to the numerical round-off which generates a small $O(10^{-15})$ imaginary part.

Of course, this differentiation example works well since the periodic boundary conditions of the FFT are essentially satisfied with the choice of function $f(x) = \text{sech}(x)$. Specifically, for the range of values $x \in [-10, 10]$, the value of $f(\pm 10) = \text{sech}(\pm 10) \approx 9 \times 10^{-5}$. For a function which does not satisfy periodic boundary conditions, the FFT routine leads to significant error. Much of this error arises from the discontinuous jump and the associated *Gibb's phenomenon* which results when approximating with a Fourier series. To illustrate this, Fig. 13.2 shows the numerically generated derivative of the function $f(x) = \tanh(x)$, which is $f'(x) = \text{sech}^2(x)$. In this example, it is clear that the FFT method is highly undesirable, whereas the finite difference method performs as well as before.

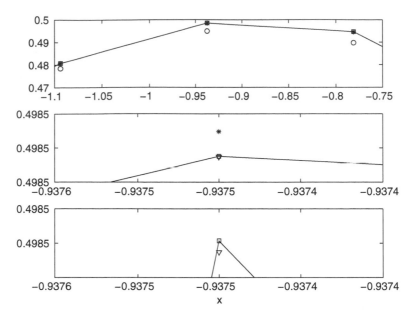

Figure 13.1: Accuracy comparison between second- and fourth-order finite difference methods and the spectral FFT method for calculating the first derivative. Note that by using the *axis* command, the exact solution (line) and its approximations can be magnified near an arbitrary point of interest. Here, the top figure shows that the second-order finite difference (circles) method is within $O(10^{-2})$ of the exact derivative. The fourth-order finite difference (star) is within $O(10^{-5})$ of the exact derivative. Finally, the FFT method is within $O(10^{-6})$ of the exact derivative. This demonstrates the spectral accuracy property of the FFT algorithm.

Higher dimensional Fourier transforms

For transforming in higher dimensions, a couple of choices in MATLAB are possible. For 2D transformations, it is recommended to use the commands **fft2** and **ifft2**. These will transform a matrix **A**, which represents data in the *x*- and *y*-direction, along the rows and then columns, respectively. For higher dimensions, the **fft** command can be modified to **fft(x,[],N)** where N is the number of dimensions. Alternatively, use **fftn(x)** for multi-dimensional FFTs.

13.2 FFT Application: Radar Detection and Filtering

FFTs and other related frequency transforms have revolutionized the field of digital signal processing and imaging. The key concept in any of these applications is to use the FFT to analyze and manipulate data in the frequency domain. There are other methods of treating the signal in the time and frequency domain, but the simplest to begin with is the Fourier transform.

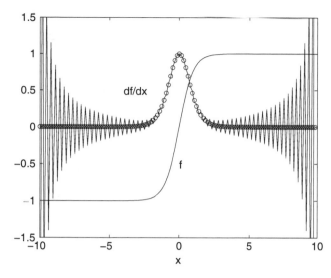

Figure 13.2: Accuracy comparison between the fourth-order finite difference method and the spectral FFT method for calculating the first derivative of $f(x) = \tanh(x)$. The fourth-order finite difference (circle) is within $O(10^{-5})$ of the exact derivative. In contrast, the FFT approximation (line) oscillates strongly and provides a highly inaccurate calculation of the derivative due to the non-periodic boundary conditions of the given function.

The Fourier transform is also used in quantum mechanics to represent a quantum wavepacket in either the spatial domain or the momentum (spectral) domain. Quantum mechanics is specifically mentioned due to the well-known *Heisenberg uncertainty principle*. Simply stated, you cannot know the exact position and momentum of a quantum particle simultaneously. This is simply a property of the Fourier transform itself. Essentially, a narrow signal in the time domain corresponds to a broadband source, whereas a highly confined spectral signature corresponds to a broad time domain source. This has significant implication for many techniques of signal analysis. In particular, an excellent decomposition of a given signal into its frequency components can render no information about *when* in time the different portions of signal actually occurred. Thus there is often a competition between trying to localize both in time and frequency a signal or stream of data. Time–frequency analysis is ultimately all about trying to resolve the time and frequency domain in an efficient and tractable manner. Various aspects of this time–frequency processing will be considered in later sections, including wavelet based methods of time–frequency analysis.

At this point, we discuss a very basic concept and manipulation procedure to be performed in the frequency domain: noise attenuation via frequency (band-pass) filtering. This filtering process is fairly common in electronics and signal detection. As a specific application or motivation for spectral analysis, consider the process of radar detection depicted in Fig. 13.3. An outgoing electromagnetic field is emitted from a source which then attempts to detect the reflections of the emitted signal. In addition to reflections coming from the desired target, reflections can also come from geographical objects (mountains, trees, etc.) and atmospheric phenomena (clouds,

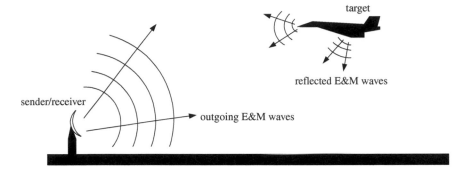

Figure 13.3: Schematic of the operation of a radar detection system. The radar emits an electromagnetic (E&M) field at a specific frequency. The radar then receives the reflection of this field from objects in the environment and attempts to determine what the objects are.

precipitation, etc.). Further, there may be other sources of the electromagnetic energy at the desired frequency. Thus the radar detection process can be greatly complicated by a wide host of environmental phenomena. Ultimately, these effects combine to give at the detector a noisy signal which must be processed and filtered before a detection diagnosis can be made.

It should be noted that the radar detection problem discussed in what follows is highly simplified. Indeed, modern day radar detection systems use much more sophisticated time–frequency methods for extracting both the position and spectral signature of target objects. Regardless, this analysis gives a high-level view of the basic and intuitive concepts associated with signal detection.

To begin we consider an ideal signal, i.e. an electromagnetic pulse, generated in the time domain.

```
clear all; close all;

L=30;    % time slot to transform
n=512;   % number of Fourier modes 2^9
t2=linspace(-L,L,n+1); t=t2(1:n);  % time discretization
k=(2*pi/(2*L))*[0:(n/2-1) -n/2:-1];  % frequency components of FFT

u=sech(t);      % ideal signal in the time domain
figure(1), subplot(3,1,1), plot(t,u,'k'), hold on
```

This will generate an ideal hyperbolic secant shape in the time domain. It is this time domain signal that we will attempt to reconstruct via denoising and filtering. Figure 13.4(a) illustrates the ideal signal.

In most applications, the signals as generated above are not ideal. Rather, they have a large amount of noise integrated within them. Usually this noise is what is called *white noise*, i.e. a

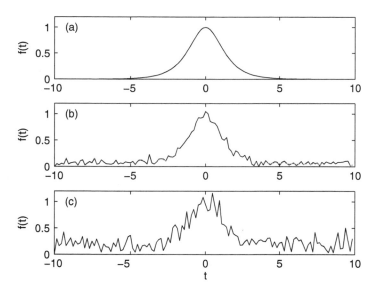

Figure 13.4: Ideal time domain pulse (a) along with two realizations of the pulse with increasing noise strength (b) and (c). The noisy signals are what are typically expected in applications.

noise that affects all frequencies the same. We can add white noise to this signal by considering the pulse in the frequency domain.

```
noise=1;
ut=fft(u);
utn=ut+noise*(randn(1,n)+i*randn(1,n));
un=ifft(utn);
figure(1), subplot(3,1,2), plot(t,abs(un),'k'), hold on
```

These lines of code generate the Fourier transform of the function along with the vector **utn** which is the spectrum of the given signal with a complex and Gaussian distributed (mean zero, unit variance) noise source term added in. Figure 13.4 shows the difference between the ideal time domain signal pulse and the more physically realistic pulse for which white noise has been added. In these figures, a clear *signal*, i.e. the original time domain pulse, is still detected even in the presence of noise.

The fundamental question in signal detection now arises: is the time domain structure received in a detector simply a result of noise, or is there an underlying signal being transmitted. For small amounts of noise, this is clearly not a problem. However, for large amounts of noise or low levels of signal, the detection becomes a nontrivial matter.

In the context of radar detection, two competing objectives are at play: for the radar detection system, an accurate diagnosis of the data is critical in determining the presence of an aircraft. Competing in this regard is, perhaps, the aircraft's objective of remaining invisible, or undetected, from the radar. Thus an aircraft attempting to remain invisible will attempt to reflect as little electromagnetic signal as possible. This can be done by, for instance, covering the aircraft with materials that absorb the radar frequencies used for detection, i.e. stealth technology. An alternative method is to fly another aircraft through the region which bombards the radar detection system with electromagnetic energy at the detection frequencies. Thus the aircraft, attempting to remain undetected, may be successful since the radar system may not be able to distinguish between the sources of the electromagnetic fields. This second option is, in some sense, like a radar jamming system. In the first case, the radar detection system will have a small signal-to-noise ratio and denoising will be critical to accurate detection. In the second case, an accurate time–frequency analysis is critical for determining the time localization and position of the competing signals.

The focus of what follows is to apply the ideas of spectral filtering to attempt to improve the signal detection by denoising. Spectral filtering is a method which allows us to extract information at specific frequencies. For instance, in the radar detection problem, it is understood that only a particular frequency (the emitted signal frequency) is of interest at the detector. Thus it would seem reasonable to filter out, in some appropriate manner, the remaining frequency components of the electromagnetic field received at the detector.

Consider a very noisy field created around the hyperbolic secant of the previous example (see Fig. 13.5):

```
noise=10;
ut=fft(u);
unt=ut+noise*(randn(1,n)+i*randn(1,n));
un=ifft(unt);

subplot(2,1,1), plot(t,abs(un),'k')
axis([-30 30 0 2])
xlabel('time (t)'), ylabel('|u|')
subplot(2,1,2)
plot(fftshift(k),abs(fftshift(unt))/max(abs(fftshift(unt))),'k')
axis([-25 25 0 1])
xlabel('wavenumber (k)', ylabel('|ut|/max(|ut|)')
```

Figure 13.5 demonstrates the impact of a large noise applied to the underlying signal. In particular, the signal is completely buried within the white noise fluctuations, making the detection of a signal difficult, if not impossible, with the unfiltered noisy signal field.

Filtering can help significantly improve the ability to detect the signal buried in the noisy field of Fig. 13.5. For the radar application, the frequency (wavenumber) of the emitted and reflected field is known, thus spectral filtering around this frequency can remove undesired frequencies and much of the white noise picked up in the detection process. There are a wide range of filters

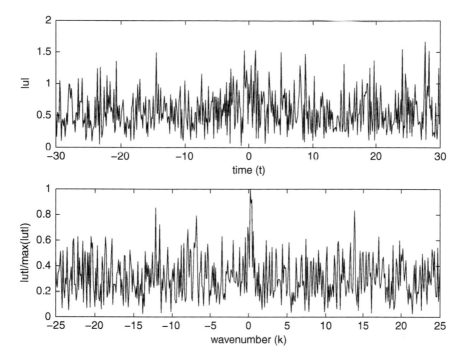

Figure 13.5: Time domain (top) and frequency domain (bottom) plots for a single realization of white noise. In this case, the noise strength has been increased to ten, thus burying the desired signal field in both time and frequency.

that can be applied to such a problem. One of the simplest filters to consider here is the Gaussian filter (see Fig. 13.6):

$$\mathcal{F}(k) = \exp\left(-\tau(k - k_0)^2\right) \tag{13.2.1}$$

where τ measures the bandwidth of the filter, and k is the wavenumber. The generic filter function $\mathcal{F}(k)$ in this case acts as a low-pass filter since it eliminates high-frequency components in the system of interest. Note that these are high frequencies in relation to the center frequency ($k = k_0$) of the desired signal field. In the example considered here $k_0 = 0$.

Application of the filter strongly attenuates those frequencies away from the center frequency k_0. Thus if there is a signal near k_0, the filter isolates the signal input around this frequency or wavenumber. Application of the filtering results in the following spectral processing in MATLAB:

```
filter=exp(-0.2*(k).^2);
unft=filter.*unt;
unf=ifft(unft);
```

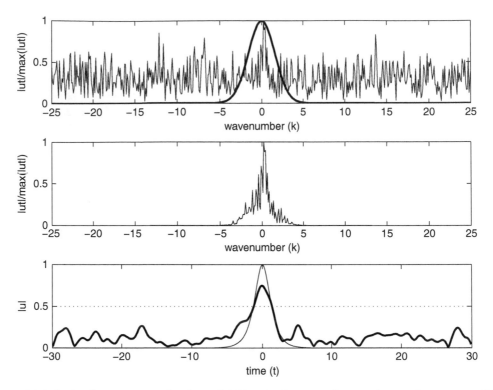

Figure 13.6: (top) White noise inundated signal field in the frequency domain along with a Gaussian filter with bandwidth parameter $\tau = 0.2$ centered on the signal center frequency. (middle) The post-filtered signal field in the frequency domain. (bottom) The time domain reconstruction of the signal field (bolded line) along with the ideal signal field (light line) and the detection threshold of the radar (dotted line).

The results of this code are illustrate in Fig. 13.6 where the white noise inundated signal field in the spectral domain is filtered with a Gaussian filter centered around the zero wavenumber $k_0 = 0$ with a bandwidth parameter $\tau = 0.2$. The filtering extracts nicely the signal field despite the strength of the applied white noise. Indeed, Fig. 13.6 (bottom) illustrates the effect of the filtering process and its ability to reproduce an approximation to the signal field. In addition to the extracted signal field, a detection threshold is placed (dotted line) on the graph in order to illustrate that the detector would read the extracted signal as a target.

As a matter of completeness, the extraction of the electromagnetic field is also illustrated when not centered on the center frequency. Figure 13.7 shows the field that is extracted for a filter centered around $k_0 = 15$. This is done easily with the MATLAB commands

```
filter=exp(-0.2*(k-15).^2);
unft=filter.*unt;
unf=ifft(unft);
```

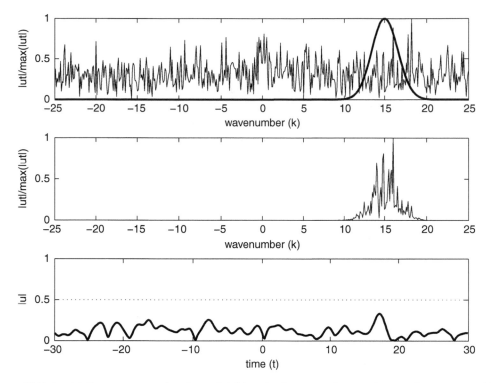

Figure 13.7: (top) White noise inundated signal field in the frequency domain along with a Gaussian filter with bandwidth parameter $\tau = 0.2$ centered on the signal frequency $k = 15$. (middle) The post-filtered signal field in the frequency domain. (bottom) The time domain reconstruction of the signal field (bolded line) along the detection threshold of the radar (dotted line).

In this case, there is no signal around the wavenumber $k_0 = 15$. However, there is a great deal of white noise electromagnetic energy around this frequency. Upon filtering and inverting the Fourier transform, it is clear from Figure 13.7 (bottom) that there is no discernible target present. As before, a decision threshold is set at $|u| = 0.5$ and the white noise fluctuations clearly produce a field well below the threshold line.

In all of the analysis above, filtering is done on a given set of signal data. However, the signal is evolving in time, i.e. the position of the aircraft is probably moving. Moreover, one can imagine that there is some statistical chance that a type I or type II error can be made in the detection. Thus a target may be identified when there is no target, or a target has been missed completely. Ultimately, the goal is to use repeated measurements from the radar of the airspace in order to try to avoid a detection error or failure. Given that you may launch a missile, the stakes are high and a high level of confidence is needed in the detection process. It should also be noted that filter design, shape and parameterization play significant roles in designing optimal filters. Thus filter design is an object of strong interest in signal processing applications.

13.3 **FFT Application: Radar Detection and Averaging**

The last section clearly shows the profound and effective use of filtering in cleaning up a noisy signal. This is one of the most basic illustrations of the power of basic signal processing. Other methods also exist to help reduce noise and generate better time–frequency resolution. In particular, the filtering example given fails to use two key facts: the noise is white, and radar will continue to produce more signal data. These two facts can be used to produce useful information about the incoming signal. In particular, it is known, and demonstrated in the last section, that white noise can be modeled adding a normally distributed random variable with zero mean and unit variance to each Fourier component of the spectrum. The key here: *with zero mean*. Thus over many signals, the white noise should, on average, add up to zero; a very simple concept, yet tremendously powerful in practice for signal processing systems where there is continuous detection of a signal.

To illustrate the power of denoising the system by averaging, the simple analysis of the last section will be considered for a time pulse. In beginning this process, a time window must be selected and its Fourier resolution decided:

```
clear all; close all; clc

L=30;   % time slot
n=512;  % Fourier modes
t2=linspace(-L,L,n+1); t=t2(1:n);
k=(2*pi/(2*L))*[0:(n/2-1) -n/2:-1]; ks=fftshift(k);
noise=10;
```

In order to be more effective in plotting our spectral results, the Fourier components in the standard shifted frame (**k**) and its unshifted counterpart (**ks**) are both produced. Further, the noise strength of the system is selected.

In the next portion of the code, the total number of data frames, or realizations of the incoming signal stream, is considered. The word realizations is used here to help underscore the fact that for each data frame, a new realization of the white noise is produced. In the following code, one, two, five and one hundred realizations are considered.

```
labels=['(a)';'(b)';'(c)';'(d)'];
realize=[1 2 5 100];
for jj=1:length(realize)
```

Continued

```
u=sech(t); ave=zeros(1,n);
ut=fft(u);
for j=1:realize(jj)
   utn(j,:)=ut+noise*(randn(1,n)+i*randn(1,n));
   ave=ave+utn(j,:);
   dat(j,:)=abs(fftshift(utn(j,:)))/max(abs(utn(j,:)));
   un(j,:)=ifft(utn(j,:));
end
ave=abs(fftshift(ave))/realize(jj);

subplot(4,1,jj)
plot(ks,ave/max(ave),'k')
set(gca,'Fontsize',[15])
axis([-20 20 0 1])
text(-18,0.7,labels(jj,:),'Fontsize',[15])
ylabel('|fft(u)|','Fontsize',[15])

end
hold on
plot(ks,abs(fftshift(ut))/max(abs(ut)),'k:','Linewidth',[2])
set(gca,'Fontsize',[15])
xlabel('frequency (k)')
```

Figure 13.8 demonstrates the averaging process in the spectral domain as a function of the number of realizations. With one realization, it is difficult to determine if there is a true signal, or if the entire spectrum is noise. Already after two realizations, the center frequency structure starts to appear. Five realizations has a high degree of discrimination between the noise and signal and 100 realizations is almost exactly converged to the ideal, noise-free signal. As expected, the noise in the averaging process is eliminated since its mean is zero. To view a few of the data realizations used in computing the averaging in Fig. 13.8, the first eight signals used to compute the average for 100 realizations are shown in Fig. 13.9. The following code is used to produce the waterfall plot used:

```
figure(2)
waterfall(ks,1:8,dat(1:8,:)), colormap([0 0 0]), view(-15,80)
set(gca,'Fontsize',[15],'Xlim',[-28 28],'Ylim',[1 8])
xlabel('frequency (k)'), ylabel('realization'),zlabel('|fft(u)|')
```

Figures 13.8 and 13.9 illustrate how the averaging process can ultimately extract a clean spectral signature. As a consequence, however, you completely loose the time domain dynamics. In other

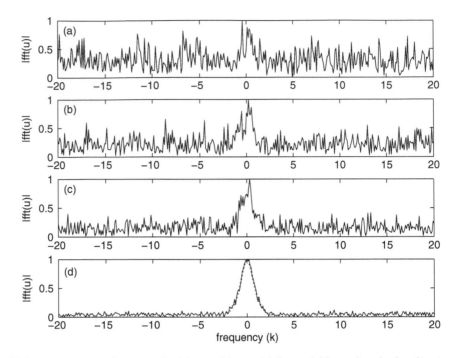

Figure 13.8: Average spectral content for (a) one, (b) two, (c) five and (d) one hundred realizations of the data. The dotted line in (d) represents the ideal, noise-free spectral signature.

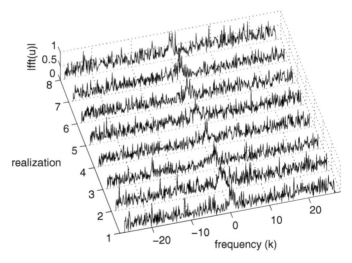

Figure 13.9: Typical time domain signals that would be produced in a signal detection system. At different times of measurement, the added white noise would alter the signal stream significantly.

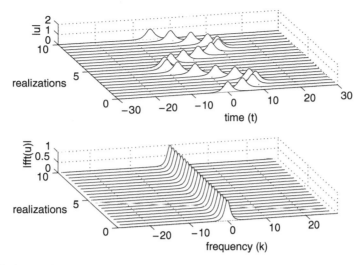

Figure 13.10: Ideal time–frequency behavior for a time domain pulse that evolves dynamically in time. The spectral signature remains unchanged as the pulse moves over the time domain.

words, one could take the cleaned up spectrum of Fig. 13.8 and invert the Fourier transform, but this would only give the averaged time domain signal, but not how it actually evolved over time. To consider this problem more concretely, consider the following bit of code that generates a time-varying signal shown in Fig. 13.10. There are a total of 21 realizations of data in this example.

```
slice=[0:0.5:10];
[T,S]=meshgrid(t,slice);
[K,S]=meshgrid(k,slice);

U=sech(T-10*sin(S)).*exp(i*0*T);
subplot(2,1,1)
waterfall(T,S,U), colormap([0 0 0]), view(-15,70)
set(gca,'Fontsize',[15],'Xlim',[-30 30],'Zlim',[0 2])
xlabel('time (t)'), ylabel('realizations'), zlabel('|u|')

for j=1:length(slice)
  Ut(j,:)=fft(U(j,:));
  Kp(j,:)=fftshift(K(j,:));
  Utp(j,:)=fftshift(Ut(j,:));
  Utn(j,:)=Ut(j,:)+noise*(randn(1,n)+i*randn(1,n));
  Utnp(j,:)=fftshift(Utn(j,:))/max(abs(Utn(j,:)));
  Un(j,:)=ifft(Utn(j,:));
end
figure(3)
subplot(2,1,2)
waterfall(Kp,S,abs(Utp)/max(abs(Utp(1,:)))), colormap([0 0 0]), view(-15,70)
set(gca,'Fontsize',[15],'Xlim',[-28 28])
xlabel('frequency (k)'), ylabel('realizations'), zlabel('|fft(u)|')
```

If this were a radar problem, the movement of the signal would represent an aircraft moving in time. The spectrum, however, remains fixed at the transmitted signal frequency. Thus it is clear even at this point that averaging over the frequency realizations produces a clean, localized signature in the frequency domain, whereas averaging over the time domain only smears out the evolving time domain pulse.

The ideal pulse evolution in Fig. 13.10 is now inundated with white noise as shown in Fig. 13.11. The noise is added to each realization, or data measurement, with the same strength as shown in Figs. 13.8 and 13.9. The following code plots the noise time domain and spectral evolution:

```
figure(4)
Subplot(2,1,1)
waterfall(T,S,abs(Un)), colormap([0 0 0]), view(-15,70)
set(gca,'Fontsize',[15],'Xlim',[-30 30],'Zlim',[0 2])
xlabel('time (t)'), ylabel('realizations'), zlabel('|u|')
subplot(2,1,2)
waterfall(Kp,S,abs(Utnp)), colormap([0 0 0]), view(-15,70)
set(gca,'Fontsize',[15],'Xlim',[-28 28])
xlabel('frequency (k)'), ylabel('realizations'), zlabel('|fft(u)|')
```

The obvious question arises about how to clean up the signal and whether something meaningful can be extracted from the data or not. After all, there are only 21 data slices to work with. The following code averages over the 21 data realizations in both the time and frequency domains.

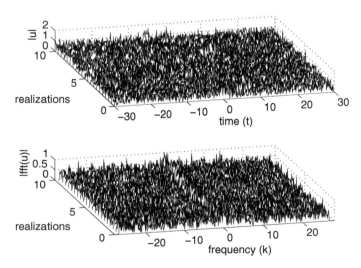

Figure 13.11: A more physically realistic depiction of the time domain and frequency spectrum as a function of the number of realizations. The goal is to extract the meaningful data buried within the noise.

```
figure(5)
Uave=zeros(1,n); Utave=zeros(1,n);
for j=1:length(slice)
  Uave=Uave+Un(j,:);
  Utave=Utave+Utn(j,:);
end
Uave=Uave/length(slice);
Utave=fftshift(Utave)/length(slice);

subplot(2,1,1)
plot(t,abs(Uave),'k')
set(gca,'Fontsize',[15])
xlabel('time (t)'), ylabel('|u|')
subplot(2,1,2)
plot(ks,abs(Utave)/max(abs(Utave)),'k'), hold on
plot(ks,abs(fftshift(Ut(1,:)))/max(abs(Ut(1,:))),'k:','Linewidth',[2])
axis([-20 20 0 1])
set(gca,'Fontsize',[15])
xlabel('frequency (k)'), ylabel('|fft(u)|')
```

Figure 13.12 shows the results from the averaging process for a nonstationary signal. The top graph shows that averaging over the time domain for a moving signal produces no discernible signal. However, averaging over the frequency domain produces a clear signature at the center

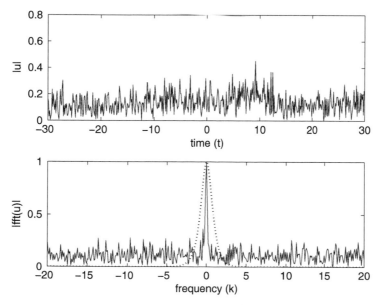

Figure 13.12: Averaged time domain and spectral profiles for the 21 realizations of data shown in Fig. 13.11. Even with limited sampling, the spectral signature at the center frequency is extracted.

frequency of interest. Ideally, if more data is collected a better average signal is produced. However, in many applications, the acquisition of data is limited and decisions must be made upon what are essentially small sample sizes.

13.4   Time–Frequency Analysis: Windowed Fourier Transforms

The Fourier transform is one of the most important and foundational methods for the analysis of signals. However, it was realized very early on that Fourier transform based methods had severe limitations. Specifically, when transforming a given time signal, it is clear that all the frequency content of the signal can be captured with the transform, but the transform fails to capture the *moment in time* when various frequencies were actually exhibited. Figure 13.13 shows a prototypical signal that may be of interest to study. The signal $S(t)$ is clearly comprised of various frequency components that are exhibited at different times. For instance, at the beginning of the signal, there are high-frequency components. In contrast, the middle of the signal has relatively low-frequency oscillations. If the signal represented music, then the beginning of the signal would produce high notes while the middle would produce low notes. The Fourier transform of the signal contains all this information, but there is no indication of *when* the high or low notes actually occur in time. Indeed, by definition the Fourier transform eliminates all time domain information since you actually integrated out all time in Eq. (13.1.8).

The obvious question to arise is this: What is the Fourier transform good for in the context of signal processing? In the previous sections where the Fourier transform was applied, the signal being investigated was fixed in frequency, i.e. a sonar or radar detector with a fixed frequency ω_0. Thus for a given signal, the frequency of interest did not shift in time. By using different measurements in time, a signal tracking algorithm could be constructed. Thus an implicit assumption

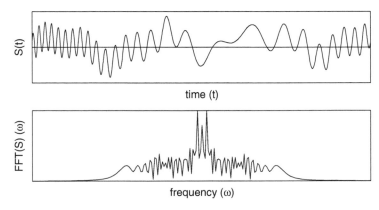

Figure 13.13: Signal $S(t)$ and its normalized Fourier transform $\hat{S}(\omega)$. Note the large number of frequency components that make up the signal.

was made about the invariance of the signal frequency. Ultimately, the Fourier transform is superb for one thing: characterizing *stationary* or periodic signals. Informally, a stationary signal is such that repeated measurements of the signal in time yield an average value that does not change in time. Most signals, however, do not satisfy this criterion. A song, for instance, changes its average Fourier components in time as the song progresses in time. Thus the generic signal $S(t)$ that should be considered, and that is plotted as an example in Fig. 13.13, is a *nonstationary signal* whose average signal value does change in time. It should be noted that in our application of radar detection of a moving target, use was made of the stationary nature of the spectral content. This allowed for a clear idea of where to filter the signal $\hat{S}(\omega)$ in order to reconstruct the signal $S(t)$.

Having established the fact that the direct application of the Fourier transform provides a nontenable method for extracting signal information, it is natural to pursue modifications of the method in order to extract time and frequency information. The most simple minded approach is to consider Fig. 13.13 and to decompose the signal over the time domain into separate time-frames. Figure 13.14 shows the original signal $S(t)$ considered but now decomposed into four smaller time windows. In this decomposition, for instance, the first time-frame is considered with the remaining three time-frames zeroed out. For each time window, the Fourier transform is applied in order to characterize the frequencies present during that time-frame. The highest frequency components are captured in Fig. 13.14(a) which is clearly seen in its Fourier transform. In contrast, the slow modulation observed in the third time-frame (c) is devoid of high-frequency

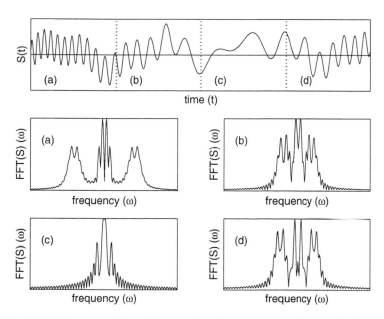

Figure 13.14: Signal $S(t)$ decomposed into four equal and separate time-frames (a), (b), (c) and (d). The corresponding normalized Fourier transform of each time-frame $\hat{S}(\omega)$ is illustrated below the signal. Note that this decomposition gives information about the frequencies present in each smaller time-frame.

components as observed in Fig. 13.14(c). This method thus exhibits the ability of the Fourier transform, appropriately modified, to extract out both time and frequency information from the signal.

The limitations of the direct application of the Fourier transform, and its inability to localize a signal in both the time and frequency domains, were realized very early on in the development of radar and sonar detection. The Hungarian physicist/mathematician/electrial engineer Gábor Dénes (Nobel Prize for Physics in 1971 for the discovery of holography in 1947) was the first to propose a formal method for localizing both time and frequency. His method involved a simple modification of the Fourier transform kernel. Thus Gábor introduced the kernel

$$g_{t,\omega}(\tau) = e^{i\omega\tau}g(\tau - t) \tag{13.4.1}$$

where the new term to the Fourier kernel $g(\tau - t)$ was introduced with the aim of localizing both time and frequency. The *Gábor transform*, also known as the *short-time Fourier transform (STFT)*, is then defined as the following:

$$\mathcal{G}[f](t,\omega) = \tilde{f}_g(t,\omega) = \int_{-\infty}^{\infty} f(\tau)\bar{g}(\tau - t)e^{-i\omega\tau}\,d\tau = \left(f, \bar{g}_{t,\omega}\right) \tag{13.4.2}$$

where the bar denotes the complex conjugate of the function. Thus the function $g(\tau - t)$ acts as a time filter for localizing the signal over a specific window of time. The integration over the parameter τ slides the time-filtering window down the entire signal in order to pick out the frequency information at each instant of time. Figure 13.15 gives a nice illustration of how the time-filtering scheme of Gábor works. In this figure, the time-filtering window is centered at τ with a width a. Thus the frequency content of a window of time is extracted and τ is modified to extract the frequencies of another window. The definition of the Gábor transform captures the entire time–frequency content of the signal. Indeed, the Gábor transform is a function of the two variables t and ω.

A few of the key mathematical properties of the Gábor transform are highlighted here. To be more precise about these mathematical features, some assumptions about commonly used $g_{t,\omega}$ are considered. Specifically, for convenience we will consider g to be real and symmetric with $\|g(t)\| = 1$ and $\|g(\tau - t)\| = 1$ where $\| \cdot \|$ denotes the L_2 norm. Thus the definition of the Gábor transform, or STFT, is modified to

$$\mathcal{G}[f](t,\omega) = \tilde{f}_g(t,\omega) = \int_{-\infty}^{\infty} f(\tau)g(\tau - t)e^{-i\omega\tau}\,d\tau \tag{13.4.3}$$

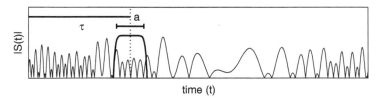

Figure 13.15: Graphical depiction of the Gábor transform for extracting the time–frequency content of a signal $S(t)$. The time-filtering window $g(\tau - t)$ is centered at τ with width a.

with $g(\tau - t)$ inducing localization of the Fourier integral around $t = \tau$. With this definition, the following properties hold

1   The energy is bounded by the Schwarz inequality so that

$$|\tilde{f}_g(t, \omega)| \leq \|f\| \|g\|. \tag{13.4.4}$$

2   The energy in the signal plane around the neighborhood of (t, ω) is calculated from

$$|\tilde{f}_g(t, \omega)|^2 = \left| \int_{-\infty}^{\infty} f(\tau) g(\tau - t) e^{-i\omega\tau} d\tau \right|^2. \tag{13.4.5}$$

3   The time–frequency spread around a Gábor window is computed from the variance, or second moment, so that

$$\sigma_t^2 = \int_{-\infty}^{\infty} (\tau - t)^2 \left| g_{t,\omega}(\tau) \right|^2 d\tau = \int_{-\infty}^{\infty} \tau^2 |g(\tau)|^2 d\tau \tag{13.4.6a}$$

$$\sigma_\omega^2 = \frac{1}{2\pi} \int_{-\infty}^{\infty} (v - \omega)^2 \left| \tilde{g}_{t,\omega}(v) \right|^2 dv = \frac{1}{2\pi} \int_{-\infty}^{\infty} v^2 \left| \tilde{g}(v) \right|^2 dv \tag{13.4.6b}$$

where $\sigma_t \sigma_\omega$ is independent of t and ω and is governed by the Heinsenberg uncertainty principle.

4   The Gábor transform is linear so that

$$\mathcal{G}[af_1 + bf_2] = a\mathcal{G}[f_1] + b\mathcal{G}[f_2]. \tag{13.4.7}$$

5   The Gábor transform can be inverted with the formula

$$f(\tau) = \frac{1}{2\pi} \frac{1}{\|g\|^2} \int_{-\infty}^{\infty} \int_{-\infty}^{\infty} \tilde{f}_g(t, \omega) g(\tau - t) e^{i\omega\tau} d\omega dt \tag{13.4.8}$$

where the integration must occur over all frequency- and time-shifting components. This double integral is in contrast to the Fourier transform which requires only a single integration since it is a function, $\hat{f}(\omega)$, of the frequency alone.

Figure 13.16 is a cartoon representation of the fundamental ideas behind a time series analysis, Fourier transform analysis and Gábor transform analysis of a given signal. In the time series method, good resolution is achieved of the signal in the time domain, but no frequency resolution is achieved. In Fourier analysis, the frequency domain is well resolved at the expense of losing all time resolution. The Gábor method, or short-time Fourier transform, trades away some measure of accuracy in both the time and frequency domains in order to give both time and frequency resolution simultaneously. Understanding this figure is critical to understanding the basic, high-level notions of time–frequency analysis.

In practice, the Gábor transform is computed by discretizing the time and frequency domains. Thus a discrete version of the transform (13.4.2) needs to be considered. Essentially, by

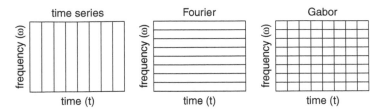

Figure 13.16: Graphical depiction of the difference between a time series analysis, Fourier analysis and Gábor analysis of a signal. In the time series method, good resolution is achieved of the signal in the time domain, but no frequency resolution is achieved. In Fourier analysis, the frequency domain is well resolved at the expense of losing all time resolution. The Gábor method, or short-time Fourier transform, is constructed to give both time and frequency resolution. The area of each box can be constructed from $\sigma_t^2 \sigma_\omega^2$.

discretizing, the transform is done on a lattice of time and frequency. Thus consider the lattice, or sample points,

$$v = m\omega_0 \tag{13.4.9a}$$

$$\tau = n t_0 \tag{13.4.9b}$$

where m and n are integers and $\omega_0, t_0 > 0$ are constants. Then the discrete version of $g_{t,\omega}$ becomes

$$g_{m,n}(t) = e^{i2\pi m\omega_0 t} g(t - n t_0) \tag{13.4.10}$$

and the Gábor transform becomes

$$\tilde{f}(m, n) = \int_{-\infty}^{\infty} f(t) \bar{g}_{m,n}(t) dt = \left(f, g_{m,n} \right). \tag{13.4.11}$$

Note that if $0 < t_0, \omega_0 < 1$, then the signal is oversampled and time-frames exist which yield excellent localization of the signal in both time and frequency. If $\omega_0, t_0 > 1$, the signal is undersampled and the Gábor lattice is incapable of reproducing the signal. Figure 13.17 shows the Gábor transform on the lattice given by Eq. (13.4.9). The overlap of the Gábor window frames ensures that good resolution in time and frequency of a given signal can be achieved.

Drawbacks of the Gábor (STFT) transform

Although the Gábor transform gives a method whereby time and frequency can be simultaneously characterized, there are obvious limitations to the method. Specifically, the method is limited by the time filtering itself. Consider the illustration of the method in Fig. 13.15. The time window filters out the time behavior of the signal in a window centered at τ with width a. Thus when considering the spectral content of this window, any portion of the signal with a wavelength longer than the window is completely lost. Indeed, since the Heinsenberg relationship must hold, the shorter the time-filtering window, the less information there is concerning the frequency content. In contrast, longer windows retain more frequency components, but this comes

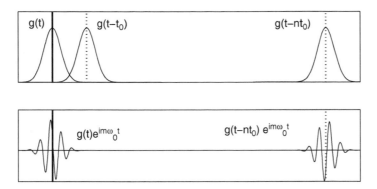

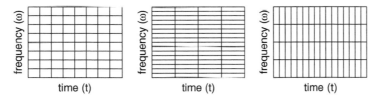

Figure 13.17: Illustration of the discrete Gábor transform which occurs on the lattice sample points of Eq. (13.4.9). In the top figure, the translation with $\omega_0 = 0$ is depicted. The bottom figure depicts both translation in time and frequency. Note that the Gábor frames (windows) overlap so that good resolution of the signal can be achieved in both time and frequency since $0 < t_0, \omega_0 < 1$.

Figure 13.18: Illustration of the resolution trade-offs in the discrete Gábor transform. The left figure shows a time-filtering window that produces nearly equal localization of the time and frequency signal. By increasing the length of the filtering window, increased frequency resolution is gained at the expense of worse time resolution (middle figure). Decreasing the time window does the opposite: time resolution is increased at the expense of poor frequency resolution (right figure).

at the expense of losing the time resolution of the signal. Figure 13.18 provides a graphical description of the failings of the Gábor transform, specifically the trade-offs that occur between time and frequency resolution, and the fact that high accuracy in one of these comes at the expense of resolution in the other parameter. This is a consequence of a fixed time-filtering window.

Other short-time Fourier transform methods

The Gábor transform is not the only windowed Fourier transform that has been developed. There are several other well-used and highly developed STFT techniques. Here, a couple of these more highly used methods will be mentioned for completeness [27].

The *Zak transform* is closely related to the Gábor transform. It is also called the *Weil–Brezin* transform in harmonic analysis. First introduced by Gelfand in 1950 as a method for characterizing eigenfunction expansions in quantum mechanical systems with periodic potentials, it has been generalized to be a key mathematical tool for the analysis of Gábor transform methods. The Zak transform is defined as

$$\mathcal{L}_a f(t, \omega) = \sqrt{a} \sum_{n=-\infty}^{\infty} f(at + an) e^{-i2\pi n\omega} \tag{13.4.12}$$

where $a > 0$ is a constant and n is an integer. Two useful and key properties of this transform are as follows: $\mathcal{L}f(t, \omega + 1) = \mathcal{L}f(t, \omega)$ (periodicity) and $\mathcal{L}f(t + 1, \omega) = \exp(i2\pi\omega)\mathcal{L}f(t, \omega)$ (quasi-periodicity). These properties are particularly important for considering physical problems placed on a lattice.

The *Wigner–Ville distribution* is a particularly important transform in the development of radar and sonar technologies. Its various mathematical properties make it ideal for these applications and provides a method for achieving great time and frequency localization. The Wigner–Ville transform is defined as

$$\mathcal{W}_{f,g}(t, \omega) = \int_{-\infty}^{\infty} f(t + \tau/2)\bar{g}(t - \tau/2) e^{-i\omega}\tau \, d\tau \tag{13.4.13}$$

where this is a standard Fourier kernel which transforms the function $f(t + \tau/2)\bar{g}(t - \tau/2)$. This transform is nonlinear since $\mathcal{W}_{f_1+f_2, g_1+g_2} = \mathcal{W}_{f_1, g_1} + \mathcal{W}_{f_1, g_2} + \mathcal{W}_{f_2, g_1} + \mathcal{W}_{f_2, g_2}$ and $\mathcal{W}_{f+g} = \mathcal{W}_f + \mathcal{W}_g + 2\Re\{\mathcal{W}_{f,g}\}$.

Ultimately, alternative forms of the STFT are developed for one specific reason: to take advantage of some underlying properties of a given system. It is rare that a method developed for radar would be broadly applicable to other physical systems unless it were operating under the same physical principles. Regardless, one can see that specialty techniques exist for time–frequency analysis of different systems.

13.5 Time–Frequency Analysis and Wavelets

The Gábor transform established two key principles for joint time–frequency analysis: *translation* of a short-time window and *scaling* of the short-time window to capture finer time resolution. Figure 13.15 shows the basic concept introduced in the theory of windowed Fourier transforms. Two parameters are introduced to handle the *translation* and *scaling*, namely τ and a. The shortcoming of this method is that it trades off accuracy in time (frequency) for accuracy in frequency (time). Thus the fixed window size imposes a fundamental limitation on the level of time–frequency resolution that can be obtained.

A simple modification to the Gábor method is to allow the scaling window (a) to vary in order to successively extract improvements in the time resolution. In other words, first the low-frequency (poor time resolution) components are extracted using a broad scaling window. The scaling window is subsequently shortened in order to extract out higher frequencies and better time resolution. By keeping a catalogue of the extracting process, both excellent time and frequency resolution of a given signal can be obtained. This is the fundamental principle of *wavelet theory.* The term wavelet means little wave and originates from the fact that the scaling window extracts out smaller and smaller pieces of waves from the larger signal.

Wavelet analysis begins with the consideration of a function known as the *mother wavelet*:

$$\psi_{a,b}(t) = \frac{1}{\sqrt{a}} \psi \left(\frac{t-b}{a} \right) \tag{13.5.1}$$

where $a \neq 0$ and b are real constants. The parameter a is the scaling parameter illustrated in Fig. 13.15 whereas the parameter b now denotes the translation parameter (previously denoted by τ in Fig. 13.15). Unlike Fourier analysis, and very much like Gábor transforms, there are a vast variety of mother wavelets that can be constructed. In principle, the mother wavelet is designed to have certain properties that are somehow beneficial for a given problem. Thus depending upon the application, different mother wavelets may be selected.

Ultimately, the wavelet is simply another expansion basis for representing a given signal or function. Thus it is not unlike the Fourier transform which represents the signal as a series of sines and cosines. Historically, the first wavelet was constructed by Haar in 1910 [28]. Thus the concepts and ideas of wavelets are a century old. However, their widespread use and application did not become prevalent until the mid-1980s. The Haar wavelet is given by the piecewise constant function

$$\psi(t) = \begin{cases} 1 & 0 \leq t < 1/2 \\ -1 & 1/2 \leq t < 1 \\ 0 & \text{otherwise.} \end{cases} \tag{13.5.2}$$

Figure 13.19 shows the Haar wavelet step function and its Fourier transform which is a sinc-like function. Note further that $\int_{-\infty}^{\infty} \psi(t)dt = 0$ and $\|\psi(t)\|^2 = \int_{-\infty}^{\infty} |\psi(t)|^2 dt = 1$. The Haar wavelet is an ideal wavelet for describing localized signals in time (or space) since it has compact support. Indeed, for highly localized signals, it is much more efficient to use the Haar wavelet basis than the standard Fourier expansion. However, the Haar wavelet has poor localization properties in the frequency domain since it decays like a sinc function in powers of $1/\omega$. This is a consequence of the Heinsenberg uncertainty principle.

To represent a signal with the Haar wavelet basis, the translation and scaling operations associated with the mother wavelet need to be considered. Depicted in Fig. 13.19 and given by

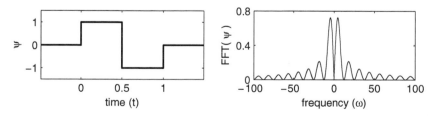

Figure 13.19: Representation of the compactly supported Haar wavelet function $\psi(t)$ and its Fourier transform $\hat{\psi}(\omega)$. Although highly localized in time due to the compact support, it is poorly localized in frequency with a decay of $1/\omega$.

Eq. (13.5.2) is the wavelet $\psi_{1,0}(t)$. Thus its translation is zero and its scaling is unity. The concept in reconstructing a signal using the Haar wavelet basis is to consider decomposing the signal into more generic $\psi_{m,n}(t)$. By appropriate selection of the m and n, finer scales and appropriate locations of the signal can be extracted. For $a < 1$, the wavelet is a compressed version of $\psi_{1,0}$ whereas for $a > 1$, the wavelet is a dilated version of $\psi_{1,0}$. The scaling parameter a is typically taken to be a power of 2 so that $a = 2^j$ for some integer j. Figure 13.20 shows the compressed and dilated Haar wavelet for $a = 0.5$ and $a = 2$, i.e. $\psi_{1/2,0}$ and $\psi_{2,0}$. The compressed wavelet allows for finer scale resolution of a given signal while the dilated wavelet captures low-frequency components of a signal by having a broad range in time.

The simple Haar wavelet already illustrates all the fundamental principles of the wavelet concept. Specifically by using scaling and translation, a given signal or function can be represented by a basis of functions which allows for higher and higher refinement in the time resolution of a signal. Thus it is much like the Gábor concept, except that now the time window is variable in order to capture different levels of resolution. Thus the large-scale structures in time are captured with broad time domain Haar wavelets. At this scale, the time resolution of the signal is very poor. However by successive rescaling in time, a finer and finer time resolution of the signal can be obtained along with its high-frequency components. The information at the low and high scales is all preserved so that a complete picture of the time–frequency domain can be constructed. Ultimately, the only limit in this process is the number of scaling levels to be considered.

The wavelet basis can be accessed via an integral transform of the form

$$(Tf)(\omega) = \int_t K(t,\omega)f(t)dt \tag{13.5.3}$$

where $K(t,\omega)$ is the kernel of the transform. This is equivalent in principle to the Fourier transform whose kernel is the oscillations given by $K(t,\omega) = \exp(-i\omega t)$. The key idea now is to define

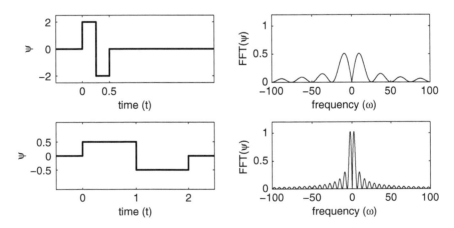

Figure 13.20: Illustration of the compression and dilation process of the Haar wavelet and its Fourier transform. In the top row, the compressed wavelet $\psi_{1/2,0}$ is shown. Improved time resolution is obtained at the expense of a broader frequency signature. The bottom row shows the dilated wavelet $\psi_{2,0}$ which allows it to capture lower frequency components of a signal.

a transform which incorporates the mother wavelet as the kernel. Thus we define the *continuous wavelet transform (CWT)*:

$$\mathcal{W}_\psi[f](a, b) = (f, \psi_{a,b}) = \int_{-\infty}^{\infty} f(t)\bar{\psi}_{a,b}(t)dt. \tag{13.5.4}$$

Much like the windowed Fourier transform, the CWT is a function of the dilation parameter a and translation parameter b. Parenthetically, a wavelet is admissible if the following property holds:

$$C_\psi = \int_{-\infty}^{\infty} \frac{\left|\hat{\psi}(\omega)\right|^2}{|\omega|} d\omega < \infty \tag{13.5.5}$$

where the Fourier transform of the wavelet is defined by

$$\hat{\psi}_{a,b} = \frac{1}{\sqrt{|a|}} \int_{-\infty}^{\infty} e^{-i\omega t} \psi\left(\frac{t-b}{a}\right) dt = \frac{1}{\sqrt{|a|}} e^{-ib\omega} \hat{\psi}(a\omega). \tag{13.5.6}$$

Thus provided the admissibility condition (13.5.5) is satisfied, the wavelet transform can be well defined.

As an example of the admissibility condition, consider the Haar wavelet (13.5.2). Its Fourier transform can be easily computed in terms of the sinc-like function:

$$\hat{\psi}(\omega) = ie^{-i\omega/2} \frac{\sin^2(\omega/4)}{\omega/4}. \tag{13.5.7}$$

Thus the admissibility constant can be computed to be

$$\int_{-\infty}^{\infty} \frac{\left|\hat{\psi}(\omega)\right|^2}{|\omega|} d\omega = 16 \int_{-\infty}^{\infty} \frac{1}{|\omega|^3} \left|\sin\frac{\omega}{4}\right|^4 d\omega < \infty. \tag{13.5.8}$$

This then shows that the Haar wavelet is in the admissible class.

Another interesting property of the wavelet transform is the ability to construct new wavelet bases. The following theorem is of particular importance.

Theorem *If ψ is a wavelet and ϕ is a bounded integrable function, then the convolution $\psi \star \phi$ is a wavelet.*

In fact, from the Haar wavelet (13.5.2) we can construct new wavelet functions by convolving with for instance

$$\phi(t) = \begin{cases} 0 & t < 0 \\ 1 & 0 \le t \le 1 \\ 0 & t \ge 1 \end{cases} \tag{13.5.9}$$

or the function

$$\phi(t) = e^{-t^2}. \tag{13.5.10}$$

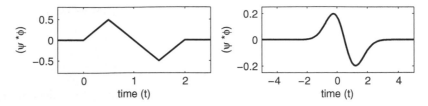

Figure 13.21: Convolution of the Haar wavelet with the functions (13.5.9) (left panel) and (13.5.10) (right panel). The convolved functions can be used as the mother wavelet for a wavelet basis expansion.

The convolutions of these functions ϕ with the Haar wavelet ψ (13.5.2) are produced in Fig. 13.21. These convolutions could also be used as mother wavelets in constructing a decomposition of a given signal or function.

The wavelet transform principle is quite simple. First, the signal is split up into a bunch of smaller signals by translating the wavelet with the parameter b over the entire time domain of the signal. Second, the same signal is processed at different frequency bands, or resolutions, by scaling the wavelet window with the parameter a. The combination of translation and scaling allows for processing of the signals at different times and frequencies. Figure 13.22 is an upgrade of Fig. 13.16 that incorporates the wavelet transform concept in the time–frequency domain. In

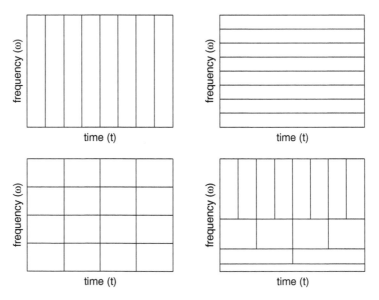

Figure 13.22: Graphical depiction of the difference between a time series analysis, Fourier analysis, Gábor analysis and wavelet analysis of a signal. This figure is identical to Fig. 13.16 but with the inclusion of the time–frequency resolution achieved with the wavelet transform. The wavelet transform starts with a large Fourier domain window so that the entire frequency content is extracted. The time window is then scaled in half, leading to finer time resolution at the expense of worse frequency resolution. This process is continued until a desired time–frequency resolution is obtained.

this figure, the standard time series, Fourier transform and windowed Fourier transform are represented along with the multi-resolution concept of the wavelet transform. In particular, the box illustrating the wavelet transform shows the multi-resolution concept in action. Starting with a large Fourier domain window, the entire frequency content is extracted. The time window is then scaled in half, leading to finer time resolution at the expense of worse frequency resolution. This process is continued until a desired time–frequency resolution is obtained. This simple cartoon is critical for understanding wavelet application to time–frequency analysis.

Example: The Mexican hat wavelet

One of the more common wavelets is the Mexican hat wavelet. This wavelet is essentially a second moment of a Gaussian in the frequency domain. The definition of this wavelet and its transform are as follows:

$$\psi(t) = (1 - t^2)e^{-t^2/2} = -d^2/dt^2\left(e^{-t^2/2}\right) = \psi_{1,0} \tag{13.5.11a}$$

$$\hat{\psi}(\omega) = \hat{\psi}_{1,0}(\omega) = \sqrt{2\pi}\,\omega^2 e^{-\omega^2/2}. \tag{13.5.11b}$$

The Mexican hat wavelet has excellent localization properties in both time and frequency due to the minimal time–bandwidth product of the Gaussian function. Figure 13.23 (top panels) shows the basic Mexican wavelet function $\psi_{1,0}$ and its Fourier transform, both of which decay in t (ω) like $\exp(-t^2)$ ($\exp(-\omega^2)$). The Mexican hat wavelet can be dilated and translated easily as is depicted in Fig. 13.23 (bottom panel). Here three wavelets are depicted: $\psi_{1,0}$, $\psi_{3/2,-3}$ and $\psi_{1/4,6}$. This shows both the scaling and translation properties associated with any wavelet function.

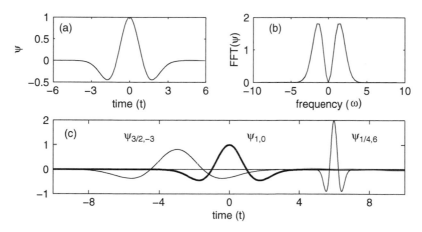

Figure 13.23: Illustration of the Mexican hat wavelet $\psi_{1,0}$ (top left panel), its Fourier transform $\hat{\psi}_{1,0}$ (top right panel), and two additional dilations and translations of the basic $\psi_{1,0}$ wavelet, namely the $\psi_{3/2,-3}$ and $\psi_{1/4,6}$ are shown (bottom panel).

To finish the initial discussion of wavelets, some of the various properties of the wavelets are listed. To begin, consider the time–frequency resolution and its localization around a given time and frequency. These quantities can be calculated from the relations:

$$\sigma_t^2 = \int_{-\infty}^{\infty} (t - \langle t \rangle)^2 \left| \psi(t) \right|^2 dt \tag{13.5.12a}$$

$$\sigma_\omega^2 = \frac{1}{2\pi} \int_{-\infty}^{\infty} (\omega - \langle \omega \rangle)^2 \left| \hat{\psi}(\omega) \right|^2 d\omega \tag{13.5.12b}$$

where the variances measure the spread of the time and frequency signal around $\langle t \rangle$ and $\langle \omega \rangle$, respectively. The Heisenberg uncertainty constrains the localization of time and frequency by the relation $\sigma_t^2 \sigma_\omega^2 \geq 1/2$. In addition, the CWT has the following mathematical properties

1   **Linearity** The transform is linear so that

$$\mathcal{W}_\psi(\alpha f + \beta g)(a, b) = \alpha \mathcal{W}_\psi(f)(a, b) + \beta \mathcal{W}_\psi(g)(a, b).$$

2   **Translation** The transform has the translation property

$$\mathcal{W}_\psi(T_c f)(a, b) = \mathcal{W}_\psi(f)(a, b - c)$$

where $T_c f(t) = f(t - c)$.

3   **Dilation** The dilation property follows

$$\mathcal{W}_\psi(D_c f)(a, b) = \frac{1}{\sqrt{c}} \mathcal{W}_\psi(f)(a/c, b/c)$$

where $c > 0$ and $D_c f(t) = (1/c)f(t/c)$.

4   **Inversion** The transform can be inverted with the definition

$$f(t) = \frac{1}{C_\psi} \int_{-\infty}^{\infty} \int_{-\infty}^{\infty} \mathcal{W}_\psi(f)(a, b) \psi_{a,b}(t) \frac{db\, da}{a^2}$$

where it becomes clear why the admissibility condition $C_\psi < \infty$ is needed.

To conclude this section, consider the idea of discretizing the wavelet transform on a computational grid. Thus the transform is defined on a lattice so that

$$\psi_{m,n}(x) = a_0^{-m/2} \psi \left(a_0^{-m} x - n b_0 \right) \tag{13.5.13}$$

where $a_0, b_0 > 0$ and m, n are integers. The *discrete wavelet transform* is then defined by

$$\begin{aligned}
\mathcal{W}_\psi(f)(m, n) &= (f, \psi_{m,n}) \\
&= \int_{-\infty}^{\infty} f(t) \bar{\psi}_{m,n}(t) dt \\
&= a_0^{-m/2} \int_{-\infty}^{\infty} f(t) \bar{\psi}(a_0^{-m} t - n b_0) dt.
\end{aligned} \tag{13.5.14}$$

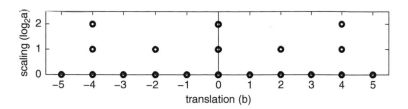

Figure 13.24: Discretization of the discrete wavelet transform with $a_0 = 2$ and $b_0 = 1$. This figure is a more formal depiction of the multi-resolution discretization shown in Fig. 13.22.

Futhermore, if $\psi_{m,n}$ are complete, then a given signal or function can be expanded in the wavelet basis:

$$f(t) = \sum_{m,n=-\infty}^{\infty} (f, \psi_{m,n}) \psi_{m,n}(t). \tag{13.5.15}$$

This expansion is in a set of wavelet frames. It still needs to be determined if the expansion is in terms of a set of basis functions. It should be noted that the scaling and dilation parameters are typically taken to be $a_0 = 2$ and $b_0 = 1$, corresponding to dilations of 2^{-m} and translations of $n2^m$. Figure 13.24 gives a graphical depiction of the time–frequency discretization of the wavelet transform. This figure is especially relevant for the computational evaluation of the wavelet transform. Further, it is the basis of fast algorithms for multi-resolution analysis.

13.6   Multi-Resolution Analysis and the Wavelet Basis

Before proceeding forward with wavelets, it is important to establish some key mathematical properties. Indeed, the most important issue to resolve is the ability of the wavelet to actually represent a given signal or function. In Fourier analysis, it has been established that any generic function can be represented by a series of cosines and sines. Something similar is needed for wavelets in order to make them a useful tool for the analysis of time–frequency signals.

The concept of a wavelet is simple and intuitive: construct a signal using a single function $\psi \in L^2$ which can be written $\psi_{m,n}$ and that represents binary dilations by 2^m and translations of $n2^{-m}$ so that

$$\psi_{m,n} = 2^{m/2} \psi \left(2^m \left(x - n/2^m \right) \right) = 2^{m/2} \psi(2^m x - n) \tag{13.6.1}$$

where m and n are integers. The use of this wavelet for representing a given signal or function is simple enough. However, there is a critical issue to be resolved concerning the *orthogonality* of the functions $\psi_{m,n}$. Ultimately, this is the primary issue which must be addressed in order to consider the wavelets as basis functions in an expansion. Thus we define the orthogonality condition as

$$(\psi_{m,n}, \psi_{k,l}) = \int_{-\infty}^{\infty} \psi_{m,n}(x) \psi_{k,l}(x) dx = \delta_{m,k} \delta_{n,l} \tag{13.6.2}$$

where δ_{ij} is the Dirac delta defined by

$$\delta_{ij} = \begin{cases} 0 & i \neq j \\ 1 & i = j \end{cases} \tag{13.6.3}$$

where i, j are integers. This statement of orthogonality is generic, and it holds in most function spaces with a defined inner product.

The importance of orthogonality should not be underestimated. It is very important in applications where a functional expansion is used to approximate a given function or solution. In what follows, two examples are given concerning the key role of orthogonality.

Fourier expansions

Consider representing an even function $f(x)$ over the domain $x \in [0, L]$ with a cosine expansion basis. By Fourier theory, the function can be represented by

$$f(x) = \sum_{n=0}^{\infty} a_n \cos \frac{n\pi x}{L} \tag{13.6.4}$$

where the coefficients a_n can be constructed by using inner product rules and orthogonality. Specifically, by multiplying both sides of the equation by $\cos(m\pi x/L)$ and integrating over $x \in [0, L]$, i.e. taking the inner product with respect to $\cos(m\pi x/L)$, the following result is found:

$$\left(f, \cos m\pi x/L\right) = \sum_{n=0}^{\infty} a_n \left(\cos n\pi x/L, \cos m\pi x/L\right). \tag{13.6.5}$$

This is where orthogonality plays a key role: the infinite sum on the right-hand side can be reduced to a single index where $n = m$ since the cosines are orthogonal to each other

$$(\cos n\pi x/L, \cos m\pi x/L) = \begin{cases} 0 & n \neq m \\ L & n = m. \end{cases} \tag{13.6.6}$$

Thus the coefficients can be computed to be

$$a_n = \frac{1}{L} \int_0^L f(x) \cos \frac{n\pi x}{L} dx \tag{13.6.7}$$

and the expansion is accomplished. Moreover, the cosine basis is complete for even functions, and any signal or function $f(x)$ can be represented, i.e. as $n \to \infty$ in the sum, the expansion converges to the given signal $f(x)$.

Eigenfunction expansions

The cosine expansion is a subset of the more general eigenfunction expansion technique that is often used to solve differential and partial differential equation problems. Consider the nonhomogeneous boundary value problem

$$Lu = f(x) \tag{13.6.8}$$

where L is a given self-adjoint linear operator. This problem can be solved with an eigenfunction expansion technique by considering the associated eigenvalue problem of the operator L:

$$Lu_n = \lambda_n u_n. \tag{13.6.9}$$

The solution of (13.6.8) can then be expressed as

$$u(x) = \sum_{n=0}^{\infty} a_n u_n \tag{13.6.10}$$

provided the coefficients a_n can be determined. Plugging in this solution to (13.6.8) yields the following calculations

$$Lu = f$$
$$L\left(\sum a_n u_n\right) = f$$
$$\sum a_n L u_n = f$$
$$\sum a_n \lambda_n u_n = f. \tag{13.6.11}$$

Taking the inner product of both sides with respect to u_m yields

$$\left(\sum a_n \lambda_n u_n, u_m\right) = \left(f, u_m\right)$$
$$\sum a_n \lambda_n \left(u_n, u_m\right) = \left(f, u_m\right)$$
$$a_m \lambda_m = \left(f, u_m\right) \tag{13.6.12}$$

where the last line is achieved by orthogonality of the eigenfunctions $(u_n, u_m) = \delta_{n,m}$. This then gives $a_m = (f, u_m)/\lambda_m$ and the eigenfunction expansion solution is

$$u(x) = \sum_{n=0}^{\infty} \frac{(f, u_n)}{\lambda_n} u_n. \tag{13.6.13}$$

Provided the u_n are a complete set, this expansion is guaranteed to converge to $u(x)$ as $n \to \infty$.

Orthonormal wavelets

The preceding examples highlight the importance of orthogonality for representing a given function. A wavelet ψ is called orthogonal if the family of functions $\psi_{m,n}$ are orthogonal as given by Eq. (13.6.2). In this case, a given signal or function can be uniquely expressed with the doubly infinite series

$$f(t) = \sum_{n,m=-\infty}^{\infty} c_{m,n} \psi_{m,n}(t) \tag{13.6.14}$$

where the coefficients are given from orthogonality by

$$c_{m,n} = (f, \psi_{m,n}). \tag{13.6.15}$$

The series is guaranteed to converge to $f(t)$ in the L^2 norm.

The above result based upon orthogonal wavelets establishes the key mathematical framework needed for using wavelets in a very broad and general way. It is this result that allows us to think of wavelets philosophically as the same as the Fourier transform.

Multi-resolution analysis (MRA)

The power of the wavelet basis is its ability to take a function or signal $f(t)$ and express it as a limit of successive approximations, each of which is a finer and finer version of the function in time. These successive approximations correspond to different resolution levels.

A *multi-resolution analysis*, commonly referred to as an MRA, is a method that gives a formal approach to constructing the signal with different resolution levels. Mathematically, this involves a sequence

$$\{V_m : \quad m \in \text{integers}\} \tag{13.6.16}$$

of embedded subspaces of L^2 that satisfies the following relations:

1   The subspaces can be embedded in each other so that

$$\textit{Coarse} \quad \cdots \subset V_{-2} \subset V_{-1} \subset V_0 \subset V_1 \subset V_2 \cdots V_m \subset V_{m+1} \cdots \quad \textit{Fine.}$$

2   The union of all the embedded subspaces spans the entire L^2 space so that

$$\cup_{m=-\infty}^{\infty} V_m$$

is dense in L^2.

3   The intersection of subspaces is the null set so that

$$\cap_{m=-\infty}^{\infty} V_m = \{0\}.$$

4   Each subspace picks up a given resolution so that $f(x) \in V_m$ if and only if $f(2x) \in V_{m+1}$ for all integers m.

5   There exists a function $\phi \in V_0$ such that

$$\left\{\phi_{0,n} = \phi(x - n)\right\}$$

is an orthogonal basis for V_0 so that

$$\|f\|^2 = \int_{-\infty}^{\infty} |f(x)|^2 \, dx = \sum_{-\infty}^{\infty} |(f, \phi_{0,n})|^2 .$$

The function ϕ is called the *scaling function* or *father wavelet*.

If $\{V_m\}$ is a multi-resolution of L^2 and if V_0 is the closed subspace generated by the integer translates of a single function ϕ, then we say ϕ generates the MRA.

One remark of importance: since $V_0 \subset V_1$ and ϕ is a scaling function for V_0 and also for V_1, then

$$\phi(x) = \sum_{-\infty}^{\infty} c_n \phi_{1,n}(x) = \sqrt{2} \sum_{-\infty}^{\infty} c_n \phi(2x - n) \tag{13.6.17}$$

where $c_n = (\phi, \phi_{1,n})$ and $\sum_{-\infty}^{\infty} |c_n|^2 = 1$. This equation, which relates the scaling function as a function of x and $2x$, is known as the *dilation equation*, or *two-scale equation*, or *refinement equation* because it reflects $\phi(x)$ in the refined space V_1 which as the finer scale of 2^{-1}.

Since $V_m \subset V_{m+1}$, we can define the orthogonal complement of V_m in V_{m+1} as

$$V_{m+1} = V_m \oplus W_m \tag{13.6.18}$$

where $V_m \perp W_m$. This can be generalized so that

$$\begin{aligned}
V_{m+1} &= V_m \oplus W_m \\
&= (V_{m-1} \oplus W_{m-1}) \oplus W_m \\
&\vdots \\
&= V_0 \oplus W_0 \oplus W_1 \oplus \cdots \oplus W_m \\
&= V_0 \oplus \left(\oplus_{n=0}^{m} W_n\right).
\end{aligned} \tag{13.6.19}$$

As $m \to \infty$, it can be found that

$$V_0 \oplus \left(\oplus_{n=0}^{\infty} W_n\right) = L^2. \tag{13.6.20}$$

In a similar fashion, the resolution can rescale upwards so that

$$\oplus_{n=-\infty}^{\infty} W_n = L^2. \tag{13.6.21}$$

Moreover, there exists a scaling function $\psi \in W_0$ (the mother wavelet) such that

$$\psi_{0,n}(x) = \psi(x - n) \tag{13.6.22}$$

constitutes an orthogonal basis for W_0 and

$$\psi_{m,n}(x) = 2^{m/2} \psi(2^m x - n) \tag{13.6.23}$$

is an orthogonal basis for W_m. Thus the mother wavelet ψ spans the orthogonal complement subset W_m while the scaling function ϕ spans the subsets V_m. The connection between the father and mother wavelet is shown in the following theorem.

Theorem *If $\{V_m\}$ is a MRA with scaling function ϕ, then there is a mother wavelet ψ*

$$\psi(x) = \sqrt{2} \sum_{-\infty}^{\infty} (-1)^{n-1} \bar{c}_{-n-1} \phi(2x - n) \tag{13.6.24}$$

where

$$c_n = (\phi, \phi_{1,n}) = \sqrt{2} \int_{-\infty}^{\infty} \phi(x)\bar{\phi}(2x - 1)dx. \tag{13.6.25}$$

That is, the system $\psi_{m,n}(x)$ is an orthogonal basis of L^2.

This theorem is critical for what we would like to do. Namely, use the wavelet basis functions as a complete expansion basis for a given function $f(x)$ in L^2. Further, it explicitly states the connection between the *scaling function* $\phi(x)$ (father wavelet) and *wavelet function* $\psi(x)$ (mother wavelet). It is only left to construct a desirable wavelet basis to use. As for wavelet construction, the idea is to build them to take advantage of certain properties of the system so that it gives an efficient and meaningful representation of your time–frequency data.

13.7 Spectrograms and the Gábor Transform in MATLAB

The aim of this section will be to use MATLAB's fast Fourier transform routines modified to handle the Gábor transform. The Gábor transform allows for a fast and easy way to analyze both the time and frequency properties of a given signal. Indeed, this windowed Fourier transform method is used extensively for analyzing speech and vocalization patterns. For such applications, it is typical to produce a *spectrogram* that represents the signal in both the time and frequency domain. Figures 13.25 and 13.26 are produced from the vocalization patterns in time–frequency of a

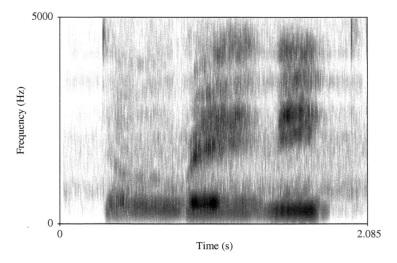

Figure 13.25: Spectrogram (time–frequency) analysis of a male human saying "do re mi." The spectrogram is created with the software program Praat, which is an open source code for analyzing phonetics.

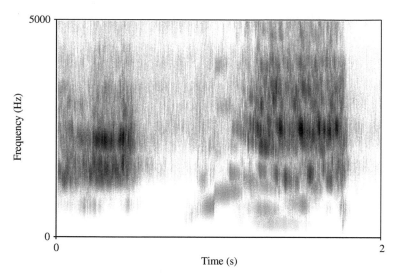

Figure 13.26: Spectrogram (time–frequency) analysis of a humpback whale vocalization over a short period of time. The spectrogram is created with the software program Praat, which is an open source code for analyzing phonetics.

human saying "do re mi" and a humpback whale vocalizing to other whales. The time–frequency analysis can be used to produce speech recognition algorithms given the characteristic signatures in the time–frequency domains of sounds. Thus spectrograms are a sort of fingerprint of sound.

To understand the algorithms which produce the spectrogram, it is informative to return to the characteristic picture shown in Fig. 13.15. This demonstrates the action of an applied time filter in extracting time localization information. To build a specific example, consider the following MATLAB code that builds a time domain (t), its corresponding Fourier domain (ω), a relatively complicated signal ($S(t)$), and its Fourier transform ($\hat{S}(\omega)$).

```
clear all; close all; clc

L=10; n=2048;
t2=linspace(0,L,n+1); t=t2(1:n);
k=(2*pi/L)*[0:n/2-1 -n/2:-1]; ks=fftshift(k);

S=(3*sin(2*t)+0.5*tanh(0.5*(t-3))+ 0.2*exp(-(t-4).^2)...
  +1.5*sin(5*t)+4*cos(3*(t-6).^2))/10+(t/20).^3;
St=fft(S);
```

The signal and its Fourier tranform can be plotted with the commands

```
figure(1)
subplot(3,1,1)   % Time domain
plot(t,S,'k')
set(gca,'Fontsize',[14]),
xlabel('Time (t)'), ylabel('S(t)')

subplot(3,1,2)   % Fourier domain
plot(ks,abs(fftshift(St))/max(abs(St)),'k');
axis([-50 50 0 1])
set(gca,'Fontsize',[14])
xlabel('frequency (\omega)'), ylabel('FFT(S)')
```

Figure 13.27 shows the signal and its Fourier transform for the above example. This signal $S(t)$ will be analyzed using the Gábor transform method.

The simplest Gábor window to implement is a Gaussian time filter centered at some time τ with width a. As has been demonstrated, the parameter a is critical for determining the level of time resolution versus frequency resolution in a time–frequency plot. Figure 13.28 shows the signal under consideration with three filter widths. The narrower the time-filtering, the better resolution in time. However, this also produces the worst resolution in frequency. Conversely, a wide window in time produces much better frequency resolution at the expense of reducing the time resolution. A simple extension to the existing code produces a signal plot along with three different filter widths of Gaussian shape.

```
figure(2)
width=[10 1 0.2];
for j=1:3
  g=exp(-width(j)*(t-4).^2);
  subplot(3,1,j)
  plot(t,S,'k'), hold on
  plot(t,g,'k','Linewidth',[2])
  set(gca,'Fontsize',[14])
  ylabel('S(t), g(t)')
end
xlabel('time (t)')
```

The key now for the Gábor transform is to multiply the time filter Gábor function $g(t)$ with the original signal $S(t)$ in order to produce a windowed section of the signal. The Fourier

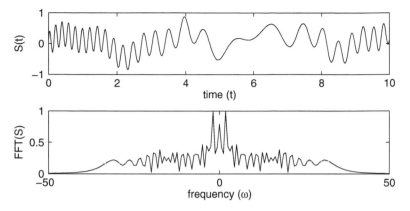

Figure 13.27: Time signal and its Fourier transform considered for a time–frequency analysis in what follows.

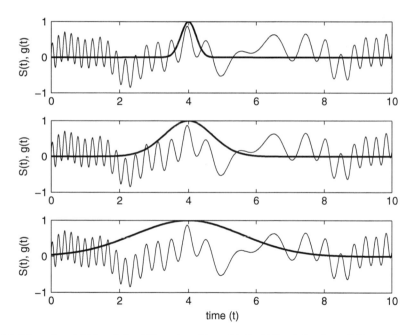

Figure 13.28: Time signal $S(t)$ and the Gábor time filter $g(t)$ (bold lines) for three different Gaussian filters: $g(t) = \exp(-10(x-4)^2)$ (top), $g(t) = \exp(-(x-4)^2)$ (middle) and $g(t) = \exp(-0.2(x-4)^2)$ (bottom). The different filter widths determine the time–frequency resolution. Better time resolution gives worse frequency resolution and vice versa due to the Heinsenberg uncertainty principle.

transform of the windowed section then gives the local frequency content in time. The following code constructs the windowed Fourier transform with the Gábor filtering function

$$g(t) = e^{-a(t-b)^2}. \tag{13.7.1}$$

The Gaussian filtering has a width parameter a and translation parameter b. The following code constructs the windowed Fourier transform using the Gaussian with $a = 2$ and $b = 4$.

```
figure(3)
g=exp(-2*(t-4).^2);
Sg=g.*S;
Sgt=fft(Sg);

subplot(3,1,1), plot(t,S,'k'), hold on
  plot(t,g,'k','Linewidth',[2])
  set(gca,'Fontsize',[14])
  ylabel('S(t), g(t)'), xlabel('time (t)')
subplot(3,1,2), plot(t,Sg,'k')
  set(gca,'Fontsize',[14])
  ylabel('S(t)g(t)'), xlabel('time (t)')
subplot(3,1,3), plot(ks,abs(fftshift(Sgt))/max(abs(Sgt)),'k')
  axis([-50 50 0 1])
  set(gca,'Fontsize',[14])
  ylabel('FFT(Sg)'), xlabel('frequency (\omega)')
```

Figure 13.29 demonstrates the application of this code and the windowed Fourier transform in extracting local frequencies of a local time window.

The key to generating a spectrogram is to now vary the position b of the time filter and produce spectra at each location in time. In theory, the parameter b is continuously translated to produce the time–frequency picture. In practice, like everything else, the parameter b is discretized. The level of discretization is important in establishing a good time–frequency analysis. Specifically, finer resolution will produce better results. The following code makes a dynamical *movie* of this process as the parameter b is translated.

```
figure(4)
Sgt_spec=[];
tslide=0:0.1:10
for j=1:length(tslide)
  g=exp(-2*(t-tslide(j)).^2); % Gabor
  Sg=g.*S; Sgt=fft(Sg);
  Sgt_spec=[Sgt_spec; abs(fftshift(Sgt))];
```

Continued

```
    subplot(3,1,1), plot(t,S,'k',t,g,'r')
    subplot(3,1,2), plot(t,Sg,'k')
    subplot(3,1,3), plot(ks,abs(fftshift(Sgt))/max(abs(Sgt)))
    axis([-50 50 0 1])
    drawnow
    pause(0.1)
end
```

This movie is particularly illustrative and provides an excellent graphical representation of how the Gábor time-filtering extracts both local time information and local frequency content. It also illustrates, as the parameter a is adjusted, the ability (or inability) of the windowed Fourier transform to provide accurate time and frequency information.

The code just developed also produces a matrix **Sgt_spec** which contains the Fourier transform at each slice in time of the parameter b. It is this matrix that produces the spectrogram of the time–frequency signal. The spectrogram can be viewed with the commands

```
pcolor(tslide,ks,Sgt_spec.'), shading interp
set(gca,'Ylim',[-50 50],'Fontsize',[14])
colormap(hot)
```

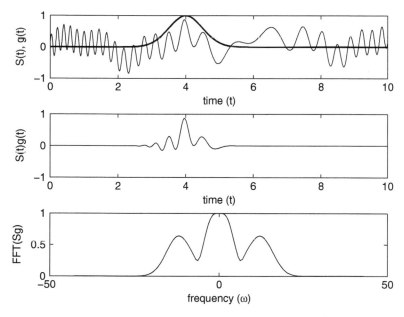

Figure 13.29: Time signal $S(t)$ and the Gábor time filter $g(t) = \exp(-2(x-4)^2)$ (bold line) for a Gaussian filter. The product $S(t)g(t)$ is depicted in the middle panel and its Fourier transform $\hat{S}g(\omega)$ is depicted in the bottom panel. Note that the windowing of the Fourier transform can severely limit the detection of low-frequency components.

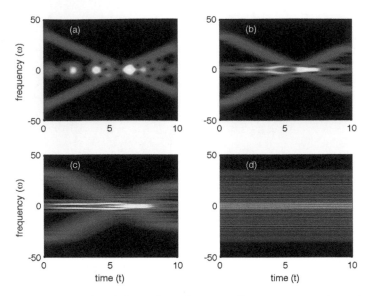

Figure 13.30: Spectrograms produced from the Gábor time-filtering equation (13.7.1) with (a) $a = 5$, (b) $a = 1$ and (c) $a = 0.2$. The Fourier transform, which has no time localization information, is depicted in (d). It is easily seen that the window width trades off time and frequency resolution at the expense of each other. Regardless, the spectrogram gives a visual time–frequency picture of a given signal.

Modifying the code slightly, a spectrogram of the signal $S(t)$ can be made for three different filter widths $a = 5, 1, 0.2$ in Eq. (13.7.1). The spectrograms are shown in Fig. 13.30 where from left to right the filtering window is broadened from $a = 5$ to $a = 0.2$. Note that for the left plot, strong localization of the signal in time is achieved at the expense of suppressing almost all the low-frequency components of the signal. In contrast, the rightmost figure with a wide temporal filter preserves excellent resolution of the Fourier domain but fails to localize signals in time. Such are the trade-offs associated with a fixed Gábor window transform.

13.8   MATLAB Filter Design and Wavelet Toolboxes

The applications of filtering and time–frequency analysis are so ubiquitous across the sciences that MATLAB has developed a suite of toolboxes that specialize in these applications. Two of these toolboxes will be demonstrated in what follows. Primarily, screenshots will give hints of the functionality and versatility of the toolboxes.

The most remarkable part of the toolbox is that it allows for entry into high-level signal processing and wavelet processing almost immediately. Indeed, one hardly needs to know anything to begin the process of analyzing, synthesizing and manipulating data to one's own ends. For the most part, each of the toolboxes allows you to begin usage once you upload your signal, image or data. The only drawback is cost. For the most part, many academic departments have access to the toolboxes. And if you are part of a university environment, or enrolled in classes, the student versions of the toolboxes can be purchased directly from MATLAB at a very reasonable price.

Filter design and analysis toolbox

The filter design toolbox is ideally suited to help create and engineer an ideal filter for whatever application you may need. The user has the option of using command line codes to directly access the filter design subroutines, or one can use the extremely efficient and useful GUI (graphical user interface) from MATLAB. Figure 13.31 shows a typical screenshot that MATLAB produces upon startup of the filter design toolbox. The command to start the toolbox is given by

```
FDAtool
```

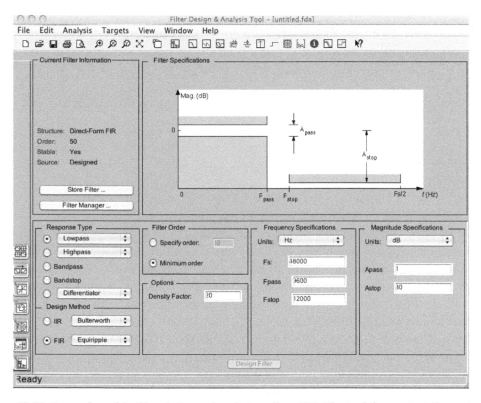

Figure 13.31: Screenshot of the filter design and analysis toolbox GUI. The top left contains information on the current filter design while the top right is a graphical representation of the filter characteristics. The lower portion of the GUI is reserved for the user interface functionality and design. Indeed, the type of filter, its spectral width, whether it is low-pass (filters high frequencies), high-pass (filters low frequencies), band-pass (filters a specified band of frequencies) or band-stop (attenuates a band of frequencies) can be chosen.

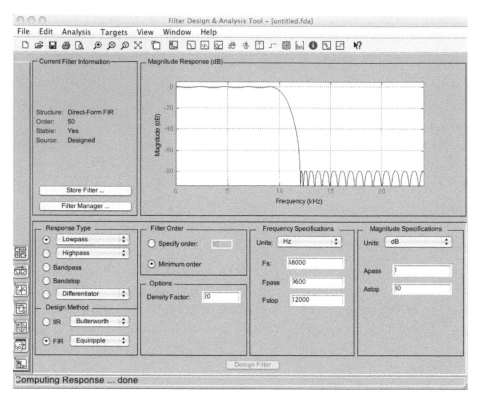

Figure 13.32: Screenshot of the magnitude response of the filter. The magnitude response, along with a host of other important characteristics of the filter, can be obtained from the drag-down menu of the **Analysis** button on the top.

This launches the highly intuitive MATLAB GUI that guides you through the process of designing your ideal filter. Figure 13.32 demonstrates, for instance, the magnitude response of the filter along with its frequency and magnitude of response. The magnitude response can be accessed by dragging down the **Analysis** button at the top of the GUI panel, in addition to the amplitude response, the phase response, a combination of phase and amplitude response, group-velocity response, etc. Depending upon your needs, complete engineering of the filter can be performed with set objectives and performance aims.

As with all toolbox design and information, it can be exported from MATLAB for use in other programs or with other parts of MATLAB. The exporting of data, as well as saving of a given filter design, can all be performed from the **file** drag-down menu at the top right of the GUI. Moreover, if you desire to skip the filter design GUI, then all the filter functions can be accessed directly from the command line window. For instance, the command

```
[B,A]=butter(N,Wn)
```

generates a Butterworth digital and analogue filter design. The parameter N denotes the Nth order low-pass digital Butterworth filter and returns the filter coefficients in length $N + 1$ vectors B (numerator) and A (denominator). The coefficients are listed in descending powers of z. The cut-off frequency Wn must be $0.0 < Wn < 1.0$, with 1.0 corresponding to half the sample rate; and besides this, you now know the butter command in MATLAB.

Wavelet toolbox

Just like the signal processing, filter design toolbox, the wavelet toolbox is an immensely powerful toolbox for signal processing, image processing, multi-resolution analysis and any manner of time–frequency analysis. The wavelet toolbox is comprised of a large number of modules that allow for various applications and methodologies. To access the toolbox, simply type

```
wavemenu
```

and a selection of menu items is generated. Figure 13.33 demonstrates the assortment of possible uses of the wavelet toolbox that are available upon launching the wavelet toolbox. In addition to such applications as signal processing and time–frequency analysis of one-dimensional signals or functions, one can proceed to denoise or manipulate image data. The program can also produce wavelet transforms and output the coefficients of a continuous or discrete wavelet transform.

As a specific example, consider the *wavelet 1-D* button and the example files it contains. The example files are accessed from the **file** button at the top left of the GUI. In this case, an example analysis of a noisy step function signal is considered. It should be noted that your own signal could have easily been imported from the **file** button with the **load** menu item. In this example, a signal is provided with the full wavelet decomposition at different levels of resolution. It is easy to see that the given signal is decomposed into its *big* features followed by its *finer* features as one descends vertically down the plots. In the standard example, the *sym* wavelet is used with five levels of resolution (see Fig. 13.34). This shows the progression, as discussed in the wavelets section, of the multi-resolution analysis. The wavelet analysis can be easily modified to consider other wavelet bases. In fact, in addition to the sym wavelet, one can choose the Haar wavelets, Daubechies wavelets (see Fig. 13.35) and Coifman wavelets among others. This remarkable packaging of the best time–frequency domain methods is one of the tremendous aspects of the toolbox. In addition to the decomposition, a histogram, the statistics, or compression of the signal can be performed by simply clicking one of the appropriate buttons on the right.

The wavelet toolbox also allows for considering a signal with a wavelet expansion. The continuous wavelet 1-D provides a full decomposition of the signal into its wavelet basis. Figure 13.36 shows a given signal along with the calculation of the wavelet coefficients $C_{a,b}$ where a and b are dilation and translation operations, respectively. This provides the basis for a time-frequency analysis. Additionally, given a chosen level of resolution determined by the parameter a, then the third panel shows the parameter $C_{a,b}$ for $a = a_{max}/2$. The local maxima of the $C_{a,b}$ function

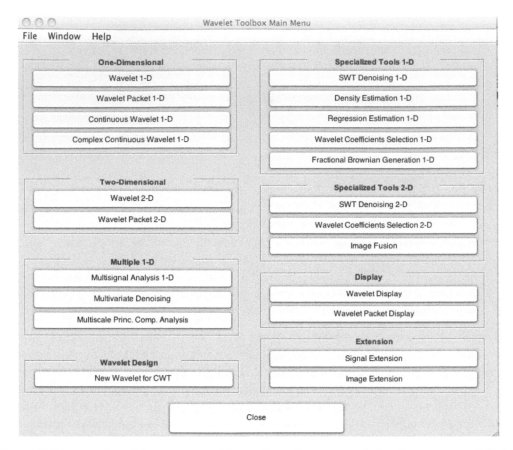

Figure 13.33: Screenshot of the entry point of the wavelet toolbox. A myriad of applications are available by simply clicking on one of the menu items. Both 1D and 2D problems can be analyzed with the toolbox.

are also shown. As before, different wavelet bases may be chosen along with different levels of resolution of a given signal.

Two final applications are demonstrated: One is an image denoising application, and the second is an image multi-resolution (compression) analysis. Both of these applications have important applications for image clean-up and size reduction; therefore they are prototypical examples of the power of the 2D wavelet tools. In the first applications, illustrated in Fig. 13.37, a noisy image is considered. The application allows you to directly important your own image or data set. Once imported, the image is decomposed at the different resolution levels. To denoise the image, *filters* are applied at each level of resolution to remove the high-frequency components of each level of resolution. This produces an exceptional, wavelet based algorithm for denoising. As before, a myriad of wavelets can be considered and full control of the denoising process is left to the user. Figure 13.38 demonstrates the application of a multi-resolution analysis to a given image. Here the image is decomposed into various resolution levels. The program allows the user complete control and flexibility in determining the level of resolution desired. By keeping fewer

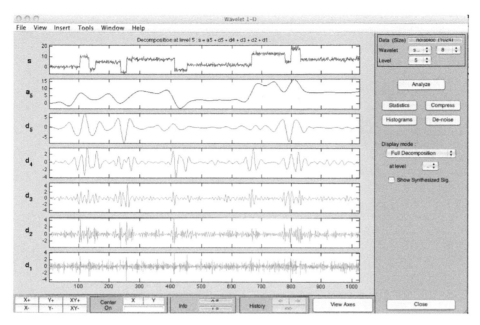

Figure 13.34: Screenshot of the wavelet 1-D with a noisy step function signal. A multi-resolution decomposition of the signal is performed with *sym* wavelets.

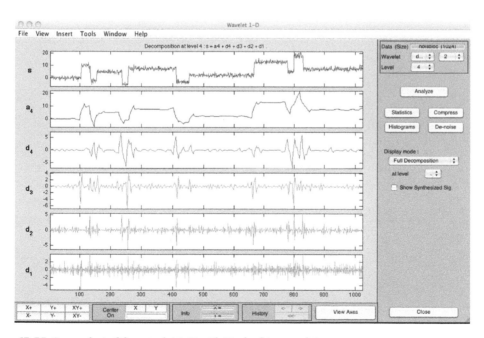

Figure 13.35: Screenshot of the wavelet 1-D with Daubechies wavelets.

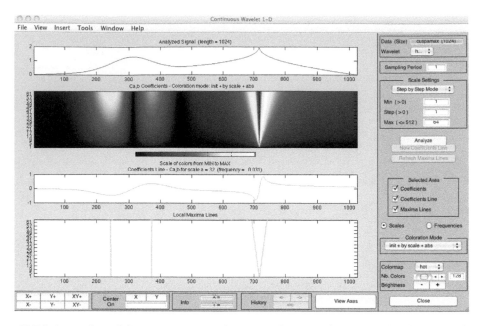

Figure 13.36: Screenshot of the continuous wavelet 1-D application where a signal is decomposed into a given wavelet basis and its wavelet coefficients are determined.

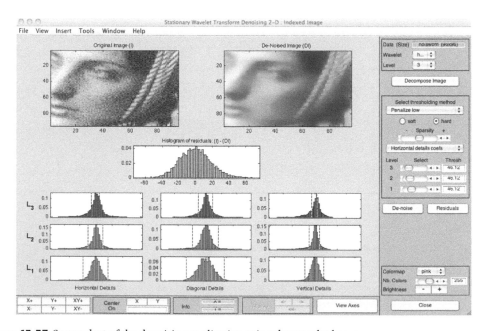

Figure 13.37: Screenshot of the denoising application using the wavelet bases.

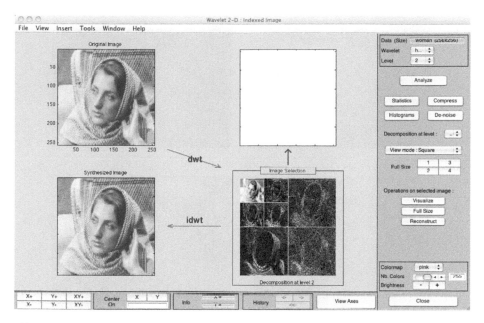

Figure 13.38: Screenshot of the decomposition of an image into different resolution levels. This application is ideal for image compression.

levels of resolution, the image quality is compromised, but the algorithm thus provides an excellent compressed image. The discrete wavelet transform (**dwt**) is used to decompose the image. The inverse discrete wavelet transform (**idwt**) is used to extract back the image at an appropriate level of multi-resolution. For this example, two levels of decomposition are applied.

The wavelets transform and its algorithms can be directly accessed from the command line, just like the filter design toolbox. For instance, to apply a continuous wavelet transform, the command

```
coeffs=cwt(S,scales,'wname')
```

can be used to compute the real or complex continuous 1D wavelet coefficients. Specifically, it computes the continuous wavelet coefficients of the vector S at real, positive *scales*, using a wavelet whose name is *wname* (for instance, haar). The signal S is real, the wavelet can be real or complex.

JPEG compression with wavelets

Wavelets can be directly applied to the arena of image compression. To illustrate the above concepts more clearly, we can consider a digital image with 600×800 pixels (rows × columns). Figure 13.39(a) shows the original image. The following MATLAB code imports the original

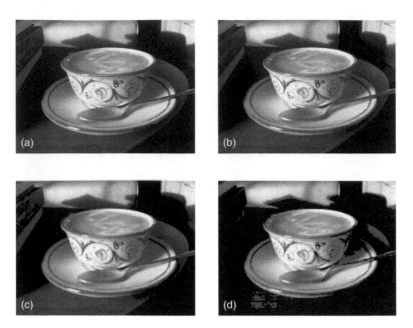

Figure 13.39: Original image (a) along with the reconstructed image using approximately (b) 5.6%, (c) 4.5% or (d) 3% of the information encoded in the wavelet coefficients. At 3%, the image reconstruction starts to fail. This reconstruction process illustrates that images are sparse in the wavelet representation.

JPEG picture and plots the results. It also constructs the matrix **Abw** which is the picture in matrix form.

```
A=imread('photo','jpeg');
Abw2=rgb2gray(A);
Abw=double(Abw2);
[nx,ny]=size(Abw);
figure(1), subplot(2,2,1), imshow(Abw2)
```

The image is well defined and of high resolution. Indeed, it contains $600 \times 800 = 480\,000$ pixels encoding the image. However, the image can actually be encoded with far less data. Specifically, it has been well-known for the past two decades that wavelets are an ideal basis in which to encode photos.

To demonstrate the nearly optimal encoding of the wavelet basis, the original photo is subjected to a two-dimensional wavelet transform:

```
[C,S]=wavedec2(Abw,2,'db1');
```

The **wavedec** command performs a two-dimensional transform and returns the wavelet coefficient vector **C** along with its ordering matrix **S**. In this example, a two-level wavelet decomposition is performed (thus the value of 2 in the **wavedec** command call) using Daubechies wavelets (thus the **db1** option). Higher level wavelet decompositions can be performed using a variety of wavelets.

In the wavelet basis, the image is expressed as a collection of weightings of the wavelet coefficients. This is identical to representing a time domain signal as a collection of frequencies using the Fourier transform. Figure 13.40(a) depicts the 480 000 wavelet coefficient values for the given picture. A zoom in of the dense looking region around 0 to 0.1 is shown in Fig. 13.41(a). In zooming in, one can clearly see that many of the coefficients that make up the image are nearly zero. Due to numerical round-off, for instance, none of the coefficients will actually be zero.

Image compression follows directly from this wavelet analysis. Specifically, one can simply choose a threshold and directly set all wavelet coefficients below this chosen threshold to be identically zero instead of nearly zero. What remains is to demonstrate that the image can be reconstructed using such a strategy. If it works, then one would only need to retain the much smaller (sparse) number of nonzero wavelets in order to store and/or reconstruct the image.

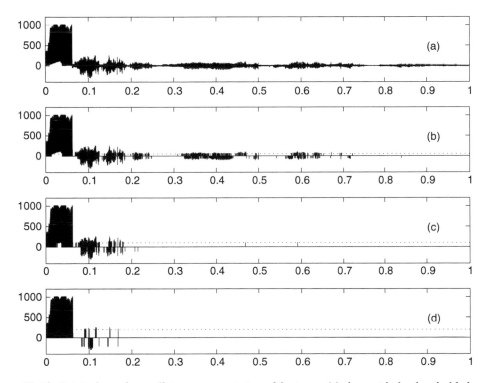

Figure 13.40: Original wavelet coefficient representation of the image (a) along with the thresholded coefficient vector for a threshold (dotted line) value of (b) 50, (c) 100 and (d) 200. The percentages of identically zero wavelet coefficients are 94.4%, 95.5% and 97.0%, respectively. Only the nonzero coefficients are kept for encoding of the image. The x-axis has been normalized to unity.

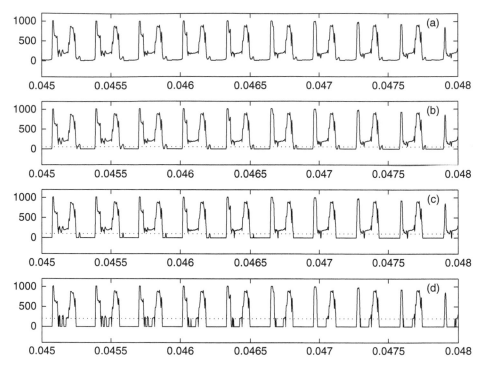

Figure 13.41: Zoom in of the original wavelet coefficient representation of the image (a) along with the thresholded coefficient vector for a threshold (dotted line) value of (b) 50, (c) 100 and (d) 200 (see Fig. 13.40). In the zoomed in region, one can clearly see that many of the coefficients are below threshold. The zoom in is over the entire wavelet domain that has been normalized to unity.

To address the image compression issue, we can apply a threshold rule to the wavelet coefficient vector **C** that is shown in Fig. 13.40(a). The following code successively sets a threshold at 50, 100 and 200 in order to generate a sparse wavelet coefficient vector **C2**. This new sparse vector is then used to encode and reconstruct the image.

```
xw=(1:nx*ny)/(nx*ny);
th=[50 100 200];
for j=1:3
  count=0;
  C2=C;
  for jj=1:length(C2);
  if abs(C2(jj)) < th(j)
    C2(jj)=0;
    count=count+1;
  end
  end
```

Continued

```
  percent=count/length(C2)*100
  figure(2), subplot(4,1,j+1),plot(xw,C2,'k',[0 1],[th(j) th(j)],'k:')
  figure(3), subplot(4,1,j+1),plot(xw,C2,'k',[0 1],[th(j) th(j)],'k:')
    set(gca,'Xlim',[0.045 0.048])

  Abw2_sparse=waverec2(C2,S,'db1');
  Abw2_sparse2=uint8(Abw2_sparse);
  figure(1), subplot(2,2,j+1), imshow(Abw2_sparse2)
end
```

The results of the compression process are illustrated in Figs. 13.39–13.41.

To understand the compression process, first consider Fig. 13.40 which shows the wavelet coefficient vector for the three different threshold values of (b) 50, (c) 100 and (d) 200. The threshold line for each of these is shown as the dotted line. Anything below the threshold line is set identically to zero, thus it can be discarded when encoding the image. Figure 13.41 shows a zoom in of the region near 0 to 0.1. In addition to setting a threshold, the above code also calculates the percentage of zero wavelet coefficients. For the three threshold values considered here, the percentage of wavelet coefficients that are zero are 94.4%, 95.5% and 97.0%, respectively. Thus only approximately 5.6%, 4.5% or 3% of the information is needed for the image reconstruction. Indeed, this is the key idea behind the JPEG2000 image compression algorithm: you only save what is needed to reconstruct the image to a desired level of accuracy.

The actual image reconstruction is illustrated in Fig. 13.39 for the three different levels of thresholding. From (b) to (d), the image is reconstructed from 5.6%, 4.5% or 3% of the original information. At 3%, the image finally becomes visibly worse and begins to fail in faithfully reproducing the original image. At 5.6% or 4.5%, it is difficult to tell much difference with the original image. This is quite remarkable and illustrates the *sparsity* of the image in the wavelet basis. Such sparsity is ubiquitous for image representation using wavelets.

14 Image Processing and Analysis

Over the past couple of decades, imaging science has had a profound impact on science, medicine and technology. The reach of imaging science is immense, spanning the range of modern day biological imaging tools to computer graphics. But the imaging methods are not limited to visual data. Instead, a general mathematical framework has been developed by which more general data analysis can be performed. The next set of sections deals with some basic tools for image processing or data analysis.

14.1 Basic Concepts and Analysis of Images

Imaging science is now ubiquitous. Indeed, it now dominates many scientific fields as a primary vehicle for achieving information and informing decisions. The following are a sample of the technologies and applications where imaging science is having a formative impact:

- **Ultrasound** Ultrasound refers to acoustic waves whose frequencies are higher than the audible range of human ears. Typically this is between 20 Hz and 20 kHz. Most medical ultrasound devices are above 1 MHz and below 20 MHz. Anybody who has had a baby will proudly show you their ultrasound pictures.

- **Infrared/thermal imaging** Imperceptible to the human eye, infrared (IR) signatures correspond to wavelengths longer than 700 nm. However, aided by signal amplification and enhancement devices, humans can perceive IR signals from above 700 nm to a few microns.

- **Tomography (CAT scans)** Using the first Nobel Prize (1895) ideas of Wilhelm Röntgen for X-rays, tomography has since become the standard in the medical industry as the wavelengths, ranging from nanometers to picometers, can easily penetrate most biological tissue. Imaging hardware and software have made this the standard for producing medical images.

- **Magnetic resonance imaging (MRI)** By applying a strong magnetic field to the body, the atomic spins in the body are aligned. High-frequency RF pulses are emitted into a slice plan of the external field. Upon turning off the RF waves, the relaxation process reveals differentials in tissue and biological material, thus giving the characteristics needed for imaging.

- **Radar and sonar imaging** Radar uses radio waves for the detection and ranging of targets. The typical wavelengths are in the microwave range of 1 cm to 1 m. Applied to imaging, radar becomes a sophisticated tool for remote sensing applications. Sonar works in a similar fashion, but with underwater acoustic waves.

- **Digital photos** Most of us are well aware of this application as we often would like to remove red-eye, undo camera blurring from out-of-focus shots or camera shakes, etc. Many software packages already exist to help you improve your image quality.

- **Computer graphics** Building virtual worlds and targeted images falls within the aegis of the computer science community. There are many mathematically challenging issues related to representation of real-life images.

In the applications above, it goes without saying that most images acquired from experiment or enhancement devices are rarely without imperfections. Often the imperfections can be ignored. However, if the noise or imperfections in the images are critical for a decision making process, i.e. detection via radar of aircraft or detection of a tumor in biological tissue, then any enhancement of the image that could contribute to a statistically better decision is of high value.

Image processing and analysis address the fundamental issue of allowing an end user to have enhanced information, or statistically more accurate data, of a given image or signal. Thus mathematical methods are the critical piece in achieving this enhancement. Before delving into specifics of mathematical methods that can be brought to bear on a given problem, it is important to classify the types of image analysis objectives that one can consider. The following gives a succinct list of image processing tasks:

- **Image contrast enhancement** Often there can be little contrast between different objects in a photo or in a given data set. Contrast enhancement is a method that tries to promote or enhance differences in an image in order to achieve sharper contrasts.

- **Image denoising** This is often of primary interest in analyzing data sets. Can noise, from either measurements or inaccuracies, be removed to give a more robust and fundamental picture of the underlying dynamics of a system?

- **Image deblurring** The leading cause of bad pictures: camera shakes and out-of-focus cameras. Image processing methods can often compensate and undo these two phenomena to produce a sharper image that is greatly deblurred. This is a remarkable application of image analysis, and is already an application that is included in many digital photo software bundles.

- **Inpainting** If data is missing over certain portions of an image or set of data, inpainting provides a mathematical framework for systematically generating the missing data. It accounts for both edges and the overall structure surrounding the missing pixels.

- **Segmentation (edge detection)** Segmentation breaks up an image into blocks of different structures. This is particularly important for biomedical applications as well as military applications. Good algorithms for determining segmentation locations are critical.

The mathematical task is then to develop methods and algorithms capable of performing a set of tasks that addresses the above list. In the sections that follow, only a small subset of these will be considered. However, a comprehensive mathematical treatment of the above issues can be performed [29].

Mathematical approaches

Given the broad appeal of imaging sciences and its applicability in such a wide variety of fields, it is not surprising that an equally diverse set of mathematical methods has been brought to bear on image processing. As with the previous lists of applications and tasks associated with image processing, the following list highlights some of the broad mathematical ideas that are present in thinking about image analysis.

Morphological approach

In this approach, one can think of the entire 2D domain of an image as a set of subdomains. Each subdomain should represent some aspect of the overall image. The fundamental idea here is to decompose the domain into fundamental shapes based upon some structure element S. The domain is then decomposed in a binary way, a small patch of the image (defined by the structure element S) is *in* the object of interest or *out* of the object of interest. This provides an efficient method for edge detection in certain sets of data or images.

Fourier analysis

As with time–frequency analysis, the key idea is to decompose the image into its Fourier components. This is easily done with a two-dimensional Fourier transform. Thus the image becomes a collection of Fourier modes. Figure 14.1 shows a photo and its corresponding 2D Fourier transform. The drawback of the Fourier transform is the fact that the Fourier transform of a sharp

Figure 14.1: Image of a beautifully made cappuccino along with its Fourier transform (log of the absolute value of the spectrum) in two dimensions. The strong vertical and horizontal lines in the spectrum correspond to vertical and horizontal structures in the photo. Looking carefully at the spectrum will also reveal diagonal signatures associated with the diagonal edges in the photo.

edged object results in a sinc-like function that decays in the Fourier modes like $1/k$ where k is the wavenumber. This slow decay means that localization of the image in both the spatial and wavenumber domain is limited. Regardless, the Fourier mode decomposition gives an alternative representation of the image. Moreover, it is clear from the Fourier spectrum that a great number of the Fourier mode components are zero or nearly so. Thus the concept of image compression can easily be seen directly from the spectrum, i.e. by saving 20% of the dominant modes only, the image can be nearly constructed. This then would compress the image five-fold. As one might also expect, filtering with the Fourier transform can also help process the image to some desired ends.

Wavelets

As with the time–frequency analysis, wavelets provide a much more sophisticated representation of an image. In particular, wavelets allow for exceptional localization in both the spatial and wavenumber domains. In the wavelet basis, the key quantities are computed from the wavelet coefficients

$$c_\alpha = (u(x, y), \psi_\alpha) \qquad (14.1.1)$$

where the wavelet coefficient c_α is equivalent to a Fourier coefficient for a wavelet ψ_α. Since the mid-1980s, the wavelet basis has taken over as the primary tool for image compression due to its excellent space–wavenumber localization properties.

Stochastic modeling

Consideration can be made in image processing of the statistical nature of the data itself. For instance, many images in nature, such as clouds, trees, sky, sandy beaches, are fundamentally statistical in nature. Thus one may not want to denoise such objects when performing image analysis tasks. In the stochastic modeling approach, hidden features are explored within a given set of observed data which accounts for the statistical nature of many of the underlying objects in an image.

Partial differential equations and diffusion

For denoising applications, or applications in which the data is choppy or highly pixelated, smoothing algorithms are of critical importance. In this case, one can take the original data as the initial conditions of a diffusion process so that the image undergoes, in the simplest case, the evolution

$$\frac{\partial u}{\partial t} = D\nabla^2 u \qquad (14.1.2)$$

where $\nabla^2 = \partial_x^2 + \partial_y^2$. This diffusion process provides smoothing to the original data file u. The key is knowing when the smoothing process should be stopped so as not to remove too much image content.

Loading images, additive noise and MATLAB

To illustrate some of the various aspects of image processing, the following example will load the ideal image, apply Gaussian white noise to each pixel and show the noisy image. In MATLAB, most standard image formats are easily loaded. The key to performing mathematical manipulations is to then transform the data into double precision numbers which can be transformed, filtered and manipulated at will. The generic data file uploaded in MATLAB is *uint8* which is an integer format ranging from 0 to 255. Moreover, color images have three sets of data for each pixel in order to specific the RGB color coordinates. The following code uploads an image file (600 × 800 pixels). Although there arc 600 × 800 pixels, the data are stored as a 600 × 800 × 3 matrix where the extra dimensions contain the color coding. This can be made into a 600 × 800 matrix by turning the image into a black-and-white picture. This is done in the code that follows:

```
clear all; close all; clc;
A=imread('photo','tiff');    % load image
Abw=rgb2gray(A);             % make image black-and-white
subplot(2,2,1), image(A); set(gca,'Xtick',[],'Ytick',[])
subplot(2,2,3), imshow(Abw);

A2=double(A);                % change form unit8 to double
Abw=double(Abw);
```

Note that at the end of the code, the matrices produced are converted to double precision numbers.

To add Gaussian white noise to the images, either the color or black-and-white versions, the **randn** command is used. In fact, noise is added to both pictures and the output is produced.

```
noise=randn(600,800,3);   % add noise to RGB image
noise2=randn(600,800);    % add noise to black-and-white

u=uint8(A2+50*noise);     % change from double to uint8
u2=uint8(Abw+50*noise2);

subplot(2,2,2), image(u); set(gca,'Xtick',[],'Ytick',[])
subplot(2,2,4), image(u2); set(gca,'Xtick',[],'Ytick',[])
```

Note that at the final step, the images are converted back to standard image formats used for JPEG or TIFF images. Figure 14.2 demonstrates the plot produced from the above code. The

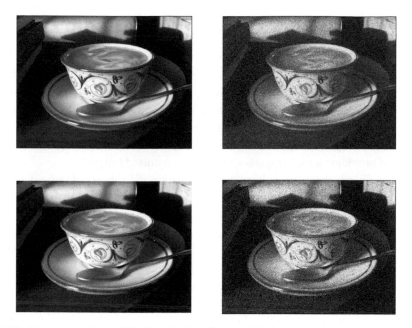

Figure 14.2: Ideal image in color and black-and-white (left panels) along with their noisy counterparts (right panels).

ability to load and manipulate the data is fundamental to all the image processing applications to be pursued in what follows.

As a final example of data manipulation, the Fourier transform of the above image can also be considered. Figure 14.1 has already demonstrated the result of this code. But for completeness, it is illustrated here

```
Abw2=Abw(600:-1:1,:);
figure(2)
subplot(2,2,1), pcolor(Abw2), shading interp,
colormap(hot), set(gca,'Xtick',[],'Ytick',[])

Abwt=abs(fftshift(fft2(Abw2)));
subplot(2,2,2), pcolor(log(Abwt)), shading interp,
colormap(hot), set(gca,'Xtick',[],'Ytick',[])
```

Note that in this set of plots, we are once again working with matrices so that commands like **pcolor** are appropriate. These matrices are the subject of image processing routines and algorithms.

14.2 **Linear Filtering for Image Denoising**

As with time–frequency analysis and denoising of time signals, many applications of image processing deal with cleaning up images from imperfections, pixelation and graininess, i.e. processing of noisy images. The objective in any image processing application is to enhance or improve the quality of a given image. In this section, the filtering of noisy images will be considered with the aim of providing a higher quality, maximally denoised image.

The power of filtering has already been demonstrated in the context of radar detection applications. Ultimately, image denoising is directly related to filtering. To see this, consider Fig. 14.1 of the last section. In this image, the ideal image is represented along with the log of its Fourier transform. Like many images, the Fourier spectrum is particularly sparse (or nearly zero) for most high-frequency components. Noise, however, tends to generate many high-frequency structures on an image. Thus it is hoped that filtering of high-frequency components might remove unwanted noise fluctuations or graininess in an image. The top two panels of Fig. 14.3 show an initial noisy image and the log of its Fourier transform. These top two panels can be compared to the ideal image and spectrum shown in Fig. 14.1. The code used to generate these top two panels is given by the following:

```
A=imread('photo','tiff');  Abw=rgb2gray(A);
Abw=double(Abw);


B=Abw+100*randn(600,800);
Bt=fft2(B);   Bts=fftshift(Bt);


subplot(2,2,1), imshow(uint8(B)), colormap(gray)
subplot(2,2,2), pcolor((log(abs(Bts)))); shading interp
colormap(gray), set(gca,'Xtick',[],'Ytick',[])
```

Note that the uploaded image is converted to a double precision number through the **double** command. It can be transformed back to the image format via the **uint8** command.

Linear filtering can be applied in the Fourier domain of the image in order to remove the high-frequency scale fluctuations induced by the noise. A simple filter to consider is a Gaussian that takes the form

$$F(k_x, k_y) = \exp(-\sigma_x(k_x - a)^2 - \sigma_y(k_y - b)^2) \qquad (14.2.1)$$

where σ_x and σ_y are the filter widths in the x- and y-directions, respectively, and a and b are the center-frequency values for the corresponding filtering. For the image under consideration, it is a 600×800 pixel image so that the center-frequency components are located at $k_x = 301$ and $k_y = 401$, respectively. To build a Gaussian filter, the following lines of code are needed:

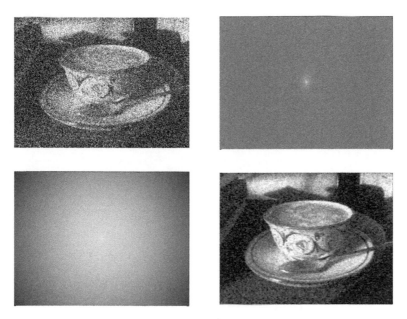

Figure 14.3: A noisy image and its Fourier transform are considered in the top panel. The ideal image is shown in Fig. 14.1. By applying a linear Gaussian filter, the high-frequency noise components can be eliminated (bottom left) and the image quality is significantly improved (bottom right).

```
kx=1:800; ky=1:600; [Kx,Ky]=meshgrid(kx,ky);
F=exp(-0.0001*(Kx-401).^2-0.0001*(Ky-301).^2);
Btsf=Bts.*F;

subplot(2,2,3), pcolor(log(abs(Btsf))); shading interp
colormap(gray), set(gca,'Xtick',[],'Ytick',[])

Btf=ifftshift(Btsf); Bf=ifft2(Btf);
subplot(2,2,4), imshow(uint8(Bf)), colormap(gray)
```

In this code, the **meshgrid** command is used to generate the two-dimensional wavenumbers in the x- and y-directions. This allows for the construction of the filter function that has a two-dimensional form. In particular, a Gaussian centered around $(k_x, k_y) = (301, 401)$ is created with $\sigma_x = \sigma_y = 0.0001$. The bottom two panels in Fig. 14.3 show the filtered spectrum followed by its inverse Fourier transform. The final image on the bottom right of this figure shows the denoised image produced by simple filtering. The filtering significantly improves the image quality.

One can also over-filter and cut out substantial information concerning the figure. Figure 14.4 shows the log of the spectrum of the noisy image for different values of filters in comparison with the unfiltered image. For narrow filters, much of the image information is irretrievably lost along

Figure 14.4: Comparison of the log of the Fourier transform for three different values of the filtering strength, $\sigma_x = \sigma_y = 0.01, 0.001, 0.0001$ (top left, top right, bottom left, respectively), and the unfiltered image (bottom right).

with the noise. Finding an optimal filter is part of the image processing agenda. In the image spatial domain, the filtering strength (or width) can be considered. Figure 14.5 shows a series of filtered images and their comparison to the unfiltered image. Strong filtering produces a smooth, yet blurry image. In this case, all fine-scale features are lost. Moderate levels of filtering produce reasonable images with significant reduction of the noise in comparison to the nonfiltered image. The following code was used to produce these last two figures.

```
% Gaussian filter
fs=[0.01 0.001 0.0001 0];
for j=1:4
  F=exp(-fs(j)*(Kx-401).^2-fs(j)*(Ky-301).^2);
  Btsf=Bts.*F; Btf=ifftshift(Btsf); Bf=ifft2(Btf);
  figure(4), subplot(2,2,j), pcolor(log(abs(Btsf)))
  shading interp,colormap(gray),set(gca,'Xtick',[],'Ytick',[])
  figure(5), subplot(2,2,j), imshow(uint8(Bf)), colormap(gray)
end
```

Note that the processed data are converted back to a graphics image before plotting with the **imshow** command.

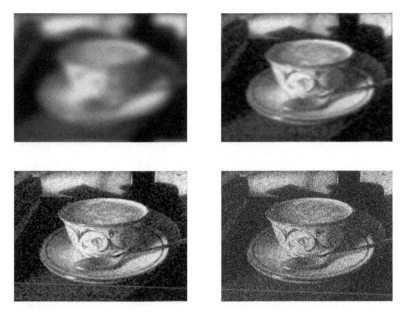

Figure 14.5: Comparison of the image quality for the three filter strengths considered in Fig. 14.4. A high degree of filtering blurs the image and no fine-scale features are observed (top left). In contrast, moderate strength filtering produces exceptional improvement of the image quality (top right and bottom left). These denoised images should be compared to the unfiltered image (bottom right).

Gaussian filtering is only one type of filtering that can be applied. There are, just like in signal processing applications, myriads of filters that can be used to process a given image, including low-pass, high-pass, band-pass, etc. As a second example filter, the Shannon filter is considered. The Shannon filter is simply a step function with value of unity within the transmitted band and zero outside of it. In the example that follows, a Shannon filter is applied of width 50 pixels around the center frequency of the image.

```
width=50;
Fs=zeros(600,800);
Fs(301-width:1:301+width,401-width:1:401+width) ...
        =ones(1:2*width+1,1:2*width+1);
Btsf=Bts.*Fs;

Btf=ifftshift(Btsf); Bf=ifft2(Btf);

subplot(2,2,3), pcolor(log(abs(Btsf))); shading interp
colormap(gray), set(gca,'Xtick',[],'Ytick',[])
subplot(2,2,4), imshow(uint8(Bf)), colormap(gray)
```

Figure 14.6: A noisy image and its Fourier transform are considered in the top panel. The ideal image is shown in Fig. 14.1. By applying a Shannon (step function) filter, the high-frequency noise components can be eliminated (bottom left) and the image quality is significantly improved (bottom right).

Figure 14.6 is a companion to Fig. 14.3. The only difference between them is the filter chosen for the image processing. The key difference is represented in the lower left panel of both figures. Note that the Shannon filter simply suppresses all frequencies outside of a given filter box. The performance difference between Gaussian filtering and Shannon filtering appears to be fairly marginal. However, there may be some applications where one or the other is more suitable. The image enhancement can also be explored as a function of the Shannon filter width. Figure 14.7 shows the image quality as the filter is widened along with the unfiltered image. Strong filtering again produces a blurred image while moderate filtering produces a greatly enhanced image quality. A code to produce this image is given by the following:

```
fs=[10 50 100 200];
for j=1:4
  Fs=zeros(600,800);
  Fs(301-fs(j):1:301+fs(j),401-fs(j):1:401+fs(j)) ...
      =ones(1:2*fs(j)+1,1:2*fs(j)+1);
  Btsf=Bts.*Fs; Btf=ifftshift(Btsf); Bf=ifft2(Btf);
  figure(7), subplot(2,2,j), imshow(uint8(Bf)), colormap(gray)
end
```

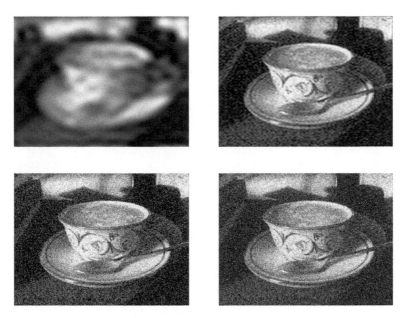

Figure 14.7: Comparison of the image quality for the three filter widths. A high degree of filtering blurs the image and no fine-scale features are observed (top left). In contrast, moderate strength filtering produces exceptional improvement of the image quality (top right and bottom left). These denoised images should be compared to the unfiltered image (bottom right).

The width is adjusted by considering the number of filter pixels around the center frequency with value unity. One might be able to argue that the Gaussian filter produces slightly better image results since the inverse Fourier transform of a step function filter produces sinc-like behavior.

 ## 14.3   Diffusion and Image Processing

Filtering is not the only way to denoise an image. Intimately related to filtering is the use of diffusion for image enhancement. Consider for the moment the simplest spatial diffusion process in two dimensions, i.e. the heat equation:

$$u_t = D\nabla^2 u \tag{14.3.1}$$

where $u(x,y)$ will represent a given image, $\nabla^2 = \partial_x^2 + \partial_y^2$, D is a diffusion coefficient, and some boundary conditions must be imposed. If for the moment we consider periodic boundary conditions, then the solution to the heat equation can be found from, for instance, the Fourier transform

$$\hat{u}_t = -D(k_x^2 + k_y^2)\hat{u} \;\; \rightarrow \;\; \hat{u} = \hat{u}_0 e^{-D(k_x^2+k_y^2)t} \tag{14.3.2}$$

where the $\hat{u}$ is the Fourier transform of $u(x, y)$ and $\hat{u}_0$ is the Fourier transform of the initial conditions, i.e. the original noisy image. The solution of the heat equation illustrates a key and critical concept: the wavenumbers (spatial frequencies) decay according to a Gaussian function. Thus linear filtering with a Gaussian is equivalent to a linear diffusion of the image for periodic boundary conditions.

The above argument establishes some equivalency between filtering and diffusion. However, the diffusion formalism provides a more general framework in which to consider image cleanup since the heat equation can be modified to

$$u_t = \nabla \cdot (D(x, y)\nabla u) \tag{14.3.3}$$

where $D(x, y)$ is now a spatial diffusion coefficient. In particular, the diffusion coefficient could be used to great advantage to target trouble spots on an image while leaving relatively noise-free patches alone.

To solve the heat equation numerically, we discretize the spatial derivative with a second-order scheme (see Table 4.1) so that

$$\frac{\partial^2 u}{\partial x^2} = \frac{1}{\Delta x^2} \left[u(x + \Delta x, y) - 2u(x, y) + u(x - \Delta x, y) \right] \tag{14.3.4a}$$

$$\frac{\partial^2 u}{\partial y^2} = \frac{1}{\Delta y^2} \left[u(x, y + \Delta y) - 2u(x, y) + u(x, y - \Delta y) \right]. \tag{14.3.4b}$$

This approximation reduces the partial differential equation to a system of ordinary differential equations. Once this is accomplished, then a variety of standard time-stepping schemes for differential equations can be applied to the resulting system.

The ODE system for 1D diffusion

Consider first diffusion in one-dimension so that $u(x, y) = u(x)$. The vector system for MATLAB can then be formulated. To define the system of ODEs, we discretize and define the values of the vector $\mathbf{u}$ in the following way:

$$u(-L) = u_1$$
$$u(-L + \Delta x) = u_2$$
$$\vdots$$
$$u(L - 2\Delta x) = u_{n-1}$$
$$u(L - \Delta x) = u_n$$
$$u(L) = u_{n+1}.$$

If periodic boundary conditions, for instance, are considered, then periodicity requires that $u_1 = u_{n+1}$. Thus the system of differential equations solves for the vector

$$\mathbf{u} = \begin{pmatrix} u_1 \\ u_2 \\ \vdots \\ u_n \end{pmatrix}. \tag{14.3.5}$$

The governing heat equation is then reformulated in discretized form as the differential equation system

$$\frac{d\mathbf{u}}{dt} = \frac{\kappa}{\Delta x^2} \mathbf{A}\mathbf{u}, \tag{14.3.6}$$

where $\mathbf{A}$ is given by the sparse matrix

$$\mathbf{A} = \begin{bmatrix} -2 & 1 & 0 & \cdots & & 0 & 1 \\ 1 & -2 & 1 & 0 & \cdots & & 0 \\ 0 & \ddots & \ddots & \ddots & & & \\ \vdots & & & & & & \vdots \\ & & & & & & 0 \\ \vdots & \cdots & 0 & 1 & -2 & 1 \\ 1 & 0 & \cdots & 0 & 1 & -2 \end{bmatrix}, \tag{14.3.7}$$

and the values of one on the upper right and lower left of the matrix result from the periodic boundary conditions.

MATLAB implementation

The system of differential equations can now be easily solved with a standard time-stepping algorithm such as *ode23* or *ode45*. The basic algorithm would be as follows

1   Build the sparse matrix $\mathbf{A}$.

```
e1=ones(n,1);   % build a vector of ones
A=spdiags([e1 -2*e1 e1],[-1 0 1],n,n);   % diagonals
A(1,n)=1;   A(n,1)=1;   % periodic boundaries
```

2   Generate the desired initial condition vector $\mathbf{u} = \mathbf{u}_0$.
3   Call an ODE solver from the MATLAB suite. The matrix $\mathbf{A}$, the diffusion constant κ and spatial step Δx need to be passed into this routine.

```
[t,y]=ode45('rhs',tspan,u0,[],k,dx,A);
```

The function *rhs.m* should be of the following form

```
function rhs=rhs(tspan,u,dummy,k,dx,A)
rhs=(k/dx^2)*A*u;
```

4   Plot the results as a function of time and space.

The algorithm is thus fairly routine and requires very little effort in programming since we can make use of the standard time-stepping algorithms already available in MATLAB.

2D MATLAB implementation

In the case of two dimensions, the calculation becomes slightly more difficult since the 2D data are represented by a matrix and the ODE solvers require a vector input for the initial data. For this case, the governing equation is

$$\frac{\partial u}{\partial t} = \kappa \left(\frac{\partial^2 u}{\partial x^2} + \frac{\partial^2 u}{\partial y^2} \right). \tag{14.3.8}$$

Provided $\Delta x = \Delta y = \delta$ are the same, the system can be reduced to the linear system

$$\frac{d\mathbf{u}}{dt} = \frac{\kappa}{\delta^2} \mathbf{A}\mathbf{u}, \tag{14.3.9}$$

where we have arranged the vector $\mathbf{u}$ by stacking slices of the data in the second dimension y. This stacking procedure is required for implementation purposes of MATLAB. Thus by defining

$$u_{nm} = u(x_n, y_m) \tag{14.3.10}$$

the collection of image (pixel) points can be arranged as follows

$$\mathbf{u} = \begin{pmatrix} u_{11} \\ u_{12} \\ \vdots \\ u_{1n} \\ u_{21} \\ u_{22} \\ \vdots \\ u_{n(n-1)} \\ u_{nn} \end{pmatrix}. \tag{14.3.11}$$

The matrix **A** is a sparse matrix and so the sparse implementation of this matrix can be used advantageously in MATLAB.

Again, the system of differential equations can now be easily solved with a standard time-stepping algorithm such as *ode23* or *ode45*. The basic algorithm follows the same course as the 1D case, but extra care is taken in arranging the 2D data into a vector.

Diffusion of an image

To provide a basic implementation of the above methods, consider the following MATLAB code which loads a given image file, converts it to double precision numbers, then diffuses the image in order to remove some of the noise content. The process begins once again with the loading of an ideal image on which noise will be projected.

```
clear all; close all; clc;

A=imread('espresso','jpeg');
Abw=rgb2gray(A); Abw=double(Abw);
[nx,ny]=size(Abw);
u2=uint8(Abw+20*randn(nx,ny));

subplot(2,2,1), imshow(A)
subplot(2,2,2), imshow(u2)
```

As before, the black-and-white version of the image will be the object consideration for diagnostic purposes. Figure 14.8 shows the ideal image along with a noisy black-and-white version. Diffusion will be used to denoise this image.

Figure 14.8: A beautiful looking image demonstrating the classic fernleaf pattern in a cappuccino (left panel). By adding noise to the ideal image, we are forced to use mathematics to clean up this image (right panel).

In what follows, a constant diffusion coefficient will be considered. To make the Laplacian operator in two dimensions, the **kron** command is used. The following code first makes the *x*- and *y*-derivatives in one dimension. The **kron** command is much like the **meshgrid** in that it converts the operators into their two-dimensional versions.

```
x=linspace(0,1,nx); y=linspace(0,1,ny); dx=x(2)-x(1); dy=y(2)-y(1);
onex=ones(nx,1); oney=ones(ny,1);
Dx=(spdiags([onex -2*onex onex],[-1 0 1],nx,nx))/dx^2; Ix=eye(nx);
Dy=(spdiags([oney -2*oney oney],[-1 0 1],ny,ny))/dy^2; Iy=eye(ny);
L=kron(Iy,Dx)+kron(Dy,Ix);
```

The generated operator **L** takes two derivatives in *x* and *y* and is the Laplacian in two dimensions. This operator is constructed to act upon data that have been stacked as in Eq. (14.3.11).

It simply remains to evolve the image through the diffusion equation. But now that the Laplacian operator has been constructed, it simply remains to reshape the image, determine a time-span for evolving, and choose a diffusion coefficient. The following code performs the diffusion of the image and produces a plot that tracks the image from its initial state to the final diffused image.

```
tspan=[0 0.005 0.02 0.04]; D=0.0005;

u3=Abw+20*noise2;
u3_2=reshape(u3,nx*ny,1);
[t,usol]=ode113('image_rhs',tspan,u3_2,[],L,D);

for j=1:length(t)
  Abw_clean=uint8(reshape(usol(j,:),nx,ny));
  subplot(2,2,j), imshow(Abw_clean);
end
```

The above code calls on the function **image_rhs** which contains the diffusion equation information. This function contains the following two lines of code:

```
function rhs=image_rhs(t,u,dummy,L,D)
rhs=D*L*u;
```

Figure 14.9 shows the evolution of the image through the diffusion process. Note that only a slight amount of diffusion is needed, i.e. a small diffusion constant as well as a short diffusion

Figure 14.9: Image denoising process for a diffusion constant of $D = 0.0005$ as a function of time for $t = 0, 0.005, 0.02$ and 0.04. Perhaps the best image is for $t = 0.02$ where a moderate amount of diffusion has been applied. Further diffusion starts to degrade the image quality. The image was rescaled to 160×240 pixels.

time, before the pixelation has been substantially reduced. Continued diffusion starts to degrade image quality. This corresponds to over-filtering the image.

Nonlinear filtering and diffusion

As mentioned previously, the linear diffusion process is equivalent to the application of a Gaussian filter. Thus it is not clear that the diffusion process is any better than simple filtering. However, the diffusion process has several distinct advantages: first, particular regions in the spatial figure can be targeted by modification of the diffusion coefficient $D(x, y)$. Second, nonlinear diffusion can be considered for enhancing certain features of the image. This corresponds to a nonlinear filtering process. In this case,

$$u_t = \nabla \cdot (D(u, \nabla u)\nabla u) \tag{14.3.12}$$

where the diffusion coefficient now depends on the image and its gradient. Nonlinear diffusion, with a properly constructed D above, can be used, for instance, as an effective method for extracting edges from an image [29]. Thus, although this section has only considered a simple linear diffusion model, the ideas are easily generalized to account for more general concepts and image processing aims.

15 Linear Algebra and Singular Value Decomposition

Linear algebra plays a central role in almost every application area of mathematics in the physical, engineering and biological sciences. It is perhaps the most important theoretical framework to be familiar with as a student of mathematics. Thus it is no surprise that it also plays a key role in data analysis and computation. In what follows, emphasis will be placed squarely on the *singular value decomposition* (SVD). This concept is often untouched in undergraduate courses in linear algebra, yet it forms one of the most powerful techniques for analyzing a myriad of application areas.

15.1 Basics of the Singular Value Decomposition (SVD)

In even the earliest experience of linear algebra, the concept of a matrix transforming a vector via multiplication was defined. For instance, the vector $\mathbf{x}$ when multiplied by a matrix $\mathbf{A}$ produces a new vector $\mathbf{y}$ that is now aligned, generically, in a new direction with a new length. To be more specific, the following example illustrates a particular transformation:

$$\mathbf{x} = \begin{bmatrix} 1 \\ 3 \end{bmatrix}, \quad \mathbf{A} = \begin{bmatrix} 2 & 1 \\ -1 & 1 \end{bmatrix} \quad \rightarrow \quad \mathbf{y} = \mathbf{A}\mathbf{x} = \begin{bmatrix} 5 \\ 2 \end{bmatrix}. \tag{15.1.1}$$

Figure 15.1(a) shows the vector $\mathbf{x}$ and its transformed version, $\mathbf{y}$, after application of the matrix $\mathbf{A}$. Thus generically, matrix multiplication will rotate and stretch (compress) a given vector as prescribed by the matrix $\mathbf{A}$ (see Fig. 15.1(a)).

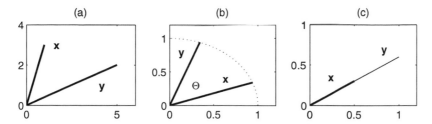

Figure 15.1: Transformation of a vector **x** under the action of multiplication by the matrix **A**, i.e. **y** = **Ax**. (a) Generic rotation and stretching of the vector as given by Eq. (15.1.1). (b) Rotation by 50° of a unit vector by the rotation matrix (15.1.2). (c) Stretching of a vector to double its length using (15.1.3) with $\alpha = 2$.

The rotation and stretching of a transformation can be precisely controlled by proper construction of the matrix **A**. In particular, it is well known that in a two-dimensional space, the rotation matrix

$$\mathbf{A} = \begin{bmatrix} \cos\theta & -\sin\theta \\ \sin\theta & \cos\theta \end{bmatrix} \tag{15.1.2}$$

takes a given vector **x** and rotates it by an angle θ to produce the vector **y**. The transformation produced by **A** is known as a *unitary transformation* since the matrix inverse is $\mathbf{A}^{-1} = \bar{\mathbf{A}}^{T}$ where the bar denotes complex conjugation [30]. Thus rotation can be directly specified without the vector being scaled. To scale the vector in length, the matrix

$$\mathbf{A} = \begin{bmatrix} \alpha & 0 \\ 0 & \alpha \end{bmatrix} \tag{15.1.3}$$

can be applied to the vector **x**. This multiplies the length of the vector **x** by α. If $\alpha = 2$ (0.5), then the vector is twice (half) its original length. The combination of the above two matrices gives arbitrary control of rotation and scaling in a two-dimentional vector space. Figure 15.1 demonstrates some of the various operations associated with the above matrix transformations.

A *singular value decomposition* (SVD) is a factorization of a matrix into a number of constitutive components all of which have a specific meaning in applications. The SVD, much as illustrated in the preceding paragraph, is essentially a transformation that stretches/compresses and rotates a given set of vectors. In particular, the following geometric principle will guide our forthcoming discussion: the image of a unit sphere under any $m \times n$ matrix is a hyper-ellipse. A hyper-ellipse in $\mathbb{R}^m$ is defined by the surface obtained upon stretching a unit sphere in $\mathbb{R}^m$ by some factors $\sigma_1, \sigma_2, \cdots, \sigma_m$ in the orthogonal directions $\mathbf{u}_1, \mathbf{u}_2, \cdots, \mathbf{u}_m \in \mathbb{R}^m$. The stretchings σ_i can possibly be zero. For convenience, consider the $\mathbf{u}_j$ to be unit vectors so that $\|\mathbf{u}_j\|_2 = 1$. The quantities $\sigma_j \mathbf{u}_j$ are then the *principal semi-axes* of the hyper-ellipse with the length σ_j. Figure 15.2 demonstrates a particular hyper-ellipse created under the matrix transformation **A** in $\mathbb{R}^2$.

A few things are worth noting at this point. First, if **A** has rank r, exactly r of the lengths σ_j will be nonzero. And if the matrix **A** is an $m \times n$ matrix where $m > n$, at most n of the σ_j will be nonzero. Consider for the moment a full rank matrix **A**. Then the n singular values of **A** are the

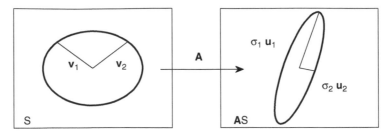

Figure 15.2: Image of a unit sphere S transformed into a hyper-ellipse AS in $\mathbb{R}^2$. The values of σ_1 and σ_2 are the singular values of the matrix $\mathbf{A}$ and represent the lengths of the semi-axes of the ellipse.

lengths of the principal semi-axes $\mathbf{A}S$ as shown in Fig. 15.2. Convention assumes that the singular values are ordered with the largest first and then in descending order: $\sigma_1 \geq \sigma_2 \geq \cdots \geq \sigma_n > 0$.

On a more formal level, the transformation from the unit sphere to the hyper-ellipse can be more succinctly stated as follows:

$$\mathbf{A}\mathbf{v}_j = \sigma_j \mathbf{u}_j \quad 1 \leq j \leq n. \tag{15.1.4}$$

Thus in total, there are n vectors that are transformed under $\mathbf{A}$. A more compact way to write all of these equations simultaneously is with the representation

$$\begin{bmatrix} \\ \mathbf{A} \\ \\ \end{bmatrix} \begin{bmatrix} \mathbf{v}_1 & \mathbf{v}_2 & \cdots & \mathbf{v}_n \end{bmatrix} = \begin{bmatrix} \mathbf{u}_1 & \mathbf{u}_2 & \cdots & \mathbf{u}_n \end{bmatrix} \begin{bmatrix} \sigma_1 & & & \\ & \sigma_2 & & \\ & & \ddots & \\ & & & \sigma_n \end{bmatrix} \tag{15.1.5}$$

so that in compact matrix notation this becomes

$$\mathbf{A}\mathbf{V} = \hat{\mathbf{U}}\hat{\boldsymbol{\Sigma}}. \tag{15.1.6}$$

The matrix $\hat{\boldsymbol{\Sigma}}$ is an $n \times n$ diagonal matrix with positive entries provided the matrix $\mathbf{A}$ is of full rank. The matrix $\hat{\mathbf{U}}$ is an $m \times n$ matrix with orthonormal columns, and the matrix $\mathbf{V}$ is an $n \times n$ unitary matrix. Since $\mathbf{V}$ is unitary, the above equation can be solved for $\mathbf{A}$ by multiplying on the right with $\mathbf{V}^*$ so that

$$\mathbf{A} = \hat{\mathbf{U}}\hat{\boldsymbol{\Sigma}}\mathbf{V}^*. \tag{15.1.7}$$

This factorization is known as the *reduced singular value decomposition*, or reduced SVD, of the matrix $\mathbf{A}$. Graphically, the factorization is represented in Fig. 15.3.

The reduced SVD is not the standard definition of the SVD used in the literature. What is typically done to augment the treatment above is to construct a matrix $\mathbf{U}$ from $\hat{\mathbf{U}}$ by adding an additional $m - n$ columns that are orthonormal to the already existing set in $\hat{\mathbf{U}}$. Thus the

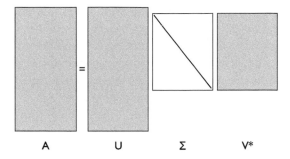

Figure 15.3: Graphical description of the reduced SVD decomposition.

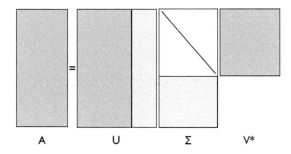

Figure 15.4: Graphical description of the full SVD decomposition where both **U** and **V** are unitary matrices. The light shaded regions of **U** and **Σ** are the silent rows and columns that are extended from the reduced SVD.

matrix **U** becomes an $m \times m$ unitary matrix. In order to make this procedure work, an additional $m - n$ rows of zeros is also added to the $\hat{\Sigma}$ matrix. These "silent" columns of **U** and rows of **Σ** are shown graphically in Fig. 15.4. In performing this procedure, it becomes evident that rank deficient matrices can easily be handled by the SVD decomposition. In particular, instead of $m - n$ silent rows and matrices, there are now simply $m - r$ silent rows and columns added to the decomposition. Thus the matrix **Σ** will have r positive diagonal entries, with the remaining $n - r$ being equal to zero.

The full SVD decomposition thus take the form

$$\mathbf{A} = \mathbf{U\Sigma V}^*$$

(15.1.8)

with the following three matrices

$$\mathbf{U} \in \mathbb{C}^{m \times m} \text{ is unitary}$$

(15.1.9a)

$$\mathbf{V} \in \mathbb{C}^{n \times n} \text{ is unitary}$$

(15.1.9b)

$$\mathbf{\Sigma} \in \mathbb{R}^{m \times n} \text{ is diagonal.}$$

(15.1.9c)

Additionally, it is assumed that the diagonal entries of Σ are nonnegative and ordered from largest to smallest so that $\sigma_1 \geq \sigma_2 \geq \cdots \geq \sigma_p \geq 0$ where $p = \min(m, n)$. The SVD decomposition of the matrix $\mathbf{A}$ thus shows that the matrix first applies a unitary transformation preserving the unit sphere via $\mathbf{V}^*$. This is followed by a stretching operation that creates an ellipse with principal semi-axes given by the matrix $\mathbf{\Sigma}$. Finally, the generated hyper-ellipse is rotated by the unitary transformation $\mathbf{U}$. Thus the statement: the image of a unit sphere under any $m \times n$ matrix is a hyper-ellipse, is shown to be true. The following is the primary theorem concerning the SVD.

Theorem *Every matrix $\mathbf{A} \in \mathbb{C}^{m \times n}$ has a singular value decomposition (15.1.8). Furthermore, the singular values $\{\sigma_j\}$ are uniquely determined, and, if $\mathbf{A}$ is square and the σ_j distinct, the singular vectors $\{\mathbf{u}_j\}$ and $\{\mathbf{v}_j\}$ are uniquely determined up to complex signs (complex scalar factors of absolute value 1).*

Computing the SVD

The above theorem guarantees the existence of the SVD, but in practice, it still remains to be computed. This is a fairly straightforward process if one considers the following matrix products:

$$
\begin{aligned}
\mathbf{A}^T\mathbf{A} &= (\mathbf{U}\mathbf{\Sigma}\mathbf{V}^*)^T(\mathbf{U}\mathbf{\Sigma}\mathbf{V}^*) \\
&= \mathbf{V}\mathbf{\Sigma}\mathbf{U}^*\mathbf{U}\mathbf{\Sigma}\mathbf{V}^* \\
&= \mathbf{V}\mathbf{\Sigma}^2\mathbf{V}^*
\end{aligned}
\tag{15.1.10}
$$

and

$$
\begin{aligned}
\mathbf{A}\mathbf{A}^T &= (\mathbf{U}\mathbf{\Sigma}\mathbf{V}^*)(\mathbf{U}\mathbf{\Sigma}\mathbf{V}^*)^T \\
&= \mathbf{U}\mathbf{\Sigma}\mathbf{V}^*\mathbf{V}\mathbf{\Sigma}\mathbf{U}^* \\
&= \mathbf{U}\mathbf{\Sigma}^2\mathbf{U}^* .
\end{aligned}
\tag{15.1.11}
$$

Multiplying (15.1.10) and (15.1.11) on the right by $\mathbf{V}$ and $\mathbf{U}$, respectively, gives the two self-consistent eigenvalue problems

$$
\mathbf{A}^T\mathbf{A}\mathbf{V} = \mathbf{V}\mathbf{\Sigma}^2
\tag{15.1.12a}
$$

$$
\mathbf{A}\mathbf{A}^T\mathbf{U} = \mathbf{U}\mathbf{\Sigma}^2 .
\tag{15.1.12b}
$$

Thus if the normalized eigenvectors are found for these two equations, then the orthonormal basis vectors are produced for $\mathbf{U}$ and $\mathbf{V}$. Likewise, the square root of the eigenvalues of these equations produces the singular values σ_j.

Example: Consider the SVD decomposition of

$$
A = \begin{bmatrix} 3 & 0 \\ 0 & -2 \end{bmatrix} .
\tag{15.1.13}
$$

The following quantities are computed:

$$\mathbf{A}^T\mathbf{A} = \begin{bmatrix} 3 & 0 \\ 0 & -2 \end{bmatrix}\begin{bmatrix} 3 & 0 \\ 0 & -2 \end{bmatrix} = \begin{bmatrix} 9 & 0 \\ 0 & 4 \end{bmatrix} \tag{15.1.14a}$$

$$\mathbf{A}\mathbf{A}^T = \begin{bmatrix} 3 & 0 \\ 0 & -2 \end{bmatrix}\begin{bmatrix} 3 & 0 \\ 0 & -2 \end{bmatrix} = \begin{bmatrix} 9 & 0 \\ 0 & 4 \end{bmatrix}. \tag{15.1.14b}$$

The eigenvalues are clearly $\lambda = \{9, 4\}$, giving singular values $\sigma_1 = 3$ and $\sigma_2 = 2$. The eigenvectors can similarly be constructed and the matrices $\mathbf{U}$ and $\mathbf{V}$ are given by

$$\begin{bmatrix} \pm 1 & 0 \\ 0 & \pm 1 \end{bmatrix} \tag{15.1.15}$$

where there is an indeterminate sign in the eigenvectors. However, a self-consistent choice of signs must be made. One possible choice gives

$$\mathbf{A} = \mathbf{U}\mathbf{\Sigma}\mathbf{V}^* = \begin{bmatrix} 1 & 0 \\ 0 & 1 \end{bmatrix}\begin{bmatrix} 3 & 0 \\ 0 & 2 \end{bmatrix}\begin{bmatrix} 1 & 0 \\ 0 & -1 \end{bmatrix}. \tag{15.1.16}$$

The SVD can be easily computed in MATLAB with the following command:

```
[U,S,V]=svd(A);
```

where the $[U, S, V]$ correspond to $\mathbf{U}$, $\mathbf{\Sigma}$ and $\mathbf{V}$, respectively. This decomposition is a critical tool for analyzing many data driven phenomena. But before proceeding to such applications, the connection of the SVD with standard and well-known techniques is elucidated.

15.2 The SVD in Broader Context

The SVD is an exceptionally important tool in many areas of applications. Part of this is due to its many mathematical properties and its guarantee of existence. In what follows, some of its more important theorems and relations to other standard ideas of linear algebra are considered with the aim of setting the mathematical framework for future applications.

Eigenvalues, eigenvectors and diagonalization

The concept of eigenvalues and eigenvectors is critical to understanding many areas of applications. One of the most important areas where it plays a role is in understanding differential equations. Consider, for instance, the system of differential equations:

$$\frac{d\mathbf{y}}{dt} = \mathbf{A}\mathbf{y} \tag{15.2.1}$$

for some vector $\mathbf{y}(t)$ representing a dynamical system of variables and where the matrix $\mathbf{A}$ determines the interaction among these variables. Assuming a solution of the form $\mathbf{y} = \mathbf{x}\exp(\lambda t)$ results in the eigenvalue problem:

$$\mathbf{A}\mathbf{x} = \lambda\mathbf{x}. \tag{15.2.2}$$

The question remains: How are the eigenvalues and eigenvectors found? To consider this problem, we rewrite the eigenvalue problem as

$$\mathbf{A}\mathbf{x} - \lambda\mathbf{I}\mathbf{x} = (\mathbf{A} - \lambda\mathbf{I})\mathbf{x} = \mathbf{0}. \tag{15.2.3}$$

Two possibilities now exist.

Option I

The determinant of the matrix $(\mathbf{A} - \lambda\mathbf{I})$ is not zero. If this is true, the matrix is *nonsingular* and its inverse, $(\mathbf{A} - \lambda\mathbf{I})^{-1}$, can be found. The solution to the eigenvalue problem (15.2.2) is then

$$\mathbf{x} = (\mathbf{A} - \lambda\mathbf{I})^{-1}\mathbf{0} \tag{15.2.4}$$

which implies that $\mathbf{x} = \mathbf{0}$. This trivial solution could have easily been guessed. However, it is not relevant as we require nontrivial solutions for $\mathbf{x}$.

Option II

The determinant of the matrix $(\mathbf{A} - \lambda\mathbf{I})$ is zero. If this is true, the matrix is *singular* and its inverse, $(\mathbf{A} - \lambda\mathbf{I})^{-1}$, cannot be found. Although there is no longer a guarantee that there is a solution, it is the only scenario which allows for the possibility of $\mathbf{x} \neq 0$. It is this condition which allows for the construction of eigenvalues and eigenvectors. Indeed, we choose the eigenvalues λ so that this condition holds and the matrix is singular.

Another important operation which can be performed with eigenvalues and eigenvectors is the evaluation of

$$\mathbf{A}^M \tag{15.2.5}$$

where M is a large integer. For large matrices $\mathbf{A}$, this operation is computationally expensive. However, knowing the eigenvalues and eigenvectors of $\mathbf{A}$ allows for a significant ease in computational expense. Assuming we have all the eigenvalues and eigenvectors of $\mathbf{A}$, then

$$\mathbf{A}\mathbf{x}_1 = \lambda_1\mathbf{x}_1$$
$$\mathbf{A}\mathbf{x}_2 = \lambda_2\mathbf{x}_2$$
$$\vdots$$
$$\mathbf{A}\mathbf{x}_n = \lambda_1\mathbf{x}_n.$$

This collection of eigenvalues and eigenvectors gives the matrix system

$$\mathbf{A}\mathbf{S} = \mathbf{S}\boldsymbol{\Lambda} \tag{15.2.6}$$

where the columns of the matrix $\mathbf{S}$ are the eigenvectors of $\mathbf{A}$,

$$\mathbf{S} = (\mathbf{x}_1 \ \mathbf{x}_2 \ \cdots \ \mathbf{x}_n),\tag{15.2.7}$$

and $\mathbf{\Lambda}$ is a matrix whose diagonals are the corresponding eigenvalues

$$\mathbf{\Lambda} = \begin{pmatrix} \lambda_1 & 0 & \cdots & & 0 \\ 0 & \lambda_2 & 0 & \cdots & 0 \\ \vdots & & \ddots & & \vdots \\ 0 & & \cdots & 0 & \lambda_n \end{pmatrix}.\tag{15.2.8}$$

By multiplying (15.2.6) on the right by $\mathbf{S}^{-1}$, the matrix $\mathbf{A}$ can then be rewritten as (note the similarity between this expression and Eq. (15.1.8) for the SVD decomposition)

$$\mathbf{A} = \mathbf{S}\mathbf{\Lambda}\mathbf{S}^{-1}.\tag{15.2.9}$$

The final observation comes from

$$\mathbf{A}^2 = (\mathbf{S}\mathbf{\Lambda}\mathbf{S}^{-1})(\mathbf{S}\mathbf{\Lambda}\mathbf{S}^{-1}) = \mathbf{S}\mathbf{\Lambda}^2\mathbf{S}^{-1}.\tag{15.2.10}$$

This then generalizes to

$$\mathbf{A}^M = \mathbf{S}\mathbf{\Lambda}^M\mathbf{S}^{-1}\tag{15.2.11}$$

where the matrix $\mathbf{\Lambda}^M$ is easily calculated as

$$\mathbf{\Lambda}^M = \begin{pmatrix} \lambda_1^M & 0 & \cdots & & 0 \\ 0 & \lambda_2^M & 0 & \cdots & 0 \\ \vdots & & \ddots & & \vdots \\ 0 & & \cdots & 0 & \lambda_n^M \end{pmatrix}.\tag{15.2.12}$$

Since raising the diagonal terms to the Mth power is easily accomplished, the matrix $\mathbf{A}$ can then be easily calculated by multiplying the three matrices in (15.2.11).

Diagonalization can also recast a given problem so as to elucidate its more fundamental dynamics. A classic example of the use of diagonalization is a two-mass spring system where the masses m_1 and m_2 are acted on by forces $F_1(t)$ and $F_2(t)$. A schematic of this situation is depicted in Fig. 15.5. For each mass, we can write down Newton's law:

$$\sum F_1 = m_1 \frac{d^2 x_1}{dt^2} \quad \text{and} \quad \sum F_2 = m_2 \frac{d^2 x_2}{dt^2}\tag{15.2.13}$$

where $\sum F_1$ and $\sum F_2$ are the sums of the forces on m_1 and m_2, respectively. Note that the equations for $x_1(t)$ and $x_2(t)$ are coupled because of the spring with spring constant k_2. The resulting governing equations are then of the form:

$$m_1 \frac{d^2 x_1}{dt^2} = -k_1 x_1 + k_2(x_2 - x_1) + F_1 = -(k_1 + k_2)x_1 + k_2 x_2 + F_1\tag{15.2.14a}$$

$$m_2 \frac{d^2 x_2}{dt^2} = -k_3 x_2 - k_2(x_2 - x_1) + F_2 = -(k_2 + k_3)x_2 + k_2 x_1 + F_2.\tag{15.2.14b}$$

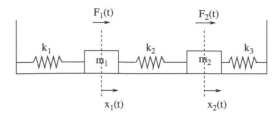

Figure 15.5: A two-mass, three-spring system. The fundamental behavior of the system can be understood by decomposition of the system via diagonalization. This reveals that all motion can be expressed as the sum of two fundamental motions: the masses oscillating in-phase, and the masses oscillating exactly out-of-phase.

This can be reduced further by assuming, for simplicity, $m = m_1 = m_2$, $K = k_1/m = k_2/m$ and $F_1 = F_2 = 0$. This results in the linear system which can be diagonalized via Eq. (15.2.9). The pairs of complex conjugate eigenvalues are produced: $\lambda_1^{\pm} = \pm i(2K + \sqrt{2K})^{1/2}$ and $\lambda_2^{\pm} = \pm i(2K - \sqrt{2K})^{1/2}$. Upon diagonalization, the full system can be understood as simply a linear combination of oscillations of the masses that are in-phase with each other or out-of-phase with each other.

Diagonalization via SVD

Like the eigenvalue and eigenvector diagonalization technique presented above, the SVD method also makes the following claim: *the SVD makes it possible for every matrix to be diagonal if the proper bases for the domain and range are used.* To consider this statement more formally, consider that since $\mathbf{U}$ and $\mathbf{V}$ are orthonormal bases in $\mathbb{C}^{m \times m}$ and $\mathbb{C}^{n \times n}$, respectively, then any vector in these spaces can be expanded in their bases. Specifically, consider a vector $\mathbf{b} \in \mathbb{C}^{m \times m}$ and $\mathbf{x} \in \mathbb{C}^{n \times n}$. Then each can be expanded in the bases of $\mathbf{U}$ and $\mathbf{V}$ so that

$$\mathbf{b} = \mathbf{U}\hat{\mathbf{b}}, \quad \mathbf{x} = \mathbf{V}\hat{\mathbf{x}} \tag{15.2.15}$$

where the vectors $\hat{\mathbf{b}}$ and $\hat{\mathbf{x}}$ give the weightings for the orthonormal bases expansion. Now consider the simple equation:

$$\mathbf{A}\mathbf{x} = \mathbf{b} \rightarrow \mathbf{U}^*\mathbf{b} = \mathbf{U}^*\mathbf{A}\mathbf{x}$$
$$\mathbf{U}^*\mathbf{b} = \mathbf{U}^*\mathbf{U}\mathbf{\Sigma}\mathbf{V}^*\mathbf{x}$$
$$\hat{\mathbf{b}} = \mathbf{\Sigma}\hat{\mathbf{x}}. \tag{15.2.16}$$

Thus the last line shows that $\mathbf{A}$ reduces to the diagonal matrix $\mathbf{\Sigma}$ when the range is expressed in terms of the basis vectors of $\mathbf{U}$ and the domain is expressed in terms of the basis vectors of $\mathbf{V}$.

Thus matrices can be diagonalized via either an eigenvalue decomposition or an SVD decomposition. However, there are three key differences in the diagonalization process.

- The SVD performs the diagonalization using two different bases, $\mathbf{U}$ and $\mathbf{V}$, while the eigenvalue method uses a single basis $\mathbf{X}$.

- The SVD method uses an orthonormal basis while the basis vectors in $\mathbf{X}$, while linearly independent, are not generally orthogonal.

- Finally, the SVD is guaranteed to exist for any matrix $\mathbf{A}$ while the same is not true, even for square matrices, for the eigenvalue decomposition.

Useful theorems of SVD

Having established the similarities and connections between eigenvalues and singular values, in what follows a number of important theorems are outlined concerning the SVD. These theorems are important for several reasons. First, the theorems play a key role in the numerical evaluation of many matrix properties via the SVD. Second, the theorems guarantee certain behaviors of the SVD that can be capitalized upon for future applications. Here is a list of important results. A more detailed account is given by Trefethen and Bau [30].

Theorem *If the rank of $\mathbf{A}$ is r, then there are r nonzero singular values.*

The proof of this is based upon the fact that the rank of a diagonal matrix is equal to the number of its nonzero entries. And because of the decomposition $\mathbf{A} = \mathbf{U\Sigma V}^*$ where $\mathbf{U}$ and $\mathbf{V}$ are full rank, then $\text{rank}(\mathbf{A}) = \text{rank}(\mathbf{\Sigma}) = r$. As a side note, the rank of a matrix can be found with the MATLAB command:

```
rank(A)
```

The standard algorithm used to compute the rank is based upon the SVD and the computation of the nonzero singular values. Thus the theorem is quite useful in practice.

Theorem *range*$(\mathbf{A}) = \langle \mathbf{u}_1, \mathbf{u}_2, \cdots, \mathbf{u}_r \rangle$ *and null* $(\mathbf{A}) = \langle \mathbf{v}_{r+1}, \mathbf{v}_{r+2}, \cdots, \mathbf{v}_n \rangle$.

Note that the range and null space come from the two expansion bases $\mathbf{U}$ and $\mathbf{V}$. The range and null space can be found in MATLAB with the commands:

```
range(A)
null(A)
```

This theorem also serves as the most accurate computation basis for determining the range and null space of a given matrix $\mathbf{A}$ via the SVD.

Theorem $\|\mathbf{A}\|_2 = \sigma_1$ *and* $\|\mathbf{A}\|_F = \sqrt{\sigma_1^2 + \sigma_2^2 + \cdots + \sigma_r^2}$.

These norms are known as the 2-norm and the Frobenius norm, respectively. They essentially measure the *energy* of a matrix. Although it is difficult to conceptualize abstractly what this means, it will become much more clear in the context of given applications. This result is established by the fact that U and V are unitary operators so that their norm is unity. Thus with $A = U\Sigma V^*$, the 2-norm is $\|A\|_2 = \|\Sigma\|_2 = \max\{|\sigma_j|\} = \sigma_1$. Similar reasoning holds for the Frobenius norm definition. The norm can be calculated in MATLAB with

```
norm(A)
```

Notice that the Frobenius norm contains the total matrix energy while the 2-norm definition contains the energy of the largest singular value. The ratio $\|A\|_2/\|A\|_F$ effectively measures the portion of the energy in the semi-axis u_1. This fact will be tremendously important for us. Furthermore, this theorem gives the standard way for computing matrix norms.

Theorem *The nonzero singular values of A are the square roots of the nonzero eigenvalues of A^*A or AA^*. (These matrices have the same nonzero eigenvalues.)*

This has already been shown in the calculation for actually determining the SVD. In particular, this process is illustrated in Eqs. (15.1.11) and (15.1.10). This theorem is important for actually producing the unitary matrices U and V.

Theorem *If $A = A^*$ (self-adjoint), then the singular values of A are the absolute values of the eigenvalues of A.*

As with most self-adjoint problems, there are very nice properties of the matrices, such as the above theorem. Eigenvalues can be computed with the command:

```
eig(A)
```

Alternatively, one can use the **eigs** to specify the number of eigenvalues desired and their specific ordering.

Theorem *For $A \in \mathbb{C}^{m \times m}$, the determinant is given by $|\det(A)| = \prod_{j=1}^{m} \sigma_j$.*

Again, due to the fact that the matrices U and V are unitary, their determinants are unity. Thus $|\det(A)| = |\det(U\Sigma V^*)| = |\det(U)||\det(\Sigma)||\det(V^*)| = |\det(\Sigma)| = \prod_{j=1}^{m} \sigma_j$. Thus even determinants can be computed via the SVD and its singular values.

Low dimensional reductions

Now comes the last, and most formative, property associated with the SVD: *low dimensional approximations* to high degree of freedom or complex systems. In linear algebra terms, this is also *low rank approximations*. The interpretation of the theorems associated with these low dimensional reductions are critical for the use and implementation of the SVD. Thus we consider the following:

Theorem **A** *is the sum of r rank-one matrices*

$$\mathbf{A} = \sum_{j=1}^{r} \sigma_j \mathbf{u}_j \mathbf{v}_j^* . \tag{15.2.17}$$

There are a variety of ways to express the $m \times n$ matrix **A** as a sum of rank-one matrices. The bottom line is this: *the Nth partial sum captures as much of the matrix* **A** *as possible.* Thus the partial sum of the rank-one matrices is an important object to consider. This leads to the following theorem:

Theorem *For any N so that $0 \leq N \leq r$, we can define the partial sum*

$$\mathbf{A}_N = \sum_{j=1}^{N} \sigma_j \mathbf{u}_j \mathbf{v}_j^* . \tag{15.2.18}$$

And if $N = \min\{m, n\}$, define $\sigma_{N+1} = 0$. Then

$$\|A - A_N\|_2 = \sigma_{N+1} . \tag{15.2.19}$$

Likewise, if using the Frobenius norm, then

$$\|A - A_N\|_N = \sqrt{\sigma_{N+1}^2 + \sigma_{N+2}^2 + \cdots + \sigma_r^2} . \tag{15.2.20}$$

Interpreting this theorem is critical. Geometrically, we can ask *what is the best approximation of a hyper-ellipsoid by a line segment? Simply take the line segment to be the longest axis, i.e. that associated with the singular value σ_1.* Continuing this idea, what is the best approximation by a two-dimensional ellipse? Take the longest and second longest axes, i.e. those associated with the singular values σ_1 and σ_2. After r steps, the total energy in **A** is completely captured. **Thus the SVD gives a type of least-square fitting algorithm, allowing us to project the matrix onto low dimensional representations in a formal, algorithmic way.** Herein lies the ultimate power of the method.

15.3 Introduction to Principal Component Analysis (PCA)

To make explicit the concept of the SVD, a simple model example will be formulated that will illustrate all the key concepts associated with the SVD. The model to be considered will be a

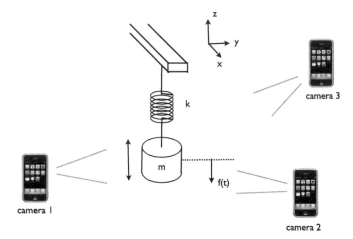

Figure 15.6: A prototypical example of how we might apply a principal component analysis, or SVD, is the simple mass–spring system exhibited here. The mass m is suspended with a spring with Hooke's constant k. Three video cameras collect data about its motion in order to ascertain its governing equations.

simple spring–mass system as illustrated in Fig. 15.6. Of course, this is a fairly easy problem to solve from basic concepts of $\mathbf{F} = m\mathbf{a}$. But for the moment, let's suppose we didn't know the governing equations. In fact, our aim in this section is to use a number of cameras (probes) to extract out data concerning the behavior of the system and then to extract empirically the governing equations of motion.

This prologue highlights one of the key applications of the SVD, or alternatively a variant of *principal component analysis* (PCA), *proper mode decomposition* (POD), *Hotelling transform*, *empirical orthogonal functions* (EOF), *reduced order modeling* (ROM), *dimensionality reduction* or *Karhunen–Loéve decomposition* as it is also known in the literature. Namely, from seemingly complex, perhaps random data, can low dimensional reductions of the dynamics and behavior be produced when the governing equations are not known? Such methods can be used to quantify low dimensional dynamics arising in such areas as turbulent fluid flows [32], structural vibrations [33, 34], insect locomotion [35], damage detection [36], and neural decision making strategies [37], to name just a few areas of application. It will also become obvious as we move forward on this topic that the breadth of applications is staggering and includes image processing and signal analysis. Thus the perspective to be taken here is clearly one in which the data analysis of an unknown, but potentially low dimensional system is to be analyzed.

Again we turn our attention to the simple experiment at hand: a mass suspended by a spring as depicted in Fig. 15.6. If the mass is perturbed or taken from equilibrium in the z-direction only, we know that the governing equations are simply

$$\frac{d^2 f(t)}{dt^2} = -\omega^2 f(t) \tag{15.3.1}$$

where the function $f(t)$ measures the displacement of the mass in the z-direction as a function of time. This has the well-known solution (in amplitude–phase form)

$$f(t) = A\cos(\omega t + \omega_0) \tag{15.3.2}$$

where the values of A and ω_0 are determined from the initial state of the system. This essentially states that the state of the system can be described by a one degree of freedom system.

In the above analysis, there are many things we have ignored. Included in the list of things we have ignored is the possibility that the initial excitation of the system actually produces movement in the $x-y$ plane. Further, there is potentially noise in the data from, for instance, shaking of the cameras during the video capture. Moreover, from what we know of the solution, only a single camera is necessary to capture the underlying motion. In particular, a single camera in the $x-y$ plane at $z = 0$ would be ideal. Thus we have oversampled the data with three cameras and have produced redundant data sets. From all of these potential perturbations and problems, it is our goal to extract out the simple solution given by the simple harmonic motion.

This problem is a good example of what kind of processing is required to analyze a realistic data set. Indeed, one can imagine that most data will be quite noisy, perhaps redundant, and certainly not produced from an optimal viewpoint. But through the process of PCA, these can be circumvented in order to extract out the ideal or simplified behavior. Moreover, we may even learn how to transform the data into the optimal viewpoint for analyzing the data.

Data collection and ordering

Assume now that we have started the mass in motion by applying a small perturbation in the z-direction only. Thus the mass will begin to oscillate. Three cameras are then used to record the motion. Each camera produces a two-dimensional representation of the data. If we denote the data from the three cameras with subscripts a, b and c, then the data collected are represented by the following:

$$\text{camera 1:} \quad (\mathbf{x}_a, \mathbf{y}_a) \tag{15.3.3a}$$

$$\text{camera 2:} \quad (\mathbf{x}_b, \mathbf{y}_b) \tag{15.3.3b}$$

$$\text{camera 3:} \quad (\mathbf{x}_c, \mathbf{y}_c) \tag{15.3.3c}$$

where each set $(\mathbf{x}_j, \mathbf{y}_j)$ is data collected over time of the position in the $x-y$ plane of the camera. Note that this is not the same $x-y$ plane of the oscillating mass system as shown in Fig. 15.6. Indeed, we should pretend we don't know the correct $x-y-z$ coordinates of the system. Thus the camera positions and their relative $x-y$ planes are arbitrary. The length of each vector $\mathbf{x}_i$ and $\mathbf{y}_i$ depends on the data collection rate and the length of time the dynamics is observed. We denote the length of these vectors as n.

All the data collected can then be gathered into a single matrix:

$$\mathbf{X} = \begin{bmatrix} \mathbf{x}_a \\ \mathbf{y}_a \\ \mathbf{x}_b \\ \mathbf{y}_b \\ \mathbf{x}_c \\ \mathbf{y}_c \end{bmatrix}. \tag{15.3.4}$$

Thus the matrix $X \in \mathbb{R}^{m \times n}$ where m represents the number of measurement types and n is the number of data points taken from the camera over time.

Now that the data have been arranged, two issues must be addressed: noise and redundancy. Everyone has an intuitive concept that noise in your data can only deteriorate, or corrupt, your ability to extract the true dynamics. Just as in image processing, noise can alter an image beyond restoration. Thus there is also some idea that if the measured data are too noisy, fidelity of the underlying dynamics is compromised from a data analysis point of view. The key measure of this is the so-called signal-to-noise ratio: $\text{SNR} = \sigma^2_{\text{signal}} / \sigma^2_{\text{noise}}$, where the ratio is given as the ratio of the variances of the signal and noise fields. A high SNR (much greater than unity) gives almost noiseless (high precision) data whereas a low SNR suggests the underlying signal is corrupted by the noise. The second issue to consider is redundancy. In the example of Fig. 15.6, the single degree of freedom is sampled by three cameras, each of which is really recording the same single degree of freedom. Thus the measurements should be rife with redundancy, suggesting that the different measurements are statistically dependent. Removing this redundancy is critical for data analysis.

The covariance matrix

An easy way to identify redundant data is by considering the covariance between data sets. Recall from our early chapters on probability and statistics that the covariance measures the statistical dependence/independence between two variables. Obviously, strongly statistically dependent variables can be considered as redundant observations of the system. Specifically, consider two sets of measurements with zero means expressed in row vector form:

$$\mathbf{a} = [a_1 \ a_2 \ \cdots \ a_n] \quad \text{and} \quad \mathbf{b} = [b_1 \ b_2 \ \cdots \ b_n] \tag{15.3.5}$$

where the subscript denotes the sample number. The variances of $\mathbf{a}$ and $\mathbf{b}$ are given by

$$\sigma^2_{\mathbf{a}} = \frac{1}{n-1} \mathbf{a}\mathbf{a}^T \tag{15.3.6a}$$

$$\sigma^2_{\mathbf{b}} = \frac{1}{n-1} \mathbf{b}\mathbf{b}^T \tag{15.3.6b}$$

while the covariance between these two data sets is given by

$$\sigma^2_{\mathbf{ab}} = \frac{1}{n-1} \mathbf{a}\mathbf{b}^T \tag{15.3.7}$$

where the normalization constant of $1/(n-1)$ is for an unbiased estimator.

We don't just have two vectors, but potentially quite a number of experiments and data that would need to be correlated and checked for redundancy. In fact, the matrix in Eq. (15.3.4) is exactly what needs to be checked for covariance. The appropriate *covariance matrix* for this case is then

$$\mathbf{C_X} = \frac{1}{n-1} \mathbf{X}\mathbf{X}^T. \tag{15.3.8}$$

This is easily computed with MATLAB from the command line:

```
cov(X)
```

The covariance matrix C_X is a square, symmetric $m \times m$ matrix whose diagonal represents the variance of particular measurements. The off-diagonal terms are the covariances between measurement types. Thus C_X captures the correlations between all possible pairs of measurements. Redundancy is thus easily captured since if two data sets are identical (identically redundant), the off-diagonal term and diagonal term would be equal since $\sigma_{ab}^2 = \sigma_a^2 = \sigma_b^2$ if $a = b$. Thus large off-diagonal terms correspond to redundancy while small off-diagonal terms suggest that the two measured quantities are close to being statistically independent and have low redundancy. It should also be noted that large diagonal terms, or those with large variances, typically represent what we might consider *the dynamics of interest* since the large variance suggests strong fluctuations in that variable. Thus the covariance matrix is the key component to understanding the entire data analysis.

Achieving the goal

The insight given by the covariance matrix leads to our ultimate aim of

(i)   removing redundancy
(ii) identifying those signals with maximal variance.

Thus in a mathematical sense we are simply asking to represent C_X so that the diagonals are ordered from largest to smallest and the off-diagonals are zero, i.e. our task is to diagonalize the covariance matrix. This is *exactly* what the SVD does, thus allowing it to becomes the tool of choice for data analysis and dimensional reduction. In fact, the SVD diagonalizes and each singular direction captures as much energy as possible as measured by the singular values σ_j.

 ## 15.4   **Principal Components, Diagonalization and SVD**

The example presented in the previous section shows that the key to analyzing a given experiment is to consider the covariance matrix

$$C_X = \frac{1}{n-1}XX^T \tag{15.4.1}$$

where the matrix X contains the experimental data of the system. In particular, $X \in \mathbb{C}^{m \times n}$ where m are the number of probes or measuring positions, and n is the number of experimental data points taken at each location.

In this setup, the following facts are highlighted:

- C_X is a square, symmetric $m \times m$ matrix.

- The diagonal terms of C_X are the variances for particular measurements. By assumption, large variances correspond to *dynamics of interest*, whereas low variances are assumed to correspond to *uninteresting dynamics*.

- The off-diagonal terms of C_X are the covariances between measurements. Indeed, the off-diagonals capture the correlations between all possible pairs of measurements. A large off-diagonal term represents two events that have a high degree of redundancy, whereas a small off-diagonal coefficient means there is little redundancy in the data, i.e. they are statistically independent.

Diagonalization

The concept of diagonalization is critical for understanding the underpinnings of many physical systems. In this process of diagonalization, the correct coordinates, or basis functions, are revealed that reduce the given system to its low dimensional essence. There is more than one way to diagonalize a matrix, and this is certainly true here as well since the constructed covariance matrix C_X is square and symmetric, both properties that are especially beneficial for standard eigenvalue/eigenvector expansion techniques.

The key idea behind the diagonalization is simply this: there exists an *ideal* basis in which the C_X can be written (diagonalized) so that in this basis, all redundancies have been removed, and the largest variances of particular measurements are ordered. In the language being developed here, this means that the system has been written in terms of its *principal components*, or in a *proper orthogonal decomposition*.

Eigenvectors and eigenvalues

The most straightforward way to diagonalize the covariance matrix is by making the observation that XX^T is a square, symmetric $m \times m$ matrix, i.e. it is self-adjoint so that the m eigenvalues are real and distinct. Linear algebra provides theorems which state that such a matrix can be rewritten as

$$XX^T = S\Lambda S^{-1} \qquad (15.4.2)$$

as stated in Eq. (15.2.9) where the matrix S is a matrix of the eigenvectors of XX^T arranged in columns. Since it is a symmetric matrix, these eigenvector columns are orthogonal so that ultimately the S can be written as a unitary matrix with $S^{-1} = S^T$. Recall that the matrix Λ is a diagonal matrix whose entries correspond to the m distinct eigenvalue of XX^T.

This suggests that instead of working directly with the matrix X, we consider working with the transformed variable, or in the principal component basis,

$$Y = S^T X. \qquad (15.4.3)$$

For this new basis, we can then consider its covariance

$$
\begin{aligned}
\mathbf{C_Y} &= \frac{1}{n-1}\mathbf{YY}^T \\
&= \frac{1}{n-1}(\mathbf{S}^T\mathbf{X})(\mathbf{S}^T\mathbf{X})^T \\
&= \frac{1}{n-1}\mathbf{S}^T(\mathbf{XX}^T)\mathbf{S} \\
&= \frac{1}{n-1}\mathbf{S}^T\mathbf{S\Lambda S}\mathbf{S}^T \\
\mathbf{C_Y} &= \frac{1}{n-1}\mathbf{\Lambda}.
\end{aligned}
\tag{15.4.4}
$$

In this basis, the *principal components* are the eigenvectors of $\mathbf{XX}^T$ with the interpretation that the jth diagonal value of $\mathbf{C_Y}$ is the variance of $\mathbf{X}$ along $\mathbf{x}_j$, the jth column of $\mathbf{S}$. The following lines of code produce the principal components of interest.

```
[m,n]=size(X);   % compute data size
mn=mean(X,2); % compute mean for each row
X=X-repmat(mn,1,n); % subtract mean

Cx=(1/(n-1))*X*X';  % covariance
[V,D]=eig(Cx);      % eigenvectors(V)/eigenvalues(D)
lambda=diag(D);     % get eigenvalues

[dummy,m_arrange]=sort(-1*lambda); % sort in decreasing order
lambda=lambda(m_arrange);
V=V(:,m_arrange);

Y=V'*X;  % produce the principal components projection
```

This simple code thus produces the eigenvalue decomposition and the projection of the original data onto the principal component basis.

Singular value decomposition

A second method for diagonalizing the covariance matrix is the SVD method. In this case, the SVD can diagonalize any matrix by working in the appropriate pair of bases $\mathbf{U}$ and $\mathbf{V}$ as outlined in the first part of this chapter. Thus by defining the transformed variable

$$
\mathbf{Y} = \mathbf{U}^*\mathbf{X}
\tag{15.4.5}
$$

where $\mathbf{U}$ is the unitary transformation associated with the SVD: $\mathbf{X} = \mathbf{U\Sigma V}^*$. Just as in the eigenvalue/eigenvector formulation, we then compute the variance in $\mathbf{Y}$:

$$\mathbf{C_Y} = \frac{1}{n-1}\mathbf{YY}^T$$
$$= \frac{1}{n-1}(\mathbf{U}^*\mathbf{X})(\mathbf{U}^*\mathbf{X})^T$$
$$= \frac{1}{n-1}\mathbf{U}^*(\mathbf{XX}^T)\mathbf{U}$$
$$= \frac{1}{n-1}\mathbf{U}^*\mathbf{U\Sigma}^2\mathbf{UU}^*$$
$$\mathbf{C_Y} = \frac{1}{n-1}\mathbf{\Sigma}^2. \tag{15.4.6}$$

This makes explicit the connection between the SVD and the eigenvalue method, namely that $\mathbf{\Sigma}^2 = \mathbf{\Lambda}$. The following lines of code produce the principal components of interest using the SVD (assume that you have the first three lines from the previous MATLAB code).

```
[u,s,v]=svd(X'/sqrt(n-1));   % perform the SVD
lambda=diag(s).^2;   % produce diagonal variances
Y=u'*X;   % produce the principal components projection
```

This gives the SVD method for producing the principal components. Overall, the SVD method is the more robust method and should be used. However, the connection between the two methods becomes apparent in these calculations.

Spring experiment

To illustrate this completely in practice, three experiments will be performed with the configuration of Fig. 15.6. The following experiments will attempt to illustrate various aspects of the PCA and its practical usefulness and the effects of noise on the PCA algorithms.

- **Ideal case** Consider a small displacement of the mass in the z-direction and the ensuing oscillations. In this case, the entire motion is in the z-direction with simple harmonic motion being observed.

- **Noisy case** Repeat the ideal case experiment, but this time, introduce camera shake into the video recording. This should make it more difficult to extract the simple harmonic motion. But if the shake isn't too bad, the dynamics will still be extracted with the PCA algorithms.

- **Horizontal displacement** In this case, the mass is released off-center so as to produce motion in the $x-y$ plane as well as the z-direction. Thus there is both a pendulum motion and a simple harmonic oscillation. See what the PCA tells us about the system.

In order to extract out the mass movement from the video frames, the following MATLAB code is needed. This code is a generic way to read in movie files to MATLAB for post-processing.

```
obj=mmreader('matlab_test.mov')

vidFrames = read(obj);
numFrames = get(obj,'numberOfFrames');

for k = 1 : numFrames
    mov(k).cdata = vidFrames(:,:,:,k);
    mov(k).colormap = [];
end

for j=1:numFrames
  X=frame2im(mov(j));
  imshow(X); drawnow
end
```

This multimedia reader command **mmreader** simply characterizes the file type and its attributes. The bulk of time in this is the **read** command which actually uploads the movie frames into the MATLAB desktop for processing. Once the frames are extracted, they can again be converted to double precision numbers for mathematical processing. In this case, the position of the mass is to be determined from each frame. This basic shell of code is enough to begin the process of extracting the spring–mass system information.

15.5   Principal Components and Proper Orthogonal Modes

Now that the basic framework of the principal component analysis and its relation to the SVD has been laid down, a few remaining issues need to be addressed. In particular, the principal component analysis seems to suggest that we are simply expanding our solution in another *orthonormal* basis, one which can always diagonalize the underlying system.

Mathematically, we can consider a given function $f(x, t)$ over a domain of interest. In most applications, the variables x and t will refer to the standard space–time variables we are familiar with. The idea is to then expand this function in some basis representation so that

$$f(x, t) \approx \sum_{j=1}^{N} a_j(t)\phi_j(x) \tag{15.5.1}$$

where N is the total number of modes, $\phi_j(x)$, to be used. The remaining function $a_j(t)$ determines the weights of the spatial modes.

The expansion (15.5.1) is certainly not a new idea to us. Indeed, we have been using this concept extensively already with Fourier transforms, for instance. Specifically, here are some of the more common expansion bases used in practice:

$$\phi_j(x) = (x - x_0)^j \qquad \text{Taylor expansion} \qquad (15.5.2a)$$

$$\phi_j(x) = \cos(jx) \qquad \text{Discrete cosine transform} \qquad (15.5.2b)$$

$$\phi_j(x) = \sin(jx) \qquad \text{Discrete sine transform} \qquad (15.5.2c)$$

$$\phi_j(x) = \exp(jx) \qquad \text{Fourier transform} \qquad (15.5.2d)$$

$$\phi_j(x) = \psi_{a,b}(x) \qquad \text{Wavelet transform} \qquad (15.5.2e)$$

$$\phi_j(x) = \phi_{\lambda_j}(x) \qquad \text{Eigenfunction expansion} \qquad (15.5.2f)$$

where $\phi_{\lambda_j}(x)$ are eigenfunctions associated with the underlying system. This places the concept of a basis function expansion in familiar territory. Further, it shows that such a basis function expansion is not unique, but rather can potentially have an infinite number of different possibilities.

As for the weighting coefficients $a_j(t)$, they are simply determined from the standard inner product rules and the fact that the basis functions are orthonormal (note that they are not written this way above):

$$\int \phi_j(x)\phi_n(x)dx = \begin{cases} 1 & j = n \\ 0 & j \neq n. \end{cases} \qquad (15.5.3)$$

This then gives for the coefficients

$$a_j(t) = \int f(x, t)\phi_j(x)dx \qquad (15.5.4)$$

and the basis expansion is completed.

Interestingly enough, the basis functions selected are often chosen for simplicity and/or their intuitive meaning for the system. For instance, the Fourier transform very clearly highlights the fact that the given function has a representation in terms of *frequency* components. This is fundamental for understanding many physical problems as there is clear and intuitive meaning to the Fourier modes.

In contrast to selecting basis functions for simplicity or intuitive meaning, the broader question can be asked: what criteria should be used for selecting the functions $\phi_j(x)$? This is an interesting question given that any complete basis can approximate the function $f(x, t)$ to any desired accuracy given N large enough. But what is desired is the best basis functions such that, with N as small as possible, we achieve the desired accuracy. The goal is then the following: find a sequence of orthonormal basis functions $\phi_j(x)$ so that the first two terms give the best two-term approximation to $f(x, t)$, or the first five terms give the best five-term approximation to $f(x, t)$. These special, ordered, orthonormal functions are called the *proper orthogonal modes* (POD) for the function $f(x, t)$. With these modes, the expansion (15.5.4) is called the POD of $f(x, t)$. The relation to the SVD will become obvious momentarily.

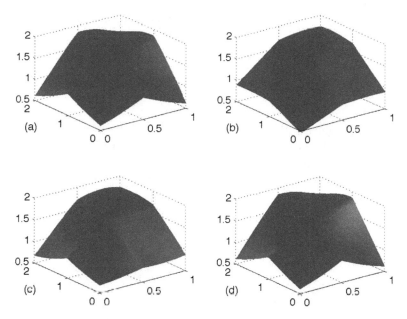

Figure 15.7: Representation of the surface (a) by a series of low rank (low-dimensional) approximations with (b) one mode, (c) two modes and (d) three modes. The energy captured in the first mode is approximately 92% of the total surface energy. The first three modes together capture 99.9% of the total surface energy, thus suggesting that the surface can be easily and accurately represented by a three-mode expansion.

Example 1: Consider an approximation to the following surface

$$f(x, t) = e^{-|(x-0.5)(t-1)|} + \sin(xt) \quad x \in [0, 1], \, t \in [0, 2]. \tag{15.5.5}$$

As listed above, there are a number of methods available for approximating this function using various basis functions. And all of the ones listed are guaranteed to converge to the surface for a large enough N in (15.5.1).

In the POD method, the surface is first discretized and approximated by a finite number of points. In particular, the following discretization will be used:

```
x=linspace(0,1,25);
t=linspace(0,2,50);
```

where x has been discretized into 25 points and t is discretized into 50 points. The surface is represented in Fig. 15.7(a) over the domain of interest. The surface is constructed by using the **meshgrid** command as follows:

```
[T,X]=meshgrid(t,x);
f=exp(-abs((X-.5).*(T-1)))+sin(X.*T);
subplot(2,2,1)
surfl(X,T,f), shading interp, colormap(gray)
```

The **surfl** produces a lighted surface that in this case is represented in gray-scale. Note that in producing this surface, the matrix **f** is a 25 × 50 matrix so that the *x*-values are given as row locations while the *t*-values are given by column locations.

The basis functions are then computed using the SVD. Recall that the SVD produces ordered singular values and associated orthonormal basis functions that capture as much energy as possible. Thus an SVD can be performed on the matrix **f** that represents the surface.

```
[u,s,v]=svd(f);   % perform SVD

for j=1:3
    ff=u(:,1:j)*s(1:j,1:j)*v(:,1:j)'; % modal projections
    subplot(2,2,j+1)
    surfl(X,T,ff), shading interp, colormap(gray)
    set(gca,'Zlim',[0.5 2])
end
subplot(2,2,1), text(-0.5,1,0.5,'(a)','Fontsize',[14])
subplot(2,2,2), text(-0.5,1,0.5,'(b)','Fontsize',[14])
subplot(2,2,3), text(-0.5,1,0.5,'(c)','Fontsize',[14])
subplot(2,2,4), text(-0.5,1,0.5,'(d)','Fontsize',[14])
```

The SVD command pulls out the diagonal matrix along with the two unitary matrices **U** and **V**. The loop above processes the sum of the first, second and third modes. This gives the POD modes of interest. Figure 15.7 demonstrates the modal decomposition and the accuracy achieved with representing the surface with one, two and three POD modes. The first mode alone captures 92% of the surface while three modes produce a staggering 99.9% of the energy. This should be contrasted with Fourier methods, for instance, which would require potentially hundreds of modes to achieve such accuracy. *As stated previously, these are the best one, two and three mode approximations of the function $f(x, t)$ that can be achieved.*

To be more precise about the nature of the POD, consider the *energy* in each mode. This can be easily computed, or has already been computed, in the SVD, i.e. they are the singular values σ_j. In MATLAB, the energy in the one mode and three mode approximation is computed from

```
energy1=sig(1)/sum(sig)
energy3=sum(sig(1:3))/sum(sig)
```

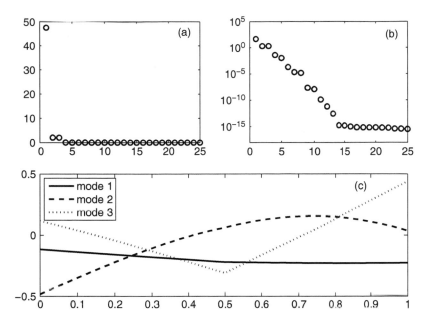

Figure 15.8: Singular values on a standard plot (a) and a log plot (b) showing the energy in each POD mode. For the surface considered in Fig. 15.7, the bulk of the energy is in the first POD mode. A three mode approximation produces 99.9% of the energy. The linear POD modes are illustrated in panel (c).

More generally, the entire *spectrum* of singular values can be plotted. Figure 15.8 shows the complete spectrum of singular values on a standard plot and a log plot. The clear dominance of a single mode is easily deduced. Additionally, the first three POD modes are illustrated: they are the first three columns of the matrix **U**. These constitute the orthonormal expansion basis of interest. The code for producing these plots is as follows:

```
sig=diag(s);
subplot(2,2,1), plot(sig,'ko','Linewidth',[1.5])
axis([0 25 0 50])
set(gca,'Fontsize',[13],'Xtick',[0 5 10 15 20 25])
text(20,40,'(a)','Fontsize',[13])
subplot(2,2,2), semilogy(sig,'ko','Linewidth',[1.5])
axis([0 25 10^(-18) 10^(5)])
set(gca,'Fontsize',[13],'Ytick',[10^(-15) 10^(-10) 10^(-5) 10^0 10^5],...
    'Xtick',[0 5 10 15 20 25]);
text(20,10^0,'(b)','Fontsize',[13])

subplot(2,1,2)
plot(x,u(:,1),'k',x,u(:,2),'k--',x,u(:,3),'k:','Linewidth',[2])
set(gca,'Fontsize',[13])
legend('mode 1','mode 2','mode 3','Location','NorthWest')
text(0.8,0.35,'(c)','Fontsize',[13])
```

Note that the key part of the code is simply the calculation of the POD modes which are weighted, in practice, by the singular values and their evolution in time as given by the columns of **V**.

Example 2: Consider the following space–time function

$$f(x, t) = [1 - 0.5\cos 2t]\text{sech}\, x + [1 - 0.5\sin 2t]\text{sech}\, x \tanh x. \tag{15.5.6}$$

This represents an asymmetric, time-periodic breather of a localized solution. Such solutions arise in a myriad of contexts are often obtained via full numerical simulations of an underlying physical system. Let's once again apply the techniques of POD analysis to see what is involved in the dynamics of this system. A discretization of the system is first applied and made into a two-dimensional representation in time and space.

```
x=linspace(-10,10,100);
t=linspace(0,10,30);
[X,T]=meshgrid(x,t);
f=sech(X).*(1-0.5*cos(2*T))+(sech(X).*tanh(X)).*(1-0.5*sin(2*T));
subplot(2,2,1), waterfall(X,T,f), colormap([0 0 0])
```

Note that, unlike before, the matrix **f** is now a 30×100 matrix so that the x-values are along the columns and t-values are along the rows. This is done simply so that we can use the **waterfall** command in MATLAB.

The singular value decomposition of the matrix **f** produces the quantities of interest. In this case, since we want the x-value in the rows, the transpose of the matrix is considered in the SVD.

```
[u,s,v]=svd(f');
for j=1:3
  ff=u(:,1:j)*s(1:j,1:j)*v(:,1:j)';
  subplot(2,2,j+1)
  waterfall(X,T,ff'), colormap([0 0 0]), set(gca,'Zlim',[-1 2])
end
subplot(2,2,1), text(-19,5,-1,'(a)','Fontsize',[14])
subplot(2,2,2), text(-19,5,-1,'(b)','Fontsize',[14])
subplot(2,2,3), text(-19,5,-1,'(c)','Fontsize',[14])
subplot(2,2,4), text(-19,5,-1,'(d)','Fontsize',[14])
```

This produces the original function along with the first three modal approximations of the function. As is shown in Fig. 15.9, the two-mode and three-mode approximations appear to be identical. In fact, they are. The singular values of the SVD show that $\approx 83\%$ of the energy is in

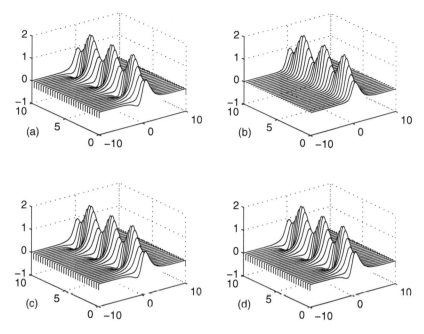

Figure 15.9: Representation of the space–time dynamics (a) by a series of low-rank (low dimensional) approximations with (b) one mode, (c) two modes and (d) three modes. The energy captured in the first mode is approximately 83% of the total energy. The first two modes together capture 100% of the surface energy, thus suggesting that the surface can be easily and accurately represented by a simple two-mode expansion.

the first mode of Fig. 15.9(a) while 100% of the energy is captured by a two mode approximation as shown in Fig. 15.9(b). Thus the third mode is completely unnecessary. The singular values are produced with the following lines of code.

```
figure(2)
sig=diag(s);
subplot(3,2,1), plot(sig,'ko','Linewidth',[1.5])
axis([0 25 0 50])
set(gca,'Fontsize',[13],'Xtick',[0 5 10 15 20 25])
text(20,40,'(a)','Fontsize',[13])
subplot(3,2,2), semilogy(sig,'ko','Linewidth',[1.5])
axis([0 25 10^(-18) 10^(5)])
set(gca,'Fontsize',[13],'Ytick',[10^(-15) 10^(-10) 10^(-5) 10^0 10^5],...
  'Xtick',[0 5 10 15 20 25]);
text(20,10^0,'(b)','Fontsize',[13])

energy1=sig(1)/sum(sig)
energy2=sum(sig(1:2))/sum(sig)
```

As before, we can also produce the first two modes, which are the first two columns of **U**, along with their time behavior, which are the first two columns of **V**.

```
subplot(3,1,2)   % spatial modes
plot(x,u(:,1),'k',x,u(:,2),'k--','Linewidth',[2])
set(gca,'Fontsize',[13])
legend('mode 1','mode 2','Location','NorthWest')
text(8,0.35,'(c)','Fontsize',[13])

subplot(3,1,3)   % time behavior
plot(t,v(:,1),'k',t,v(:,2),'k--','Linewidth',[2])
text(9,0.35,'(d)','Fontsize',[13])
```

Figure 15.10 shows the singular values along with the spatial and temporal behavior of the first two modes. It is not surprising that the SVD characterizes the total behavior of the system as an exactly (to numerical precision) two mode behavior. Recall that is exactly what we started with! Note, however, that our original two modes are different from the SVD modes. Indeed, the first SVD mode is asymmetric unlike our symmetric hyperbolic secant.

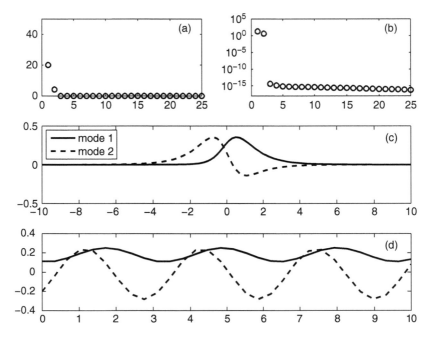

Figure 15.10: Singular values on a standard plot (a) and a log plot (b) showing the energy in each POD mode. For the space–time surface considered in Fig. 15.9, the bulk of the energy is in the first POD mode. A two mode approximation produces 100% of the energy. The linear POD modes are illustrated in panel (c) which their time evolution behavior in panel (d).

Summary and limits of SVD/PCA/POD

The methods of PCA and POD, which are essentially related to the idea of diagonalization of any matrix via the SVD, are exceptionally powerful tools and methods for the evaluation of data driven systems. Further, they provide the most precise method for performing low dimensional reductions of a given system. A number of key steps are involved in applying the method to experimental or computational data sets.

- Organize the data into a matrix $\mathbf{A} \in \mathbb{C}^{m \times n}$ where m is the number of measurement types (or probes) and n is the number of measurements (or trials) taken from each probe.
- Subtract off the mean for each measurement type or row $\mathbf{x}_j$.
- Compute the SVD and singular values of the covariance matrix to determine the principal components.

If considering a fitting algorithm, then the POD method is the appropriate technique to consider and the SVD can be applied directly to the $\mathbf{A} \in \mathbb{C}^{m \times n}$ matrix. This then produces the singular values as well as the POD modes of interest.

Although extremely powerful, and seemingly magical in its ability to produce insightful quantification of a given problem, these SVD methods are not without limits. Some fundamental assumptions have been made in producing the results thus far, the most important being the assumption of *linearity*. Recently some efforts to extend the method using nonlinearity prior to the SVD have been explored [39, 40, 41]. A second assumption is that larger variances (singular values) represent more important dynamics. Although generally this is believed to be true, there are certainly cases where this assumption can be quite misleading. Additionally, in constructing the covariance matrix, it is assumed that the mean and variance are sufficient statistics for capturing the underlying dynamics. It should be noted that only exponential (Gaussian) probability distributions are completely characterized by the first two moments of the data. This will be considered further in the next section on independent component analysis. Finally, it is often the case that PCA may not produce the optimal results. For instance, consider a point on a propeller blade. Viewed from the PCA framework, two orthogonal components are produced to describe the circular motion. However, this is really a one-dimensional system that is solely determined by the phase variable. The PCA misses this fact unless use is made of the information already known, i.e. that a polar coordinate system with fixed radius is the appropriate context to frame the problem. Thus blindly applying the technique can also lead to nonoptimal results that miss obvious facts.

15.6 Robust PCA

The singular value decomposition and its various guises, most notably principal component analysis, are perhaps the most widely used statistical tools for generating dimensionality reduction of data. As has been demonstrated and discussed, the SVD produces characteristic features (principal components) that are determined by the covariance matrix of the data. Fundamental in producing the SVD/PCA/POD modes is the L^2 norm for data fitting. Although the L^2 norm is highly appealing due to its centrality (and physical interpretation) in most applications, it does have potential flaws that can severely limit its applicability. Specifically, if corrupt data or large noise fluctuations are present in the data matrix, the higher dimensional least-square fitting performed by the SVD algorithm squares this pernicious error, leading to significant deformation of the singular values and PCA modes describing the data. Although many physical systems and/or computations are relatively free of data corruption, many modern applications in image processing, PIV fluid measurements, web data analysis, and bioinformatics, for instance, are rife with arbitrary corruption of the measured data, thus ensuring that standard application of the SVD will be of limited use.

Ideally, in performing dimensionality reduction one should not allow the corrupt or large noise fluctuations to so grossly influence one's results. In order to limit the impact of such *data outliers*, a different measure, or norm, can be envisioned in measuring the best data fit (SVD/PCA/POD modes). One measure that has recently produced great success in this arena is the L^1 norm [42] which no longer squares the distance of the data to the best-fit mode. Indeed, the L^1 norm has been demonstrated to promote sparsity and is the underlying theoretical construct in *compressive sensing* or *sparse sensing* algorithms (see Section 17.3 for further details). Here, the L^1 norm is simply used as an alternative measure (norm) for performing the equivalent of the least-square fit to the data. As a result, a more robust method is achieved of computing PCA components, i.e. the so-called *robust PCA* method.

Matrix decomposition: Low-rank plus sparse

To illustrate the entanglement of low-rank data with corrupt (sparse) measurement/matrix error, we can construct a new matrix which is composed of these two elements together

$$\mathbf{M} = \mathbf{L} + \mathbf{S} \qquad (15.6.1)$$

where $\mathbf{M}$ is the data matrix of interest that is composed of a low-rank structure $\mathbf{L}$ along with some sparse data $\mathbf{S}$. Our objective is the following: given observations $\mathbf{M}$, can the low-rank matrix $\mathbf{L}$ and sparse matrix $\mathbf{S}$ be recovered? Adding to the difficulty of this task is the following: we do not know the rank of the matrix $\mathbf{L}$, nor do we know how many nonzero (sparse) elements of the matrix $\mathbf{S}$ exist.

Remarkably, Candés and co-workers [42] have recently proved that this can indeed be solved through the formulation of a convex optimization problem. At first sight, the separation problem seems impossible to solve since the number of unknowns to infer for $\mathbf{L}$ and $\mathbf{S}$ is twice as many as the given measurements in $\mathbf{M}$. Furthermore, it seems even more daunting that we expect to reliably obtain the low-rank matrix $\mathbf{L}$ with errors in $\mathbf{S}$ of arbitrarily large magnitude. But not only can this problem be solved, it can be solved by tractable convex optimization [42].

Of course, in real applications, it is rare that the matrix decomposition is as ideal as (15.6.1). More generally, the decomposition is of the form

$$\mathbf{M} = \mathbf{L} + \mathbf{S} + \mathbf{N} \qquad (15.6.2)$$

where the matrix **N** is a dense, small perturbation accounting for the fact that the low-rank component is only approximately low-rank and that small errors can be added to all the entries (in some sense, this model unifies the classical PCA and the robust PCA by combining both sparse gross errors and dense small noise). But even in this case, the convex optimization algorithm formulated by Candés and co-workers [42] seems to do a remarkable job at separating low-rank from sparse data. This can be used to great advantage when certain types of data are considered, namely those with corrupt data or large, sparse noisy perturbations.

The details of the method and its proof are beyond the scope of this book. However, we will utilize the algorithms developed in the robust PCA literature in order to separate data matrices into low-rank and sparse components. A number of MATLAB algorithms can be downloaded from the web for this purpose (http://perception.csl.illinois.edu/matrix-rank/sample_code.html). Indeed, a wide variety of techniques have been developed for specifically providing a framework for robust PCA, including the augmented Lagrange multiplier (ALM) method [43], accelerated proximal gradient method [44], dual method [44], singular value thresholding method [45] and the alternating direction method [46]. For our purposes, we will use the MATLAB program **inexact_alm_rpca** since it is extremely simple to use and exceedingly fast [43].

Example: As an example, consider the function from the previous section

$$f(x, t) = [1 - 0.5\cos 2t]\operatorname{sech} x + [1 - 0.5\sin 2t]\operatorname{sech} x \tanh x. \tag{15.6.3}$$

Previously, the SVD was applied to this function and a two-mode dominance was clearly demonstrated (see Figs. 15.9 and 15.10). To illustrate the problem to be considered, the above function is plotted along with a corrupted version in Fig. 15.11. In this example, the data were corrupted by the addition of sparse, but large, noise fluctuations to the ideal data. The following MATLAB code produces the ideal figure.

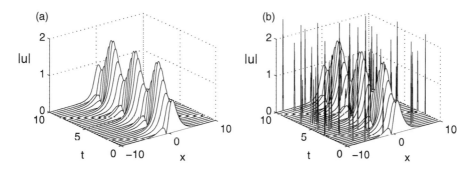

Figure 15.11: Representation of the space–time dynamics considered in Figs. 15.9 and 15.10. In (a), the ideal dynamics is demonstrated while in (b), a corrupted (noisy) image is assumed to be generated by the data collection process, for instance. The addition of noise brings into question the ability of the SVD to produce meaningful results. For this example, the low-rank matrix **L** has two modes of significance while the sparse matrix **S** has 60 nonzero entries.

```
n=200;
x=linspace(-10,10,n);
t=linspace(0,10,30);
[X,T]=meshgrid(x,t);
usol=sech(X).*(1-0.5*cos(2*T))+(sech(X).*tanh(X)).*(1-0.5*sin(2*T));
subplot(2,2,1), waterfall(x,t,abs(usol)); colormap([0 0 0]);
```

To add sparse noise, the **randintrlv** is used to produce noise spikes (60 spikes in total) in certain matrix/pixel locations, thus corrupting the data matrix.

```
sam=60;
Atest2=zeros(length(t),n);
Arand1=rand(length(t),n);
Arand2=rand(length(t),n);
r1 = randintrlv([1:length(t)*n],793);
r1k= r1(1:sam);
for j=1:sam
    Atest2(r1k(j))=-1;
end
Anoise=Atest2.*(Arand1+i*Arand2);
unoise=usol+2*Anoise;
subplot(2,2,2), waterfall(x,t,abs(unoise)); colormap([0 0 0]);
```

Figure 15.11 shows the ideal data along with the corrupt, or more physically relevant, data that might be collected in practice. The complex white noise fluctuations are, of course, the problematic part of the dimensionality reduction issue. Recall that without such noise fluctuations, the ideal data can be faithfully approximated with a low-rank, two-mode approximation.

The PCA reduction of the data matrices can now be compared by using the singular value decomposition. The following code produces the SVD decomposition of both data matrices

```
A1=usol; A2=unoise;
[U1,S1,V1]=svd(A1);
[U2,S2,V2]=svd(A2);
```

Both the singular values and the modal structures are represented in Figs. 15.12 and 15.13, respectively. Specifically, the top panel of Fig. 15.12 shows the two-mode dominance (top panel of Fig. 15.13) of the ideal data while the middle panels of these figures show the singular value

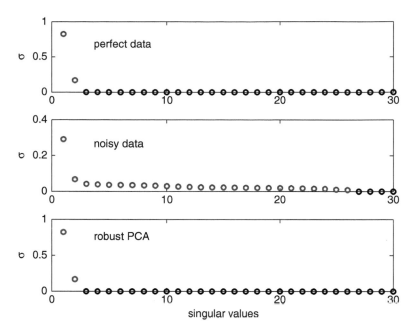

Figure 15.12: Singular values of the data matrices for the ideal data (top panel), the corrupted data (middle panel), and the low-rank data computed through robust PCA (bottom panel). The robust PCA construction produces almost a perfect match to the original, ideal data. The magenta dots represent the number of modes necessary to construct 99% of the original data. In this case, the ideal and robust PCA require two modes while the corrupt data require 26 modes.

distribution and the first four modes, respectively. Note that the addition of the corrupt data activates many more modes. In fact, it takes 26 modes to capture 99% of the energy of the data compared to two modes previously. The modes are also highly perturbed from their ideal states due to the sparse (corrupt) noise that was added.

To apply the robust PCA algorithm, the corrupt data matrix is decomposed into real and imaginary parts before applying the **inexact_alm_rpca** algorithm. This algorithm effectively separates the data into low-rank and sparse components as given by (15.6.1). The low-rank portion is kept in order to then acquire the PCA modes necessary for reconstruction.

```
ur=real(unoise);
ui=imag(unoise);

lambda=0.2;
[R1r,R2r]=inexact_alm_rpca(real(ur.'),lambda);
[R1i,R2i]=inexact_alm_rpca(real(ui.'),lambda);
```

Continued

```
R1=R1r+i*R1i;
R2=R2r+i*R2i;

[U3,S3,V3]=svd(R1.');

subplot(2,2,1), waterfall(x,t,abs(R1)'), colormap([0 0 0])
subplot(2,2,2), waterfall(x,t,abs(R2)'), colormap([0 0 0])
```

Note here that the real and imaginary portions are put back together at the end of the algorithm in order to SVD the low-rank portion. Further, the two matrices corresponding to the low-rank (*R*1) and sparse (*R*2) portions of the data matrix are plotted. The parameter *lambda* used in the **inexact_alm_rpca** algorithm can be tuned to best separate the sparse from low-rank matrices in (15.6.1). Here a value of 0.2 is used, which produces an almost ideal separation as is shown in Figs. 15.12 (bottom panel), 15.13 (bottom panel) and 15.14. Indeed, an almost perfect separation is achieved as advertised (guaranteed) by Candés and co-workers [42].

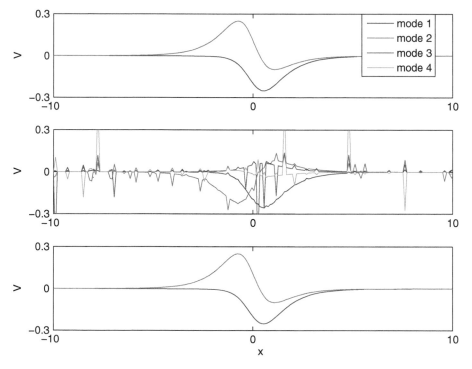

Figure 15.13: Dominant modes for the ideal data (top panel), the corrupted data (middle panel), and the low-rank data computed through robust PCA (bottom panel). The ideal and robust PCA produce the same dominant two modes while the corrupt data produce highly erratic modes.

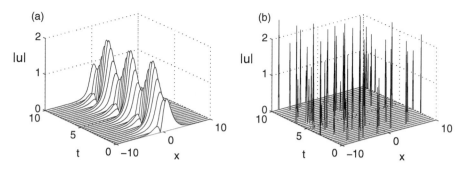

Figure 15.14: Separation of the original corrupt data matrix shown in Fig. 15.11(b) into a low-rank component (a) and a sparse component (b). The separation is almost perfect, with the low-rank matrix being dominated by two modes, i.e. they contain greater than 99% of the energy.

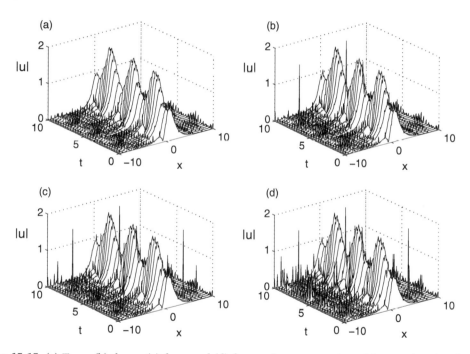

Figure 15.15: (a) Two-, (b) three-, (c) four- and (d) five-mode reconstruction of the matrix using the corrupted (sparse plus low-rank) data. Note that the addition of the higher modes adds sparse data spikes as is expected since they have not been filtered out of the data matrix.

To further investigate the robust PCA algorithm and its ability to extract the meaningful, low-rank matrix from the full matrix corrupted by sparse fluctuations, a series of low-rank approximations is made of the data matrices using the PCA modes generated by the SVD (see Figs. 15.15 and 15.16). The low-rank approximations show how effective the robust PCA algorithm is in separating out the low-rank from sparse components.

```
% low-rank approximation of corrupt data matrix
figure(1)
for jj=1:4
  for j=1:jj+1
    ff=U2(:,1:j)*S2(1:j,1:j)*V2(:,1:j).';
    subplot(2,2,jj), waterfall(x,t,abs(ff)), colormap([0 0 0])
  end
end

% approximation of low-rank matrix from robust PCA
figure(2)
for jj=1:4
  for j=1:jj+1
    ff=U3(:,1:j)*S3(1:j,1:j)*V3(:,1:j).';
    subplot(2,2,jj), waterfall(x,t,abs(ff)), colormap([0 0 0])
  end
end
```

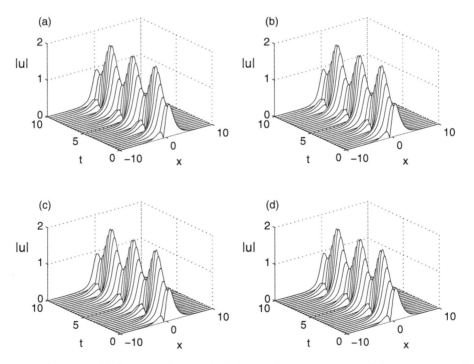

Figure 15.16: (a) Two-, (b) three-, (c) four- and (d) five-mode reconstruction of the matrix using the robust PCA algorithm for extracting the corrupted (sparse) components. Note that a two-mode expansion is sufficient to capture all the features of interest.

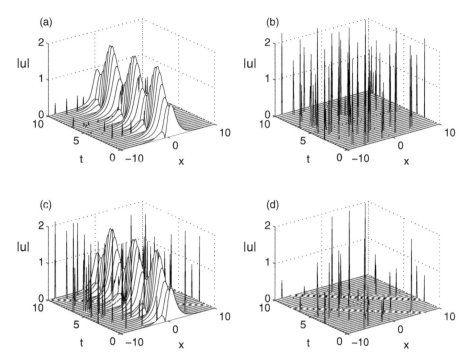

Figure 15.17: Separation of the original corrupt data matrix shown in Fig. 15.11(b) into a low-rank component (a) and (c) and a sparse component (b) and (d). The top panels are for *lambda* = 0.5 in the **inexact_alm_rpca** algorithm while the bottom panels are for *lambda* = 0.8. The separation deteriorates from the almost perfect separation illustrated previously with *lambda* = 0.2.

Finally, to end this discussion, the effects of the parameter *lambda* in the **inexact_alm_rpca** algorithm are illustrated. This parameter takes on values of zero to unity and can be tuned to achieve better or worse performance. The default, if *lambda* is not included, does a reasonable job generically in separating the low-rank from the sparse matrices. As *lambda* goes to zero, the computed low-rank matrix picks up the entire initial data matrix, i.e. the resulting sparse matrix is computed to be zero while the resulting sparse matrix contains the original data matrix. As *lambda* goes to unity, the opposite happens with the computed low-rank matrix containing nothing while the sparse matrix contains virtually the entire original matrix. Figure 15.17 depicts the results of the **inexact_alm_rpca** algorithm for *lambda* equal to 0.5 and 0.8. Note that as *lambda* is increased, the fidelity of the low-rank matrix is compromised and corrupted by the sparsity.

As a final comment, the robust PCA algorithm advocated for here is quite remarkable in its ability to separate sparse corruption from low-rank structure in a given data matrix. The application of robust PCA essentially acts as an advanced filter for cleaning up a data matrix. Indeed, one can potentially use this as a filter which is very unlike the time–frequency filtering schemes considered previously.

16 Independent Component Analysis

The concept of principal components or of a proper orthogonal decomposition are fundamental to data analysis. Essentially, there is an assertion, or perhaps a hope, that underlying seemingly complex and unordered data, there is in fact, some characteristic, and perhaps low dimensional ordering of the data. The singular value decomposition of a matrix, although a mathematical idea, was actually a formative tool for the interpretation and ordering of the data. In this section on *independent component analysis* (ICA), the ideas turn to data sets for which there is more than one statistically independent object in the data. ICA provides a foundational mathematical method for extracting interleaved data sets in a fairly straightforward way. Its applications are wide and varied, but our primary focus will be application of ICA in the context of image analysis.

16.1 The Concept of Independent Components

Independent component analysis (ICA) is a powerful tool that extends the concepts of PCA, POD and SVD. Moreover, you are already intuitively familiar with the concept as some of the examples here will show. A simple way to motivate thinking about ICA is by considering the *cocktail party problem*. Thus consider two conversations in a room that are happening simultaneously. How is it that the two different acoustic signals of conversation one and two can be separated out? Figure 16.1 depicts a scenario where two groups are conversing. Two microphones are placed in the room at different spatial locations and from the two signals $s_1(t)$ and $s_2(t)$ a mathematical attempt is made to separate the signals that have been mixed at each of the microphone locations. Provided that the noise level is not too large or that the conversation volumes are sufficiently large, humans can perform this task with remarkable ease. In our case, the two microphones are our two ears. Indeed, this scenario and its mathematical foundations are foundational to the concept of eavesdropping on a conversation.

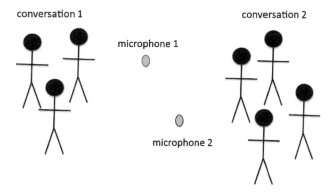

Figure 16.1: Envisioned cocktail party problem. Two groups of people are having separate conversations which are recorded from microphones one and two. The signals from the two conversations are mixed together at microphones one and two according to the placement relative to the conversations. From the two microphone recordings, the individual conversations can then be extracted.

From a mathematical standpoint, this problem can be formulated with the following mixing equation

$$x_1(t) = a_{11}s_1 + a_{12}s_2 \tag{16.1.1a}$$

$$x_2(t) = a_{21}s_1 + a_{22}s_2 \tag{16.1.1b}$$

where $x_1(t)$ and $x_2(t)$ are the mixed, recorded signals at microphones one and two, respectively. The coefficients a_{ij} are the mixing parameters that are determined by a variety of factors including the placement of the microphones in the room, the distance to the conversations, and the overall room acoustics. Note that we are omitting time-delay signals that may reflect off the walls of the room. This problem also resembles quite closely what may happen in a large number of applications. For instance, consider the following

- **Radar detection** If there are numerous targets that are being tracked, then there is significant mixing in the scattered signal from all of the targets. Without a method for clear separation of the targets, the detector becomes useless for identifying location.

- **Electroencephalogram (EEG)** EEG readings are electrical recordings of brain activities. Typically these EEG readings are from multiple locations of the scalp. However, at each EEG reading location, all the brain activity signals are mixed, thus preventing a clear understanding of how many underlying signals are contained within. With a large number of EEG probes, ICA allows for the separation of the brain activity readings and a better assessment of the overall neural activity.

- **Terminator salvation** If you remember in the movie, John Connor and the resistance found a hidden signal (ICA) embedded on the normal signals in the communications sent between the terminators and skynet vehicles and ships. Although not mentioned in the movie, this was clearly somebody in the future who is reading this book now. Somehow that bit of sweet math only made it to the cutting room floor. Regardless, one shouldn't underestimate how awesome data analysis skills are in the real world of the future.

The ICA method is closely related to the *blind source separation* (BSS) (or blind signal separation) method. Here the sources refer to the original signals, or independent components, and blind refers to the fact that the mixing matrix coefficients a_{ij} are unknown.

A more formal mathematical framework for ICA can be established by considering the following: *Given N distinct linear combinations of N signals (or data sets), determine the original N images*. This is the succinct statement of the mathematical objective of ICA. Thus to generalize (16.1.1) to the N dimensional system we have

$$x_j(t) = a_{j1}s_1 + a_{j2}s_2 + \cdots + a_{jN}s_N \quad 1 \leq j \leq N. \tag{16.1.2}$$

Expressed in matrix form, this results in

$$\mathbf{x} = \mathbf{As} \tag{16.1.3}$$

where $\mathbf{x}$ are the mixed signal measurements, $\mathbf{s}$ are the original signals we are trying to determine, and $\mathbf{A}$ is the matrix of coefficients which determines how the signals are mixed as a function of the physical system of interest. At first, one might simply say that the solution to this is trivial and it is given by

$$\mathbf{s} = \mathbf{A}^{-1}\mathbf{x} \tag{16.1.4}$$

where we are assuming that the placement of our measurement devices is such that $\mathbf{A}$ is nonsingular and its inverse exists. Such a solution makes the following assumption: we know the matrix cofficients a_{ij}. The point is, we don't know them. Nor do we know the signal $\mathbf{s}$. Mathematically then, the aim of ICA is to approximate the coefficients of the matrix $\mathbf{A}$, and in turn the original signals $\mathbf{s}$, under as general assumptions as possible.

The statistical model given by Eq. (16.1.3) is called *independent component analysis*. ICA is a *generative model*, which means that it describes how the observed data are generated by a process of mixing in the components $s_j(t)$. The independent components are *latent variables*, meaning they are never directly observed. The key step in the ICA models is to assume the following: *the underlying signals $s_j(t)$ are statistically independent with probability distributions that are not Gaussian*. Statistical independence and higher order moments of the probably distribution (thus requiring non-Gaussian distributions) are the critical mathematical tools that will allow us to recover the signals.

Image separation and SVD

To make explicit the mathematical methodology to be pursued here, a specific example of image separation will be used. Although there are a variety of mathematical alternatives for separating the independent components [47], the approach considered here will be based upon PCA and SVD. To illustrate the concept of ICA, consider the example data represented in Fig. 16.2. The three panels are fundamental to understanding the concept of ICA. In the left panel (a), measurements are taken of a given system and are shown to project nicely onto a dominant direction whose leading principal component is denoted by the red vector. This red vector would be the principal component with a length σ_1 determined by the largest singular value. It is clear that the singular value σ_2 corresponding to the orthogonal direction of the second principal component

 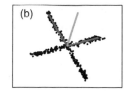 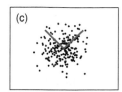

(a)   (b)   (c)

Figure 16.2: Illustration of the principles of PCA (a), ICA (b) and the failure of Gaussian distributions to distinguish principal components (c). The red vectors show the principal directions, while the green vector in the middle panel shows what would be the principal direction if a direct SVD where applied rather than an ICA. The principal directions in (c) are arbitrarily chosen since no principal directions can be distinguished.

would be considerably smaller. In the middle panel (b), the measurements indicate that there are two principal directions in the data fluctuations. If an SVD is applied directly to the data, then the dominant singular direction would be approximately the green vector. Yet it is clear that the green vector does not accurately represent the data. Rather, the data seems to be organized *independently* along two different directions. An SVD of the two independent data sets would produce two principal components, i.e. two *independent component analysis* vectors. Thus it is critical to establish a method for separating out data of this sort into their independent directions. Finally, in the third panel (c), Gaussian distributed data is shown where no principal components can be identified with certainty. Indeed, there are an infinite number of possible orthogonal projections that can be made. Two arbitrary directions have been shown in panel (c). This graphic shows why Gaussian distributed random variables are disastrous for ICA; no projections onto the independent components can be accomplished.

We now turn our attention to the issue of image separation. Figure 16.3 demonstrates a prototypical example of what can occur when photographing an image behind glass. The glass encasing

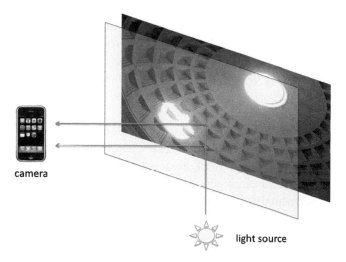

camera

light source

Figure 16.3: Illustration of the image separation problem. A piece of glass reflects part of the background light source while some of the light penetrates to the image and is reflected to the camera. Thus the camera sees two images: the image (painting) and the reflection of the background light source.

of the artwork, or image, produces a reflection of the backlit source along with the image itself. Amazingly, the human eye can very easily compensate for this and *see through* much of the reflection to the original image. Alternatively, if we choose, we can also focus on the reflected image of the backlit source. Ultimately, there is some interest in training a computer to do this image processing automatically. Indeed, in artificial vision, algorithms such as these are important for mimicking natural human behaviors and abilities.

The light reflected off the glass is partially polarized and provides the crucial degree of freedom necessary for our purposes. Specifically, to separate the images, a photo is taken as illustrated in Fig. 16.3 but now with a linear polarizer placed in front of the camera lens. Two photos are taken where the linear polarizer is rotated 90 degrees from each other. This is a simple home experiment essentially. But you must have some control over the focus of your camera in terms of focusing and flash. The picture to be considered is hanging in my office: **The Judgement of Paris** by John Flaxman (1755–1826). Flaxman was an English sculptor, draughtsman and illustrator who enjoyed a long and brilliant career that included a stint at Wedgwood in England. In fact, at the age of 19, Josiah Wedgwood hired Flaxman as a modeler of classic and domestic friezes, plaques and ornamental vessels and medallion portraits. You can still find many Wedgwood ornamental vessels today that have signature Flaxman works molded on their sides. It was during this period that the manufactures of that time perfected their style and earned great reputations. But what gained Flaxman his general fame was not his work in sculpture proper, but his designs and engravings for the **Iliad**, **Odyssey** and **Dante's Inferno**. All of his works on these mytholological subjects were engraved by Piroli with considerable loss to Flaxman's own lines. The greatest collection of Flaxman's work is housed in the Flaxman Gallery at the University College London.

In the specific Flaxman example considered here, one of the most famous scenes of mythology is depicted: **The Judgement of Paris**. In this scene, Hermes watches as young Paris plays the part of judge. His task is to contemplate giving the golden apple with the inscription *To the Fairest* to either Venus, Athena or Hera. This golden apple was placed at the wedding reception of Peleus and Thetis, father and mother of peerless Achilles, by Discord since she did not receive a wedding invitation. Each lovely goddess in turn promises him something for the apple: Athena promises that he will become the greatest warrior in history, Juno promises him to be greatest ruler in the land, ruling over a vast and unmatched empire. Last comes the lovely Venus who promises him that the most beautiful woman in the world will be his. Of course, to sweeten up her offer, she is wearing next to nothing as usual. Lo and behold, the hot chick wins! And as a result we have the Trojan War for which Helen was fought for. I could go on here, but what else is to be expected of a young 20-something year old male (an older guy would have taken the vast kingdom while a prepubescent boy would have chosen to be the great warrior). It is comforting to know that very little has changed since in some 4000 years. Figures 16.4 and 16.5 represent this Flaxman scene along with the reflected image of my office window. The two pictures are taken with a linear polarizer in front of a camera that has been rotated by 90 degrees between the photos.

SVD method for ICA

Given the physical problem, how is SVD used then to separate the two images observed in Figs. 16.4 and 16.5? Consider the action of the SVD on the image mixing matrix $\mathbf{A}$ of Eq. (16.1.3).

Figure 16.4: The Judgement of Paris by John Flaxman (1755–1826). This is a plate made by Flaxman for an illustrated version of Homer's Iliad. The picture was taken in my office and so you can see both a faint reflection of my books as well as strong signature of my window which overlooks Drumheller Fountain.

Figure 16.5: This is an identical picture to Fig. 16.4 with a linear polarizer rotated 90 degrees. The polarization state of the reflected field is affected by the linear polarizer, thus allowing us to separate the images.

In this case, there are two images S_1 and S_2 that we are working with. This gives us six unknowns $(S_1, S_2, a_{11}, a_{12}, a_{21}, a_{22})$ with only the two constraints from Eq. (16.1.3). Thus the system cannot be solved without further assumptions being made. The first assumption will be that the two images are statistically independent. If the pixel intensities are denoted by S_1 and S_2, then statistical independence is mathematically expressed as

$$P(\mathbf{S}_1, \mathbf{S}_2) = P(\mathbf{S}_1)P(\mathbf{S}_2) \tag{16.1.5}$$

which states that the joint probability density is separable. No specific form is placed upon the probability densities themselves aside from this: they cannot be Gaussian distributed. In practice, this is a reasonable constraint since natural images are rarely Gaussian. A second assumption is that the matrix **A** is full rank, thus assuming that the two measurements of the image have indeed been filtered with different polarization states. These assumptions yield an estimation problem which is the ICA.

Consider what happens under the SVD process to the matrix **A**

$$\mathbf{A} = \mathbf{U}\boldsymbol{\Sigma}\mathbf{V}^* \tag{16.1.6}$$

where again **U** and **V** are unitary matrices that simply lead to rotation and $\boldsymbol{\Sigma}$ stretches (scales) an image as prescribed by the singular values. A graphical illustration of this process is shown in Fig. 16.6 for distributions that are uniform, i.e. they form a square. The mixing matrix **A** can be thought to first rotate the square via unitary matrix $\mathbf{V}^*$, then stretch it into a parallelogram via the diagonal matrix $\boldsymbol{\Sigma}$ and then rotate the parallelogram via the unitary matrix **U**. This is now the mixed image **X**. *The estimation, or ICA, of the independent images thus reduces to finding how to transform the rotated parallelogram back into a square. Or mathematically, transforming the mixed image back into a separable product of one-dimensional probability distributions.* This is the defining mathematical goal of the ICA image analysis problem, or any general ICA reduction technique.

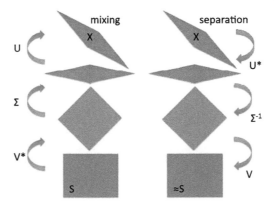

Figure 16.6: Graphical depiction of the SVD process of mixing the two images. The reconstruction of the images is accomplished by approximating the inversions of the SVD matrices so as to achieve a separable (statistically independent) probability distribution of the two images.

16.2   Image Separation Problem

The method proposed here to separate two images relies on reversing the action of the SVD on the two statistically independent images. Again, the matrix **A** in (16.1.6) is not known, so a direct computation of the SVD cannot be performed. However, each of the individual matrices can be approximated by considering its net effect on the assumed uniformly distributed images.

Three specific computations must be made: (i) the rotation of the parallelogram must be approximated, (ii) the scaling of the parallelogram according to the variances must be computed, and (iii) the final rotation back to a separable probability distribution must be obtained. Each step is outlined below.

Step 1: Rotation of the parallelogram

To begin, consider once again Fig. 16.6. Our first objective is to undo the rotation of the unitary matrix **U**. Thus we will ultimately want to compute the inverse of this matrix which is simply **U***. In a geometrical way of thinking, our objective is to align the long and short axes of the parallelogram with the primary axis as depicted in the two top right shaded boxes of Fig. 16.6. The angle of the parallelogram relative to the primary axes will be denoted by θ, and the long and short axes correspond to the axes of the maximal and minimal variance, respectively. From the image data itself, then, the maximal and minimal variance directions will be extracted. Assuming mean-zero measurements, the variance at an arbitrary angle of orientation is given by

$$Var(\theta) = \sum_{j=1}^{N} \left\{ [x_1(j) \quad x_2(j)] \begin{bmatrix} \cos\theta \\ \sin\theta \end{bmatrix} \right\}^2. \tag{16.2.1}$$

The maximal variance is determined by computing the angle θ that maximizes this function. It will be assumed that the corresponding angle of minimal variance will be orthogonal to this at $\theta - \pi/2$. These axes are essentially the principal component directions that would be computed if we actually knew components of the matrix **A**.

The maximum of (16.2.1) with respect to θ can be found by differentiating $Var(\theta)$ and setting it equal to zero. To do this, Eq. (16.2.1) is rewritten as

$$\begin{aligned} Var(\theta) &= \sum_{j=1}^{N} \left\{ [x_1(j) \quad x_2(j)] \begin{bmatrix} \cos\theta \\ \sin\theta \end{bmatrix} \right\}^2 \\ &= \sum_{j=1}^{N} [x_1(j)\cos\theta + x_2(j)\sin\theta]^2 \\ &= \sum_{j=1}^{N} x_1^2(j)\cos^2\theta + 2x_1(j)x_2(j)\cos\theta\sin\theta + x_2^2(j)\sin^2\theta. \end{aligned} \tag{16.2.2}$$

Differentiating with respect to θ then gives

$$\frac{d}{d\theta} Var(\theta) = 2 \sum_{j=1}^{N} -x_1^2(j) \sin \theta \cos \theta + x_1(j)x_2(j)\left[\cos^2 \theta - \sin^2 \theta\right] + x_2^2(j) \sin \theta \cos \theta$$

$$= 2 \sum_{j=1}^{N} \left[x_2^2(j) - x_1^2(j)\right] \sin \theta \cos \theta + x_1(j)x_2(j)\left[\cos^2 \theta - \sin^2 \theta\right]$$

$$= \sum_{j=1}^{N} \left[x_2^2(j) - x_1^2(j)\right] \sin 2\theta + 2x_1(j)x_2(j) \cos 2\theta \tag{16.2.3}$$

which upon setting this equal to zero gives the value of θ desired. Thus taking $d(Var(\theta))/d\theta = 0$ gives

$$\frac{\sin 2\theta}{\cos 2\theta} = \frac{-2 \sum_{j=1}^{N} x_1(j)x_2(j)}{\sum_{j=1}^{N} x_2^2(j) - x_1^2(j)}, \tag{16.2.4}$$

or in terms of θ alone

$$\theta_0 = \frac{1}{2} \tan^{-1} \left[\frac{-2 \sum_{j=1}^{N} x_1(j)x_2(j)}{\sum_{j=1}^{N} x_2^2(j) - x_1^2(j)}\right]. \tag{16.2.5}$$

In the polar coordinates $x_1(j) = r(j) \cos \psi(j)$ and $x_2(j) = r(j) \sin \psi(j)$, the expression reduces further to

$$\theta_0 = \frac{1}{2} \tan^{-1} \left[\frac{\sum_{j=1}^{N} r^2(j) \sin 2\psi(j)}{\sum_{j=1}^{N} r^2(j) \cos 2\psi(j)}\right]. \tag{16.2.6}$$

The rotation matrix, or unitary transformation, associated with the rotation of the parallelogram back to its aligned position is then

$$\mathbf{U}^* = \begin{bmatrix} \cos \theta_0 & \sin \theta_0 \\ -\sin \theta_0 & \cos \theta_0 \end{bmatrix} \tag{16.2.7}$$

with the angle θ_0 computed directly from the experimental data.

Step 2: Scaling of the parallelogram

The second task is to undo the principal component scaling achieved by the singular values of the SVD decomposition. This process is illustrated as the second step in the right column of Fig. 16.6. This task, however, is rendered straightforward now that the principal axes have been determined from step 1. In particular, the assumption was that along the direction θ_0, the maximal variance is achieved, while along $\theta_0 - \pi/2$, the minimal variance is achieved. Thus the components, or singular values, of the diagonal matrix $\mathbf{\Sigma}^{-1}$ can be computed.

The variances along the two principal component axes are given by

$$
\sigma_1 = \sum_{j=1}^{N} \left\{ [x_1(j) \; x_2(j)] \begin{bmatrix} \cos\theta_0 \\ \sin\theta_0 \end{bmatrix} \right\}^2
\tag{16.2.8a}
$$

$$
\sigma_2 = \sum_{j=1}^{N} \left\{ [x_1(j) \; x_2(j)] \begin{bmatrix} \cos(\theta_0 - \pi/2) \\ \sin(\theta_0 - \pi/2) \end{bmatrix} \right\}^2.
\tag{16.2.8b}
$$

This then gives the diagonal elements of the matrix $\mathbf{\Sigma}$. To undo this scaling, the inverse of $\mathbf{\Sigma}$ is constructed so that

$$
\mathbf{\Sigma}^{-1} = \begin{bmatrix} 1/\sqrt{\sigma_1} & 0 \\ 0 & 1/\sqrt{\sigma_2} \end{bmatrix}.
\tag{16.2.9}
$$

This matrix, in combination with that of step 1, easily undoes the principal component direction of rotation and its associated scaling. However, this process has only decorrelated the images, and a separable probability distribution has not yet been produced.

Step 3: Rotation to produce a separable probability distribution

The final rotation to separate the probability distributions is a more subtle and refined issue, but critical to producing nearly separable probability distributions. This separation process typically relies on the higher moments of the probability distribution. Since the mean has been assumed to be zero and there is no reason to believe that there is an asymmetry in the probability distributions, i.e. higher order odd moments (such as the skewness) are negligible, the next dominant statistical moment to consider is the fourth moment, or the kurtosis of the probability distribution. The goal will be to minimize this fourth-order moment, and by doing so we will determine the appropriate rotation angle. Note that the second moment has already been handled through steps 1 and 2. Said in a different mathematical way, minimizing the kurtosis will be another step in trying to approximate the probability distribution of the images as separable functions so that $P(\mathbf{S}_1\mathbf{S}_2) \approx P(\mathbf{S}_1)P(\mathbf{S}_2)$.

The kurtosis of a probability distribution is given by

$$
kurt(\phi) = K(\phi) = \sum_{j=1}^{N} \left\{ [\bar{x}_1(j) \; \bar{x}_2(j)] \begin{bmatrix} \cos\phi \\ \sin\phi \end{bmatrix} \right\}^4
\tag{16.2.10}
$$

where ϕ is the angle of rotation associated with the unitary matrix $\mathbf{U}$ and the variables $\bar{x}_1(j)$ and $\bar{x}_2(j)$ represent the image that has undergone the two steps of transformation as outlined previously.

For analytic convenience, a normalized version of the above definition of kurtosis will be considered. Specifically, a normalized version is computed in practice. The expedience of the normalization will become clear through the algebraic manipulations. Thus consider

$$\bar{K}(\phi) = \sum_{j=1}^{N} \frac{1}{\bar{x}_1^2(j) + \bar{x}_2^2(j)} \left\{ [\bar{x}_1(j) \quad \bar{x}_2(j)] \begin{bmatrix} \cos\phi \\ \sin\phi \end{bmatrix} \right\}^4. \tag{16.2.11}$$

As in our calculation regarding the second moment, the kurtosis will be written in a more natural form for differentiation

$$\bar{K}(\phi) = \sum_{j=1}^{N} \frac{1}{\bar{x}_1^2(j) + \bar{x}_2^2(j)} \left\{ [\bar{x}_1(j) \quad \bar{x}_2(j)] \begin{bmatrix} \cos\phi \\ \sin\phi \end{bmatrix} \right\}^4$$

$$= \sum_{j=1}^{N} \frac{1}{\bar{x}_1^2(j) + \bar{x}_2^2(j)} \left[\bar{x}_1(j)\cos\phi + \bar{x}_2(j)\sin\phi \right]^4 \tag{16.2.12}$$

$$= \sum_{j=1}^{N} \frac{1}{\bar{x}_1^2(j) + \bar{x}_2^2(j)} \left[\bar{x}_1^4(j)\cos^4\phi + 4\bar{x}_1^3(j)\bar{x}_2(j)\cos^3\phi\sin\phi + 6\bar{x}_1^2(j)\bar{x}_2^2(j)\cos^2\phi\sin^2\phi \right.$$

$$\left. + 4\bar{x}_1(j)\bar{x}_2^3(j)\cos\phi\sin^3\phi + \bar{x}_2^4(j)\sin^4\phi \right]$$

$$= \sum_{j=1}^{N} \frac{1}{\bar{x}_1^2(j) + \bar{x}_2^2(j)} \left[\frac{1}{8}\bar{x}_1^4(j)(3 + 4\cos 2\phi + \cos 4\phi) + \bar{x}_1^3(j)\bar{x}_2(j)(\sin 2\phi + (1/2)\sin 4\phi) \right.$$

$$+ \frac{3}{4}\bar{x}_1^2(j)\bar{x}_2^2(j)(1 - \cos 4\phi) + \bar{x}_1(j)\bar{x}_2^3(j)(\sin 2\phi - (1/2)\sin 4\phi)$$

$$\left. + \frac{1}{8}\bar{x}_2^4(j)(3 - 4\cos 2\phi + \cos 4\phi) \right].$$

This quantity needs to be minimized with an appropriate choice of ϕ. Thus the derivative $d\bar{K}/d\phi$ must be computed and set to zero:

$$\frac{d\bar{K}(\phi)}{d\phi} = \sum_{j=1}^{N} \frac{1}{\bar{x}_1^2(j) + \bar{x}_2^2(j)} \left[\frac{1}{8}\bar{x}_1^4(j)(-8\sin 2\phi - 4\sin 4\phi) + \bar{x}_1^3(j)\bar{x}_2(j)(2\cos 2\phi + 2\cos 4\phi) \right.$$

$$+ 3\bar{x}_1^2(j)\bar{x}_2^2(j)\sin 4\phi + \bar{x}_1(j)\bar{x}_2^3(j)(2\cos 2\phi - 2\cos 4\phi)$$

$$\left. + \frac{1}{8}\bar{x}_2^4(j)(8\sin 2\phi - 4\sin 4\phi) \right]$$

$$= \sum_{j=1}^{N} \frac{1}{\bar{x}_1^2(j) + \bar{x}_2^2(j)} \left\{ \left[\bar{x}_2^4(j) - \bar{x}_1^4(j) \right]\sin 2\phi + \left[2\bar{x}_1(j)\bar{x}_2(j)^3 + 2\bar{x}_1^3(j)\bar{x}_2(j) \right]\cos 2\phi \right.$$

$$+ \left[2\bar{x}_1^3(j)\bar{x}_2(j) - 2\bar{x}_1(j)\bar{x}_2^3(j) \right]\cos 4\psi$$

$$\left. + \left[3\bar{x}_1(j)^2\bar{x}_2^2(j) - (1/2)\bar{x}_1^4(j) - (1/2)\bar{x}_2^4(j) \right]\sin 4\psi \right\}$$

$$= \sum_{j=1}^{N} \frac{1}{\bar{x}_1^2(j) + \bar{x}_2^2(j)} \left\{ \left[(\bar{x}_1^2(j) + \bar{x}_2^2(j))(\bar{x}_2^2(j) - \bar{x}_1^2(j)) \right]\sin 2\phi \right.$$

$$+ \left[2\bar{x}_1(j)\bar{x}_2(j)(\bar{x}_1^2(j) + \bar{x}_2^2(j)) \right]\cos 2\phi$$

$$+ \left[2\bar{x}_1^3(j)\bar{x}_2(j) - 2\bar{x}_1(j)\bar{x}_2^3(j) \right]\cos 4\psi$$

$$\left. + \left[3\bar{x}_1(j)^2\bar{x}_2^2(j) - (1/2)\bar{x}_1^4(j) - (1/2)\bar{x}_2^4(j) \right]\sin 4\psi \right\}$$

$$= \sum_{j=1}^{N} \left[\bar{x}_2^2(j) - \bar{x}_1^2(j) \right] \sin 2\phi + 2\bar{x}_1(j)\bar{x}_2(j)\cos 2\phi$$

$$+ \frac{1}{\bar{x}_1^2(j) + \bar{x}_2^2(j)} \left\{ \left[2\bar{x}_1^3(j)\bar{x}_2(j) - 2\bar{x}_1(j)\bar{x}_2^3(j) \right] \cos 4\psi \right.$$

$$\left. + \left[3\bar{x}_1(j)^2\bar{x}_2^2(j) - (1/2)\bar{x}_1^4(j) - (1/2)\bar{x}_2^4(j) \right] \sin 4\psi \right\}$$

$$= \sum_{j=1}^{N} \frac{1}{\bar{x}_1^2(j) + \bar{x}_2^2(j)} \left\{ \left[2\bar{x}_1^3(j)\bar{x}_2(j) - 2\bar{x}_1(j)\bar{x}_2^3(j) \right] \cos 4\psi \right.$$

$$\left. + \left[3\bar{x}_1(j)^2\bar{x}_2^2(j) - (1/2)\bar{x}_1^4(j) - (1/2)\bar{x}_2^4(j) \right] \sin 4\psi \right\}, \qquad (16.2.13)$$

where the terms proportional to $\sin 2\phi$ and $\cos 2\phi$ add to zero since they are minimized when considering the second moment or variance, i.e. we would like to minimize both the variance and the kurtosis, and this calculation does both. By setting this to zero, the angle ϕ_0 can be determined to be

$$\phi_0 = \frac{1}{4} \tan^{-1} \left[\frac{ -\sum_{j=1}^{N} \left[2\bar{x}_1^3(j)\bar{x}_2(j) - 2\bar{x}_1(j)\bar{x}_2^3(j) \right] / \left[\bar{x}_1^2(j) + \bar{x}_2^2(j) \right] }{ \sum_{j=1}^{N} \left[3\bar{x}_1(j)^2\bar{x}_2^2(j) - (1/2)\bar{x}_1^4(j) - (1/2)\bar{x}_2^4(j) \right] / \left[\bar{x}_1^2(j) + \bar{x}_2^2(j) \right] } \right]. \qquad (16.2.14)$$

When converted to polar coordinates, this reduces very nicely to the following expression:

$$\phi_0 = \frac{1}{4} \tan^{-1} \left[\frac{ \sum_{j=1}^{N} r^2 \sin 4\psi(j) }{ \sum_{j=1}^{N} r^2 \cos 4\psi(j) } \right]. \qquad (16.2.15)$$

The rotation back to the approximately statistically independent square is then given by

$$V = \begin{bmatrix} \cos\phi_0 & \sin\phi_0 \\ -\sin\phi_0 & \cos\phi_0 \end{bmatrix} \qquad (16.2.16)$$

with the angle θ_0 computed directly from the experimental data. At this point, it is unknown whether the angle ϕ_0 is a minimum or maximum of the kurtosis and this should be checked. Sometimes this is a maximum, especially in cases when one of the image histograms has long tails relative to the other [48].

Completing the analysis

Recall that our purpose was to compute, through the statistics of the images (pixels) themselves, the SVD decomposition. The method relied on computing (approximating) the principal axes and scaling of the principal components. The final piece was to try and make the probability distribution as separable as possible by minimizing the kurtosis as a function of the final rotation angle.

To reconstruct the image, the following linear algebra operation is performed:

$$\mathbf{s} = \mathbf{A}^{-1}\mathbf{x} = \mathbf{V}\boldsymbol{\Sigma}^{-1}\mathbf{U}^*\mathbf{x}$$
$$= \begin{bmatrix} \cos\phi_0 & \sin\phi_0 \\ -\sin\phi_0 & \cos\phi_0 \end{bmatrix} \begin{bmatrix} 1/\sqrt{\sigma_1} & 0 \\ 0 & 1/\sqrt{\sigma_2} \end{bmatrix} \begin{bmatrix} \cos\theta_0 & \sin\theta_0 \\ -\sin\theta_0 & \cos\theta_0 \end{bmatrix}\mathbf{x}. \tag{16.2.17}$$

A few things should be noted about the inherent ambiguities in the recovery of the two images. The first is the ordering ambiguity. Namely, the two matrices are indistinguishable:

$$\begin{bmatrix} a_{11} & a_{12} \\ a_{21} & a_{22} \end{bmatrix} \begin{bmatrix} x_1 \\ x_2 \end{bmatrix} = \begin{bmatrix} a_{21} & a_{22} \\ a_{11} & a_{12} \end{bmatrix} \begin{bmatrix} x_2 \\ x_1 \end{bmatrix}. \tag{16.2.18}$$

Thus images one and two are arbitrary to some extent. In practice this does not matter since the aim was simply to separate, not label, the data measurements. There is also an ambiguity in the scaling since

$$\begin{bmatrix} a_{11} & a_{12} \\ a_{21} & a_{22} \end{bmatrix} \begin{bmatrix} x_1 \\ x_2 \end{bmatrix} = \begin{bmatrix} a_{11}/\alpha & a_{12}/\delta \\ a_{21}/\alpha & a_{22}/\delta \end{bmatrix} \begin{bmatrix} \alpha x_1 \\ \delta x_2 \end{bmatrix}. \tag{16.2.19}$$

Again this does not matter since the two separated images can then be rescaled to their full intensity range afterwards. Regardless, this shows the basic process that needs to be applied to the system data in order to construct a self-consistent algorithm capable of image separation.

16.3 Image Separation and MATLAB

Using MATLAB, the previously discussed algorithm will be implemented. However, before proceeding to try and extract the images displayed in Figs. 16.4 and 16.5, a more idealized case is considered. Consider the two ideal images of Sicily in Fig. 16.7. These two images will be mixed

Figure 16.7: Two images of Sicily: an ocean view from the hills of Erice and an interior view of the Capella Palatina in Palermo.

with two different weights to produce two mixed images. Our objective will be to use the algorithm outlined in the last section to separate out the images. Upon separation, the images can be compared with the ideal representation of Fig. 16.7. To load in the images and start the processing, the following MATLAB commands are used.

```
S1=imread('sicily1','jpeg');
S2=imread('sicily2','jpeg');

subplot(2,2,1), imshow(S1);
subplot(2,2,2), imshow(S2);
```

The two images are quite different with one overlooking the Mediterranean from the medieval village of Erice on the northwest corner, and the second is of the Capella Palatina in Palermo. Recall that what is required is statistical independence of the two images. Thus the pixel strengths and colors should be largely decorrelated throughout the image.

The two ideal images are now mixed with the matrix $\mathbf{\Lambda}$ in Eq. (16.1.3). Specifically, we will consider the arbitrarily chosen mixing of the form

$$\mathbf{A} = \begin{bmatrix} a_{11} & a_{12} \\ a_{21} & a_{22} \end{bmatrix} = \begin{bmatrix} 4/5 & \beta \\ 1/2 & 2/3 \end{bmatrix} \tag{16.3.1}$$

where in what follows we will consider the values $\beta = 1/5, 3/5$. This mixing produces two composite images with the strengths of images one and two altered between the two composites. This can be easily accomplished in MATLAB with the following commands.

```
A=[0.8 0.2; 1/2 2/3];   % mixing matrix
X1=double(A(1,1)*S1+A(1,2)*S2);
X2=double(A(2,1)*S1+A(2,2)*S2);

subplot(2,2,1), imshow(uint8(X1));
subplot(2,2,2), imshow(uint8(X2));
```

Note that the coefficients of the mixing matrix $\mathbf{A}$ will have a profound impact on our ability to extract one image from another. Indeed, just switching the parameter β from 1/5 to 3/5 will show the impact of a small change to the mixing matrix. It is these images that we wish to reconstruct by numerically computing an approximation to the SVD. The top rows of Figs. 16.8 and 16.9 demonstrate the mixing that occurs with the two ideal images given the mixing matrix (16.3.1) with $\beta = 1/5$ (Fig. 16.8) and $\beta = 3/5$ (Fig. 16.9).

Figure 16.8: The top two figures show the mixed images produced from Eq. (16.3.1) with $\beta = 1/5$. The bottom two figures are the reconstructed images from the image separation process and approximation to the SVD. For the value of $\beta = 1/5$, the Capella Palatina has been exceptionally well reconstructed while the view from Erice has quite a bit of residual from the second image.

Image statistics and the SVD

The image separation method relies on the statistical properties of the image for reconstructing an approximation to the SVD decomposition. Three steps have been outlined in the last section that must be followed: first the rotation of the parallelogram must be computed by finding the maximal and minimal directions of the variance of the data. Second, the scaling of the principal component directions is evaluated by calculating the variance, and third, the final rotation is computed by minimizing both the variance and kurtosis of the data. This yields an approximately separable probability distribution. The three steps are each handled in turn.

Step 1: Maximal/minimal variance angle detection

This step is illustrated in the top right of Fig. 16.6. The actual calculation that must be performed for calculating the rotation angle is given by either Eq. (16.2.5) or (16.2.6). Once computed, the desired rotation matrix $\mathbf{U}^*$ given by Eq. (16.2.7) can be constructed. Recall the underlying assumption that a mean-zero distribution is considered. In MATLAB, this computation takes the following form.

Figure 16.9: The top two figures show the mixed images produced from Eq. (16.3.1) with $\beta = 3/5$. The bottom two figures are the reconstructed images from the image separation process and approximation to the SVD. For the value of $\beta = 3/5$, the view from Erice has been well reconstructed aside from the top right corner of the image while the Capella Palatina remains of poor quality.

```
[m,n]=size(X1);
x1=reshape(X1,m*n,1);
x2=reshape(X2,m*n,1);
x1=x1-mean(x1); x2=x2-mean(x2);

theta0=0.5*atan( -2*sum(x1.*x2) / sum(x1.^2-x2.^2) )
Us=[cos(theta0) sin(theta0); -sin(theta0) cos(theta0)];
```

This completes the first step in the SVD reconstruction process.

Step 2: Scaling of the principal components

The second step is to undo the scaling/compression that has been performed by the singular values along the principal component directions θ_0 and $\theta_0 - \pi/2$. The rescaling matrix thus is computed as follows:

```
sig1=sum( (x1*cos(theta0)+x2*sin(theta0)).^2 )
sig2=sum( (x1*cos(theta0-pi/2)+x2*sin(theta0-pi/2)).^2 )
Sigma=[1/sqrt(sig1) 0; 0 1/sqrt(sig2)];
```

Note that the variable called sigma above is actually the inverse of the diagonal matrix constructed from the SVD process. This completes the second step in the SVD process.

Step 3: Rotation to separability

The final rotation is aimed towards producing, as best as possible, a separable probability distribution. The analytic method used to do this is to minimize both the variance and kurtosis of the remaining distributions. The angle that accomplishes this task is computed analytically from either (16.2.14) or (16.2.15) and the associated rotation matrix $\mathbf{V}$ is given by (16.2.16). Before computing this final rotation, the image after steps 1 and 2 must be produced, i.e. we compute the $\bar{\mathbf{X}}_1$ and $\bar{\mathbf{X}}_2$ used in (16.2.14) and (16.2.15). The MATLAB code used to produce the final rotation matrix is then

```
X1bar= Sigma(1,1)*(Us(1,1)*X1+Us(1,2)*X2);
X2bar= Sigma(2,2)*(Us(2,1)*X1+Us(2,2)*X2);

x1bar=reshape(X1bar,m*n,1);
x2bar=reshape(X2bar,m*n,1);

phi0=0.25*atan( -sum(2*(x1bar.^3).*x2bar-2*x1bar.*(x2bar.^3)) ...
   / sum(3*(x1bar.^2).*(x2bar.^2)-0.5*(x1bar.^4)-0.5*(x2bar.^4)) )

V=[cos(phi0) sin(phi0); -sin(phi0) cos(phi0)];
S1bar=V(1,1)*X1bar+V(1,2)*X2bar;
S2bar=V(2,1)*X1bar+V(2,2)*X2bar;
```

This final rotation completes the three steps of computing the SVD matrix. However, the images needed to be rescaled back to the full image brilliance. Thus the data points need to be made positive and scaled back to values ranging from 0 (dim) to 255 (bright). This final bit of code rescales and plots the separated images.

```
min1=min(min(min(S1bar)));
S1bar=S1bar-min1;
max1=max(max(max(S1bar)));
S1bar=S1bar*(255/max1);
```

Continued

```
min2=min(min(min(S2bar)));
S2bar=S2bar-min2;
max2=max(max(max(S2bar)));
S2bar=S2bar*(255/max2);

subplot(2,2,3), imshow(uint8(S1bar))
subplot(2,2,4), imshow(uint8(S2bar))
```

Note that the successive use of the **min/max** command is due to the fact that the color images are $m \times n \times 3$ matrices where the three levels give the color mapping information.

Image separation from reflection

The algorithm above can be applied to the original problem presented of the **Judgement of Paris** by Flaxman. The original images are again illustrated in the top row of Fig. 16.10. The difference

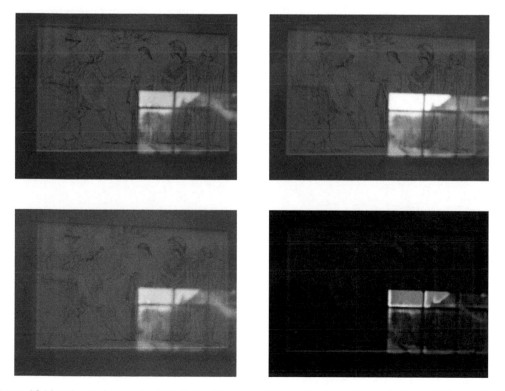

Figure 16.10: The Judgement of Paris by Flaxman with a reflection of my window which overlooks Drumheller Fountain. In the top row are the two original photos taken with the linear polarizer rotated 90 degrees between shots. The bottom row represents the separated images. The brightness of the reflected image greatly affects the quality of the image separation.

between the two figures was the application of a polarizer whose angle was rotated 90 degrees between shots. The bottom two figures shows the extraction process that occurs for the two images. Ultimately, the technique applied here did not do such a great job. This is largely due to the fact that the reflected image was so significantly brighter than the original photographed image itself, thus creating a difficult image separation problem. Certainly experience tells us that we also have this same difficulty: when it is very bright and sunny out, reflected images are much more difficult for us to see through to the true image.

Image Recognition: Basics of Machine Learning

One of the most intriguing mathematical developments of the past decade concerns the ideas of image recognition. In particular, mathematically plausible methods are being rapidly developed to mimic real biological processes towards automating computers to recognize people, animals and places. Indeed, there are many applications, many of them available on smart phones, that can perform sophisticated algorithms for doing the following: tag all images in a photo library with a specific person, and automatically identify the location and name of a set of mountain peaks given a photo of the mountains. These image recognition methods are firmly rooted in the basic mathematical techniques to be described in the following sections. The pace of this technology and its potential for scientific impact is remarkable. The methods here will bring together statistics, time–frequency analysis and the SVD.

Taken as a whole, the dimensionality reduction and statistical methods presented here are a simple example of so-called *machine learning*, otherwise known as *statistical learning*. More broadly, such data classification schemes fall under pattern theory and are becoming a dominant paradigm in computer science for handling *big data*. In this example, a *supervised learning* technique is implemented in which a prescribed set of data is known and classified beforehand. Alternatively, one could modify the code to do *unsupervised learning* so that the code itself learns to classify and group the data in an appropriate fashion. We will not dwell on these issues here, but will simply present an intuitive example of the machine learning architecture.

17.1 Recognizing Dogs and Cats

A basic and plausible methodology will be formulated in the context of image recognition. Specifically, we will generate a mathematically plausible mechanism by which we can

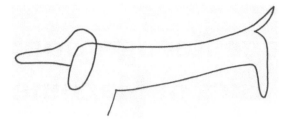

Figure 17.1: Author's sketch of a dog inspired by Picasso's own sketch of a dog. Note the identification of the image occurs simply from drawing the edges of the animal.

discriminate between dogs and cats. Before embarking on this pursuit, however, it is useful to consider our own perception and image recognition capabilities. To some extent, the art world has been far ahead of the mathematicians and scientists on this, often asking the question about how a figure can be minimally represented. For instance, consider the Picasso inspired image represented in Fig. 17.1. With very simple lines, Picasso was able to represent a dog. This example is particular striking given that not only do we recognize a dog, but we recognize the specific dog represented: a dauschung. It is clear from these examples that the concept of edge detection must play a critical role in the image recognition problem. The question to ask is this: could a computer be trained the recognize the figures and animals represented by Picasso in these works.

Picasso certainly pushed much further in the representation of figures and images. Indeed, many of his paintings challenge our ability to perceive the figures within them. But the fact remains, we often do perceive the figures represented even if they are remarkably abstract and distorted from our normal perception of what a human figure should look like, and yet there remains clear indications to our senses of the figures within the abstractions. This highlights the amazingly robust imaging system available to humans.

Wavelet analysis of images and the SVD and LDA

As discussed previously in the time–frequency analysis, wavelets represent an ideal way to represent multi-scale information. As such, wavelets will be used here to represent images of dogs and cats. Specifically, wavelets are tremendously efficient in detecting and highlighting edges in images. In the image detection of cats versus dogs, the edge detection will be the dominant aspect of the discrimination process.

The mathematical objective is to develop an algorithm by which we can train the computer to distinguish between dogs and cats. The algorithm will consist of the following key steps:

- **Step 1** Decompose images of dogs and cats into wavelet basis functions. The motivation is simply to provide an effective method for doing edge detection. The training sets for dogs and cats will be 80 images each.

- **Step 2** From the wavelet expanded images, find the principal components associated with the dogs and cats, respectively.

- **Step 3** Design a statistical decision threshold for discrimination between the dogs and cats. The method to be used here will be a *linear discrimination analysis* (LDA).

- **Step 4** Test the algorithm efficiency and accuracy by running through it 20 pictures of cats and 20 pictures of dogs.

The focus of the remainder of this section will simply be on step 1 of the algorithm and the appliation of a wavelet analysis.

Wavelet decomposition

To begin the analysis, a sample of the dog pictures is presented. The file **dogData.mat** contains a matrix **dog** which is a 4096 × 80 matrix. The 4096 data points correspond to a 64 × 64 resolution image of a dog while the 80 columns are 80 different dogs for the training set of dogs. A similar file exists for the cat data called **catData.mat**. The following command lines load the data and plot the first nine dog faces.

```
load dogData
for j=1:9
subplot(3,3,j)
  dog1=reshape(dog(:,j),64,64);
  imshow(dog1)
end
```

Figure 17.2 demonstrates nine typical dog faces that are used in training the algorithm to recognize dogs versus cats. Note that high-resolution images will not be required to perform this task. Indeed, a fairly low resolution can accomplish the task quite nicely.

These images are now decomposed into a wavelet basis using the discrete wavelet transform in MATLAB. In particular, the **dwt2** command performs a single-level discrete 2D wavelet transform, or decomposition, with respect to either a particular wavelet or particular wavelet filters that you specify. The command

```
[cA,cH,cV,cD]  = DWT2(X,'wname')
```

computes the approximation coefficient matrix **cA** and details coefficient matrices **cH**, **cV**, **cD**, obtained by a wavelet decomposition of the input image matrix **X**. The input **'wname'** is a string containing the wavelet name. Thus the large-scale features are in **cA** and the fine-scale

Figure 17.2: A sampling of nine dog faces to be used in the training set for the dog versus cat recognition algorithm.

features in the horizontal, vertical and diagonal directions are in **cH**, **cV**, **cD**, respectively. The command lines

```
figure(2)
X=double(reshape(dog(:,6),64,64));
[cA,cH,cV,cD]=dwt2(X,'haar');
subplot(2,2,1), imshow(uint8(cA));
subplot(2,2,2), imshow(uint8(cH));
subplot(2,2,3), imshow(uint8(cV));
subplot(2,2,4), imshow(uint8(cD));
```

produce the Haar wavelet decomposition in four 32×32 matrices. Figure 17.3 shows the wavelet decomposition matrices for the sixth dog in the data set. Part of the reason the images are so dark in Fig. 17.3 is that the images have not been rescaled to the appropriate pseudo-color scaling used by the image commands. The following lines of code note only rescale the images,

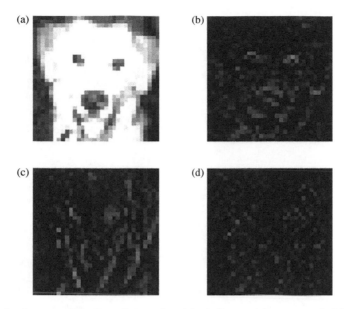

Figure 17.3: Wavelet decomposition into the matrices (a) **cA** (large-scale structures), (b) **cH** (fine scales in the horizontal direction), (c) **cV** (fine scales in the vertical direction), and (d) **cD** (fine scales in the diagonal direction).

but also combine the edge detection wavelets in the horizontal and vertical direction into a single matrix.

```
nbcol = size(colormap(gray),1);
cod_cH1 = wcodemat(cH,nbcol);
cod_cV1 = wcodemat(cV,nbcol);
cod_edge=cod_cH1+cod_cV1;

figure(3)
subplot(2,2,1), imshow(uint8(cod_cH1))
subplot(2,2,2), imshow(uint8(cod_cV1))
subplot(2,2,3), imshow(uint8(cod_edge))
subplot(2,2,4), imshow(reshape(dog(:,6),64,64))
```

The new matrices have been appropriately scaled to the correct image brightnesses so that the images now appear sharp and clean in Fig. 17.4. Indeed, the bottom two panels of this figure show the ideal dog along with the dog represented in its wavelet basis constructed by weighting the vertical and horizontal directions.

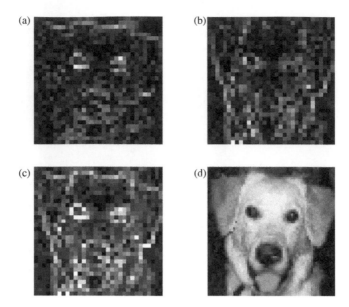

Figure 17.4: This figure represents the (a) horizontal and (b) vertical representations of the wavelet transform in the appropriately scaled image scale. The overall wavelet transformation process is represented in (c) where the edge detection in the horizontal and vertical directions is combined to represent the ideal dog in (d).

The cat data set can similarly be analyzed. Figures 17.5 and 17.6 demonstrate the wavelet decomposition of a cat. Specifically, the bottom two panels of Fig. 17.6 show the ideal cat along with the cat represented in its wavelet basis constructed by weighting the vertical and horizontal directions. Just as before, the analysis was done for the sixth cat in the data set. This choice of cat (and dog) was arbitrary, but it does illustrate the general principle quite nicely. Indeed the bottom right figures of both Fig. 17.4 and 17.6 will form the basis of our statistical decision making process.

17.2   The SVD and Linear Discrimination Analysis

Having decomposed the images of dogs and cats into a wavelet representation, we now turn our attention to understanding the mathematical and statistical distinction between dogs and cats. In particular, steps 2 and 3 in the image recognition algorithm of the previous section will be addressed here. In step 2, our aim will be to decompose our dog and cat data sets via the SVD, i.e. dogs and cats will be represented by principal components or by a proper orthogonal mode decomposition. Following this step, a statistical decision based upon this

Figure 17.5: A sampling of nine cat faces to be used in the training set for the dog versus cat recognition algorithm.

representation will be made. The specific method used here is called a *linear discrimination analysis* (LDA).

To begin, the following main function file is created:

```
load dogData
load catData

dog_wave = dc_wavelet(dog);
cat_wave = dc_wavelet(cat);

feature=20;  %  1<feature<80
[result,w,U,S,V,threshold] = dc_trainer(dog_wave,cat_wave,feature);
```

In this algorithm, the dog and cat data are loaded and run through the wavelet algorithm of the last section. As a subroutine, this takes on the following form

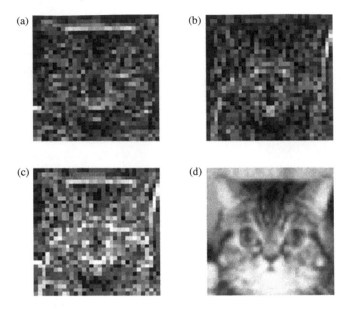

Figure 17.6: This figure represents the (a) horizontal and (b) vertical representations of the wavelet transform in the appropriately scaled image scale. The overall wavelet transformation process is represented in (c) where the edge detection in the horizontal and vertical directions is combined to represent the ideal cat in (d).

```
function dcData = dc_wavelet(dcfile)
[m,n]=size(dcfile);   % 4096 x 80
nw=32*32;   % wavelet resolution
nbcol = size(colormap(gray),1);

for i=1:n
    X=double(reshape(dcfile(:,i),64, 64));
    [cA,cH,cV,cD]=dwt2(X,'haar');
    cod_cH1 = wcodemat(cH,nbcol);
    cod_cV1 = wcodemat(cV,nbcol);
    cod_edge=cod_cH1+cod_cV1;
    dcData(:,i)=reshape(cod_edge,nw,1);
end
```

where is should be recalled that **cA**, **cH**, **cV** and **cD** are the wavelet decompositions with a resolution of 32×32. Upon generating these data, the subroutine **dc_trainer.m** is called. This subroutine will perform and SVD and LDA in order to train itself to recognize dogs and cats. The variable **feature** determines the number of principal components that will be used to classify the dog and cat features.

The SVD decomposition

The first step in setting up the statistical analysis is the SVD decomposition of the wavelet generated data. The subroutine **dc_trainer.m** accepts as input the wavelet decomposed dog and cat data along with the number of PCA/POD components to be considered with the variable **feature**. Upon entry into the subroutine, the dog and cat data are concatenated and the SVD is performed. The following code performs with the reduced SVD to extract the key components **U**, Σ and **V**.

```
function [result,w,U,S,V,th]=dc_trainer(dog0,cat0,feature)
nd=length(dog0(1,:)); nc=length(cat0(1,:));


[U,S,V] = svd([dog0,cat0],0); % reduced SVD


animals = S*V';
U = U(:,1:feature);
dogs = animals(1:feature,1:nd);
cats = animals(1:feature,nd+1:nd+nc);
```

Recall that only a limited number of the principal components are considered in the feature detection as determined by the variable **feature**. Thus upon completion of the SVD step, a limited number of columns of **U** and rows of $\Sigma\mathbf{V}^*$ are extracted. Further, new variables **cats** and **dogs** are generated giving the strength of the cat and dog projections on the PCA/POD modes generated by **U**.

To get a better feeling of what has just happened, the SVD matrices **U**, Σ and **V** are all returned in the subroutine to the main file. In the main file, the decomposition can be visualized. What is of most interest is the PCA/POD that is generated from the cat and dog data. This can be reconstructed with the following code

```
for j=1:4
  subplot(2,2,j)
  ut1=reshape(U(:,j),32,32);
  ut2=ut1(32:-1:1,:);
  pcolor(ut2)
  set(gca,'Xtick',[],'Ytick',[])
end
```

Thus the first four columns, or PCA/POD modes, are plotted by reshaping them back into 32×32 images. Figure 17.7 shows the first four POD modes associated with the dog and cat data. The strength of each of these modes in a given cat or dog can be computed by the projection onto $\Sigma\mathbf{V}^*$. In particular, the singular (diagonal) elements of Σ give the strength of the projection of

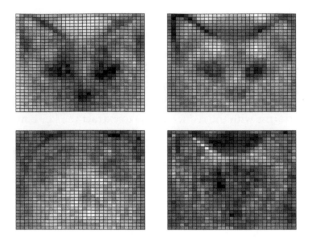

Figure 17.7: Representation of the first four principal components, or proper orthogonal modes, of the cat and dog data. Note the dominance of the triangular ears in the first two modes, suggesting a strong signature of a cat. These modes are used as the basis for classifying a cat or dog image.

each cat onto the POD modes. Figures 17.8 and 17.9 demonstrate the singular values associated with the cat and dog data along with the strength of each cat projected onto the POD modes. These two plots are created from the main file with the commands

```
figure(2)
subplot(2,1,1)
plot(diag(S),'ko','Linewidth',[2])
set(gca,'Fontsize',[14],'Xlim',[0 80])
subplot(2,1,2)
semilogy(diag(S),'ko','Linewidth',[2])
set(gca,'Fontsize',[14],'Xlim',[0 80])

figure(3)
for j=1:3
    subplot(3,2,2*j-1)
    plot(1:40,V(1:40,j),'ko-')
    subplot(3,2,2*j)
    plot(81:120,V(81:120,j),'ko-')
end
subplot(3,2,1), set(gca,'Ylim',[-.15 0],'Fontsize',[14]), title('dogs')
subplot(3,2,2), set(gca,'Ylim',[-.15 0],'Fontsize',[14]), title('cats')

subplot(3,2,3), set(gca,'Ylim',[-.2 0.2],'Fontsize',[14])
subplot(3,2,4), set(gca,'Ylim',[-.2 0.2],'Fontsize',[14])
subplot(3,2,5), set(gca,'Ylim',[-.2 0.2],'Fontsize',[14])
subplot(3,2,6), set(gca,'Ylim',[-.2 0.2],'Fontsize',[14])
```

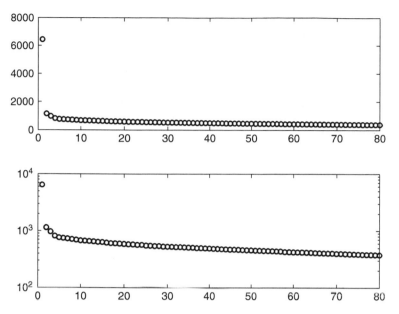

Figure 17.8: Singular values of the SVD decomposition of the dog and cat data. Note the dominance of a single singular value and the heavy-tail distribution (non-Gaussian) fall-off of the singular values.

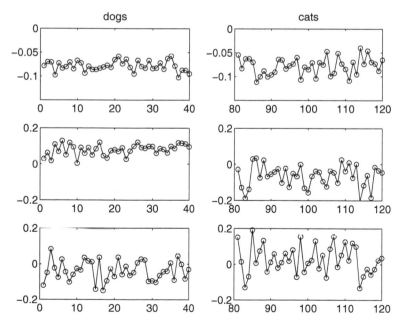

Figure 17.9: Projection of the first 40 individual cats and dogs onto the first three POD modes as described by the SVD matrix **V**. The left column is the first 40 dogs while the right column is the first 40 cats. The three rows of figures are the projections on to the first three POD modes.

Note the dominance of a single mode and the heavy-tail distribution of the remaining singular values. The singular values in Fig. 17.8 are plotted on both a normal and log scale for convenience. Figure 17.9 demonstrates the projection strength of the first 40 individual cats and dogs onto the first three POD modes. Specifically, the first three mode strengths correspond to the rows of the figures. Note the striking difference in cats and dogs as shown in the second mode. Namely, for dogs, the first and second mode are of opposite sign and "cancel" whereas for cats, the signs are the same and the features of the first and second mode add together. Thus for cats, the triangular ears are emphasized. This starts to give some indication of how the discrimination might work.

Linear discrimination analysis

Thus far, two decompositions have been performed: wavelets and SVD. The final piece in the training algorithm is to use statistical information about the wavelet/SVD decomposition in order to make a decision about whether the image is indeed a dog or cat. Figure 17.10 gives a cartoon of the key idea involved in the LDA. The idea of the LDA was first proposed by Fisher [49] in the context of taxonomy. In our example, two data sets are considered and projected onto new bases. In the left figure, the projection shows the data to be completely mixed, not allowing for any reasonable way to separate the data from each other. In the right figure, which is the ideal caricature for LDA, the data are well separated with the means μ_1 and μ_2 being well apart when projected onto the chosen subspace. Thus the goal of LDA is two-fold: *find a suitable projection that maximizes the distance between the inter-class data while minimizing the intra-class data.*

For a two-class LDA, the above idea results in consideration of the following mathematical formulation. Construct a projection $\mathbf{w}$ such that

$$\mathbf{w} = \arg\max_{\mathbf{w}} \frac{\mathbf{w}^T \mathbf{S}_B \mathbf{w}}{\mathbf{w}^T \mathbf{S}_W \mathbf{w}} \tag{17.2.1}$$

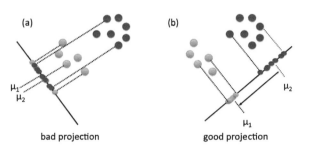

Figure 17.10: Graphical depiction of the process involved in a two-class linear discrimination analysis. Two data sets are projected onto a new basis function. In the left figure, (a), the projection produces highly mixed data and very little distinction or separation can be drawn between data sets. In the right figure, (b), the projection produces the ideal, well-separated statistical distribution between the data sets. The goal is to mathematically construct the optimal projection basis which separates the data most effectively.

where the scatter matrices for between-class $\mathbf{S}_B$ and within-class $\mathbf{S}_W$ data are given by

$$\mathbf{S}_B = (\mu_2 - \mu_1)(\mu_2 - \mu_1)^T \qquad (17.2.2)$$

$$\mathbf{S}_W = \sum_{j=1}^{2} \sum_{\mathbf{x}} (\mathbf{x} - \mu_j)(\mathbf{x} - \mu_j)^T. \qquad (17.2.3)$$

These quantities essentially measure the variance of the data sets as well as the variance of the difference in the means. There are a number of tutorials available online with details of the method, thus the mathematical details will not be presented here. The criterion given by Eq. (17.2.1) is commonly known as the generalized Rayleigh quotient whose solution can be found via the generalized eigenvalue problem

$$\mathbf{S}_B \mathbf{w} = \lambda \mathbf{S}_W \mathbf{w} \qquad (17.2.4)$$

where the maximum eigenvalue λ and its associated eigenvector give the quantity of interest and the projection basis. Thus once the scatter matrices are constructed, the generalized eigenvectors can easily be constructed with MATLAB.

Once the SVD decomposition has been performed in the **dc_trainer.m** subroutine, the LDA is applied. The following code produces the LDA projection basis as well as the decision threshold level associated with the dog and cat recognition.

```
md = mean(dogs,2);
mc = mean(cats,2);

Sw=0;   % within class variances
for i=1:nd
    Sw = Sw + (dogs(:,i)-md)*(dogs(:,i)-md)';
end
for i=1:nc
    Sw = Sw + (cats(:,i)-mc)*(cats(:,i)-mc)';
end

Sb = (md-mc)*(md-mc)';   % between class

[V2,D] = eig(Sb,Sw);   % linear discriminant analysis
[lambda,ind] = max(abs(diag(D)));
w = V2(:,ind);   w = w/norm(w,2);

vdog = w'*dogs; vcat = w'*cats;
result = [vdog,vcat];
```

Continued

```
if mean(vdog)>mean(vcat)
    w = -w;
    vdog = -vdog;
    vcat = -vcat;
end
% dog < threshold < cat

sortdog = sort(vdog);
sortcat = sort(vcat);

t1 = length(sortdog);
t2 = 1;
while sortdog(t1)>sortcat(t2)
    t1 = t1-1;
    t2 = t2+1;
end
threshold = (sortdog(t1)+sortcat(t2))/2;
```

This code ends the subroutine **dc_trainer.m** since the statistical threshold level has been determined along with the appropriate LDA basis for projecting a new image of a dog or cat. For **feature = 20** as carried through the codes presented here, a histogram can be constructed of the dog and cat statistics.

```
figure(4)
subplot(2,2,1)
hist(sortdog,30); hold on, plot([18.22 18.22],[0 10],'r')
set(gca,'Xlim',[-200 200],'Ylim',[0 10],'Fontsize',[14])
title('dog')
subplot(2,2,2)
hist(sortcat,30,'r'); hold on, plot([18.22 18.22],[0 10],'r')
set(gca,'Xlim',[-200 200],'Ylim',[0 10],'Fontsize',[14])
title('cat')
```

The above code not only produces a histogram of the statistical distributions associated with the dog and cat properties projected to the LDA basis, it also shows the decision threshold that is computed for the statistical decision making problem of categorizing a given image as a dog or cat (see Fig. 17.11). Note that as is the case in most statistical processes, the threshold level shows that some cats are mistaken as dogs and vice versa. The more features that are used, the fewer "mistakes" are made in the recognition process.

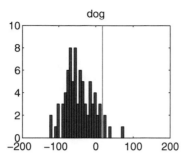
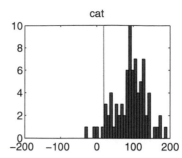

Figure 17.11: Histogram of the statistics associated with the dog and cat data once projected onto the LDA basis. The red line is the numerically computed decision threshold. Thus a new image would be projected onto the LDA basis and a decision would be made concerning the image depending upon which side of the threshold line it sits.

17.3   Implementing Cat/Dog Recognition in MATLAB

Up to now, we have performed three key steps in training or organizing our dog and cat data images: we have (i) wavelet decomposed the images, (ii) considered this wavelet representation and its corresponding SVD/PCA/POD, and (iii) projected these results onto the subspace provided by the linear discrimination analysis. The three steps combined are the basis of the training set of the dog versus cat recognition problem.

It remains now to test the training set (and machine learning) on a sample of dog and cat pictures outside of the training data set. The data set **PatternRecAns.mat** contains 38 new pictures of dogs and cats for us to test our recognition algorithm on. Specifially, two files are contained within this data set: **TestSet** which is a 4096 × 38 matrix containing the 38 pictures of resolution 64 × 64, and **hiddenlabels** which is a 1 × 38 vector containing either 1s (cat) or 0s (dog) identifying the image as a true dog or cat. To see part of the sample set of test pictures, the following lines of code are used.

```
load PatternRecAns

figure(1)
for j=1:9
 subplot(3,3,j)
 test=reshape(TestSet(:,j+5),64,64);
 imshow(uint8(test))
end
```

Figure 17.12: Nine sample images of dogs and cats that will be classified by the wavelet, SVD, LDA scheme. The second dog in this set (sixth picture) will be misclassified due to its ears.

Figure 17.12 shows nine or the 38 test images of dogs and cats to be processed by our wavelet, SVD, LDA algorithm.

To classify the images as dogs or cats, the new data set must now undergo the same process as the training images. Thus the following will occur

- Decompose the new image into the wavelet basis.
- Project the images onto the SVD/PCA/POD modes selected from the training algorithm.
- Project this two-level decomposition onto the LDA eigenvector.
- Determine the projection value relative to the LDA threshold determined in the training algorithm.
- Classify the image as dog or cat.

The following lines of code perform the first three tasks above.

```
Test_wave = dc_wavelet(TestSet);  % wavelet transformation
TestMat = U'*Test_wave;           % SVD projection
pval = w'*TestMat;                % LDA projection
```

Note that the subroutine **dc_wavelet.m** is considered in the previous section. Further, it is assumed that the training algorithm of the last section has been run so that both the SVD matrix **U** has been computed, and the first 20 columns (features) have been preserved, and the eigenvector **w** of the LDA has been computed. Note that the above code performs the three successive transformations necessary for this problem: wavelet, SVD, LDA.

It only remains to see how the recognition algorithm worked. The variable **pval** contains the final projection to the LDA subspace and it must be compared with the threshold determined by the LDA. The following code produces a new vector **ResVec** containing 1s and 0s depending upon wether **pval** is above or below threshold. The vector **ResVec** is then compared with **hiddenlabels** which has the true identification of the dog and cat data. These are then compared for accuracy.

```
%cat = 1, dog = 0
ResVec = (pval>threshold)

disp('Number of mistakes');
errNum = sum(abs(ResVec - hiddenlabels))

disp('Rate of success');
sucRate = 1-errNum/TestNum
```

In addition to identifying and classifying the test data, the number of mistakes, or misidentified, images is reported and the percentage rate of success is given. In this algorithm, a nearly 95% chance of success is given which is quite remarkable given the low-resolution, 2D images and the simple structure of the algorithm.

The mistaken images in the 38 test images can now be considered. In particular, by understanding why mistakes where made in identification, better algorithms can be constructed. The following code identifies which images where mistaken and plots the results.

```
k = 1;
TestNum = length(pval);
figure(2)
for i = 1:TestNum
    if ResVec(i)~=hiddenlabels(i)
        S = reshape(TestSet(:,i),64,64);
        subplot(2,2,k)
        imshow(uint8(S))
        k = k+1;
    end
end
```

Figure 17.13: Of the 38 images tested, two mistakes are made. The two dogs above were both misclassified as cats. Looking at the POD of the dog/cat decomposition gives a clue to why the mistake was made: they have triangular ears. Indeed, the image recognition algorithm developed here favors, in some sense, recognition of cats versus dogs simply due to the strong signature of ears that cats have. This is the primary feature picked out in the edge detection of the wavelet expansion.

Figure 17.13 shows the two errors that were made in the identification process. These two dogs have the quintessential features of cats: triangular ears. Recall that even in the training set, the decision threshold was set at a value for which some cats and dogs where misclassified. It is these kinds of features that can give the misleading results, suggesting that edge detection is not the only mechanism we use to identify cats versus dogs. Indeed, one can see that much more sophisticated algorithms could be used based upon the size of the animal, color, coat pattern, walking cadence, etc. Of course, the more sophisticated the algorithm, the more computationally expensive. This is a fast, low-resolution routine that produces 95% accuracy and can be easily implemented.

Basics of Compressed Sensing

In most dimensionality reduction applications, a least-square fitting of the data in higher dimensions is traditionally applied due to both (i) well-developed factorization algorithms like the SVD and (ii) the intuitive appeal offered by an *energy* norm. However, this does not necessarily make such an L^2 based reduction the best choice. Indeed, recent progress in the field of *compressed sensing* has shown that there is great advantage in reduction techniques based upon the L^1 norm instead. This will be explored in a series of three sections where sparse sampling and an L^1 data projection can be used quite effectively for signal and image reconstruction.

18.1 Beyond Least-Square Fitting: The L^1 Norm

Understanding of compressed sensing comes from building an intuitive feel for data fitting and the norms we use to measure our error in such a process. In typical applications, a least-square fitting algorithm is used as the standard measure of error. However, other options are available beyond this standard choice. It should be noted that this standard choice is often chosen since the L^2 norm represents an *energy* norm and often it has a clear physical interpretation.

Curve fitting: L^2 versus L^1

To begin understanding, let us first consider a standard curve fitting problem. Specifically, consider a best fit curve through a set of data points (x_i, y_i) where $i = 1, 2, \ldots, N$. This is a fairly standard problem that involves an underlying optimization routine. Section 2.6 outlines the basic ideas which are reviewed here.

To solve this problem, it is desired to minimize some error between the line fit

$$y(x) = Ax + B \qquad (18.1.1)$$

and the data points (x_i, y_i). Thus an error can be defined as follows:

$$E_p = \left(\frac{1}{N} \sum_{n=1}^{N} |y(x_i) - y_i|^p \right)^{1/p} \qquad (18.1.2)$$

where p denotes the error norm to consider. For $p = 2$, this gives the standard least-square error E_2. However, other choices are possible and we will consider here $p = 1$ as well.

In order to solve this problem, an optimization routine must be performed to determine the coefficients A and B in the line fit. What is specifically required is to determine A and B from the minimization constraints: $\partial E_p / \partial A = 0$ and $\partial E_p / \partial B = 0$. For a line fit, these are trivial to calculate for either the L^2 or L^1 norms. However, when higher-degree of freedom fits are involved, the optimization routine can be costly. For L^2, the higher-degree of freedom fit is generated in $O(N^3)$ via a the singular value decomposition. For L^1, a different optimization routine is required. In the end, these curve fitting techniques are simply methods for solving overdetermined systems of equations, i.e. there are N constraints/equations and two unknowns in A and B. Learning how to solve such overdetermined systems is of generic importance.

The difference in L^2 and L^1 norms is most readily apparent when considering data with large variance or many outliers. The L^2 norm squares the distance to these outlying points so that they have a big impact on the curve fit. In contrast, the L^1 norm is much less susceptible to outliers in data as the distance to these points is not squared, but rather only the absolute distance is considered. Thus the L^1 curve fit is highly robust to outliers in the data.

To demonstrate the difference between L^2 and L^1 line fitting, consider the following data set of (x_i, y_i) pairs

```
x=[0.1 0.4 0.7 1.2 1.3 1.7 2.2 2.8 3.0 4.0 4.3 4.4 4.9];
y=[0.5 0.9 1.1 1.5 1.5 2.0 2.2 2.8 2.7 3.0 3.5 3.7 3.9];
```

A line fit can be performed for this data set using both L^2 and L^1 norm minimization using the **fminsearch** command. By generically guessing values of $A = B = 1$, the following two **fminsearch** routines will converge to optimal values of A and B that minimize the L^2 and L^1 norm, respectively.

```
coeff_L2=fminsearch('line_L2_fit',[1,1],[],x,y)
coeff_L1=fminsearch('line_L1_fit',[1,1],[],x,y)
```

Note that these two function calls return 2×1 vectors with the updated values of A and B. The functions look like the following:

line_L2_fit.m

```
function E=line_L2_fit(x0,x,y)
E=sum( ( x0(1)*x+x0(2)  - y ).^2 )
```

line_L1_fit.m

```
function E=line_L1_fit(x0,x,y)
E=sum( abs( x0(1)*x+x0(2)  - y ) )
```

Note that in each of these, $p = 2$ or $p = 1$, respectively. This makes clear that **fminsearch** is simply an optimization routine and can be used for far more than curve fitting.

Figure 18.1 (top panel) demonstrates the line fitting process to the data proposed here. In this first example, which is devoid of outliers in the data, both the L^2 and L^1 fit look reasonably similar. Indeed, hardly any difference is seen between the dashed and dotted lines. However, if additional outliers are added to the data set (two outlying points):

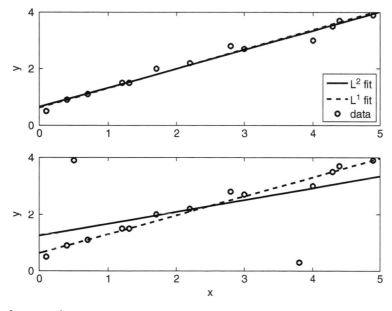

Figure 18.1: L^2 versus L^1 data fitting for data without outliers (top) and with outliers (bottom). Note the significant skewing of the data that occurs with the L^2 fitting algorithm (solid line) versus the L^1 algorithm (dotted line). The raw data are denoted with circles.

```
x=[0.5 3.8];
y=[3.9 0.3];
```

then significant differences can be seen between the L^2 and L^1 fits. This is demonstrated in Fig. 18.1 (bottom panel). Note how the L^2 fit is significantly skewed by the outlying data while the L^1 data remain robust and locked along the bulk of the data. One should keep this example in mind for future reference as the L^2 and L^1 minimization process is effectively what is underlying SVD and compressed sensing in higher dimensions.

Underdetermined systems

Continuing on this theme, consider now solving an underdetermined system of equations. As a simple example, solve $\mathbf{Ax} = \mathbf{b}$ with

```
A=[1/2 1/4 3 7 -2 10];
b=[3];
```

Of course, this has an infinite number of solutions. So one may ask how would an algorithm decide on what solution to pick. Two options that are immediately available to generate a solution are the backslash command and the pseudo-inverse (for underdetermined systems)

```
x1=A\b
x2=pinv(A)*b
```

The solutions for both these two cases are strikingly different. For the backslash, the solution produced is sparse, i.e. it is almost entirely all zeros.

```
x1=
         0
         0
         0
         0
         0
    0.3000
```

In contrast, the solution produced with the pseudo-inverse produces the following solution vector

```
x2=
   0.0092
   0.0046
   0.0554
   0.1294
  -0.0370
   0.1848
```

Thus no sparsity is exhibited in the solution structure.

This example can be scaled up to a much more complex situation. In particular, consider the 100×500 system where there are 100 equations and 500 unknowns. As before, these can be solved using the backslash or pseudo-inverse commands.

```
m=100; n=500;
A=randn(m,n);
b=randn(m,1);

x1=A\b;
x2=pinv(A)*b;
```

In addition to the backslash and pseudo-inverse, L^2 or L^1 optimization can also be applied to the problem. That is, solve the system subject to minimizing the L^2 or L^1 norm of the solution vector. There are a number of open source convex optimization codes that can be downloaded from the Internet. In the example here, a convex optimization package is used that can be directly implemented with MATLAB: **http://cvxr.com/cvx/**. In the implementation of the code, the following lines of code generate a solution that minimizes the L^1 residual and L^2 residual, respectively

```
cvx_begin;
   variable x3(n);
   minimize( norm(x3,1) );
   subject to
     A*x3 == b;
cvx_end;
```

Continued

```
cvx_begin;
  variable x4(n);
  minimize( norm(x4,2) );
  subject to
    A*x4 == b;
cvx_end;
```

Figure 18.2 demonstrates the use of all four methods for solving this underdetermined system, i.e. backslash, pseudo-inverse, L^1 minimization and L^2 minimization. What is plotted in this figure is a histogram of the values of the solution vector. Note that L^1 optimization and the backslash command both produce solutions that are sparse, i.e. there are a large number of zeros in the solution. Specifically, both these techniques produce 400 zeros and a remaining 100 nonzero elements that satisfy the remaining equations. In contrast, the L^2 optimization and pseudo-inverse produce identical, nonsparse solution vectors.

Figure 18.3 shows a detail of the backslash and L^1 optimization routines. The purpose of this figure is to show that the backslash and L^1 do indeed produce different results. Note that the backslash histogram is still much more tightly clustered around the zero solution while the L^1 minimization allows for larger variance in the nonzero solutions. It is this solution sparsity that will play a critical role in the ability to reconstruct signals and/or images from sparse sampling.

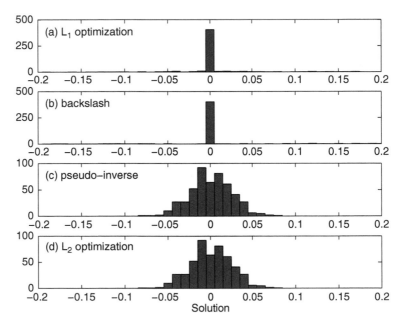

Figure 18.2: Histogram of the solution of an underdetermined system with 100 equations and 500 unknowns. The backslash and L^1 minimization produce sparse solution vectors while the pseudo-inverse and L^2 minimization do not.

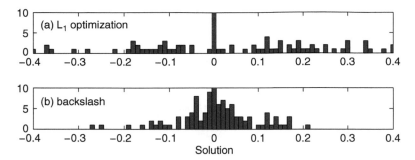

Figure 18.3: Blow-up of the L^1 minimization and backslash solution histograms demonstrating the difference the two solution methods produce.

Overdetermined systems

This example can also be modified to consider overdetermined systems of equations. In particular, consider the 500 × 150 system where there are 500 equations and 150 unknowns. As before, these can be solved using the backslash or pseudo-inverse commands.

```
m=500; n=150;
A=randn(m,n);
b=randn(m,1);

x1=A\b;
x2=pinv(A)*b;
```

In addition to the backslash and pseudo-inverse, L^2 or L^1 optimization can also be applied to the problem. That is, solve the system subject to minimizing the L^2 or L^1 norm of the solution vector. In the implementation of the code, the following lines of code generate a solution that minimizes the L^1 and L^2 norm, respectively

```
cvx_begin;
    variable x3(n);
    minimize( norm(A*x3-b,1) );
cvx_end;
cvx_begin;
    variable x4(n);
    minimize( norm(A*x4-b,2) );
cvx_end;
```

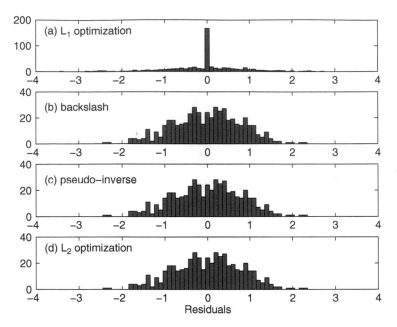

Figure 18.4: Histogram of the solution residual of an overdetermined system with 500 equations and 150 unknowns. The L^1 minimization produces sparse solution vectors while the pseudo-inverse, backslash and L^2 minimization do not.

As before, the sparsity of the resulting solution can be illustrated using all four solution methods. Figure 18.4 demonstrates the use of all four methods for solving this overdetermined system, i.e. backslash, pseudo-inverse, L^1 minimization and L^2 minimization. What is plotted in this figure is a histogram of the solution residual vector. Note that L^1 optimization produces solutions that are sparse, i.e. there are a large number of zeros in the residual. In contrast, the L^2 optimization, backslash and pseudo-inverse produce identical, nonsparse solution vectors. Sparsity again plays a critical role in the ability to solve the resulting overdetermined system. The critical idea in compressed sensing is to take advantage of sparsity of signals or images in order to perform reconstruction. Indeed, the L^1 norm should be thought of as a proxy for sparsity as it promotes sparsity in its solution technique of over- or underdetermined systems.

 ## Signal Reconstruction and Circumventing Nyquist

To illustrate the power of the L^1 norm, and ultimately of compressive sensing, it is informative to consider the reconstruction of a temporal signal. This example [50] makes connection directly to signal processing. However, it is fundamentally the same issue that was considered in the last section. Therefore, consider the "A" key on a touch-tone telephone which is the sum of two sinusoids with incommensurate frequencies [50]:

$$f(t) = \sin(1394\pi t) + \sin(3266\pi t). \tag{18.2.1}$$

To begin, the signal will be sampled over 1/8 of a second at 40 000 Hz. Thus a vector of length 5000 will be generated. The signal can also be represented using the discrete cosine transform (DCT). In MATLAB, the following lines are sufficient to produce the signal and its DCT:

```
n=5000;
t=linspace(0,1/8,n);
f=sin(1394*pi*t)+sin(3266*pi*t);
ft=dct(f);
```

As can be seen from the form of the signal, there should be dominance by two Fourier modes in (18.2.1). However, since the frequencies are incommensurate and do not fall within the frequencies spanned by the DCT, there are a number of nonzero DCT coefficients. Figures 18.5 and 18.6 show the original signal over 1/8 of a second and a blow-up of the signal in both time and frequency. The key observation to make: *The signal is highly sparse in the frequency domain.*

The objective now will be to randomly *sample* the signal at a much lower frequency (i.e. sparse sampling) and try to reconstruct the original signal, in both frequency and time, as best as possible. Such an exercise is at the heart of compressive sensing. Namely, given that the signal is sparse (in frequency) to begin with, can we sample sparsely and yet faithfully reconstruct the desired signal [51, 52, 53]? Indeed, compressive sensing algorithms are data acquisition protocols which perform as if, where possible, to directly acquire just the important information about the signal, i.e. the sparse signal representation in the DCT domain. In effect, the idea is to acquire only the important information and ignore that part of the data which effectively makes no contribution to the signal. Figure 18.5 clearly shows that most of the information in the original signal can be neglected when considered from the perspective of the DCT domain. Thus the sampling algorithm must find a way to key in on the prominent features of the signal in the frequency domain. Since the L^1 norm promotes sparsity, this will be the natural basis in which to work to reconstruct the signal from our sparse sampling [51, 52, 53].

The matrix problem

Before proceeding too far along, it is important to relate the current exercise to the linear algebra problems considered in the last section. Specifically, this signal reconstruction problem is nothing more than a large underdetermined system of equations. To be more precise, the conversion from the time domain to frequency domain via the DCT can be thought of as a linear transformation

$$\psi \mathbf{c} = \mathbf{f} \tag{18.2.2}$$

where $\mathbf{f}$ is the signal vector in the time domain (plotted in the top panels of Figs. 18.5 and 18.6) and $\mathbf{c}$ are the cosine transform coefficients representing the signal in the DCT domain (plotted in the bottom panels of Figs. 18.5 and 18.6). The matrix ψ represents the DCT transform itself.

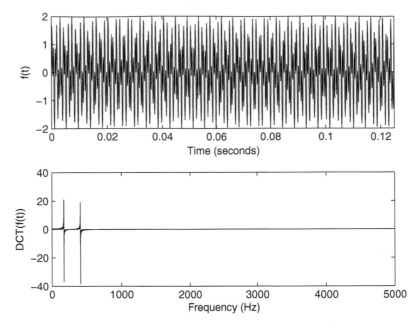

Figure 18.5: Original "A"-tone signal and its discrete cosine transform sampled at 40 000 Hz for 1/8 second.

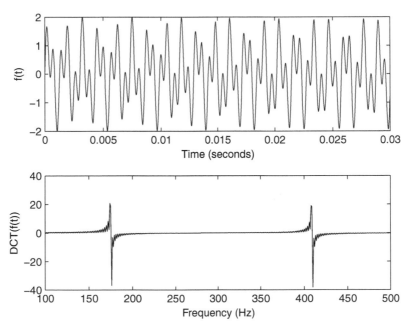

Figure 18.6: Zoom-in of the original "A"-tone signal and its discrete cosine transform sampled at 40 000 Hz for 1/8 second.

The key observation is that most of the coefficients of the vector **c** are zero, i.e. it is sparse as clearly demonstrated by the dominance of the two cosine coefficients at $\approx$175 Hz and $\approx$410 Hz. Note that the matrix ψ is of size $n \times n$ while **f** and **c** are $n \times 1$ vectors. For the example plotted, $n = 5000$.

The choice of basis functions is critical in carrying out the compressed sensing protocol. In particular, the signal must be sparse in the chosen basis. For the example here of a cosine basis, the signal is clearly sparse, allowing us to accurately reconstruct the signal using sparse sampling. The idea is to now sample the signal randomly (and sparsely) so that

$$\mathbf{b} = \phi \mathbf{f} \tag{18.2.3}$$

where **b** is a few (m) random samples of the original signal **f** (ideally $m \ll n$). Thus ϕ is a subset of randomly permuted rows of the identity operator. More complicated sampling can be performed, but this is a simple example that will illustrate all the key features. Note here that **b** is an $m \times 1$ vector while the matrix ϕ is of size $m \times n$.

Approximate signal reconstruction can then be performed by solving the linear system

$$\mathbf{A}\mathbf{x} = \mathbf{b} \tag{18.2.4}$$

where **b** is an $m \times 1$ vector, **x** is $n \times 1$ vector and

$$\mathbf{A} = \phi \psi \tag{18.2.5}$$

is a matrix of size $m \times n$. Here the **x** is the sparse approximation to the full DCT coefficient vector. Thus for $m \ll n$, the resulting linear algebra problem is highly underdetermined as in the last section. The idea is then to solve the underdetermined system using an appropriate norm constraint that best reconstructs the original signal. As already demonstrated, the L^1 norm promotes sparsity and is highly appropriate given the sparsity already demonstrated. The signal reconstruction is performed by using

$$\mathbf{f} \approx \psi \mathbf{x}. \tag{18.2.6}$$

If the original signal had exactly m nonzero coefficients, the reconstruction could be made exact.

Signal reconstruction

To begin the reconstruction process, the signal must first be randomly sampled. In this example, the original signal **f** with $n = 5000$ points will be sampled at 10% so that the vector **b** above will have $m = 500$ points. The following code randomly rearranges the integers 1 to 5000 (**randintrlv**) so that the vector **perm** retrieves 500 random sample points.

```
m=500;
r1 = randintrlv([1:n],793);
perm=r1(1:m);
```

Continued

```
f2=f(perm);
t2=t(perm);
```

In this example, the resulting vectors **t2** and **f2** are the 500 random point locations (out of the 5000 original). The $m \times n$ matrix **A** is then constructed by constructing $\mathbf{A} = \phi\psi$.

```
D=dct(eye(n,n));
A=D(perm,:);
```

Recall that the matrix ψ was the DCT transform while the matrix ϕ was the permutation matrix of the identity yielding the sampling points for **f2**.

Although the L^1 norm is of primary importance, the underdetermined system will be solved in three ways: with the pseudo-inverse, the backslash and with L^1 optimization. It was already illustrated in the last section that the backslash promoted sparsity for such systems, but as will be illustrated, all sparsity is not the same! The following commands generate all three solutions for the underdetermined system.

```
x=pinv(A)*f2';
x2=A\f2';
cvx_begin;
    variable x3(n);
    minimize( norm(x3,1) );
    subject to
    A*x3 == f2';
cvx_end;
```

The above code generates three vectors approximating the DCT coefficients, i.e. these solutions are used for the reconstruction (18.2.6). Figure 18.7 illustrates the results of the reconstructed signal in the DCT domain over the range of 100–500 Hz. The top panel illustrates the original signal in the DCT domain while the next three panels represent the approximation to the DCT coefficients using the L^1 norm optimization, the pseudo-inverse and the backslash, respectively. As expected, both the L^1 and backslash methods produce sparse representations. However, the backslash produces a result that looks nothing like the original DCT signal. In contrast, the pseudo-inverse does not produce a sparse result. Indeed, the spectral content is quite full and only the slightest indication of the importance of the cosine coefficients at $\approx$175 Hz and $\approx$410 Hz is given.

To more clearly see the results of the cosine coefficient reconstruction, Fig. 18.8 zooms in near the $\approx$410 Hz peak of the spectrum. The top panel shows the excellent reconstruction yielded by

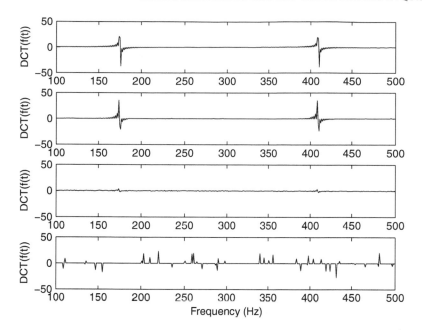

Figure 18.7: Original signal in the DCT domain (top panel) and the reconstructions using L^1 optimization (second panel), the pseudo-inverse (third panel) and the backslash (fourth panel), respectively. Note that the L^1 optimization does an exceptional job at reconstructing the spectral content with only 10% sampling. In contrast, both the pseudo-inverse and backslash fail miserably at the task.

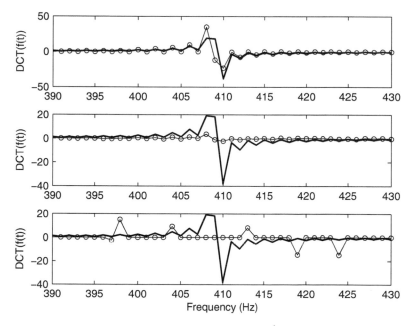

Figure 18.8: Zoom-in of the DCT domain reconstructions using L^1 optimization (top panel), the pseudo-inverse (middle panel) and the backslash (bottom panel), respectively. Here a direct comparison is given to the original signal (bold line). Although both the L^1 optimization and backslash produce sparse results, it is clear the backslash produces a horrific reconstruction of the signal.

the L^1 optimization routine. The pseudo-inverse (middle panel) and backslash fail to capture any of the key features of the spectrum. This example shows the almost "magical" ability of the L^1 optimization to reconstruct the original signal in the cosine domain.

It remains to consider the signal reconstruction in the time domain. To perform this task, the DCT domain data must be transformed to the time domain via the discrete cosine transform:

```
sig1=dct(x);
sig2=dct(x2);
sig3=dct(x3);
```

As before, the L^1 optimization will be compared to the standard L^2 schemes of the pseudo-inverse and backslash. Figure 18.9 illustrates the original signal (top panel) along with the L^1 optimization, pseudo-inverse and backslash methods, respectively. Even at 10% sampling, the L^1 does a very nice job in reconstructing the signal. In contrast, both L^2 based methods fail severely in the reconstruction. Note, however, that in this case the pseudo-inverse does promote

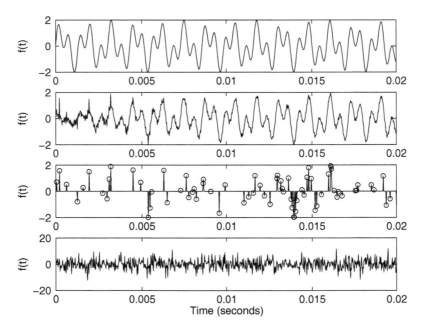

Figure 18.9: Original time domain signal (top panel) and the reconstructions using L^1 optimization (second panel), the pseudo-inverse (third panel) and the backslash (fourth panel), respectively. Note that the L^1 optimization does an exceptional job at reconstructing the signal with only 10% sampling. In contrast, both the pseudo-inverse and backslash fail. Note that in the pseudo-inverse reconstruction (third panel), the original sparse sampling points are included, demonstrating that the method does well in keeping the L^2 error in check for these points. However, the L^2 methods do poorly elsewhere.

sparsity in the time domain. Indeed, when the original sparse sampling points are overlaid on the reconstruction, it is clear that the reconstruction does a very nice job of minimizing the L^2 error for these points. In contrast, the backslash produces a solution which almost resembles white noise. However, as with the pseudo-inverse, if considering only the sampling points, the L^2 error is quite low.

Finally, Fig. 18.10 demonstrates the effect of sampling on the signal reconstruction. In what was considered in the previous plots, the sampling was $m = 500$ points out of a possible $n = 5000$, thus 10% sparse sampling was sufficient to approximate the signal. The sampling is lowered to see the deterioration of the signal reconstructions. In particular, Fig. 18.10 demonstrates the signal reconstruction using $m = 100$ (second panel), 300 (third panel) and 500 (bottom panel) sampling points. In the top panel, the original signal is included for reference. In subsequent panels, the original signal is included as the dotted line. Also included is the sparse sample points (circles). This provides a naked eye measure of the effectiveness of the compressed sensing algorithm and L^1 optimization method. At $m = 500$ (bottom panel), the reconstruction is quite faithful to the original signal. As m is lowered, the signal recovery becomes worse. But remarkably at even 2% sampling ($m = 100$), many of the key features are still visible in the signal reconstruction. This is especially remarkable given the amazing sparsity in the sampling. For instance, between ≈ 0.006 and 0.012 Hz there are no sampling points, yet the compressed sensing

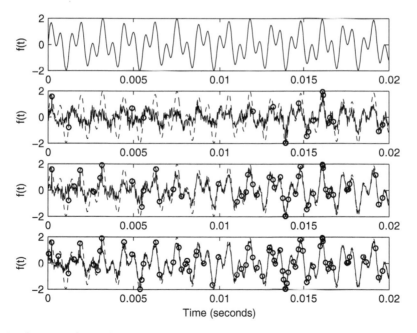

Figure 18.10: The original signal (top panel) and the reconstruction using $m = 100$ (second panel), 300 (third panel) and 500 (bottom panel) sampling points. In the top panel, the original signal is included for reference. In subsequent panels, the original signal is included as the dotted line. Also included is the sparse sample points (circles).

scheme still picks up the key oscillatory features in this range. This is in direct contrast to the thinking and intuition established by Nyquist. Indeed, Nyquist (or Nyquist–Shannon sampling theory) states that to completely recover a signal, one must sample at twice the rate of the highest frequency of the signal. This is the gold standard of the information theory community. Here, however, the compressed sensing algorithm seems to be Nyquist. The apparent contradiction is overcome by one simple idea: sparsity. Specifically, Nyquist–Shannon theory assumes in its formulation that the signal is dense in the frequency domain. Thus it is certainly true that an accurate reconstruction can only be obtained by sampling at twice the highest frequency. Here, however, compressed sensing worked on a fundamental assumption about the sparsity of the signal. This is a truly amazing idea that is having revolutionary impact across many disciplines of science [51, 52, 53].

Data (Image) Reconstruction from Sparse Sampling

Compressed sensing can now be brought to the arena of image reconstruction. Much like the signal recovery problem, the idea is to exploit the sparse nature of images in order to reconstruct them using greatly reduced sampling. These concepts also make natural connection to image compression algorithms. What is of particular importance is choosing a basis representation in which images are, in fact, sparse.

To illustrate the above concepts more clearly, we can consider a digital image with 600×800 pixels (rows × columns). Figure 18.11(a) shows the original image. The following MATLAB code imports the original JPEG picture and plots the results. It also constructs the matrix **Abw** which is the picture in matrix form.

```
A=imread('photo','jpeg');
Abw2=rgb2gray(A);
Abw=double(Abw2);
[nx,ny]=size(Abw);
figure(1), subplot(2,2,1), imshow(Abw2)
```

The image is well defined and of high resolution. Indeed, it contains $600 \times 800 = 480\,000$ pixels encoding the image. However, the image can actually be encoded with far less data as illustrated with the JPEG wavelet compression algorithm (see Figs. 13.39–13.41). It has also been shown that the image is highly compressed (or sparse) in the Fourier basis as well (See Fig. 14.1). In what follows, the compressed sensing algorithm will be applied using the Fourier mode basis.

To demonstrate the nearly optimal encoding of the picture in the Fourier domain, the original photo is subjected to a two-dimensional Fourier transformation. The Fourier coefficients are reshaped into a single vector where the dominant portion of the signal is plotted on both a regular axis and semi-log axis:

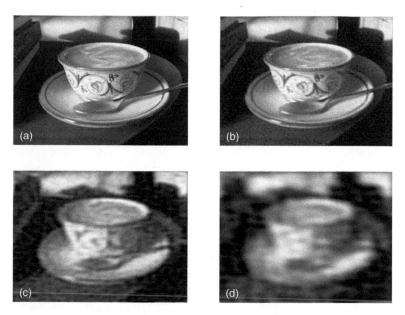

Figure 18.11: Original image (800×600 pixels) (a) along with the reconstructed image using approximately (b) 1.4%, (c) 0.2% or (d) 0.06% of the information encoded in the Fourier coefficients. At 0.06%, the image reconstruction starts to fail. This reconstruction process illustrates that images are sparse in the Fourier representation.

```
At=fftshift(fft2(Abw));
figure(2)
subplot(2,1,1)
plot(reshape(abs(At),nx*ny,1),'k','Linewidth',[2])
set(gca,'Xlim',[2.3*10^5 2.5*10^5],'Ylim',[-0.5*10^7 5*10^7])

subplot(2,1,2)
semilogy(reshape(abs(At),nx*ny,1),'k','Linewidth',[2])
set(gca,'Xlim',[2.3*10^5 2.5*10^5],'Ylim',[-0.5*10^7 5*10^7])
```

The **fft2** command performs a two-dimensional transform and returns the Fourier coefficient matrix **At** which is reshaped into an $nx \times ny$ length vector.

Figure 18.12 shows the Fourier coefficients over the dominant range of Fourier mode coefficients. In addition to the regular y-axis scale, a semi-log plot is presented showing the low-amplitude (nonzero) energy in each of the Fourier mode coefficients. One can clearly see that many of the coefficients that make up the image are nearly zero. Due to numerical round-off, for instance, none of the coefficients will actually be zero.

Image compression follows directly from this Fourier analysis. Specifically, one can simply choose a threshold and directly set all wavelet coefficients below this chosen threshold to be

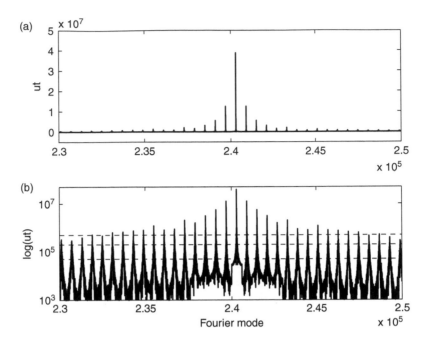

Figure 18.12: (a) Amplitude of the Fourier mode coefficients (in vector form) around the dominant Fourier modes. Due to the small amplitude of many of the Fourier modes, the amplitudes are also shown on a semi-log scale (b). In panel (b), the three threshold lines (dotted) are demonstrated for reconstructing the images in Fig. 18.11(b)–(d). The percentage of nonzero Fourier coefficients for the three threshold lines is approximately 1.4%, 0.2% and 0.06%, respectively. Only the nonzero coefficients are kept for encoding of the image.

identically zero instead of nearly zero. What remains is to demonstrate that the image can be reconstructed using such a strategy. If it works, then one would only need to retain the much smaller (sparse) number of nonzero wavelets in order to store and/or reconstruct the image.

To address the image compression issue, we can apply a threshold rule to the Fourier coefficient vector **At** that is shown in Fig. 18.12. The following code successively sets a threshold at three different levels (dotted lines in Fig. 18.12) in order to generate a sparse Forier coefficient vector **At2**. This new sparse vector is then used to encode and reconstruct the image.

```
count_pic=2;
for thresh=[0.005*10^7 0.02*10^7 0.05*10^7];
  At2=reshape(At,nx*ny,1);
  count=0;
  for j=1:length(At2);
      if abs(At2(j)) < thresh
          At2(j)=0;
          count=count+1;
```

Continued

```
      end
   end
   percent=100-count/length(At2)*100

   Atlow=fftshift(reshape(At2,nx,ny)); Alow=uint8(ifft2(Atlow));
   figure(1), subplot(2,2,count_pic), imshow(Alow);
   count_pic=count_pic+1;
end
```

The results of the compression process are illustrated in Figs. 18.11(b)–(d) where only 1.4%, 0.2% and 0.06% of the modes are retained, respectively. The variable **percent** in the code keeps track of the percentage of nonzero modes in the sparse vector **At2**. Remarkably, the image can be very well reconstructed with only 1.4% of the modes, thus illustrating the sparse nature of the image. Such sparsity is ubiquitous for image representation using wavelets or Fourier coefficients. Thus the idea is to take advantage of this fact. *Much like representing dynamics in a POD basis, the aim is always to take advantage of low dimsional structure. This is exactly what compressed sensing does!*

Compressed sensing

Compressed sensing aims to take advantage of sparsity in information. In the wavelet or Fourier basis, images are clearly sparse and thus compressed sensing can play a definitive role. Specifically, note what the image compression algorithm does: (i) first the image is taken, (ii) a wavelet/Fourier transform is applied in order to apply a threshold rule, and (iii) the majority of information is promptly discarded. In the example here, 98.6% of the information is directly discarded and a faithful reconstruction can be produced. Overall, it seems quite astonishing to go through all the effort of acquiring the image only to discard 98.6% of what was procured. Could we instead sparse sample the image and faithfully reconstruct the image instead? In particular, of the 480 000 pixels sampled, could we instead randomly sample, let's say 5% (24 000 pixels), of the pixels and still reconstruct the image? Knowing that the image is sparse allows us to do just that. Specifically, we already know that the L^1 optimization routines already considered allow for a faithful reconstruction of sparse data.

To make the problem of a more manageable size, and to simply illustrate the process of compressed sensing, a greatly reduced image resolution will be considered since the processing time for the 800 × 600 image is significant. Thus the image if first resized to 100 × 75:

```
B=imresize(A,[75 100]);

Abw2=rgb2gray(B);
Abw=double(Abw2);
[ny,nx]=size(Abw)
```

At this size, the image is not nearly as sparse as the high-resolution picture, thus the compressed sensing algorithm is not as ideal as it would be in a more realistically applied problem. Regardless, the algorithm developed here is generic and allows for a signal reconstruction using sparse information.

The algorithm begins by first deciding how many samples of the image to take. Thus much like the signal problem of the last section, a random number of pixels will be sampled and an attempt will be made to completely reconstruct the image. The following code samples the image in 2500 randomly chosen pixel locations. The matrix **Atest2** shows the sample mask. At 2500 random points, only 1/3 of the image is sampled for the reconstruction.

```
k=2500;  % number of sparse samples
Atest2=zeros(ny,nx);
r1 = randintrlv([1:nx*ny],793);
r1k= r1(1:k);
for j=1:k
   Atest2(r1k(j))=-1;
end
imshow(Atest2), caxis([-1 0])
```

The matrix **Atest2** is what is illustrated in Fig. 18.13. The black dots represent the randomly chosen sampling points on the original image.

As was already stated, by using a low-resolution image with only 100×75 pixels, the image is no longer remarkably sparse as with the 800×600 original. Thus more Fourier coefficients are

Figure 18.13: Typical realization of a random sampling of the image with 2500 random pixel points (black). The white pixels, which comprise 2/3 of the pixels, are not sampled. It should be noted that for images, the sampling mask can be extremely important in the image reconstruction [54].

required for reconstructing the image. The following code works much as the signal construction problem of the last section. Indeed, the image will now be represented in the Fourier domain using the two-dimensional discrete transform. Thus the randomly chosen sampling points, which are treated as delta functions, are responsible for producing the matrix leading to the highly undetermined system.

```
Atest=zeros(ny,nx);
for j=1:k
    Atest(r1k(j))=1;
    Adel=reshape(idct2(Atest),nx*ny,1);
    Adelta(j,:)=Adel;
    Atest(r1k(j))=0;
end
b=Abw(r1k).';
```

This code produces two key quantities: the matrix **Adelta** (which is 2500 × 7500) and the vector **b** (which is 2500 × 1). The objective is then to solve the underdetermined system **Adelta y = b**, where the vector **y** (which is 7500 × 1) is the sparse recovery of the discrete cosine coefficients. The sparse recovery of this vector is done using the L^1 optimization.

```
n=nx*ny;
cvx_begin;
    variable y(n);
    minimize( norm(y,1) );
    subject to
        Adelta*y == b;
cvx_end;

Alow=uint8((dct2(reshape(y,ny,nx))));
figure(1), subplot(2,2,2), imshow(Alow)
```

As before, convex optimization is now performed to recover the discrete cosine coefficients. These are in turn used to reconstruct the original image.

Figure 18.14 demonstrates the image recovery using the compressed sensing format. Ultimately, the image recovery is not stellar given the sizeable number of pixels used. However, recall that the signal considered is already of low-resolution, thus losing some of its sparsity. Furthermore, the random sampling is not optimal for recovering the image. Specifically, it is often desirable to apply a sampling mask in order to enhance the compressed sensing algorithm [54]. Regardless, the above code demonstrates the key features required in the compressed sensing algorithm.

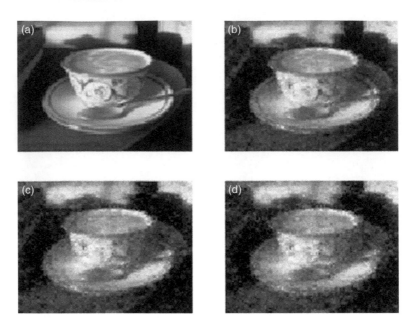

Figure 18.14: Compressed sensing recovery of the original image (a) which has 7500 pixels. Panels (b)–(d) demonstrate the image recovery using $k = 4000, 3000$ and 2500 random sample points of the image, respectively.

Sampling locations

The above example can be greatly improved by sampling with an appropriate sampling mask. As is the case with most images expressed in the Fourier basis, the dominant Fourier modes correspond to low-frequency content. Indeed, most high-frequency content can be easily neglected without much loss to the image quality. Thus when sampling, it would be highly advantageous to sample in the frequency domain with heavier sampling near the zero wavenumber. Candes and Romberg [54] have built such a sampling mask in their L^1-magic MATLAB routines. Specifically, they sample along diagonals in the frequency domain, thus allowing for dense sampling near the zero wavenumbers. Figure 18.15 demonstrates an efficient sampling of a particular image along diagonal lines in the Fourier domain. Here the construction is done by sampling only a small fraction of the total pixels.

Wavelet basis

Although we demonstrated the compressed sensing algorithm with standard Fourier techniques, largely because most people have easy access to Fourier transforms via MATLAB, the compressive sensing is perhaps even more effective using wavelets. However, when using wavelets, the sampling must be performed in a very different way. Unlike the sampling using Fourier modes, which takes a delta function in the image space in order to create global spanning of the Fourier

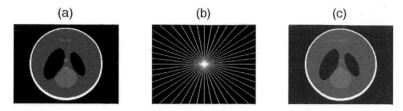

Figure 18.15: Compressed sensing recovery of the original image (a) which has 256×256 pixels. The sampling mask in the Fourier domain (22 diagonals) is shown in (b) and the reconstructed image sampled sparsely along the Fourier diagonals is shown in (c).

space, wavelets require something different since they are highly localized both in space and time. In particular, global and noisy sampling modes are required to perform the task in the most efficient manner [51, 56]. It is in this context that the compressive sensing can make serious performance gains in relation to what is demonstrated in the current example.

The one-pixel camera

One can imagine taking this idea of compressed sensing further. Indeed, various research groups in image processing coupled with compressive sensing have proposed the idea of the one-pixel camera [55]. The single-pixel camera compresses the image data via its hardware before the pixels are recorded. As a result, it is able to capture an image with only thousands of pieces of information rather than millions.

The key benefit of such a camera is that it needs much less information to assemble an image. Massive CCD arrays collect millions of pixels worth of data, which are typically compressed to keep file sizes manageable. It's an approach the Rice researchers describe as "acquire first, ask questions later" [55]. Many pictures, however, have portions that contain relatively little information, such as a clear blue sky or a snowy white background. Conventional cameras record every pixel and later eliminate redundancy with compression algorithms. The idea is to eliminate the redundancy ahead of time.

Currently, the single-pixel camera is simply not competitive with standard digital cameras, mostly because of the processing (L^1 optimization) that must be performed in acquiring the image. Specifically, it takes about 15 minutes to record a reasonable photo with the single pixel. However, one potential payoff of this sort of compressed sensing camera is that it may make conventional digital cameras much better. If a single pixel can do the job of an array of pixels, then you could potentially get each of the pixels in a megapixel camera to do extra duty as well. Effectively, you can multiply the resolution of your camera with the techniques developed in a single-pixel camera.

19 Dimensionality Reduction for Partial Differential Equations

One of the most important uses of computational methods and techniques is in its applications to modeling of physical, engineering or biological systems that demonstrate spatio-temporal dynamics and patterns. For instance, the field of fluid dynamics fundamentally revolves around being able to predict the time and space dynamics of a fluid through some potentially complicated geometry. Understanding the interplay of spatial patterns in time is indeed often the central focus of the field of partial differential equations. In systems where the underlying spatial patterns exhibit low dimensional (pattern forming) dynamics, then the application of the proper orthogonal decomposition can play a critical role in predicting the resulting low dimensional dynamics.

 ## 19.1 Modal Expansion Techniques for PDEs

Partial differential equations (PDEs) are fundamental to the understanding of the underlying physics, biophysics, etc. of complex dynamical systems. In particular, a partial differential equation gives the governing evolution of a system in both time and space. Abstractly, this can be represented by the following PDE system:

$$\mathbf{U}_t = N(\mathbf{U}, \mathbf{U}_x, \mathbf{U}_{xx}, \cdots, x, t) \tag{19.1.1}$$

where $\mathbf{U}$ is a vector of physically relevant quantities and the subscripts t and x denote partial differentiation. The function $N(\cdot)$ captures the space–time dynamics that is specific to the system being considered. Along with this PDE are some prescribed boundary conditions and initial conditions. In general, this can be a complicated, nonlinear function of the quantity $\mathbf{U}$, its derivatives and both space and time.

Figure 19.1: Basic solution technique for solving a partial differential equation. The first step is to reduce it to an appropriate set of ordinary differential equations and then algebra. If the reduction can be undone, then a complete analytic solution can be produced. The starred quantity, i.e. algebra, is the goal of any PDE or ODE reduction.

In general, there are no techniques for building analytic solutions for (19.1.1). However for specific forms of $N(\cdot)$, such as the case where the function is linear in $\mathbf{U}$, time-indenpendent and constant coefficient, then standard PDE solution techniques may be used to characterize the dynamics analytically. Such solution methods are the core of typical PDE courses at both the graduate and undergraduate level. The specific process for solving PDE systems is illustrated in Fig. 19.1 where a reduction from a PDE to ODE is first performed and then a reduction from an ODE to algebra is enacted. Provided such reductions can be achieved and undone, then an analytic solution is produced.

To be more specific about the solution method concept in Fig. 19.1, a number of standard solution techniques can be discussed. The first to be highlighted is the technique of self-similarity reduction. This method simply attempts to extract a fundamental relationship between time and space by making a change of variable to, for instance, a new variable such as

$$\xi = x^{\beta} t^{\alpha} \tag{19.1.2}$$

where β and α are determined by making the change of variable in (19.1.1). This effectively performs the first step in the reduction process by reducing the PDE to an ODE in the variable ξ alone. A second, and perhaps the most common introductory technique, is the method of separation of variables where the assumption

$$\mathbf{U}(x, t) = \mathbf{F}(x)\mathbf{G}(t) \tag{19.1.3}$$

is made, thus effectively separating time and space in the solution. When applied to (19.1.1) with $N(\cdot)$ being linear in $\mathbf{U}$, two ODEs result: one for the time dynamics $\mathbf{G}(t)$ and one for the spatial dynamics $\mathbf{F}(x)$. If the function $N(\cdot)$ is nonlinear in $\mathbf{U}$, then separation cannot be achieved.

As a specific example of the application of the method of separation, we consider here the eigenfunction expansion technique. Strictly speaking, to make analytic progress with the method of separation of variables, one must require that separation occurs. However, no such requirement needs to be made if we are simply using the method as the basis of a computational technique. In particular, the eigenfunction expansion technique assumes a solution of the form

$$u(x, t) = \sum_{n=1}^{\infty} a_n(t)\phi_n(x) \tag{19.1.4}$$

where the $\phi_n(x)$ are an orthogonal set of eigenfunctions and we have assumed that the PDE is now described by a scalar quantity $u(x, t)$. The $\phi_n(x)$ can be any orthogonal set of functions such that

$$(\phi_j(x), \phi_k(x)) = \delta_{jk} = \begin{cases} 1 & j = k \\ 0 & j \neq q \end{cases} \tag{19.1.5}$$

where δ_{jk} is the Dirac function and the notation $(\phi_j, \phi_k) = \int \phi_j \phi_k^* dx$ gives the inner product. For a given physical problem, one may be motivated to use a specified set of eigenfunctions such as those special functions that arise for specific geometries or symmetries. More generally, for computational purposes, it is desired to use a set of eigenfunctions that produce accurate and rapid evaluation of the solutions of (19.1.1). Two eigenfunctions immediately come to mind: the Fourier modes and Chebyshev polynomials. This is largely in part due to their spectral accuracy properties and tremendous speed advantages such as $O(N \log N)$ speed in projection to the spectral basis.

To give a specific demonstration of this technique, consider the nonlinear Schrödinger (NLS) equation

$$iu_t + \frac{1}{2}u_{xx} + |u|^2 u = 0 \tag{19.1.6}$$

with the boundary conditions $u \rightarrow 0$ as $x \rightarrow \pm\infty$. If not for the nonlinear term, this equation could be solved easily in closed form. However, the nonlinearity mixes the eigenfunction components in the expansion (19.1.4) making a simple analytic solution not possible.

To solve the NLS computationally, a Fourier mode expansion is used. Thus use can be made of the standard fast Fourier transform. The following code formulates the PDE solution as an eigenfunction expansion technique (19.1.4) of the NLS (19.1.6). The first step in the process is to define an appropriate spatial and time domain for the solution along with the Fourier frequencies present in the system. The following code produces both the time and space domain of interest:

```
% space
L=40; n=512;
x2=linspace(-L/2,L/2,n+1); x=x2(1:n);
k=(2*pi/L)*[0:n/2-1 -n/2:-1].';
% time
t=0:0.1:10;
```

It now remains to consider a specific spatial configuration for the initial condition. For the NLS, there are a set of special initial conditions called solitons where the initial conditions are given by

$$u(x, 0) = N\text{sech}(x) \tag{19.1.7}$$

where N is an integer. We will consider the soliton dynamics with $N = 1$ and $N = 2$, respectively. In order to do so, the initial condition is projected onto the Fourier modes with the fast Fourier transform. Rewriting (19.1.6) in the Fourier domain, i.e. Fourier transforming, gives the set of differential equations

$$\hat{u}_t = -\frac{i}{2}k^2\hat{u} + i|u|^2 u \tag{19.1.8}$$

where the Fourier mode mixing occurs due to the nonlinear mixing in the cubic term. This gives the system of differential equations to be solved for in order to evaluate the NLS behavior. The following code solves the set of differential equations in the Fourier domain.

```
% initial conditions
N=1;
u=N*sech(x);
ut=fft(u);
[t,utsol]=ode45('nls_rhs',t,ut,[],k);
for j=1:length(t)
    usol(j,:)=ifft(utsol(j,:));   % bring back to space
end

subplot(2,2,1), waterfall(x,t,abs(usol)), colormap([0 0 0])
subplot(2,2,2), waterfall(fftshift(k),t,abs(fftshift(utsol)))
```

In the solution code, the following right-hand side function **nls_rhs.m** is called that effectively produces the differential equation (19.1.8):

```
function rhs=nls_rhs(t,ut,dummy,k)
u=ifft(ut);
rhs=-(i/2)*(k.^2).*ut+i*fft( (abs(u).^2).*u );
```

This gives a complete code that produces both the time–space evolution and the time–frequency evolution.

The dynamics of the $N = 1$ and $N = 2$ solitons are demonstrated in Figs. 19.2 and 19.3, respectively. During evolution, the $N = 1$ soliton only undergoes phase changes while its amplitude remains stationary. In contrast, the $N = 2$ soliton undergoes periodic oscillations. In both cases, a large number of Fourier modes, about 50 and 200, respectively, are required to model the simple behaviors illustrated.

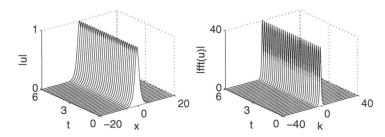

Figure 19.2: Evolution of the $N = 1$ soliton. Although it appears to be a very simple evolution, the phase is evolving in a nonlinear fashion and approximately 50 Fourier modes are required to model this behavior accurately.

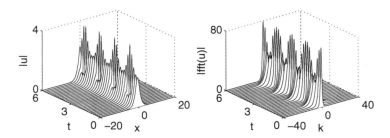

Figure 19.3: Evolution of the $N = 2$ soliton. Here periodic dynamics is observed and approximately 200 Fourier modes are required to model the behavior accurately.

The potentially obvious question to ask in light of our dimensionality reduction thinking is this: is the soliton dynamics really a 50 or 200 degree-of-freedom system as implied by the Fourier mode solution technique? The answer is no. Indeed, with the appropriate basis, i.e. the POD modes generated from the SVD, it can be shown that the dynamics is a simple reduction to one or two modes, respectively. Indeed, it can easily be shown that the $N = 1$ and $N = 2$ are truly low dimensional by computing the singular value decomposition of the evolutions shown in Figs. 19.2 and 19.3, respectively. Indeed, just as in the example of Section 4.5, the simulations can be aligned in a matrix **A** where the columns are different segments of time and the rows are the spatial locations.

Figures 19.4 and 19.5 demonstrate the low dimensional nature explicitly by computing the singular values of the numerical simulations along with the modes to be used in our new eigenfunction expansion. What is clear is that for both of these cases, the dynamics is truly low dimensional with the $N = 1$ soliton being modeled exceptionally well by a single POD mode while the $N = 2$ dynamics is modeled quite well with two POD modes. Thus in performing the expansion (19.1.4), the modes chosen should be the POD modes generated from the simulations themselves. In the next section, we will derive the dynamics of the modal interaction for these two cases and show that quite a bit of analytic progress can then be made within the POD framework.

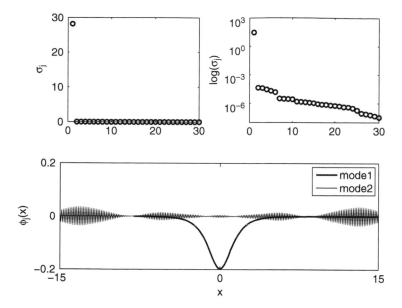

Figure 19.4: Projection of the $N = 1$ evolution to POD modes. The top two figures are the singular values σ_j of the evolution demonstrated in Fig. 19.2 on both a regular scale and log scale. This demonstrates that the $N = 1$ soliton dynamics is primarily a single mode dynamics. The first two modes are shown in the bottom panel. Indeed, the second mode is meaningless and is generated from noise and numerical round-off.

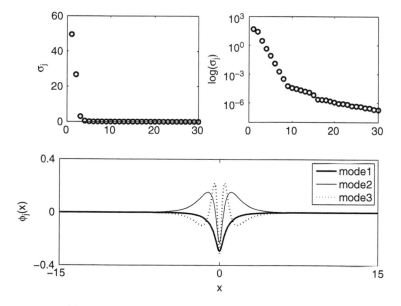

Figure 19.5: Projection of the $N = 2$ evolution to POD modes. The top two figures are the singular values σ_j of the evolution demonstrated in Fig. 19.3 on both a regular scale and log scale. This demonstrates that the $N = 2$ soliton dynamics is primarily a two-mode dynamics as these two modes contain approximately 95% of the evolution energy. The first three modes are shown in the bottom panel.

19.2   PDE Dynamics in the Right (Best) Basis

The last section demonstrated the $N = 1$ and $N = 2$ soliton dynamics of the NLS equation (19.1.6). In particular, Figs. 19.2 and 19.3 showed the evolution dynamics in the Fourier mode basis that was used in the computation of the evolution. The primary motivation in performing a POD reduction of the dynamics through (19.1.4) is the observation in Figs. 19.4 and 19.5 that both dynamics are truly low dimensional with only one or two modes playing a dominant role, respectively.

N = 1 soliton reduction

To take advantage of the low dimensional structure, we first consider the $N = 1$ soliton dynamics. Figure 19.2 shows that a single mode in the SVD dominates the dynamics. This is the first column of the **U** matrix (see Section 4.5). Thus the dynamics is recast in a single mode so that

$$u(x, t) = a(t)\phi(x) \tag{19.2.1}$$

where now there is a sum in (19.1.4) since there is only a single mode $\phi(x)$. Plugging in this equation into the NLS equation (19.1.6) yields the following:

$$ia_t\phi + \frac{1}{2}a\phi_{xx} + |a|^2 a|\phi|^2\phi = 0. \tag{19.2.2}$$

The inner product is now taken with respect to ϕ which gives

$$ia_t + \frac{\alpha}{2}a + \beta|a|^2 a = 0 \tag{19.2.3}$$

where

$$\alpha = \frac{(\phi_{xx}, \phi)}{(\phi, \phi)} \tag{19.2.4a}$$

$$\beta = \frac{(|\phi|^2\phi, \phi)}{(\phi, \phi)} \tag{19.2.4b}$$

and the inner product is defined as integration over the computational interval so that $(\phi, \psi) = \int \phi\psi^* dx$. Note that the integration is over the computational domain and the $*$ denotes the complex conjugate.

The differential equation for $a(t)$ given by (19.2.3) can be solved explicitly to yield

$$a(t) = a(0) \exp\left[i\frac{\alpha}{2}t + \beta|a(0)|^2 t\right] \tag{19.2.5}$$

where $a(0)$ is the initial value condition for $a(t)$. To find the initial condition, recall that

$$u(x, 0) = \text{sech}(x) = a(0)\phi(x). \tag{19.2.6}$$

Taking the inner product with respect to $\phi(x)$ gives

$$a(0) = \frac{(\text{sech}(x), \phi)}{(\phi, \phi)}.$$ (19.2.7)

Thus the one-mode expansion gives the approximate PDE solution

$$u(x, t) = a(0) \exp\left[i\frac{\alpha}{2}t + \beta|a(0)|^2 t\right]\phi(x).$$ (19.2.8)

This solution is the low dimensional POD approximation of the PDE expanded in the best basis possible, i.e. the SVD determined basis.

For the $N = 1$ soliton, the spatial profile remains constant while its phase undergoes a non-linear rotation. The POD solution (19.2.8) can be solved exactly to give a characterization of this phase rotation. Figure 19.6 shows both the full PDE dynamics along with its one-mode approximation. In this example, the parameters α, β and $a(0)$ were computed using the following code:

```
% one soliton match
phi_xx=ifft(-(k.^2).*fft(U(:,1)));
norm=trapz(x,U(:,1).*conj(U(:,1)));
A0=trapz(x,(sech(x).').*conj(U(:,1)))/norm;
alpha=trapz(x,U(:,1).*phi_xx)/norm;
beta=trapz(x,(U(:,1).*conj(U(:,1))).^2)/norm;
```

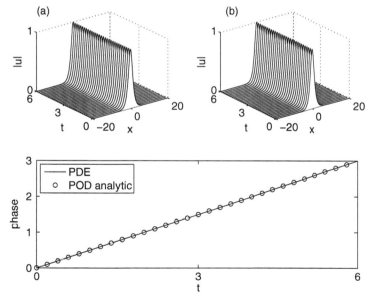

Figure 19.6: Comparison of (a) full PDE dynamics and (b) one-mode POD dynamics. In the bottom figure, the phase of the pulse evolution at $x = 0$ is plotted for the full PDE simulations and the one-mode analytic formula (circles) given in (19.2.8).

This suggest that the $N = 1$ behavior is indeed a single mode dynamics. This is also known to be true from other solution methods, but it is a critical observation that the POD method for PDEs reproduces all of the expected dynamics.

$N = 2$ soliton reduction

The case of the $N = 2$ soliton is a bit more complicated and interesting. In this case, two modes clearly dominate the behavior of the system. These two modes are the first two columns of the matrix $\mathbf{U}$ and are now used to approximate the dynamics observed in Fig. 19.3. In this case, the two-mode expansion takes the form

$$u(x, t) = a_1(t)\phi_1(x) + a_2(t)\phi_2(x) \tag{19.2.9}$$

where the ϕ_1 and ϕ_2 are simply taken from the first two columns of the $\mathbf{U}$ matrix in the SVD. Inserting this approximation into the governing equation (19.1.6) gives

$$i\left(a_{1t}\phi_1 + a_{2t}\phi_2\right) + \frac{1}{2}\left(a_1\phi_{1xx} + a_2\phi_{2xx}\right) + (a_1\phi_1 + a_2\phi_2)^2(a_1^*\phi_1^* + a_2^*\phi_2^*) = 0. \tag{19.2.10}$$

Multiplying out the cubic term gives the equation

$$\begin{aligned}
& i\left(a_{1t}\phi_1 + a_{2t}\phi_2\right) + \frac{1}{2}\left(a_1\phi_{1xx} + a_2\phi_{2xx}\right) \\
& + \left(|a_1|^2 a_1 |\phi_1|^2 \phi_1 + |a_2|^2 a_2 |\phi_2|^2 \phi_2 + 2|a_1|^2 a_2 |\phi_1|^2 \phi_2 + 2|a_2|^2 a_1 |\phi_2|^2 \phi_1 \right. \\
& \left. + a_1^2 a_2^* \phi_1^2 \phi_2^* + a_2^2 a_1^* \phi_2^2 \phi_1^* \right).
\end{aligned} \tag{19.2.11}$$

All that remains is to take the inner product of this equation with respect to both $\phi_1(x)$ and $\phi_2(x)$. Recall that these two modes are orthogonal, thus the resulting 2×2 system of nonlinear equations results

$$\begin{aligned}
& ia_{1t} + \alpha_{11}a_1 + \alpha_{12}a_2 + \left(\beta_{111}|a_1|^2 + 2\beta_{211}|a_2|^2\right)a_1 \\
& \quad + \left(\beta_{121}|a_1|^2 + 2\beta_{221}|a_2|^2\right)a_2 + \sigma_{121}a_1^2 a_2^* + \sigma_{211}a_2^2 a_1^* = 0
\end{aligned} \tag{19.2.12a}$$

$$\begin{aligned}
& ia_{2t} + \alpha_{21}a_1 + \alpha_{22}a_2 + \left(\beta_{112}|a_1|^2 + 2\beta_{212}|a_2|^2\right)a_1 \\
& \quad + \left(\beta_{122}|a_1|^2 + 2\beta_{222}|a_2|^2\right)a_2 + \sigma_{122}a_1^2 a_2^* + \sigma_{212}a_2^2 a_1^* = 0
\end{aligned} \tag{19.2.12b}$$

where

$$\alpha_{jk} = (\phi_{j_{xx}}, \phi_k)/2 \tag{19.2.13a}$$

$$\beta_{jkl} = \left(|\phi_j|^2 \phi_k, \phi_l\right) \tag{19.2.13b}$$

$$\sigma_{jkl} = \left(\phi_j^2 \phi_k^*, \phi_l\right) \tag{19.2.13c}$$

and the initial values of the two components are given by

$$a_1(0) = \frac{(2\text{sech}(x), \phi_1)}{(\phi_1, \phi_1)} \tag{19.2.14a}$$

$$a_2(0) = \frac{(2\text{sech}(x), \phi_2)}{(\phi_2, \phi_2)}. \tag{19.2.14b}$$

This gives a complete description of the two-mode dynamics predicted from the SVD analysis.

The 2×2 system (19.2.12) can be easily simulated with any standard numerical integration algorithm (e.g. fourth-order Runge–Kutta). Before computing the dynamics, the inner products given by α_{jk}, β_{jkl} and σ_{jkl} must be calculated along with the initial conditions $a_1(0)$ and $a_2(0)$. Starting from the code of the previous section and the SVD decomposition, these quantities can be computed as follows:

```
% compute the second derivatives
phi_1_xx=  ifft( -(k.^2).*fft(U(:,1))  );
phi_2_xx=  ifft( -(k.^2).*fft(U(:,2))  );

% compute the norms of the SVD modes
norm1=trapz(x,U(:,1).*conj(U(:,1)));
norm2=trapz(x,U(:,2).*conj(U(:,2)));

% compute the initial conditions
A0_1=trapz(x,(2.0*sech(x).').*conj(U(:,1)))/norm1;
A0_2=trapz(x,(2.0*sech(x).').*conj(U(:,2)))/norm2;

% compute inner products alpha, beta, sigma
alpha11=trapz(x,conj(U(:,1)).*phi_1_xx)/norm1;
alpha12=trapz(x,conj(U(:,1)).*phi_2_xx)/norm1;
beta11_1=trapz(x,(U(:,1).*conj(U(:,1))).^2)/norm1;
beta22_1=trapz(x,(U(:,2).*(abs(U(:,2)).^2).*conj(U(:,1))))/norm1;
beta21_1=trapz(x,(U(:,1).*(abs(U(:,2)).^2).*conj(U(:,1))))/norm1;
beta12_1=trapz(x,(U(:,2).*(abs(U(:,1)).^2).*conj(U(:,1))))/norm1;
sigma12_1=trapz(x,(conj(U(:,2)).*(U(:,1).^2).*conj(U(:,1))))/norm1;
sigma21_1=trapz(x,(conj(U(:,1)).*(U(:,2).^2).*conj(U(:,1))))/norm1;

alpha21=trapz(x,conj(U(:,2)).*phi_1_xx)/norm2;
alpha22=trapz(x,conj(U(:,2)).*phi_2_xx)/norm2;
beta11_2=trapz(x,(U(:,1).*(abs(U(:,1)).^2).*conj(U(:,2))))/norm2;
beta22_2=trapz(x,(U(:,2).*(abs(U(:,2)).^2).*conj(U(:,2))))/norm2;
beta21_2=trapz(x,(U(:,1).*(abs(U(:,2)).^2).*conj(U(:,2))))/norm2;
beta12_2=trapz(x,(U(:,2).*(abs(U(:,1)).^2).*conj(U(:,2))))/norm2;
sigma12_2=trapz(x,(conj(U(:,2)).*(U(:,1).^2).*conj(U(:,2))))/norm2;
sigma21_2=trapz(x,(conj(U(:,1)).*(U(:,2).^2).*conj(U(:,2))))/norm2;
```

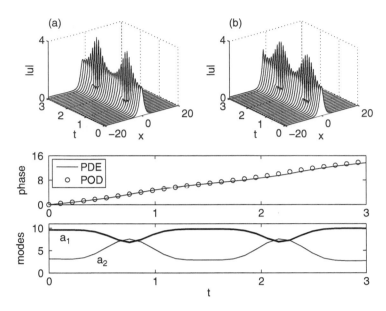

Figure 19.7: Comparison of (a) full PDE dynamics and (b) two-mode POD dynamics. In the bottom two figures, the phase of the pulse evolution at $x = 0$ (middle panel) is plotted for the full PDE simulations and the two-mode dynamics along with the nonlinear oscillation dynamics of $a_1(t)$ and $a_2(t)$ (bottom panel).

Upon completion of these calculations, *ode45* can then be used to solve the system (19.2.12) and extract the dynamics observed in Fig. 19.7. Note that the two-mode dynamics does a good job in approximating the solution. However, there is a phase drift that occurs in the dynamics that would require higher precision in both taking time slices of the full PDE and more accurate integration of the inner products for the coefficients. Indeed, the simplest trapezoidal rule has been used to compute the inner products and its accuracy is somewhat suspect. Higher order schemes could certainly help improve the accuracy. Additionally, incorporation of the third mode could also help. In either case, this demonstrates sufficiently how one would in practice use the low dimensional structures for approximating PDE dynamics.

 ## Global Normal Forms of Bifurcation Structures in PDEs

The preceding two sections illustrate quite effectively the role of POD in reducing the dynamics onto the optimal basis functions for producing analytically tractable results. Specifically, given the observation of low dimensionality, the POD provides the critical method necessary for projecting the complicated system dynamics onto a limited set of physically meaningful modes. This technique can be used to characterize the underlying dynamical behavior, or bifurcation structure, of a given PDE. In what follows, a specific example will be considered where the POD reduction technique is utilized. In this example, a specific physical behavior, which is modified via a bifurcation parameter, is considered. This allows the POD method to reconstruct the *global normal form* for the evolution equations, thus providing a technique similar to normal form reduction for differential equations.

The specific problem considered here to illustrate the construction of a global normal form for PDEs arises in the arena of laser physics and the generation of high-intensity, high-energy laser (optical) pulses [57]. Ultrashort lasers have a large variety of applications, ranging from small-scale problems such as ocular surgeries and biological imaging to large-scale problems such as optical communication systems and nuclear fusion. In the context of telecommunications and broadband sources, the laser is required to robustly produce pulses in the range of 10 femtoseconds to 50 picoseconds. The generation of such short pulses is often referred to as mode-locking [58, 59]. One of the most widely used mode-locked lasers developed to date is a ring cavity laser with a combination of waveplates and a passive polarizer. The combination of such components acts as an effective saturable absorber, providing an intensity discriminating mechanism to shape the pulse. It was first demonstrated experimentally in the early 1990s that ultrashort pulse generation and robust laser operation could be achieved in such a laser cavity. Since then, a number of theoretical models have been proposed to characterize the mode-locked pulse evolution and its stability, including the present work which demonstrates that the key phenomenon of multi-pulsing can be completely characterized by a low dimensional, four-mode interaction generated by a POD expansion.

The master mode-locking equation proposed by H.A. Haus [58, 59], which is the complex Ginzburg–Landau equation with a bandwidth-limited gain, was the first model used to describe the pulse formation and mode-locking dynamics in the ring cavity laser. The Ginzburg–Landau equation was originally developed in the context of particle physics as a model of superconductivity, and has since been widely used as a prototypical model for nonlinear wave propagation and pattern-forming systems. A large number of stationary solutions are supported by this equation. These stationary solutions can be categorized as single-pulse, double-pulse and in general n-pulse solutions depending on the strength of the cavity gain. The phenomenon for which a single mode-locked pulse solution splits into two or more pulses is known as multi-pulsing [58, 59], and has been observed in a wide variety of experimental and theoretical configurations. Specifically, when the pulse energy is increased, the spectrum of the pulse experiences a broadening effect. Once the spectrum exceeds the bandwidth limit of the gain, the pulse becomes energetically unfavorable and a double-pulse solution is generated which has a smaller bandwidth.

A number of analytical and numerical tools have been utilized over the past 15 years to study the mode-locking stability and multi-pulsing transition. One of the earliest theoretical approaches was to consider the energy rate equations derived by Namiki *et al.* that describe the averaged energy evolution in the laser cavity [60]. Other theoretical approaches involved the linear stability analysis of stationary pulses. These analytical methods were effective in characterizing the stability of stationary pulses, but were unable to describe the complete pulse transition that occurs during multi-pulsing. In addition to the limitation of the analytic tools, the transition region is also difficult to characterize computationally even with the fastest and most accurate algorithms available. Moreover, there is no efficient (algorithmic) way to find bifurcations by direct simulations. Solutions that possess a small basin of attraction are very difficult to find. This difficulty has led to the consideration of reduction techniques that simplify the full governing equation and allow for a deeper insight into the multi-pulsing instability. The POD method provides a generic, analytic tool that can be used efficiently, unlike the direct methods.

To describe the evolution of the electric field in the laser, the dominant physical effects in the laser cavity must be included. Specifically, the leading order dynamics must account for the chromatic dispersion, fiber birefringence, self- and cross-phase modulations for the two orthogonal components of the polarization vector in the fast and slow fields, bandwidth-limited gain saturation, cavity attenuation, and the discrete application of the saturable absorber element after each cavity round-trip. Distributing the discrete saturable absorption over the entire cavity and using a low-intensity approximation, one obtains (in dimensionless form) the single-field evolution equation [59]

$$i\frac{\partial u}{\partial z} + \frac{D}{2}\frac{\partial^2 u}{\partial t^2} + |u|^2 u - \nu|u|^4 u = ig(z)\left(1 + \tau\frac{\partial^2}{\partial t^2}\right)u - i\delta u + i\beta|u|^2 u - i\mu|u|^4 u \qquad (19.3.1)$$

where

$$g(z) = \frac{2g_0}{1 + \|u\|^2/e_0} = \frac{2g_0}{1 + \frac{1}{e_0}\int_{-\infty}^{\infty}|u|^2 dt}. \qquad (19.3.2)$$

The above equation is known as the cubic–quintic complex Ginzburg–Landau equation (CQGLE), and is a valid description of the averaged mode-locking dynamics when the intracavity fluctuations are small. Here u represents the complex envelope of the electric field propagating in the fiber. The independent variables z (the time-like variable) and t (the space-like variable) denote the propagating distance (number of cavity round-trips) and the time in the rest frame of the pulse, respectively. All the parameters are assumed to be positive throughout this section, which is usually the case for physically realizable laser systems. The parameter D measures the chromatic dispersion, ν is the quintic modification of the self-phase modulation, and β (cubic nonlinear gain) and μ (quintic nonlinear loss) arise directly from the averaging process in the derivation of the CQGLE.

For the linear dissipative terms (the first three terms on the right-hand side of the CQGLE), δ is the distributed total cavity attenuation. The gain $g(z)$, which is saturated by the total cavity energy (L^2-norm) $\|u\|^2$, has two control parameters g_0 (pumping strength) and e_0 (saturated pulse energy). In principle, the energy supplied by the gain can be controlled by adjusting either g_0 or e_0. In practice, however, one cannot change one of the parameters without affecting the other one. In what follows, we will assume without loss of generality that $e_0 = 1$ so that the cavity gain is totally controlled by g_0. The parameter τ characterizes the parabolic bandwidth of the saturable gain. Unless specified otherwise, the parameters are assumed to be $D = 0.4$, $\nu = 0.01$, $\tau = 0.1$, $\delta = 1$, $\beta = 0.3$ and $\mu = 0.02$ throughout the rest of this section. These parameters are physically achievable in typical ring cavity lasers.

In the appropriate parameter regime, the CQGLE supports single mode-locked pulses which are robust with respect to the initial conditions, as confirmed by experiments and numerical studies [59]. These mode-locked pulses are self-starting, and can be formed from low-amplitude white noise. When the gain parameter g_0 is increased, the stable pulse may undergo a multi-pulsing transition which is commonly observed in mode-locking systems [58, 59] and is illustrated in Fig. 19.8. At $g_0 = 3$ the initial condition quickly mode-locks into a single stationary pulse. When

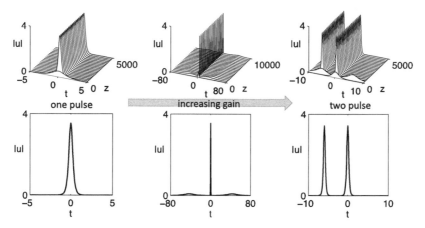

Figure 19.8: Multi-pulsing instability observed in the CQGLE (19.3.1). Top: As the gain is increased, the stable single mode-locked pulse (left, $g_0 = 3$) bifurcates to a periodic solution with small oscillations which persist indefinitely over z (middle, $g_0 = 3.24$), and finally to a stable double-pulse solution when the gain is too large (right, $g_0 = 3.6$). Bottom: The corresponding pulse profiles when $z \to \infty$.

g_0 is increased, this stationary pulse eventually loses its stability at $g_0 \approx 3.15$ and a periodic solution is formed. The periodic solution consists of a tall central pulse which is inherited from the previous stationary pulse, and one low-amplitude pulse on each side of the central pulse. The amplitudes of the pulses undergo small oscillations (on the order of 10^{-4}) which persist indefinitely over the propagation distance z. As g_0 keeps increasing, the two side pulses gradually move inwards. When the gain exceeds $g_0 \approx 3.24$, the periodic solution becomes unstable. At this gain value the central pulse experiences a little dip in intensity, and the system randomly chooses one of the small side pulses to be amplified and the other one to be suppressed, thus forming a double-pulse solution. The two pulses in the system have the same amount of energy, and the separation between them is fixed throughout the evolution. Additional pulses can be obtained by further increasing the value of g_0. The main focus of this section is to develop a low dimensional analytical model that describes the transition from one to two pulses as a function of the gain parameter g_0.

Figure 19.9 shows the POD modes obtained from a typical single-pulse evolution using a pulse-like initial condition (see the left panel of Fig. 19.8). As mentioned before, the modal energy E_j of the jth mode in a particular decomposition can be measured by the normalized eigenvalue λ_j, i.e.

$$E_j = \frac{\lambda_j}{\lambda_1 + \cdots + \lambda_M} . \qquad (19.3.3)$$

We find that the first mode in the decomposition dominates the dynamics of the evolution by taking up over 99.9% of the total energy. All the other POD modes have negligible energy, and they represent the transient behavior in the evolution that shapes the initial pulse into the final form. One can also compute the POD modes by using random initial conditions. The higher order modes may look different and they may carry higher energy as the transient effect is different, but the fundamental mode ϕ_1 will still dominate the entire evolution.

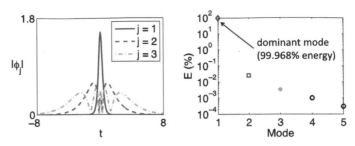

Figure 19.9: The POD modes and the corresponding normalized modal energies of a single-pulse evolution with $g_0 = 3$. The first three modes contain 99.968% (blue diamond), 0.0267% (red square), and 0.0035% (green asterisk) of the total energy, respectively.

To extend the reduced model to the periodic and double-pulse region we consider a combination of POD modes from different regions. The underlying idea is to use the dominating POD modes from different attractors of the CQGLE so that the resulting low dimensional system carries as much information as possible, and thus approximates the mode-locking dynamics better. We illustrate this by combining the first m modes from the single-pulse region with the first n modes from the double-pulse region.

Denote S as the union of the two sets of orthonormal basis mentioned above, i.e.

$$S = \left\{ \phi_1^{(1)}, \cdots \phi_m^{(1)}, \phi_1^{(2)}, \cdots, \phi_n^{(2)} \right\} . \tag{19.3.4}$$

Here $\phi_j^{(1)}$ and $\phi_j^{(2)}$ are the POD modes computed from the single-pulse and double-pulse region, respectively. The original field u can be projected onto the elements of S as

$$u = e^{i\theta_1} \left(a_1 S_1 + \sum_{j=2}^{m+n} a_j e^{i\psi_j} S_j \right) , \tag{19.3.5}$$

where S_j represent the elements of the combined basis S, a_j are the modal amplitudes, θ_1 is the phase of the first mode, and ψ_j denote the phase differences with respect to θ_1. One can obtain a low-dimensional system governing the evolutions of the modal amplitudes a_k and the phase differences ψ_k by substituting (19.3.5) into the CQGLE, multiplying the resulting equation by the complex conjugate of S_k ($k = 1, \cdots, m + n$), integrating over all t, and then separating the result into real and imaginary parts. The reduced system has, however, a complicated form due to the fact the elements in the combined basis S are not all orthonormal to each other. To address this issue, one can orthogonalize the combined basis S prior to the projection to obtain a new basis $\left\{ \Phi_j \right\}_{j=1}^{m+n}$. This can be achieved by the Gram–Schmidt algorithm given by

$$\begin{cases} \Phi_1 = S_1 , \\ \Phi_j = \chi_j / \sqrt{\langle \chi_j, \chi_j \rangle} , \, j = 2, \cdots, m + n \end{cases} \tag{19.3.6}$$

where

$$\chi_j = S_j - \sum_{q=1}^{j-1} \langle S_j, \Phi_q \rangle \Phi_q , \tag{19.3.7}$$

and the inner product is defined in the standard way. The reduced model derived from the new basis Φ_j has a simpler form than the one derived directly from S, since it does not have terms that are due to nonorthogonality of the basis functions. This reduced model will be called the $(m + n)$-model, which indicates the number of modes taken from the single-pulse (m) and double-pulse region (n).

Before moving on to the investigation of the mode-locking dynamics governed by the $(m + n)$-model, we first present the typical POD modes of a double-pulse evolution of the CQGLE in Fig. 19.10. The dominating modes in this case are $\phi_1^{(2)}$, $\phi_2^{(2)}$ and $\phi_3^{(2)}$, which takes up 99.9% of the total energy. As a first approximation to the CQGLE dynamics, we consider the (1+3)-model since the dominating mode in the single-pulse region is $\phi_1^{(1)}$ (see Fig. 19.9). The orthonormal basis $\{\Phi_j\}_{j=1}^4$ obtained using the Gram–Schmidt algorithm (19.3.6) is shown in Fig. 19.11. By construction, the first mode Φ_1 in the new basis is identical to $\phi_1^{(1)}$. The second mode Φ_2 contains a tall pulse which is identical to the first mode at $t = -6$ and a small bump at the origin. In general this tall pulse can be located on either the left or the right of the origin, depending on the data set U for which the POD modes are computed from. The other two modes have complicated temporal structures, and their oscillatory nature is mainly due to the orthogonalization. The (1+3)-model is derived from these four basis functions.

Classification of solutions of the (1+3)-model

The four-mode (1+3) model is seven-dimensional in the sense that the dynamic variables are the four modal amplitudes a_1, a_2, a_3, a_4, and the three phase differences ψ_2, ψ_3, ψ_4. To effectively characterize any periodic behavior in the system, it is customary to use the new set of variables

$$\begin{cases} x_1 = a_1 , \\ x_j = a_j \cos \psi_j , \quad y_j = a_j \sin \psi_j \end{cases} \tag{19.3.8}$$

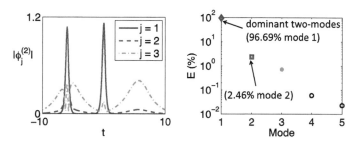

Figure 19.10: The POD modes and the corresponding normalized modal energies of a double-pulse evolution of the CQGLE with $g_0 = 3.6$ (cf. right column of Fig. 19.8). The first three modes contain 96.69% (blue diamond), 2.46% (red square), and 0.74% (green asterisk) of the total energy, respectively.

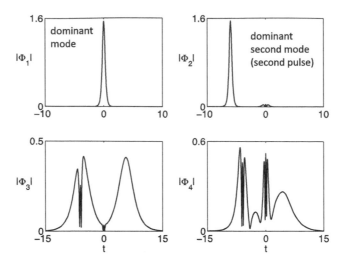

Figure 19.11: Basis functions used in the (1+3)-model. These functions are obtained by applying the Gram–Schmidt algorithm (19.3.6) to the set $S = \left\{ \phi_1^{(1)}, \phi_1^{(2)}, \phi_2^{(2)}, \phi_3^{(2)} \right\}$.

for $j = 2, 3, 4$ so that the formal projection of the field is given by

$$u = e^{i\theta_1} \left(x_1 \Phi_1 + \sum_{j \geq 2} \left(x_j + i y_j \right) \Phi_j \right). \tag{19.3.9}$$

Figure 19.12 shows the solution of the (1+3)-model expressed in terms of the above variables and the reconstructed field u at different cavity gains g_0. At $g_0 = 3$ (top panel), the (1+3)-model supports a stable fixed point with $x_1 = 2.2868$ while the other variables are nearly zero, i.e. the steady state content of u mainly comes from Φ_1, and thus a single mode-locked pulse centered at $t = 0$ is expected. Increasing g_0 eventually destabilizes the fixed point and a periodic solution is formed (middle panel) which corresponds to a limit cycle in the seven-dimensional phase space. The reconstructed field has a tall pulse centered at the origin due to the significant contribution of x_1 (and hence Φ_1). The small-amplitude side pulses result from the linear combination of the higher order modes present in the system, and their locations are completely determined by the data matrix U representing the double-pulse evolution of the CQGLE. Further increasing g_0 causes the limit cycle to lose its stability and bifurcate to another type of stable fixed point. Unlike the first fixed point, this new fixed point has significant contributions from both x_1 and y_2 which are associated to Φ_1 and Φ_2. Since the other variables in the system become negligible when $z \to \infty$, a double-pulse solution is formed.

The (1+3)-model is able to describe both the single-pulse and double-pulse evolution of the CQGLE, which is a significant improvement compared to the two-mode model considered in the previous section. The small-amplitude side pulses in the reconstructed periodic solution do not appear at the expected locations predicted by the CQGLE (see the middle column of Fig. 19.8), since we are using the data with a particular gain ($g_0 = 3.6$) to characterize the mode-locking

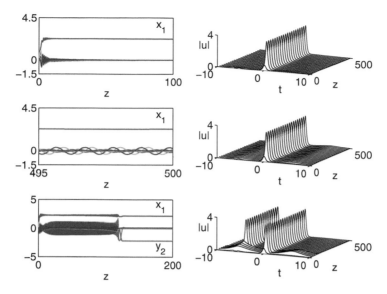

Figure 19.12: Left: The solution of the (1+3)-model at $g_0 = 3$ (top), $g_0 = 3.24$ (middle) and $g_0 = 3.8$ (bottom). Here only the significant variables (in absolute value) are labeled. Right: The corresponding reconstructed field u from the variables of the (1+3)-model. [From E. Ding, E. Shlizerman and J. N. Kutz, Modeling multipulsing transition in ring cavity lasers with proper orthogonal decomposition, Physical Review A **82**, 023823 (2010) [57] ©APS.]

behavior in the whole range of gain. Nevertheless, the two key features of the periodic solution of the CQGLE are explicitly captured: (i) the existence of side pulses, and (ii) the small-amplitude oscillations of the entire structure. These two features are not demonstrated by the reduced models considered before. The (1+3)-model also captures the amplification and suppression of the side pulses in the formation of the double-pulse solution in the CQGLE. Since the POD modes in the double-pulse region are computed from the numerical data where the left side pulse is chosen over the right side pulse (see Fig. 19.8), the second pulse in the reconstructed field always forms on the left. The case where the second pulse is formed on the right is not of particular interest since the underlying dynamics is essentially the same.

We construct a global bifurcation diagram in Fig. 19.13 to characterize the transition between the single-pulse and double-pulse solution of the (1+3)-model. The diagram shows the total cavity energy of the reconstructed field u as a function of the cavity gain g_0, which is obtained using MATCONT [63]. In terms of the dynamic variables, the energy of the field is given by

$$\|u\|^2 = x_1^2 + \sum_{j\geq 2} \left(x_j^2 + y_j^2 \right) . \tag{19.3.10}$$

In the diagram, the branch with the lowest energy corresponds to the single mode-locked pulses of the (1+3)-model. With increasing g_0, the reconstructed mode-locked pulse readjusts its amplitude and width accordingly to accommodate the increase in the cavity energy. The branch of the single-pulse solutions is stable until $g_0 = 3.181$ where a Hopf bifurcation occurs. The fluctuation

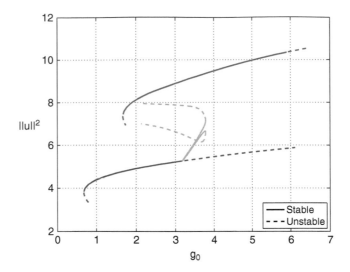

Figure 19.13: Bifurcation diagram of the total cavity energy of the reconstructed field u as a function of g_0. Here the bottom branch (blue) represents the single-pulse solution and the top branch (red) represents the double-pulse solution. The middle branch (green) denotes the extrema of the energy of the periodic solution. [From E. Ding, E. Shlizerman and J. N. Kutz, Modeling multipulsing transition in ring cavity lasers with proper orthogonal decomposition, Physical Review A **82**, 023823 (2010) [57] ©APS.]

in the cavity energy as a result of the periodic solution is small at first, but increases gradually with g_0. At $g_0 = 3.764$ the periodic solution becomes unstable by a fold bifurcation of the limit cycle. In this case a discrete jump is observed in the bifurcation diagram and a new branch of solutions is reached. This new branch represents the double-pulse solution of the system and has higher energy than the single-pulse branch and the periodic branch.

The bifurcation diagram also illustrates that in the region $1.681 \leq g_0 \leq 3.181$ the single-pulse and the double-pulse solutions are stable simultaneously. This bistability is not restricted to the (1+3)-model as it also happens in the CQGLE as well. Given a set of parameters, the final form of the solution is determined by the basin of attraction of the single-pulse and double-pulse branches rather than the initial energy content. When the initial condition is selected such that it is "close" to the double-pulse branch (characterized by the Euclidean distance between them), the result of linear analysis holds and a double-pulse mode-locked state can be achieved. For random initial data that is far away from both branches, the system tends to settle into the single-pulse branch as it is more energetically favorable.

Comparison of different $(m + n)$-models

To demonstrate that the (1+3)-model is the optimal $(m + n)$-model to characterize the transition from a single mode-locked pulse into two pulses that occurs in the CQGLE, we consider the (1+2)-model whose results are shown in the top row of Fig. 19.14. This model can be derived

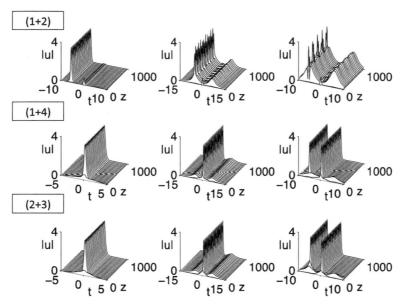

Figure 19.14: Top: The reconstructed field of the (1+2)-model at $g_0 = 3$ (left), $g_0 = 4.3$ (middle) and $g_0 = 7$ (right). Middle: The reconstructed field of the (1+4)-model at $g_0 = 3$ (left), $g_0 = 3.4$ (middle) and $g_0 = 3.8$ (right). Bottom: The reconstructed field of the (2+3)-model at $g_0 = 3$ (left), $g_0 = 3.4$ (middle) and $g_0 = 3.7$ (right). [From E. Ding, E. Shlizerman and J. N. Kutz, Modeling multipulsing transition in ring cavity lasers with proper orthogonal decomposition, Physical Review A **82**, 023823 (2010) [57] ©APS.]

by setting the variables x_4 and y_4 in the (1+3)-model to zero. At $g_0 = 3$ a tall stable pulse in formed at $t = -6$ together with a small residual at the origin. The entire structure has a close resemblance to Φ_2 from the Gram–Schmidt procedure. Except for the location, the tall pulse agrees with the single-pulse solution of the CQGLE quantitatively. One can think of this as the result of the translational invariance of the CQGLE. In the full simulation of the CQGLE, the small residual is automatically removed by the intensity discriminating mechanism (saturable absorption). When the structure loses its stability, a periodic solution is formed which persists even at unrealistically large values of cavity gain such as $g_0 = 7$.

The middle row of the same figure shows the numerical simulations of the (1+4)-model, which is derived from the combination of the first POD mode from the single-pulse region with the first four modes from the double-pulse region. This model is able to reproduce all the key dynamics observed during the multi-pulsing phenomenon shown in Fig. 19.13. The difference between the (1+4)-model and the (1+3)-model is the locations at which the bifurcations occur. In the previous section we showed that the single-pulse solution described by the (1+3)-model loses stability at $g_0 = 3.181$ (Hopf bifurcation), and the subsequent periodic solution bifurcates into the double-pulse solution at $g_0 = 3.764$. For the (1+4)-model these two bifurcations occur at $g_0 = 3.173$ and $g_0 = 3.722$, respectively, which are closer to the values predicted using the CQGLE ($g_0 \approx 3.15$ and $g_0 \approx 3.24$). The (2+3)-model (bottom row of Fig. 19.14) produces similar results to the (1+3)- and the (1+4)-model, with pulse transitions occurring at $g_0 = 3.177$ and 3.678. The comparison here shows that the third POD mode $\phi_3^{(2)}$ from the double-pulse region is essential for getting the right pulse transition. Once this mode is included, further addition of the POD modes has only a minor effect on the accuracy of the low dimensional model.

19.4 The POD Method and Symmetries/Invariances

The POD reduction technique can be a powerful tool in projecting seemingly complex dynamics to a low dimensional manifold where all the *important* dynamics occurs. Of course, it is not without its drawbacks and limitations. In particular, the method can often fail if the data sampled is simply put through the SVD algorithm without considering its basic structure and potential invariances. Invariances can arise due to various symmetry considerations in a given physical system, for instance, translational or rotational invariance.

As an example, we can once again consider the N-soliton dynamics as illustrated in Figs. 19.2–19.5. In this example, the N-soliton ($N = 1$ and $N = 2$) solution of the nonlinear Schrödinger equation was integrated using a spectral method in order to generate the data for sampling. A small change to this simulation can make a tremendous difference in the results of the POD reduction scheme. Specifically, consider integrating once again the nonlinear Schrödinger equation (19.1.6) but with the initial condition

$$u(x, 0) = N\text{sech}(x + x_0) \exp(i\Omega x) \tag{19.4.1}$$

where Ω represents a center-frequency shift of the soliton and x_0 is a center-position offset variable. The center-frequency perturbation to the N-soliton solution causes the soliton solution to move at a prescribed group velocity. Indeed, such a group-velocity movement of the pulse, when the center-frequency is driven by noise, leads to a fundamental limit in fiber optical communication systems known as the Gordon–Haus jitter [61].

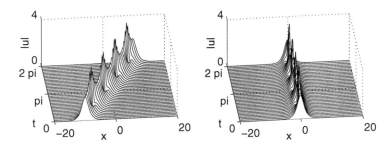

Figure 19.15: Evolution dynamics of the $N = 2$ soliton solution when subject to a center-frequency shift (left panel) and when the translational dynamics is factored out (right panel). The translating two-soliton generates a high-dimensional singular value distribution. In contrast, by factoring out the translational mode, the distribution once again becomes dominated by two modes as previously shown in Figs. 19.2–19.5.

To observe the impact of the center-frequency on the soliton propagation, consider numerically integrating the nonlinear Schrödinger equation with the above initial conditions and $N = 2$, $\Omega = \pi$ and $x_0 = 10$. Figure 19.15 (left panel) shows the soliton evolution. It is clearly seen that the center frequency causes a group-velocity drift in the pulse. Other than that, however, the $N = 2$ soliton dynamics appears to be identical to what is observed and characterized in Figs. 19.2–19.5. As before, we could arrange our numerical simulations into a data matrix and perform an SVD reduction. The bottom left panel of Fig. 19.16 shows the singular values that would result from such a reduction. It is clear in this case that such a method would not result in a reduction of the dynamics as the translating data, when correlated across time and space, do not produce redundant data. Indeed, there is a very slow decay of the singular values and the POD method cannot be used to generate a low dimensional manifold for the dynamics.

The failure of the method in this case is due simply to the translational invariance. If the invariance is *removed*, or factored out [62], before a data reduction is attempted, then the POD method can once again be used. In order to remove the invariance, the invariance must first be identified and an auxiliary variable defined. Thus we consider the dynamics rewritten as

$$u(x, t) \rightarrow u(x - c(t)) \tag{19.4.2}$$

where $c(t)$ corresponds to the translational invariance in the system responsible for limiting the POD method. The parameter c can be found by a number of methods. Rowley and Marsden [62]

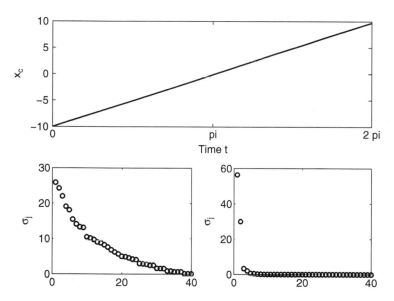

Figure 19.16: Dynamics of the center position $x_c(t)$ (top panel) showing the constant translation of the two-soliton to the right in the computational domain. The distributions of singular values for the two data sets demonstrated in Fig. 19.15 are exhibited in the bottom panels. Without factoring out the translation, the SVD does not produce low dimensional behavior (bottom left) whereas appropriate factoring of the data reveals the dominant two-mode dynamics (bottom right).

propose a template based technique for factoring out the invariance. Here, a simple center-of-mass calculation will be used to compute the location of the soliton.

The following code can be implemented to find the center position of the pulse as it evolves in time

```
for j=1:length(t)
    com=(x.*abs(usol(j,:)).^2);
    com2=(abs(usol(j,:)).^2);
    c(j)=trapz(x,com)/trapz(x,com2);
end
```

The end result of this computation is the production of the vector or function **c** which prescribes the location of the pulse. The center position can then be factored out by the following lines of code:

```
for j=1:length(t)
    [mn,jj]=min(abs(x-c(j)));
    ns=n/2-jj;
    usift=[usol(j,:) usol(j,:) usol(j,:)];
    usol_shift(j,:)=usift(1,n+1-ns:2*n-ns);
end
```

The new matrix **usol_shift** has now had the group-velocity drift factored out. Figure 19.15 (right panel) shows the net result of this. The center position vector as a function of time is demonstrated in the top panel of Fig. 19.16. Once the translation is factored out, the SVD reduction of the remaining data set produces the singular value in the bottom right panel of Fig. 19.16. This singular value distribution is identical to what was previously demonstrated in the simple $N = 2$ case. Thus the dynamics is now three dimensional: two modes interact nonlinearly to describe the pulse changes while one mode describes the translation. This example serves to demonstrate that if the invariance is handled properly, only a single degree of freedom extra is required to account for its action in the dynamics.

Advection–Diffusion equations

A more sophisticated example of handling such an invariance is provided by the advection–diffusion equations (see Sections 10.3 and 25.4):

$$\frac{\partial \omega}{\partial t} + [\psi, \omega] = \nu \nabla^2 \omega \qquad (19.4.3a)$$

$$\nabla^2 \psi = \omega \qquad (19.4.3b)$$

where

$$[\psi, \omega] = \frac{\partial \psi}{\partial x} \frac{\partial \omega}{\partial y} - \frac{\partial \psi}{\partial y} \frac{\partial \omega}{\partial x} \qquad (19.4.4)$$

and $\nabla^2 = \partial_x^2 + \partial_y^2$ is the two-dimensional Laplacian. A spectral method has been developed for solving this problem in Section 10.3. Here, the initial data applied to this problem will be a dipole-type initial condition

$$\omega(x, y, 0) = \exp[-(x + 8)^2 - (y - 2)^2] - \exp[-(x + 8)^2 - (y + 2)^2]. \qquad (19.4.5)$$

Such an initial condition leads to the formation of a dipole pair that translates to the right as depicted in Fig. 19.17. The evolution over nine time-frames is shown using the time-span variable

```
tspan=linspace(0,135,9);
```

A nearly constant velocity is displayed as the dipole pair is advected to the right of the computational domain.

As previously noted, the dimensionality of the dynamics can be assessed by using the SVD on the computational data. In this case, the solution is first allowed to settle to the dipole state before

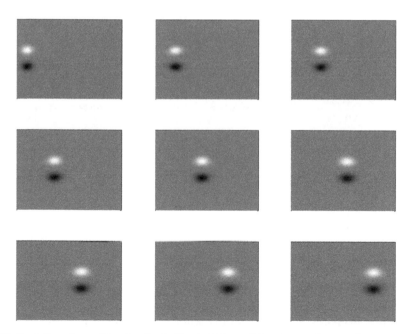

Figure 19.17: Evolution dynamics of a dipole-like initial condition in evenly spaced increments over the time-frame $t \in [0, 135]$. The dipole is seen to generate a translation of the structure to the right of the computational domain.

sampling is applied. For this case, the time domain and resulting sampling in the advection–diffusion dynamics is given by

```
tspan=linspace(0,135,60);
```

where 60 slices of data are taken, but only the last 40 slices of this data will be used in our POD projection. Given that the data are two-dimensional, they must first be reshaped into vector form before applying the SVD.

```
for j=21:length(t)
  w=real(ifft2(reshape(wtsol(j,:),nx,ny)));
  A(j-20,:)=reshape(w,1,nx*ny);
end
[U,S,V]=svd(A);
```

The top left panel of Fig. 19.18 shows the distribution of singular values after sampling the dynamics. Again, without taking care of the translation, the dynamics appears to be high

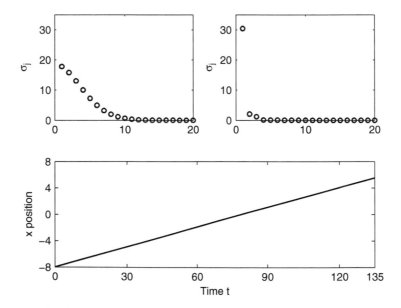

Figure 19.18: Dynamics of the center position (bottom panel) showing the constant translation of the dipole solution to the right in the computational domain. The distribution of singular values for the raw data and factored out data are exhibited in the top panels. Without factoring out the translation, the SVD does not produce low dimesional behavior (top left) whereas appropriate factoring of the data reveals a dominant one-mode dynamic (top right).

dimensional. However, it is clear from Fig. 19.17 that the dynamics should indeed by low dimensional with a dominant dipole mode translating from left to right.

As previously, a computation must be made of the translation of the dipole mode. Here, since the data are symmetric about $y = 0$, it only remains to find the center-of-mass via a standard moment method. The following code finds the center position variable corresponding to the translation:

```
for j=1:length(t)
    w=ifft2(reshape(wtsol(j,:),nx,ny));
    for jj=1:nx
        wy(jj)=trapz(y,abs(w(:,jj)));
    end
    com=(x.*wy);
    com2=(wy);
    c(j)=trapz(x,com)/trapz(x,com2);
end
```

The resulting translation vector is plotted in the bottom panel of Fig. 19.18. This gives the auxiliary variable that is to be factored out of the dynamics. The following code factors out the translation in order to align the data for the SVD and dimensionality reduction step.

```
for j=21:length(t)
    w=ifft2(reshape(wtsol(j,:),nx,ny));
    [mn,jj]=min(abs(x-c(j)));
    ns=nx/2-jj;
    usift=[w w w];
    w_shift=usift(:,nx+1-ns:2*nx-ns);
    A(j-20,:)=reshape(w_shift,nx*ny,1);
end
[U,S,V]=svd(A);
```

The top right panel of Fig. 19.18 shows the singular value distribution for this factored case. Note that the dynamics, as expected, is dominated by a single mode, i.e. the dipole mode. Figure 19.19 shows the first four modes of the SVD reduction. The first mode is clearly the dominant dipole solution. Modes two through four show the effects of the advection at the back of the dipole, i.e. fluid is ejected out the back of the dipole as it moves towards the left. Note that 88% of the energy is in the first dipole mode while 98% is in the first four modes.

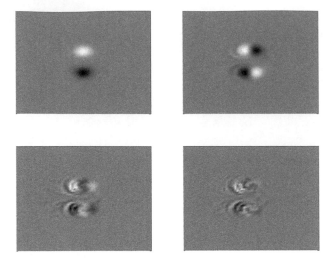

Figure 19.19: The four most dominant modes of the SVD decomposition for the data which has factored out the translational invariance. The first mode (top right) contains 88% of the modal energy while the first four contains 98%.

A one-mode decomposition of the data can be performed by assuming a solution of the form

$$\omega(x, y, t) = a(t)\phi(x, y) \tag{19.4.6}$$

where $\phi(x, y)$ is the dominant dipole mode. Once the vorticity is determined, then the stream-function can be computed from

$$\nabla^2 \psi = \omega = a(t)\phi(x, y) \quad \rightarrow \quad \psi = a(t)L^{-1}\phi(x, y) \tag{19.4.7}$$

where the operator L^{-1} inverts the Laplacian operator. Plugging the one-mode expansion into the governing equations and taking the inner product of both sides of the equation for vorticity yields

$$\frac{da}{dt} = \alpha a + \beta a^2 \tag{19.4.8}$$

where

$$\alpha = \nu \frac{(\nabla^2 \phi, \phi)}{(\phi, \phi)} \tag{19.4.9a}$$

$$\beta = \frac{((L^{-1}\phi)_y \phi_x, \phi)}{(\phi, \phi)} - \frac{((L^{-1}\phi)_x \phi_y, \phi)}{(\phi, \phi)} \tag{19.4.9b}$$

where the coefficients α and β are determined from inner products once the mode structures are computed from the SVD. Due to symmetry considerations of the dipole solution, $\beta = 0$, thus reducing everything to the calculation of α alone. The calculation of α can be performed with the following code:

```
w1=reshape(V(1,:),nx,ny);   w1t=fft2(w1);
w1lap=-real(ifft2(K.*w1t));

Q2=w1lap.*w1;
Q3=w1.*w1;
for j=1:nx
   s2(j)=trapz(y,Q2(j,:));
   s3(j)=trapz(y,Q3(j,:));
end
alpha=trapz(x,s2)/trapz(x,s3)
```

This yields a value of $\alpha = -1.771\nu$. Thus the leading order (dominant) POD solution for the dipole is simply exponential decay given by

$$a(t) = a(0)\exp(-1.771\nu t),\qquad(19.4.10)$$

where $a(0) = (\omega(x,y,0),\phi(x,y))$. In total then, the overall solution has been deconstructed into its constitute parts. To reconstruct, one simply needs to put the three pieces together: the time dynamics of the mode amplitude, the dominant mode structure, and the translational dynamics. This gives the one-mode expansion for the dipole dynamics:

$$\omega(x,y,t) = a(0)\exp(-1.771\nu t)\phi(x-c(t),y).\qquad(19.4.11)$$

Of course, a more complicated dynamics would result if a higher order mode expansion was assumed. However, 88% of the energy is in this single mode which was handled nicely by factoring out the translation.

To conclude this discussion, one can easily imagine factoring out a myriad of other symmetries. The most obvious is translation, but perhaps the second most obvious is factoring out a rotation. This might arise, for example, when considering spiral wave dynamics in reaction–diffusion systems. Ultimately, this is simply a data preprocessing step which must be performed prior to the dimensionality reduction step.

 ## 19.5 POD Using Robust PCA

The POD technique is ideal for sampling data, determining the low-dimensional structure therein, and exploiting the low-rank structure in the context of dynamical systems. The examples considered to this point demonstrate this by using the POD basis in combination with Galerkin projection in order to reduce the dynamics of a given complex system to a low dimensional manifold on which the dynamics occurs. At the heart of this dimension reduction is the singular value decomposition and its various guises, most notably principal component analysis or proper orthogonal decomposition. The SVD produces characteristic features (principal components)

that are determined by the covariance matrix of the data. Fundamental in producing the SVD/PCA/POD modes is the L^2 norm for data fitting. Although the L^2 norm is highly appealing due to its centrality (and physical interpretation) in most applications as an *energy*, it does have potential flaws that can severely limit its applicability when trying to integrate it with dynamical systems theory. In particular, if the data gathered through measurement is corrupt or suffers from large and sparse noise fluctuations, the higher dimensional least-square fitting to the data matrix by the SVD algorithm squares this pernicious error; leading to significant deformation of the singular values and PCA/POD modes describing the data. Although many physical systems and/or computations are relatively free of data corruption, many modern applications, such as PIV fluid measurements, are rife with arbitrary corruption of the measured data due to the sensitivity of measuring *derivative*-type data. This ensures that the standard application of the SVD will be of limited use.

Ideally, in performing dimensionality reduction one should not allow the corrupt or large noise fluctuations to so strongly influence one's results. In order to limit the impact of such *data outliers*, a different measure, or norm, can be envisioned in measuring the best data fit (SVD/PCA/POD modes). One measure that has recently produced great success in this arena is the L^1 norm [51] which no longer squares the distance of the data to the best-fit mode. Indeed, the L^1 norm has been demonstrated to promote sparsity and is the underlying theoretical construct in *compressive sensing* or *sparse sensing* algorithms (see Section 17.3 for further details). Here, the L^1 norm is simply used as an alternative measure (norm) for performing the equivalent of the least-square fit to the data. As a result, a more robust method is achieved for computing PCA components, i.e. the so-called *robust PCA* method.

The robust PCA technique has already been considered in Section 15.6. Briefly, the idea of the robust PCA is to provide an algorithm capable of separating in an efficient manner low-rank data with corrupt (sparse) measurement/matrix error. Thus the idea would be to consider a matrix which is composed of these two elements together

$$\mathbf{M} = \mathbf{L} + \mathbf{S} \tag{19.5.1}$$

where $\mathbf{M}$ is the data matrix of interest that is composed of a low-rank structure $\mathbf{L}$ along with some sparse data $\mathbf{S}$. The data-analysis objective is the following: given measurements or observations $\mathbf{M}$, can the low-rank matrix $\mathbf{L}$ and sparse matrix $\mathbf{S}$ be recovered? Adding to the difficulty of this task is the following: neither the rank of the matrix $\mathbf{L}$, nor the number of nonzero (sparse) elements of the matrix $\mathbf{S}$ are known.

The separation problem seems impossible to solve since the number of unknowns to infer for $\mathbf{L}$ and $\mathbf{S}$ is twice as many as the given measurements in $\mathbf{M}$. However, Candés and co-workers [51] have recently proved that this can indeed be solved through the formulation of a tractable convex optimization problem. In real applications, however, it is rare that the matrix decomposition is as ideal as (19.5.1). More generally, the decomposition is of the form

$$\mathbf{M} = \mathbf{L} + \mathbf{S} + \mathbf{N} \tag{19.5.2}$$

where the matrix $\mathbf{N}$ is a dense, small perturbation accounting for the fact that the low-rank component is only approximately low rank and that small errors can be added to all the entries. But even in this case, the convex optimization algorithm for generating robust PCA seems to do a remarkable job at separating low-rank from sparse data. This can be used to great advantage when certain types of data are considered, namely those with corrupt data or large, sparse noisy perturbations.

POD mode selection via robust PCA

To illustrate the role of robust PCA in dimensionality reduction of dynamical systems, we once again revisit the two-soliton example considered throughout this chapter. The two-soliton dynamics are illustrated in Figs. 19.2–19.5. It is once again demonstrated in Fig. 19.20 along with a modified two-soliton data matrix which has been corrupted by sparse noise fluctuations. Assuming the two-soliton (nonlinear Schrödinger solver) has been executed with the initial condition $u(x, 0) = 2$ sech x and over the interval $t \in [0, 2\pi]$, then the resulting complex solution is contained in the matrix **usol**. This matrix is corrupted by 60 randomly chosen, sparse noise fluctuations with the following code:

```
sam=60; % corrupt data points
Atest2=zeros(length(t),n);
Arand1=rand(length(t),n);
Arand2=rand(length(t),n);
r1 = randintrlv([1:length(t)*n],793);
r1k= r1(1:sam);
for j=1:sam
    Atest2(r1k(j))=-1;
end
Anoise=Atest2.*(Arand1+i*Arand2);
unoise=usol+5*Anoise;

subplot(2,2,1), waterfall(x,t,abs(usol)); colormap([0 0 0]);
subplot(2,2,2), waterfall(x,t,abs(unoise)); colormap([0 0 0]);
```

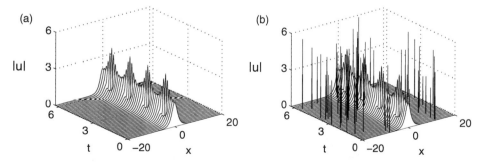

Figure 19.20: (a) Ideal evolution dynamics of the two-soliton solution of the nonlinear Schrödinger equation, and (b) the ideal data matrix corrupted by a sparse set of 60 measurements.

Recall that the noise added has both real and imaginary parts as is the case for the complex evolution field given by the nonlinear Schrödinger equation.

Previously, the data matrix was used in its raw (and perfect) form to generate the low dimensional manifold spanned by a reduced set of POD modes. However, with the sparse corruption of the data, it seems unlikely that the desired low dimensional manifold can be computed. Application of the SVD to the ideal data and corrupt data shows significant differences. The following code produces the singular values and POD modes for each.

```
A1=usol; A2=unoise;
[U1,S1,V1]=svd(A1);
[U2,S2,V2]=svd(A2);
```

Figures 19.21 and 19.22 show the singular values and POD modes for both reductions. In the top panel of both figures are the results for the ideal data, showing that a clear two- to three-mode dominance exists in the data, i.e. three modes (magenta dots) captures more than 99% of the data matrix. Indeed, this was directly exploited in Section 19.2 for the construction of a low

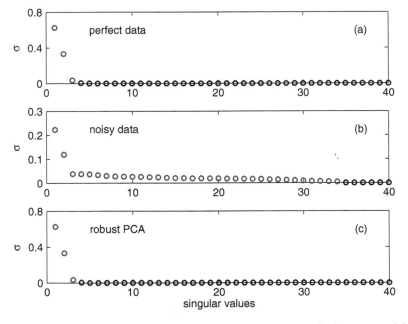

Figure 19.21: Singular values of the data matrices for the ideal data (top panel), the corrupted data (middle panel), and the low-rank data computed through robust PCA (bottom panel). The robust PCA construction produces almost a perfect match to the original, ideal data. The magenta dots represent the number of modes necessary to construct 99% of the original data. In this case, the ideal and robust PCA require three modes while the corrupt data require 34 modes.

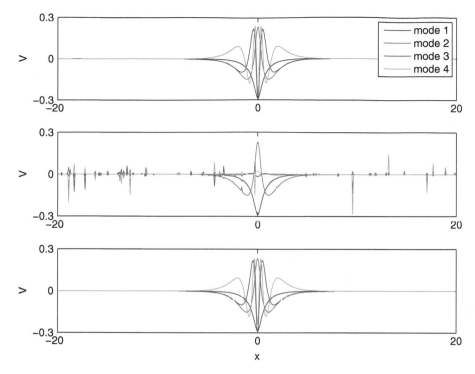

Figure 19.22: Dominant modes for the ideal data (top panel), the corrupted data (middle panel), and the low-rank data computed through robust PCA (bottom panel). The ideal and robust PCA produce the same dominant two to three modes while the corrupt data produce highly erratic modes.

dimensional manifold for the dynamical evolution of the system. In contrast, the middle panels show that the corrupted data matrix activates a significant number of modes. Specifically, it now takes 34 modes (magenta circles) in order to capture 99% of the corrupt data matrix. The POD modes are also now severely misshaped from their ideal, noisy free state.

To remove the corrupted data, the **inexact_alm_rpca** algorithm is used which is based upon the alternating direction method [46]. For our purposes, we will use this MATLAB program since it is extremely simple to use and exceedingly fast. The following code separates the low-rank from the sparse data as in (19.5.1)

```
ur=real(unoise);
ui=imag(unoise);

lambda=0.2;
[R1r,R2r]=inexact_alm_rpca(real(ur.'),lambda);
[R1i,R2i]=inexact_alm_rpca(real(ui.'),lambda);
```

Continued

```
R1=R1r+i*R1i;
R2=R2r+i*R2i;

[U3,S3,V3]=svd(R1.');
```

Note that the use of **inexact_alm_rpca** requires the data be real. Additionally, there is a parameter *lambda* that can adjusted to best separate the low-rank from sparse data. This is considered in more detail in Section 15.6.

The real and imaginary portions are put back together at the end of the algorithm in order to SVD the low-rank portion of interest, i.e. the portion of the data considered to be without corrupt data. The two matrices corresponding to the low-rank ($R1$) and sparse ($R2$) portions of the data matrix are plotted in Fig. 19.23. The parameter *lambda* used in the **inexact_alm_rpca** algorithm can be tuned to best separate the sparse from low-rank matrices in (15.6.1). Here a value of 0.2 is used, which produces an almost ideal separation as is shown in Figs. 19.21(bottom panel), 19.22(bottom panel) and 19.23. Indeed, an almost perfect separation is achieved as advertised (guaranteed) by Candés and co-workers [51].

Finally, the POD dynamics are compared for a two-mode truncation of the corrupt data versus the low-rank approximation of the data achieved through robust PCA. Figure 19.24 shows the results of both two-mode expansions. The robust PCA modes produce almost identical results to those achieved with the standard POD reduction using the ideal data. In contrast, the POD reduction using the corrupt data produces inner products in (19.2.12) of Section 19.2 that greatly skew their value and their ability to predict accurately the true underlying evolution. Indeed, the two POD mode dynamics is significantly deteriorated from its robust PCA counterpart.

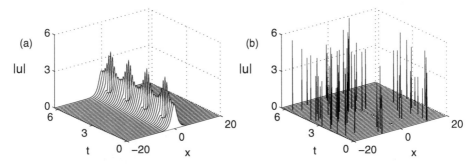

Figure 19.23: Separation of the original corrupt data matrix shown in Fig. 19.20(b) into a low-rank component (a) and a sparse component (b). The separation is almost perfect, with the low-rank matrix being dominated by two to three modes, i.e. the three modes contain greater than 99% of the energy.

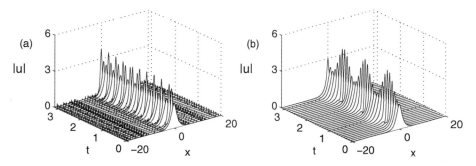

Figure 19.24: Comparison of the two-mode POD evolution given by (19.2.12) of Section 19.2 using the (a) SVD reduction of the corrupt data matrix and (b) using the low-rank data generated from the robust PCA algorithm. The use of the robust PCA algorithm renders the POD results almost identical to those with the ideal data, thus suggesting it can potentially be a filter for corrupt data before applying POD reductions.

As was already pointed out in Section 15.6 the robust PCA algorithm advocated for here is quite remarkable in its ability to separate sparse corruption from low-rank structure in a given data matrix. The application of robust PCA essentially acts as an advanced filter for cleaning up a data matrix. Indeed, one can potentially use this as a filter which is very unlike the time–frequency filtering schemes considered previously. This has the potential to greatly enhance the POD reduction schemes advocated here when they are subject to corrupt data measurements.

Dynamic Mode Decomposition

20

Exploiting low dimensionality in complex systems has already been demonstrated to be an effective method for either computationally or theoretically reducing a given system to a more tractable form. In what has been proposed so far with POD reductions, we have used *model-based* algorithms. Thus a given set of governing equations dictate the dynamics of the underlying system. These equations construe the *model*. However, an alternative to this approach exists when either the model equations are beyond their range of validity or the governing equations simply are not known or well-formulated. In this alternative approach, experimental data itself drives the understanding of the system, without model equations being prescribed. Thus *data-based* algorithms can be constructed to understand, mimic and control the complex system. *Dynamic mode decomposition* (DMD) is a relatively new data-based algorithm that requires no underlying governing equations, rather snapshots of experimental measurements are used to predict and control a given system [64, 65, 66, 67]. In atmospheric sciences, a version of the DMD method is used to fit a set of data to an underlying (best statistical fit) linear stochastic model. This is known as *linear inverse models* (LIMs) [68, 69, 70]. Thus the methodology presented here uses data alone as the source for informing us of the state and dimensionality of the system.

20.1 Theory of Dynamic Mode Decomposition (DMD)

Exploitation of low dimensionality: that is the overarching goal of data-driven modeling methods. In the case of the technique of dynamic mode decomposition, it is the aim of the method to take advantage of low dimensionality in the experimental data itself without having to rely on a given set of governing equations.

The DMD method provides a decomposition of experimental data into a set of dynamic modes that are derived from snapshots of the data in time. The mathematics underlying the extraction of dynamic information from time-resolved snapshots is closely related to the idea of the Arnoldi algorithm, one of the workhorses of fast computational solvers. To set the stage, we begin with the data collection process. To important parameters are required:

$$N = \text{number of spatial points saved per time snapshot} \qquad (20.1.1a)$$

$$M = \text{number of snapshots taken}. \qquad (20.1.1b)$$

In this case, it is imperative to collect data at regularly spaced intervals of time

$$\text{data collection times}: \quad t_{m+1} = t_m + \Delta t \qquad (20.1.2)$$

where the collection time starts at t_1 and ends at t_M, and the interval between data collection times is Δt.

The data can then be arranged into an $N \times M$ matrix

$$\mathbf{X} = \begin{bmatrix} U(\mathbf{x}, t_1) & U(\mathbf{x}, t_2) & U(\mathbf{x}, t_3) & \cdots & U(\mathbf{x}, t_M) \end{bmatrix} \qquad (20.1.3)$$

where $\mathbf{x}$ is a vector of data collection points of length N. A limitation of the DMD technique, at least applied in a straightforward manner, is apparent from the start: sampling must be done in equally spaced time intervals. Be that as it may, the objective is to mine the data matrix $\mathbf{X}$ for important dynamical information. For the purposes of the DMD method, the following matrix is also defined:

$$\mathbf{X}_j^k = \begin{bmatrix} U(\mathbf{x}, t_j) & U(\mathbf{x}, t_{j+1}) & \cdots & U(\mathbf{x}, t_k) \end{bmatrix}. \qquad (20.1.4)$$

Thus this matrix includes columns j through k of the original data matrix.

The DMD method approximates the modes of the so-called *Koopman operator*. The Koopman operator is a linear, infinite dimensional operator that represents nonlinear, infinite dimensional dynamics without linearization [71, 72], and is the adjoint of the Perron–Frobenius operator. The method can be viewed as computing, from the experimental data, the eigenvalues and eigenvectors (low dimensional modes) of a linear model that approximates the underlying dynamics, even if the dynamics is nonlinear. Since the model is assumed to be linear, the decomposition gives the growth rates and frequencies associated with each mode. If the underlying model is linear, then the DMD method recovers the leading eigenvalues and eigenvectors normally computed using standard solution methods for linear differential equations.

Mathematically, the Koopman operator $\mathbf{A}$ is a linear, time-independent operator $\mathbf{A}$ such that

$$\mathbf{x}_{j+1} = \mathbf{A}\mathbf{x}_j \qquad (20.1.5)$$

where j indicates the specific data collection time and $\mathbf{A}$ is the linear operator that maps the data from time t_j to t_{j+1}. The vector $\mathbf{x}_j$ is an N-dimensional vector of the data points collected at time j. The computation of the Koopman operator is at the heart of the DMD methodology. As already stated, the mapping over Δ is linear even though the underlying dynamics that generated $\mathbf{x}_j$ may be nonlinear. It should be noted that this is different from linearizing the dynamics.

To construct the appropriate Koopman operator that best represents the data collected, the matrix $\mathbf{X}_1^{M-1}$ is considered:

$$\mathbf{X}_1^{M-1} = \begin{bmatrix} \mathbf{x}_1 & \mathbf{x}_2 & \mathbf{x}_3 & \cdots & \mathbf{x}_{M-1} \end{bmatrix}. \qquad (20.1.6)$$

Making use of (20.1.5), this matrix reduces to

$$\mathbf{X}_1^{M-1} = \begin{bmatrix} \mathbf{x}_1 & \mathbf{A}\mathbf{x}_1 & \mathbf{A}^2\mathbf{x}_1 & \cdots & \mathbf{A}^{M-2}\mathbf{x}_1 \end{bmatrix}. \tag{20.1.7}$$

Here is where the DMD method connects to Krylov subspaces and the Arnoldi algorithm. Specifically, the columns of $\mathbf{X}_1^{M-1}$ are each elements in a Krylov space. This matrix attempts to fit the first $M-1$ data collection points using the Koopman operator (matrix) $\mathbf{A}$. In the DMD technique, the final data point $\mathbf{x}_M$ is represented, as best as possible, in terms of this Krylov basis, thus

$$\mathbf{x}_M = \sum_{m=1}^{M-1} b_m \mathbf{x}_m + \mathbf{r} \tag{20.1.8}$$

where the b_m are the coefficients of the Krylov space vectors and $\mathbf{r}$ is the residual (or error) that lies outside (orthogonal to) the Krylov space. Ultimately, this best fit to the data using this DMD procedure will be done in an L^2 sense using a pseudo-inverse.

Before proceeding further, it is at this point that the data matrix $\mathbf{X}_1^{M-1}$ in (20.1.7) should be considered further. In particular, our dimensionality reduction methods look to take advantage of any low dimensional structures in the data. To exploit this, the SVD of (20.1.7) is computed:

$$\mathbf{X}_1^{M-1} = \mathbf{U}\boldsymbol{\Sigma}\mathbf{V}^* \tag{20.1.9}$$

where $*$ denotes the conjugate transpose, $\mathbf{U} \in \mathbb{C}^{N \times K}$, $\boldsymbol{\Sigma} \in \mathbb{C}^{K \times K}$ and $\mathbf{V} \in \mathbb{C}^{M-1 \times K}$. Here K is the reduced SVD's approximation to the rank of $\mathbf{X}_1^{M-1}$. If the data matrix is full rank and the data have no suitable low dimensional structure, then the DMD method fails immediately. However, if the data matrix can be approximated by a low-rank matrix, then DMD can take advantage of this low dimensional structure to project a future state of the system. Thus once again, the SVD plays the critical role in the methodology.

Armed with the reduction (20.1.9) to (20.1.7), we can return to the results of the Koopman operator and Krylov basis (20.1.8). Specifically, generalizing (20.1.5) to its matrix form yields

$$\mathbf{A}\mathbf{X}_1^{M-1} = \mathbf{X}_2^M. \tag{20.1.10}$$

But by using (20.1.8), the right-hand side of this equation can be written in the form

$$\mathbf{X}_2^M = \mathbf{X}_1^{M-1}\mathbf{S} + \mathbf{r}e_{M-1}^* \tag{20.1.11}$$

where e_{M-1} is the $(M-1)$th unit vector and

$$\mathbf{S} = \begin{bmatrix} 0 & \cdots & & 0 & b_1 \\ 1 & \ddots & & 0 & b_2 \\ 0 & \ddots & \ddots & & \vdots \\ & \ddots & \ddots & 0 & b_{M-2} \\ 0 & \cdots & 0 & 1 & b_{M-1} \end{bmatrix}. \tag{20.1.12}$$

Recall that the b_j are the unknown coefficients in (20.1.8).

The key idea now is the observation that the eigenvalues of **S** approximate some of the eigenvalues of the unknown Koopman operator **A**, making the DMD method similar to the Arnoldi algorithm and its approximations to the Ritz eigenvalues. Schmid [64] showed that rather than computing the matrix **S** directly, we can instead compute the *lower-rank* matrix

$$\tilde{\mathbf{S}} = \mathbf{U}^*\mathbf{X}_2^M\mathbf{V}\boldsymbol{\Sigma}^{-1} \tag{20.1.13}$$

which is related to **S** via a similarity transformation. Recall that the matrices **U**, $\boldsymbol{\Sigma}$ and **V** arise from the SVD reduction of $\mathbf{X}_1^{M-1}$ in (20.1.9).

Consider then the eigenvalue problem associated with $\tilde{\mathbf{S}}$:

$$\tilde{\mathbf{S}}\mathbf{y}_k = \mu_k\mathbf{y}_k \qquad k = 1, 2, \cdots, K \tag{20.1.14}$$

where K is the rank of the approximation we are choosing to make. The eigenvalues μ_k capture the time dynamics of the discrete Koopman map **A** as a Δt step is taken forward in time. These eigenvalues and eigenvectors can be related back to the similarity transformed original eigenvalues and eigenvectors of **S** in order to construct the DMD modes:

$$\psi_k = \mathbf{U}\mathbf{y}_k . \tag{20.1.15}$$

With the low-rank approximations of both the eigenvalues and eigenvectors in hand, the projected future solution can be constructed for all time in the future. By first rewriting for convenience $\omega_k = \ln(\mu_k)/\Delta t$ (recall that the Koopman operator time dynamics is linear), then the approximate solution at all future times, $\mathbf{x}_{DMD}(t)$, is given by

$$\mathbf{x}_{DMD}(t) = \sum_{k=1}^{K} b_k(0)\psi_k(\mathbf{x})\exp(\omega t) = \boldsymbol{\Psi}\text{diag}(\exp(\omega t))\mathbf{b} \tag{20.1.16}$$

where $b_k(0)$ is the initial amplitude of each mode, $\boldsymbol{\Psi}$ is the matrix whose columns are the eigenvectors ψ_k, diag(ωt) is a diagonal matrix whose entries are the eigenvalues $\exp(\omega_k t)$, and **b** is a vector of the coefficients b_k.

It only remains to compute the initial coefficient values $b_k(0)$. If we consider the initial snapshot ($\mathbf{x}_1$) at time zero, let's say, then (20.1.16) gives $\mathbf{x}_1 = \boldsymbol{\Psi}\mathbf{b}$. This generically is not a square matrix so that its solution

$$\mathbf{b} = \boldsymbol{\Psi}^+\mathbf{x}_1 \tag{20.1.17}$$

can be found using a pseudo-inverse. Indeed, $\boldsymbol{\Psi}^+$ denotes the Moore–Penrose pseudo-inverse that can be accessed in MATLAB via the **pinv** command. As already discussed in the compressive sensing chapter, the pseudo-inverse is equivalent to finding the best solution **b** in the least-squares (best fit) sense. This is equivalent to how DMD modes were derived originally.

Overall then, the DMD algorithm presented here takes advantage of low dimensionality in the data in order to make a low-rank approximation of the linear mapping that best approximates the nonlinear dynamics of the data collected for the system. Once this is done, a prediction of the future state of the system is achieved for all time. Unlike the POD method, which requires solving a low-rank set of dynamical quantities to predict the future state, no additional work is required for the future state prediction outside of plugging in the desired future time into (20.1.16). Thus the advantages of DMD revolve around the fact that (i) no equations are needed, and (ii) the future state is known for all time (of course, provided the DMD approximation holds).

The algorithm is as follows:

(i) Sample data at N prescribed locations M times. The data snapshots should be evenly spaced in time by a fixed Δt. This gives the data matrix $\mathbf{X}$.

(ii) From the data matrix $\mathbf{X}$, construct the two submatrices $\mathbf{X}_1^{M-1}$ and $\mathbf{X}_2^{M}$.

(iii) Compute the SVD decomposition of $\mathbf{X}_1^{M-1}$.

(iv) The matrix $\tilde{\mathbf{S}}$ can then be computed and its eigenvalues and eigenvectors found.

(v) Project the initial state of the system onto the DMD modes using the pseudo-inverse.

(vi) Compute the solution at any future time using the DMD modes along with their projection to the initial conditions and the time dynamics computed using the eigenvalue of $\tilde{\mathbf{S}}$.

20.2 Dynamics of DMD Versus POD

To illustrate the application of the DMD method, a simple example is chosen in order to highlight the implementation of the algorithm advocated at the end of the last section. In fact, it is the same example used to illustrate the POD reduction technique, namely, the $N = 2$ soliton dynamics of the nonlinear Schrödinger (NLS) equation (see Section 18.3). For completeness, the NLS will be given here again:

$$iu_t + \frac{1}{2}u_{xx} + |u|^2 u = 0 \tag{20.2.1}$$

with initial conditions $N\mathrm{sech}(x)$. Thus the data to be used will be generated through simulations of this equation. The PDE solution can be computed using FFTs with the following code. First, the initialization of the space and time discretization is computed:

```
% space
L=40; n=512;
x2=linspace(-L/2,L/2,n+1); x=x2(1:n);
k=(2*pi/L)*[0:n/2-1 -n/2:-1].';
% time
slices=20;
t=linspace(0,2*pi,slices+1); dt=t(2)-t(1);
```

Note that the parameter **slices** will be critical in what follows ($M = $ **slices**+1). Specifically, this determines the number of data samples taken over the time interval $t \in [0, 2\pi]$. The initial conditions and time-stepping routines are as follows

```
u=2*sech(x).';    % initial conditions
ut=fft(u);
[t,utsol]=ode45('nls_rhs',t,ut,[],k);
for j=1:length(t)
    usol(j,:)=ifft(utsol(j,:));   % bring back to space
end
```

This implementation of the solver was also demonstrated in Section 18.3. Note that a call is made to the right-hand side function **nls_rhs.m** which is the following

```
function nls_rhs=nls_rhs(z,ut,dummy,k)
u=ifft(ut);
nls_rhs= -(i/2)*(k.^2).*ut + i*fft( (abs(u).^2).*u );
```

Having executed the above code, the matrix **usol** results which arranges the simulation data into a matrix with M time slices and N data points at each slice. Thus the matrix **usol** is the desired data matrix $\mathbf{X}$ required in the first step of the DMD algorithm.

The top left panel of Fig. 20.1 shows the result of the simulation of the NLS equation over the interval $t \in [0, 2\pi]$ with $M = 21$ and $N = 512$. Our goal is to sample from this set of data in order

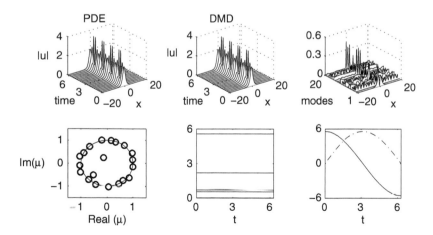

Figure 20.1: The top left panels represent the full PDE simulation and DMD reconstruction from sampling for $t \in [0, 2\pi]$ and $M = 21$ slices of data. The DMD modes used for the linear reconstruction are shown in the top right panel with its eigenvalue distribution μ_k shown in the bottom left panel. The solid line of the bottom left panel represents the unit circle for which eigenvalues outside of it are growth modes. The bottom right panels show the absolute value of the time dynamics (bottom middle) for each of the 20 DMD modes and the real (solid) and imaginary (dotted) amplitude of the time evolution of the first mode (bottom right).

to (i) find a low dimensional framework to exploit, and (ii) project the future state of the system in time without relying on computations of the PDE.

Our first objective is to then construct the data matrices in the second part of the algorithm: $\mathbf{X}_1^{M-1}$ and $\mathbf{X}_2^M$. Thus we have

```
X = usol.'; X1 = X(:,1:end-1); X2 = X(:,2:end);
```

Armed with these subsets of the full data matrix, the SVD decomposition of $\mathbf{X}_1^{M-1}$ can be computed and the eigenvalues and eigenvectors of $\tilde{\mathbf{S}}$ found. The following code performs these operations.

```
[U,Sigma,V] = svd(X1, 'econ');
S = U'*X2*V*diag(1./diag(Sigma));
[eV,D] = eig(S);
mu = diag(D);
omega = log(mu)/(dt);
Phi = U*eV;
```

Note that the vector **vals** contains the eigenvalues μ of interest to the DMD dynamics and **Phi** contains the DMD modes which are the basis function used to predict the future state of the system.

The final steps in the DMD algorithm both project the initial conditions, via the pseudo-inverse, to the DMD modes and then predicts the future state using the DMD modes and their linear time dynamics dictated by the μ_k. The following gives a reconstruction of the dynamics over the same interval as used in collecting the data.

```
y0 = Phi\u;  % pseudo-inverse initial conditions

u_modes = zeros(size(v1,2),length(t));
for iter = 1:length(t)
    u_modes(:,iter) =(y0.*exp(omega*t(iter)));
end
u_dmd = Phi*u_modes;
```

Figure 20.1 gives a summary of the DMD method, including demonstrating the PDE versus DMD dynamics. In fact, in the top panels of this figure, the PDE (exact solution) and DMD reconstruction are shown. In the top right panel, the 20 DMD modes constructed from the data reduction are illustrated. Thus these modes are used as the basis for reconstructing the best

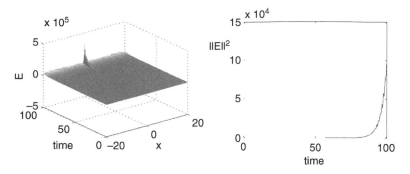

Figure 20.2: Evolution of the error, i.e. the absolute value of the difference between the DMD solution and the PDE solution, over the time interval $t \in [0, 100]$ (left panel) and the evolution of the norm of the error (right panel). Due to the DMD approximation having eigenvalues outside the unit circle, growth modes appear which blow-up for long time projections of the dynamics. However, the error remains small for times less than $t \approx 50$, giving an upper range for when the DMD approximation holds.

possible linear fit to the nonlinear evolution. Their time evolution is determined by the eigenvalues μ_k which are given in the bottom left panel. Also plotted on this panel is the unit circle. Those eigenvalues that lie outside the unit circle represent growth modes in the system. The absolute value of the time dynamics for each mode is given in the bottom middle panel while the first mode time dynamics (the real and imaginary parts) is given in the bottom right panel.

Although the $N = 2$ soliton dynamics is oscillatory (nonlinear) in time, the DMD reduction shows that a few of the eigenvalues μ_k sit slightly outside the unit circle when $M = 21$. This will ultimately lead to a long-term blow-up of the predicted solution using the DMD method. Figure 20.2 shows the error as a function of time. Here, a comparison is made between the DMD predicted solution, using data only in $t \in [0, 2\pi]$, and the full numerical solution for the interval $t \in [0, 100]$. The agreement is quite good until about $t \approx 40$ when the solution starts to diverge exponentially from the true solution. This is simply a result of sampling error as will be shown in what follows. Either way, it is important to recognize that how one samples and how frequently one samples can make a big impact on the DMD method. Moreover, any noise in the system and/or data can push an eigenvalue outside of the unit circle and limit the range (in time) of the DMD algorithm and prediction. Thus to continue to use it in this context, resampling must occur in order to re-project the data and avoid eventual (unphysical) blow-up of the DMD predictions.

As already stated, the computation of the eigenvalues μ_k changes as a function of the sampling rate. To illustrate this, two new figures are produced. Figure 20.3 shows the eigenvalue distribution μ_k for $M = 16, 21, 31$ and 41 when sampling the same PDE dynamics over the interval $t \in [0, 2\pi]$. Note that as the sampling rate gets higher, the eigenvalues outside the unit circle draw closer to the unit circle, thus producing DMD predictions that are valid for a longer window in time since the growth rates of these eigenvalues is smaller. The modes ψ_k associated with these eigenvalues μ_k are shown in Fig. 20.4 for the corresponding number of samplings. These modes are used in linear combination with their time dynamics given by the μ_k to give the approximation to the true PDE dynamics.

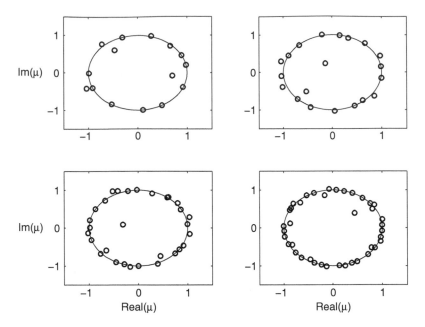

Figure 20.3: Spectral distribution of the eigenvalues μ_k as a function of the sampling rate over the interval $t \in [0, 2\pi]$. From top left to bottom right, the values are $M = 16, 21, 31$ and 41. Note that for higher sampling, the eigenvalues outside of the unit circle collapse to the unit circle, allowing for longer time dynamical predictions using the DMD modes. The corresponding eigenvectors are shown in Fig. 20.4.

An advantage of the DMD algorithm is that it actually can easily identify growth modes of a given physical system. In the example above, the growth modes were spurious. However, in many physical systems, there may indeed be a growth mode that one wishes to suppress. Thus the DMD framework is a natural way to apply the ideas of control theory, i.e. the growth modes are identified and a control algorithm is placed on the complex physical system with the aim of suppressing the pernicious growth mode(s). Such an application might be for the control of turbulent flow fields, for instance, where full simulations not only disagree with experimental observations quantitatively, but where such simulations are prohibitively expensive to do in real time. With the DMD framework, direct measurements can be performed in order to understand the underlying dynamics and growth modes for short windows of time into the future, thus allowing for control and manipulation of the flow field.

Finally, a comment should be made to conclude the section about the DMD method in comparison with the POD method. In the POD method, a low dimensional basis is selected from the data matrix **X** by use of the SVD. The dynamics of the governing equations are then projected onto this reduced basis and the resulting nonlinear dynamical system for the mode amplitudes evolved forward in time. With DMD, the dynamics are assumed to be linear so that an exact solution can be constructed. Thus there is no need to evolve a dynamical system forward in time for the mode amplitudes, rather, once the reduction is done and the DMD modes and their eigenvalues computed, then the future state of the system can be predicted without any addition work.

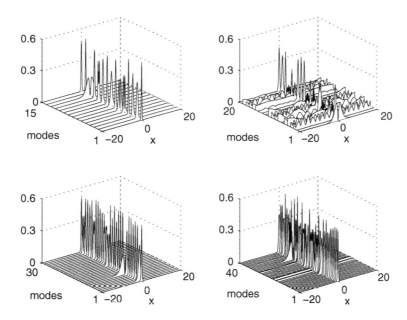

Figure 20.4: Eigenvectors associated with the eigenvalue distribution μ_k of Fig. 20.3. For $M = 16, 21, 31$ and 41, there are 15, 20, 30 and 40 eigenvectors generated, respectively, for reconstruction or approximation of the PDE solution using these DMD basis modes.

Thus the POD method retains the *nonlinear dynamics* of the system through its projection of the low dimensional modes onto the governing equations. DMD tries to construct the Koopman operator that best approximates the nonlinear dynamics directly without the need of a governing equation or additional evolution of any equation. Both have strengths and weaknesses, so that the method of choice usually boils down to taking maximal advantage of specific aspects of the problem at hand. But if the equations are unknown, hard to parameterize, overly approximated, etc, then DMD provides a viable method for extracting and potentially controlling some underlying complex system.

20.3 Applications of DMD

The example given in the previous section is nice in the sense that the computations for the PDE can be performed quite easily and an intuitive understanding of the dynamics is already present. Thus the comparison of the POD, DMD and full PDE solution techniques can be made and each of their advantages and disadvantages compared. But just like the POD method considered in Section 18.3, the primary goal is to apply the DMD method to complex physical systems where full advantage can be taken of its strengths. In this section, two examples will be given for how one might use the DMD method for the modeling of physical and/or computational systems.

Reduced order integrator using DMD/POD

Modern methods in scientific computing aim to exploit any underlying mathematical structure to make computations more efficient. Adaptive time-stepping schemes for ordinary and partial differential equations are an example of a numerical technique that has continued to have significant impact on almost any field of the mathematical, physical, biological and engineering sciences. Adaptive time-stepping with Runge–Kutta methods are used so universally that they are now an integral component in most commercial and open source scientific computing software packages. Indeed, adaptive time-stepping is the default in every numerical integrator in MATLAB including **ode45**, **ode23**, **ode 113**, etc. The advantage of these adaptive integrators is particularly profound in time-dependent multi-scale physics problems where large steps can be taken in certain regimes but more refinement is needed to accurately capture fast time-scales in other regimes.

Motivated by the increase in performance that can be obtained by adapting the integrators, it makes sense to consider hybrid numerical integration schemes that exploit, when possible, dimensionality reduction of a given system. To have maximal impact, the low dimensional basis must be adaptively selected when the dynamics of the system change, i.e. an initial low dimensional basis will be unable to accommodate dynamical changes orthogonal to the reduced basis. The POD method outlined in Section 18.3 is a standard method for generating a reduced order model. Improvements to the POD technique include nonlinear methods based on manifold learning techniques [74, 75], balanced POD (BPOD) reductions [76, 77, 78], traveling POD [79], trust-region POD [80] and sequential POD (SPOD) [81]. for instance. All these extensions to the basic POD methodology are aimed to improve the ability of the technique to model complex systems and/or to compute its underlying solution branches and bifurcation structure.

Although there are some exceptions, the generation of the POD modes is typically performed in a static and offline fashion where data for the modes are collected prior to generation of the model, and the POD basis itself remains unchanging despite any new incoming information. In order to make the numerical integrator more robust while retaining some of the computational benefits of the reduced model, we adaptively switch between the accurate but expensive PDE simulations, to collect data or evolve the solutions in regimes that are high dimensional, and reduced order simulations that, while inexpensive to integrate, require representative data in order to be constructed and lose accuracy should the dynamics of the system travel too far away from the location in phase space where data were obtained.

Here, exploitation is made of the predictive and equation-free nature of the dynamic mode decomposition (DMD) in conjunction with the potentially accurate POD technique [73]. If the predictions of the two are close, then we will continue to use the reduced model. If not, we revert to the full PDE. The result is a robust yet simple integration scheme that can, for appropriate and nontrivial types of system dynamics, retain many of the computational benefits of reduced order models while avoiding spurious results. Such a *dimensionally adaptive* scheme has been shown to be successful in reducing computational costs of complex systems [73].

As a specific example, consider application of these ideas to the complex Ginzburg–Landau equation (CQGLE) of mathematical physics.

$$u_t = \left(\frac{i}{2} - \tau\right) u_{xx} + \kappa u_{xxxx} + (i + \beta)|u|^2 u + (i\nu + \sigma)|u|^4 u + \gamma u \qquad (20.3.1)$$

where $\tau, \kappa, \beta, \nu, \sigma$ and γ are related to the specific application from which the equation arises. If these parameters are all zero, then the NLS equation results. The CQGLE can be used, for instance, to model so-called mode-locked lasers [59]. For our purposes, the parameters are chosen to be $\tau = 0.08, \kappa = 0, \beta = 0.66, \nu = -0.1$ and $\sigma = -0.1$ with γ varying in time as a step function. Specifically, $\gamma = -0.1$ aside from the interval $t \in [500/3, 1000/3]$ where $\tau = -0.2$. This causes the solution transitions and forces the POD/DMD solver back to the full PDE simulations.

The idea is to simulate (20.3.1). Methods for doing so are presented in previous chapters of this book, thus it is not a complicated thing to do. However, when observing the resulting dynamics, much of it is low dimensional in nature as there are a variety of *attracting steady states* in the system. Thus if the evolution becomes low dimensional during the evolution, the goal is to switch to the appropriate POD basis expansion and simulate the system at a fraction of the cost. Checking for low dimensional behavior is relatively easy and there are a variety of ways to do so [73]. The more difficult task is to assess wether the low dimensional simulation needs to revert back to the full PDE simulation. This happens, for instance, when one of many parameters of the CQGLE changes and the POD modes are incapable of representing the new (bifurcated) behavior. Two techniques are developed in Ref. [73] that set criteria for evaluating whether or not to continue in the POD basis or whether to return to the full PDE simulation.

Figure 20.5 illustrates the hybrid DMD/POD time-stepper that takes advantage of low dimensionality. Here, the parameter γ is a step function in time, causing the solution to change steady state attractors in the evolution. In the top left panel, the full, high-resolution PDE simulation is shown for the time interval $t \in [0, 500]$. The remaining top panels show the results of the DMD/POD stepper for the two different low dimensional criteria developed in the paper. The red region is where full PDE simulations occur while the black region represents the low dimensional POD evolution. The bottom panels represent the energy (L^2 norm) in the CQGLE laser cavity for the two methods proposed. As can be seen, aside from the transition regions, the dynamics can almost always be represented in a low dimensional POD basis, thus allowing for significant computational savings. Indeed, the simulation times drop by an order of magnitude in comparison with full PDE simulations [73].

DMD for a fluid jet

The previous example tied together the two concepts of DMD and POD. In this second example, DMD is used to extract the modal structure of a complex flow field and predict the dynamics of the modes. In this case, no governing equations, or underlying model, is assumed. Thus it is a purely data-driven approach to the problem. Details of the experiment can be found in

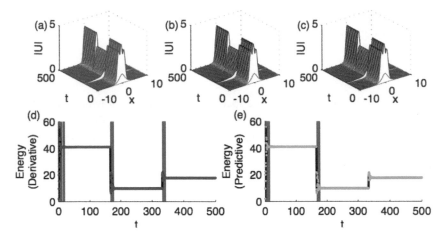

Figure 20.5: Comparison of the PDE evolution with the two proposed hybrid DMD/POD reductions. The full CQGLE PDE simulations are shown in the top left panel (a) while the two hybrid DMD/POD steppers are shown in the remaining top two panels (b) and (c). The red regions are where full PDE simulations occur while the black region is where low dimensional POD stepping occurs. The energy (L^2 norm) evolution for the two DMD/POD method are shown in the bottom panels (d) and (e). Note that because of the changes of γ in time, the DMD/POD method must return to the full PDE simulations. [from M. Williams, P. Schmid and J. N. Kutz, Hybrid reduced order integration with the proper orthogonal decomposition and dynamic mode decomposition, SIAM Journal of Multiscale Modeling and Simulation (2013) [73] ©SIAM]

Ref. [65]. Basically, data for the flow of a helium jet is visualized using Schlieren snapshots in time. Figure 20.6 shows the visualization of the flow field. This constitutes the data. Note that only a small region of the flow field will be used to extract modal structures and their dynamics.

Once the data collection process is finished, the data matrix $\mathbf{X}$ can be constructed and the DMD algorithm as outlined in the previous section can be implemented. Schmid *et al.* [65] uses a slightly modified implementation of the algorithm. Specifically, once the data preparation is accomplished and the two matrices $\mathbf{X}_1^{M-1}$ and $\mathbf{X}_2^M$ constructed, a QR algorithm is used instead of the SVD. These, of course, are closely related and ultimately give the same thing, i.e. they both generate the DMD modes and their eigenvalues μ_k.

Figure 20.7 shows the DMD spectrum for the helium jet for the Schlieren snapshots of the data. Both the real and imaginary parts of the eigenvalue are shown along with a colored ball of a given size. The marker size indicates the spatial coherence of the associated DMD modes. Thus larger balls correspond to large-scale structures in the fluid flow while smaller balls correspond to small-scale structures. The DMD spectrum shows the typical spectrum for an advection–diffusion dominated flow. Specifically, the eigenvalues show a strong alignment along the horizontal neutral stability line, but are overall damped. An eigenvalue at the origin (labelled a) is present which represents the steady state; its corresponding eigenvector displays the mean flow field. This feature is also present in a POD analysis of the same flow field: the first, most dominant, right singular vector of the snapshot basis consists of the mean flow. The least stable modes then fall

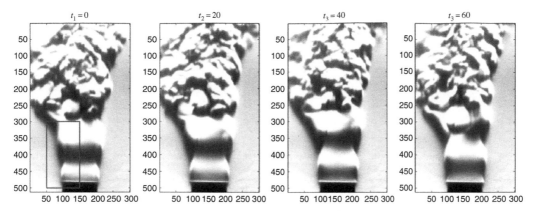

Figure 20.6: A selection of Schlieren snapshots of a helium fluid jet at Reynolds number $Re = 940$. The blue boxed region in the left panel indicates the data collection region for which time-resolved data are extracted for the DMD reduction. [From P. J. Schmid, L. Li, M. P. Juniper and O. Pust, Applications of the dynamic mode decomposition, Theoretical and Computational Fluid Dynamics **25**, 249–259 (2011) [65] ©Springer.]

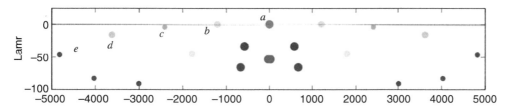

Figure 20.7: DMD spectrum for the flow field using Schlieren snapshots. Both the real and imaginary parts of the eigenvalue are shown along with a colored ball of a given size. The marker size indicates the spatial coherence of the associated DMD modes. Thus larger balls correspond to large-scale structures in the fluid flow while smaller balls correspond to small-scale structures. [From P. J. Schmid, L. Li, M. P. Juniper and O. Pust, Applications of the dynamic mode decomposition, Theoretical and Computational Fluid Dynamics **25**, 249–259 (2011) [65] ©Springer.]

on parabolic arcs, denoted by b through e. Since over the temporal observation period the flow is in an equilibrated state, we do not expect and observe any unstable eigenvalues. For even longer observation periods and data sequences, the damping rates of the dominant dynamic modes will approach the neutral line. The color coding and symbol size of the eigenvalues represent a measure of coherence of the associated modes (similar to the energy content of POD modes) and help in the separation of relevant structures from noise-contaminated ones. This coherence measure is computed by projecting individual dynamic modes onto energy-ranked proper orthogonal modes (except the mean flow); the resulting coefficients of this projection give a ratio of large-scale to fine-scale structures. This feature adds information about spatial coherence and scales to the temporal information contained in the spectrum. Figure 20.8 presents five dynamic modes associated with the labeled eigenvalues in Fig. 20.7. As mentioned before the mode corresponding

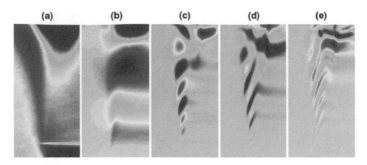

Figure 20.8: Five DMD modes generated from Schlieren snapshots. These five modes (a)–(e) correspond to the eigenvalues *a* through *e* in Fig. 20.7. Note that the modes show both large-scale and small-scale structures in the flow field. [From P. J. Schmid, L. Li, M. P. Juniper and O. Pust, Applications of the dynamic mode decomposition, Theoretical and Computational Fluid Dynamics **25**, 249–259 (2011) [65] ©Springer.]

to label *a* represents the mean flow, while the higher modes *b* through *e* increasingly show the presence of small-scale (but coherent) structures. The support of the dynamic modes is clearly located at the diverging outer edge of the jet where vortex roll-up, mixing and entrainment processes are prevalent. The combination of the spectral content and DMD mode structures allows for prediction of the future state of the system. Further, if one were interested in flow control, such predictions could be used in a control algorithm for controlling the fluid jet nozzle or other parameters in the physical system. Much like the POD method, the DMD reduction highlights the key features for consideration in the complex system of interest. It is yet another tool that allows for understanding of a given system.

Data Assimilation Methods 21

Most of the data-driven techniques presented in this book were applied to systems where the underlying governing equations were prescribed. However, in the DMD method (or in the equation-free method), no governing equations were required to extract meaningful information about the dynamics of the complex system under consideration. The method of *data assimilation* is a hybrid method that uses both data measurements collected in time about the system in conjunction with a prescribed set of governing equations. The fact is, both the measurements and simulations are in practice heavily influenced by uncertainty and/or noise fluctuations. Thus neither the experiment nor the theoretical model can be fully trusted. However, combining the two so that the experimental measurements helps inform the model and vice versa can greatly improve the predictive powers of the model, or analysis of the state of the system [83, 84, 85, 86]. Thus data is assimilated into the model predictions and hence the name. Data assimilation is potentially one of the most useful data-driven modeling techniques as we are rarely without some underlying governing equations or without experimental measurements. And to make optimal use of both, data assimilation techniques are ideal.

21.1 Theory of Data Assimilation

To begin thinking about data assimilation, we will first consider a given complex system and its underlying dynamical evolution. Specifically, let us begin by assuming that the system under consideration has the following evolution equation

$$\frac{d\mathbf{y}}{dt} = f(t, \mathbf{y}) \tag{21.1.1}$$

with the initial state of the system given by

$$\mathbf{y}(0) = \mathbf{y}_0. \tag{21.1.2}$$

Of course, techniques for solving such a system have been considered extensively throughout this book. Indeed, provided that $f(t, \mathbf{y})$ is well behaved, i.e. continuous and differentiable, for instance, then a solution to the prescribed problem can be solved for uniquely. In fact, for an N-dimensional vector $\mathbf{y}$, N initial conditions are prescribed by $\mathbf{y}_0$ so that the number of unknowns and constraints match.

Up to this point, we have ignored and/or denied a simple and obvious truth regarding the evolution equation (21.1.1) and its initial conditions (21.1.2). Specifically, we are assuming that we can perfectly prescribe both! In practice, this is surely an impossible task. And for strongly nonlinear systems (21.1.1), even small changes in the initial conditions and/or slight noise perturbations to the system can lead to large changes in the dynamical behavior, stability and predictability of (21.1.1). To be more precise then about our formulation of the true system, (21.1.1) and (21.1.2) should be modified to

$$\frac{d\mathbf{y}}{dt} = f(t, \mathbf{y}) + \mathbf{q}_1(t) \tag{21.1.3}$$

with the initial conditions

$$\mathbf{y}(0) = \mathbf{y}_0 + \mathbf{q}_2 \tag{21.1.4}$$

where $\mathbf{q}_1$ represents unknown model errors due to either noise fluctuations in the system or perhaps truncation of higher order effects in the system that are thought to be negligible. Similarly, the vector $\mathbf{q}_2$ represents the error in the prescribed initial conditions either because we cannot accurately measure them in practice or prescribe them in a realistic physical system.

Even in the presence of the perturbations $\mathbf{q}_1$ and $\mathbf{q}_2$, the system of equations (21.1.3) and (21.1.4) remains well-posed with a unique solution specifying the dynamics in time. However, for a strongly nonlinear system, the presence of $\mathbf{q}_1$ and $\mathbf{q}_2$ make it virtually impossible to use the governing equations (21.1.3) and the initial conditions (21.1.4) for an accurate prediction of the future state of the system. This is largely due to the concept of *sensitivity to initial conditions* that is displayed in many complex systems. Of course, this then gives rise to the following fundamental question: Is modeling a worthwhile exercise if it cannot predict the future state of a realistic system? In practice, great engineering is often about eliminating large unknowns in a system by essentially suppressing, as much as possible, the vectors $\mathbf{q}_1$ and $\mathbf{q}_2$. But for highly complex systems, such as climate modeling and weather prediction, the system simply cannot be engineered and one must find effective strategies for dealing with the inherent effect of $\mathbf{q}_1$ and $\mathbf{q}_2$.

Data assimilation is a technique that attempts to mitigate the problems associated with having $\mathbf{q}_1$ and $\mathbf{q}_2$ present in a given system. As the name suggests, the idea is to assimilate experimental measurements directly into the theoretical model in order to inform the dynamics. Thus a set of measurements are taken so that

$$g(t, \mathbf{y}) + \mathbf{q}_3 = 0 \tag{21.1.5}$$

where $g(t, \mathbf{y})$ is a certain set of measurements, let's say M of them, on some quantities related to the state vector $\mathbf{y}$, and the vector $\mathbf{q}_3$ is the error measurement associated with the data collection process.

In principle, it is a great idea to incorporate real experimental measurements into the modeling process. In practice, the addition of (21.1.5) now makes for an overdetermined system

(N unknowns and $N + M$ constraints) for **y** for which no solution exists in general. Thus the tradeoff of using the experimental data is that we went from a well-posed system with a unique solution to an overdetermined system with no general solution.

To deal with the overdetermined system, the following *quadratic form* is introduced:

$$J(\mathbf{y}) = \int_0^T \int_0^T \mathbf{q}_1^T(t_1)\mathbf{W}_1\mathbf{q}_1(t_2)dt_1 dt_2 + \mathbf{q}_2^T\mathbf{W}_2\mathbf{q}_2 + \mathbf{q}_3^T\mathbf{W}_3\mathbf{q}_3 \tag{21.1.6}$$

where the error vectors $\mathbf{q}_j$ are all directly included in the quadratic form. The matrices $\mathbf{W}_j$ are the inverse of the error covariance for the model, initial conditions and measurements, respectively. The introduction of such a quadratic form is motivated by one primary purpose: optimization. In particular, one potential solution to the overdetermined system is to find the solution that minimizes the weighted error, weighted with respect to the model, initial condition and measurement error, as given by the quadratic form $J(\mathbf{y})$. Since it is a quadratic form, standard convex optimization methods can be directly applied to find a solution. The least-square error model defined by $J(\mathbf{y})$ is only one potential choice. However, this choice is particularly attractive when considering Gaussian statistics for the error vectors. In this case, minimizing $J(\mathbf{y})$ is equivalent to maximizing the probability density function $P(\mathbf{y}) = C\exp[-J(\mathbf{y})]$. Thus the minimum of (21.1.6) is also the maximum-likelihood estimate.

Thus to summarize the data assimilation method, we consider a theoretical model under the influence of some error (21.1.3) subject to initial conditions which are also subject to error (21.1.4). A number of experimental measurements (21.1.5), also subject to error, are then made to inform the model. The combined errors of the resulting overdetermined system are cast as a quadratic form for which convex optimization techniques can be used to find the best-fit solution (in the L^2 sense). As a result, the experimental measurements are assimilated with the model predictions to generate better predictions of the dynamical behavior in time. The development of these key ideas will be carried forward in the next two sections.

Data assimilation for a single random variable

To illustrate the ideas of data assimilation, the simplest toy example will be considered (see, for example, the nice arguments by Holton and Hakim [82] in the context of atmospheric sciences). Specifically, consider a model that generates some prediction about the state of the system, thus effectively $\mathbf{q}_1$ and $\mathbf{q}_2$ would be combined in this model prediction process. Also, consider some experimental measurements on the systems that are also subject to error. Thus for the variable x, there are two predictions about its true value: one from the model and one from experiment. The idea is to combine these two measurements to arrive at a better prediction of the true value of x.

To begin this calculation, consider the following conditional probability statement from Bayes's formula (see Section 12.2):

$$p(x|y) = \frac{p(y|x)p(x)}{p(y)} \tag{21.1.7}$$

where $p(x)$ represents the probability density for predicting the variable x. Here, y will represent the observation (experimental assimilation) that will be made. Thus $p(x|y)$ is the probability of finding x conditioned on having measured y. At this point, we will not concern ourselves with $p(y)$ since it will simply represent a scaling factor for (21.1.7). The probability density function $p(y|x)$ is a *likelihood function* since it is a function of the variable x.

As is the case with almost all problems in probability theory, great simplifications can be made by assuming Gaussian distributed random variables. In fact, our treatment of the data assimilation problem for higher dimensional problems relies explicitly on this assumption in order to derive something tractable. It also helps simplify the current one-variable problem if this assumption is made. Thus consider the following probability density distributions

$$p(y|x) = c_1 \exp\left[-\frac{1}{2}\left(\frac{y-x}{\sigma_y}\right)^2\right] \tag{21.1.8}$$

$$p(x) = c_2 \exp\left[-\frac{1}{2}\left(\frac{x-x_0}{\sigma_0}\right)^2\right] \tag{21.1.9}$$

where σ_y is the error variance for the observation, and x_0, σ_0 are the predicted model mean and its associated error variance. The constants c_1 and c_2 are normalization constants ensuring that probability density functions integrate to unity. Thus the errors, both in measurement (σ_y) and theory (σ_0), are characterized by the variance parameters. Note that without any data assimilation, the model would predict the value x_0. Data assimilation attempts to give a correction to this value using the measurement data.

Using the conditional probability statement, which integrates in information about the observation y, along with the assumed Gaussian probability distributions (21.1.9) yields the following prediction (what is x given observation y?) for the state of the system

$$p(x|y) = c_3 \exp\left[-\frac{1}{2}\left(\frac{y-x}{\sigma_y}\right)^2\right] \exp\left[-\frac{1}{2}\left(\frac{x-x_0}{\sigma_0}\right)^2\right] \tag{21.1.10}$$

where c_3 is another constant used for convenience.

Our goal now is to construct a quadratic form like (21.1.6) for this problem in order to determine how to modify x from its default prediction value of x_0 in light of our observation y. A very simple quadratic form to construct from the Gaussian distributions is as follows

$$J(x) = -\log[p(x|y)] + \log(c_3) = \frac{1}{2}\left(\frac{y-x}{\sigma_y}\right)^2 + \frac{1}{2}\left(\frac{x-x_0}{\sigma_0}\right)^2. \tag{21.1.11}$$

But this quadratic form can be easily minimized as it is simply a parabola in one dimension, i.e. optimization can be performed in closed form unlike the general formulation (21.1.6). To find its minimum, $dJ(\bar{x})/dx = 0$ is computed. This yields the following value for $\bar{x}$ at the minimum

$$\bar{x} = \left(\frac{\sigma_y^2}{\sigma_y^2 + \sigma_0^2}\right)x_0 + \left(\frac{\sigma_0^2}{\sigma_y^2 + \sigma_0^2}\right)y. \tag{21.1.12}$$

Thus $\bar{x}$, which is a weighted linear superposition of the model prediction x_0 and the observation y, is the new and improved prediction for the outcome of x given the observation y. Such is the change of value when using the conditional probability argument.

Equation (21.1.12) has some very intuitive features. First and foremost: if there is no error in the experimental observation, then $\sigma_y = 0$ and $\bar{x} = y$. This is essentially a statement of self-consistency. It would be very bad if the data assimilation method failed to choose the observational data value if it was error free. The error variance for $\bar{x}$ can also be computed to be

$$\bar{\sigma}^2 = \frac{\sigma_0^2}{1 + \left(\sigma_0^2/\sigma_y^2\right)} = \frac{\sigma_y^2}{1 + \left(\sigma_y^2/\sigma_0^2\right)} < \sigma_0^2, \sigma_y^2 . \tag{21.1.13}$$

This is also a self-consistency check. Namely, the error in the data assimilation prediction is always better than the error of either the model alone or observation alone. Again, the fact that you are using the combination of model and data should always improve your predictions and error.

Figure 21.1 shows the three probability densities of interest. In particular, what is shown is the probability density given by the model, $p(x)$, the observation, $p(y|x)$ and the data assimilation, $p(\bar{x}) = p(y|x)p(x)$. For this figure the following were assumed $x_0 = -0.5$, $\sigma_0 = 0.5$, $y = 0.2$ and $\sigma_y = 0.3$. This allows us to compute the distribution of the assimilated prediction through (21.1.12) and (21.1.13). Note that the variance of the assimilated distribution is quite narrow and shifted strongly towards the observational distribution. This is largely because the observation has a much smaller variance than the model data.

Another way to express these results is by noting that

$$\bar{x} = x_0 + K(y - x_0) \tag{21.1.14}$$

with $\bar{\sigma}^2 = (1 - K)\sigma_0^2$ and where

$$K = \frac{\sigma_0^2}{\sigma_0^2 + \sigma_y^2} \leq 1 . \tag{21.1.15}$$

The second term in the right-hand side of (21.1.14), $K(y - x_0)$, is the so-called *innovation* since it brings in new information (an observation) to the prediction of x. The predicted value of x is a linear combination of its model prediction x_0 and the innovation. If there is no innovation, i.e. no new information, then the prediction remains x_0. The parameter $0 < K < 1$ is the gain weighting factor and is essentially the so-called *Kalman filter* or *Kalman gain*. This will be considered in more detail in the next section.

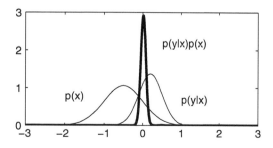

Figure 21.1: Distributions of the model (left Gaussian, $p(x)$), the observation (right Gaussian, $p(y|x)$), and the data assimilated distribution $p(\bar{x}) = p(y|x)p(x)$. Note that the error variance of the assimilated distribution is much narrower than either the model or observational data as shown in (21.1.13).

21.2   Data Assimilation, Sampling and Kalman Filtering

The preceding section outlined the high-level view of the data assimilation method and showed how to implement it on the simplest example possible. What is desired now is to develop the theory further for implementation in realistic systems. Of particular note will be the role of the *Kalman filter*, or as already hinted at in the last section, the method of incorporating *innovation* to the model predictions.

Relating observations to the state vector

Before proceeding to derive the Kalman filter for a general vector system, we address the issue of how observations (21.1.5) project onto the state variable **y** of the governing equation (21.1.1). For instance, the governing equation (21.1.1) may be the simulation of some underlying PDE system that is solved on a rectangular grid of a prescribed domain. Observations, however, may be made anywhere in the domain and will, in general, not be aligned with the grid used for (21.1.1). In weather prediction, the data are collected at observation points which may be on the coast, on top of mountains and/or in metropolitan areas. Certainly the observation points are not aligned on a regular grid. The simulation of a geographic region, however, is most likely accomplished using a given discretization, perhaps in tens of meters to kilometers. The question is how to overlay the irregularly spaced observations onto the regularly spaced state variables. The simplest thing to do is to use a linear interpolation algorithm to generate values of the state variables at each grid point of the model.

The mapping of the experimental observations can be accomplished mathematically in the following way

$$\mathbf{y}(t) = \mathbf{Hx}(t) + \mathbf{q}_3 \tag{21.2.1}$$

where $\mathbf{y}(t)$ are the observations at a given time t of the state vector $\mathbf{x}$, $\mathbf{H}$ is a matrix that maps the state vector to the observations and $\mathbf{q}_3$ is the observational error. This is a linear version of the more general form (21.1.5). This step is always necessary once a data collection has been applied at a time t.

The one-dimensional Kalman filter

The derivation of the projection operator $\mathbf{H}$ in (21.2.1) is necessary for deriving the Kalman filter. To start the derivation process, we once again return to one-dimensional considerations and a highly idealized version of the dynamics in discrete form. In particular, following Miller [84], we can consider the mapping from a time t_k to a time t_{k+1}. The full dynamics, without approximation, is assumed to be given by

$$x_{k+1} = f(x_k) + q_{k+1} \tag{21.2.2}$$

where x_{k+1} and x_k are the state of the one-dimensional system at time t_{k+1} and t_k, respectively, and q_{k+1} is a Gaussian white noise sequence. The model approximation to this system is given by

$$x_{0_{k+1}} = f\left(x_{0_k}\right) \qquad (21.2.3)$$

where x_{0_k} is the best estimate of the state of the system at time t_k, sometimes called the *analysis*, and $x_{0_{k+1}}$ is the forecast of the dynamics using this estimate at time t_{k+1}.

The error between the truth and the forecast at time t_{k+1} is then given by

$$x_{k+1} - x_{0_{k|1}} = f(x_k) - f\left(x_{0_k}\right) + q_{k+1}. \qquad (21.2.4)$$

By Taylor expanding $f(x_k)$ around x_{0_k}, the above expression can be written as follows

$$x_{k+1} - x_{0_{k+1}} = (x_k - x_{0_k})f'(x_{0_k}) + \frac{1}{2}(x_k - x_{0_k})^2 f''(x_{0_k})$$
$$+ \frac{1}{6}(x_k - x_{0_k})^3 f'''(x_{0_k}) + \cdots + q_{k+1}. \qquad (21.2.5)$$

To find the error variance between the correct solution x_{k+1} and our prediction $x_{0_{k+1}}$, the expectation value of the quantity $(x_{k+1} - x_{0_{k+1}})^2$ is computed. Denoting this as $E[(x_{k+1} - x_{0_{k+1}})^2]$, we find, by squaring the above expression and taking the expectation,

$$E\left[(x_{k+1} - x_{0_{k+1}})^2\right] = E\left[(x_k - x_{0_k})^2\right](f'(x_{0_k}))^2 + \text{h.o.t} + E\left[q_{k+1}^2\right] \qquad (21.2.6)$$

where h.o.t. denotes all higher order moment terms, i.e. the skewness and kurtosis, for instance, of the error. Keeping these terms represents a closure problem for the error. However, as a first approximation, it is often the case that the higher order moments are neglected. Thus we can discard them from the calculation at this point. However, it should be noted that there is a great deal of work that investigates the effect of preserving the higher order moments in the calculation. One might imagine that the more information that is retained, the better the approximation should work.

To simplify the notation, we define the following two quantities

$$P_{k+1} = E\left[(x_{k+1} - x_{0_{k+1}})^2\right] \qquad (21.2.7a)$$
$$P_k = E\left[(x_k - x_{0_k})^2\right] \qquad (21.2.7b)$$

which are measures of the error variance between the true solution and the prediction at t_{k+1}, and the true solution and our best estimate of it at t_k, respectively. This notation gives

$$P_{k+1} = P_k(f'(x_{0_k}))^2 + E\left[q_{k+1}^2\right]. \qquad (21.2.8)$$

Note that P_{k+1} accounts for the dynamics (and errors associated with it) while P_k accounts for errors in estimating the initial state.

To make a data assimilated prediction, $\bar{x}_{k+1}$, of the state of the system at time t_{k+1} using an observation y_{k+1}, we then apply the following formula

$$\bar{x}_{k+1} = x_{0_{k+1}} + K_{k+1}(y_{k+1} - x_{0_{k+1}}) \qquad (21.2.9)$$

where the innovation is again given by the quantity $y_{k+1} - x_{0_{k+1}}$ and the Kalman gain K_{k+1} is given by

$$K_{k+1} = \frac{P_{k+1}}{P_{k+1} + R} \tag{21.2.10}$$

where R is the observation error variance. If there is no observational error, then $R = 0$ which forces $K_{k+1} = 1$. This in turn gives the assimilated prediction to be $\bar{x}_{k+1} = y_{k+1}$, i.e. the assimilated prediction is exactly the error-free experimental observation. The variance associated with the error of our prediction $\bar{x}_{k+1}$ is given by

$$\bar{P}_{k+1} = (1 - K_{k+1})P_{k+1}. \tag{21.2.11}$$

Taken together, the data assimilation results, which now directly tie in the dynamics of the system through (21.2.2), define what is called the *extended Kalman filter* (EKF).

The vector EKF

The vector version of the above one-dimensional argument follows in a straightforward manner [84, 87]. The dynamics given by (21.2.2) and (21.2.3) are now written

$$\mathbf{x}_{k+1} = f(\mathbf{x}_k) + \mathbf{q}_{k+1} \tag{21.2.12}$$

and

$$\mathbf{x}_{0_{k+1}} = f(\mathbf{x}_{0_k}), \tag{21.2.13}$$

respectively. Once again Taylor expanding now yields the covariance evolution

$$\mathbf{P}_{k+1} = \mathbf{J}(\mathbf{f})\mathbf{P}_k\mathbf{J}(\mathbf{f})^T + \mathbf{Q} \tag{21.2.14}$$

where $\mathbf{J}(\mathbf{f})$ is the Jacobian generated from the multi-dimensional Taylor expansion and higher order moments have been neglected. This formula is equivalent to (21.2.8).

Given a data observation $\mathbf{y}_{k+1}$ at time t_{k+1}, the vector case requires an appropriate mapping of the observation space to the state space. But this was already discussed and is mathematically transcribed in (21.2.1). Finally, the data assimilated prediction of the correct state is given by

$$\bar{\mathbf{x}}_{k+1} = \mathbf{x}_{0_{k+1}} + \mathbf{K}_{k+1}\left(\mathbf{y}_{k+1} - \mathbf{H}\mathbf{x}_{0_{k+1}}\right) \tag{21.2.15}$$

where $\mathbf{K}_{k+1}$ is the Kalman gain matrix given by

$$\mathbf{K}_{k+1} = \mathbf{P}_{k+1}\mathbf{H}^T(\mathbf{H}\mathbf{P}_{k+1}\mathbf{H}^T + \mathbf{R})^{-1} \tag{21.2.16}$$

and $\mathbf{R}$ is the noise covariance matrix. The error covariance of the updated state vector is given by

$$\bar{\mathbf{P}}_{k+1} = (\mathbf{I} - \mathbf{K}_{k+1}\mathbf{H})\mathbf{P}_{k+1} \tag{21.2.17}$$

where $\mathbf{I}$ is the identity matrix. In vector form, the innovation is given by $\mathbf{y}_{k+1} - \mathbf{H}\mathbf{x}_{0_{k+1}}$. Thus the matrix $\mathbf{H}$ serves an important role in overlaying the data measurement locations with the underlying grid used for computationally evolving forward the model dynamics. This is the EKF for generic complex systems.

With the EKF now established, its strengths and weaknesses are briefly considered. Its strengths are obvious: the EKF provides a systematic way to integrate observational data into the modeling process, thus improving the model predictions. Its most obvious drawbacks are computational. Specifically, for complex systems where the state vector is defined by potentially millions or billions of variables on a large computational grid, computing the innovation becomes unwieldy, especially as these large matrices need to be inverted and Jacobians found. Such enormous computational expense can render the method useless from a practical point of view. One potential way to deal with such computational complexity is to minimize (21.1.6) directly using, for instance, gradient descent algorithms. This is ultimately faster than computing Jacobians and inverses of the large matrices under consideration. Another potential way to solve the problem is to use ensemble Kalman filtering (EnKF) techniques which render the problem tractable by processing observations one at a time or by breaking the domain into smaller subdomains where the matrices are tractable. Such computational considerations are necessary to consider in the types of systems (weather prediction and climate modeling [82]) where data assimilation has becomes a standard method of analysis.

21.3 Data Assimilation for the Lorenz Equation

To demonstrate the data assimilation algorithm in practice, consideration will be given to the Lorenz equations

$$x' = \sigma(y - x) \tag{21.3.1a}$$

$$y' = rx - y - xz \tag{21.3.1b}$$

$$z' = xy - bz. \tag{21.3.1c}$$

Thus the vector $\mathbf{x} = [x\ y\ z]^T$ is the three-degree of freedom state vector whose nonlinear evolution $f(\mathbf{x})$ is specified by the right-hand side (Lorenz model) in (21.3.1). The Lorenz equations are a highly simplified model of convective-driven atmospheric motion (see Section 24.3 for a derivation of the system).

To begin, a perfect simulation will be performed of the Lorenz system, i.e. a simulation that includes specified initial conditions without noise and an evolution which is free of stochastic forcing. This simulation will be the *truth* that the data assimilation will try to reproduce. The parameters to be simulated here, and in what follows, are standard in many examples: $\sigma = 10$, $b = 8/3$ and $r = 28$. The large value of r puts the system in a dynamical state that is highly sensitive to initial conditions. The perfect initial conditions will be $\mathbf{x}_0 = [5\ 5\ 5]^T$. The following MATLAB code solves this problem for the time domain $t \in [0, 20]$ and plots it in a parametric way in 3D.

```
t=0:0.01:20;
sigma=10; b=8/3; r=28;

x0=[5 5 5];
[t,xsol]=ode45('lor_rhs',t,x0,[],sigma,b,r)

x_true=xsol(:,1); y_true=xsol(:,2); z_true=xsol(:,3);
figure(1), plot3(x_true,y_true,z_true)
```

The right-hand side of the differential equation includes the dynamics through $f(\mathbf{x})$. In this case, the function **lor_rhs** is given by

```
function rhs=lor_rhs(t,x,dummy,sigma,b,r)
rhs=[sigma*(-x(1)+x(2))
     -x(1)*x(3)+r*x(1)-x(2)
     x(1)*x(2)-b*x(3)];
```

The results of simulating the Lorenz equation are shown in Figs. 21.2 and 21.3. The first of these figures demonstrates the standard *butterfly* pattern of evolution of the strange attractor in three

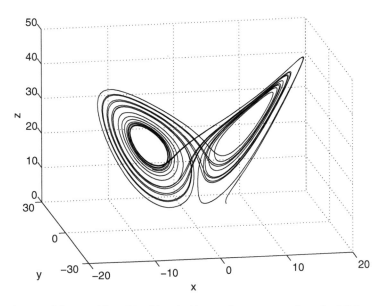

Figure 21.2: Evolution of the variables $x(t), y(t)$ and $z(t)$ over the time period $t \in [0, 20]$ for $\sigma = 10, b = 8/3$ and $r = 28$. The evolution is shown parametrically as a function of time.

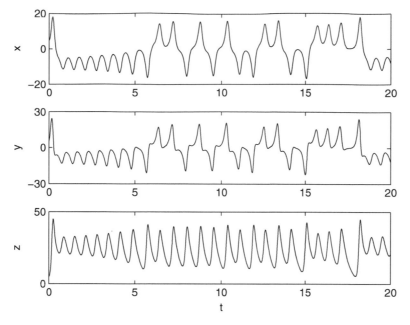

Figure 21.3: Time evolution of the variables $x(t)$, $y(t)$ and $z(t)$ over the time period $t \in [0, 20]$ for $\sigma = 10$, $b = 8/3$, $r = 28$ and $\mathbf{x}_0 = [5\ 5\ 5]^T$. In this parameter regime, specifically for this large value of r, the dynamics of the Lorenz equations are highly sensitive to initial conditions.

dimensions. In this graph, time is a parametric quantity. The second of these figures illustrates the evolution of the variables $x(t)$, $y(t)$ and $z(t)$ over the time period $t \in [0, 20]$. In what follows, instead of tracking and comparing all the variables, we will focus on $x(t)$ for illustrative purposes only.

Sensitivity to initial conditions

The first thing that will be investigated is sensitivity of the evolution to small changes in the initial conditions. The mathematical statement of this problem is given in (21.1.2). Thus the effect of $\mathbf{q}_2$ on the dynamics will be considered. And in particular, it is the perturbation of the initial conditions that compromises the predictive power of the theoretical model. To simplify this, we will assume that the error has a Gaussian distribution so that

$$\mathbf{x}(0) = \mathbf{x}_0 + \sigma_2 \mathbf{q}(0, 1) \tag{21.3.2}$$

where $\mathbf{x}_0$ is the perfect initial conditions and $\mathbf{q}(0, 1)$ is a Gaussian distributed random variable with mean zero and unit variance. Thus σ_2 is chosen to make the error variance either larger or smaller.

With this error, the evolution of (21.3.1) can once again be explored. In Fig. 21.4, eight realizations of the evolution are given using the initial conditions (21.3.2). The exact evolution which we

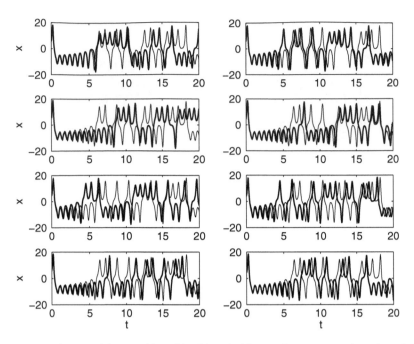

Figure 21.4: Time evolution of the variables $x(t)$, $y(t)$ and $z(t)$ over the time period $t \in [0, 20]$ for $\sigma = 10$, $b = 8/3$, $r = 28$ and $\mathbf{x}_0 = [5 \ 5 \ 5]^T$. Here, eight realizations are shown for the perturbed initial conditions given by (21.3.2) with $\sigma_2 = 1$. Note that for all simulations, after $t \approx 5$ the true dynamics (light line) differ from the dynamics with perturbed initial conditions (bold line). Thus prediction of the dynamics beyond this time is virtually impossible given such perturbations (errors) in the initial data.

are trying to model (with $\sigma_2 = 0$) is the thinner solid line while the initial conditions with error (with $\sigma_2 = 1$) is the bolded line. The following MATLAB code generates the eight realizations:

```
sigma2=1;   % error variance
for j=1:8
  xic=x0+sigma2*randn(1,3); % perturb initial conditions
  [t,xsol]=ode45('lor_rhs',t,xic,[],sigma,b,r);
  x=xsol(:,1);   % projected x values
  subplot(4,2,j), plot(t,x_true,'k'), hold on
  plot(t,x,'k','Linewidth',[2])
end
```

For all these realizations, the projected state fails to model the *true* dynamics after $t \approx 5$. Indeed, after this time, there is almost no correlation of closeness between the truth and our projection based upon noisy initial data. This illustrates the need for data assimilation. Specifically, it is hoped that occasional measurements of the data would allow for an accurate prediction of the true future state far beyond $t \approx 5$.

Data assimilation for the Lorenz equations

The goal in this example will be to simply illustrate the EKF algorithm. Thus a simple case will be taken which can be completely coded in MATLAB in a fairly straightforward manner. In this simple example, no stochastic forcing of the differential equations will be considered so that the dynamics are *exactly* as specified in (21.3.1). Said another way, the error vector $\mathbf{q}_1$ in (21.1.6) is zero. However, there will be error both in the measurements ($\mathbf{q}_3$) and the initial conditions ($\mathbf{q}_2$). Knowing full well about the sensitivity to initial conditions in the Lorenz equations, the initial noise will greatly compromise the ability of the model to predict the future state of the system. The hope is that data assimilation will mitigate this problem to some extent and allow for much more accurate, and longer time, predictions for the future state of the system.

In our example, the data collection points will be at simulation time points already specified by our differential equation solver. Moreover, data will be taken from all variables. Thus the mapping matrix is $\mathbf{H} = \mathbf{I}$. To illustrate the data collection process, consider then the modification of (21.2.1) to

$$\mathbf{y}(t_n) = \mathbf{x}(t_n) + \sigma_3 \mathbf{q}(0, 1) \tag{21.3.3}$$

where t_n is a measurement time point and $\mathbf{q}(0, 1)$ is a Gaussian distributed random variable with mean zero and unit variance. Thus σ_3 is chosen to make the error variance either larger or smaller in the data measurements. Figure 21.5 shows the true dynamics (line) and data collected every half-unit of time with $\sigma_3 = 4$ (circles). As is clearly seen, the data are collected under error and do not match up perfectly with the true dynamics. However, they do follow the true dynamics

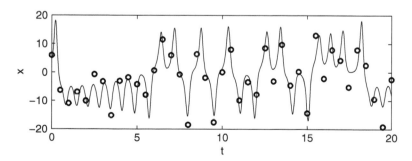

Figure 21.5: True time dynamics of the variable $x(t)$ over the time period $t \in [0, 20]$ for $\sigma = 10$, $b = 8/3$, $r = 28$ and $\mathbf{x}_0 = [5 \; 5 \; 5]^T$. The circles represent experimental measurements at every half-unit of time of the true dynamics with error given by (21.3.3) with $\sigma_3 = 4$. The dynamics of the variables $y(t)$ and $z(t)$ are similar. Data assimilation makes use of the experimental measurements to keep the model predictions closer to the true dynamics.

fairly closely over the time period of integration. To compute these data points in MATLAB, the following code is used in conjunction with the code that generates the true dynamics:

```
% noisy obserations every t=0.5
tdata=t(1:50:end);
n=length(tdata)
xn=randn(n,1); yn=randn(n,1); zn=randn(n,1);
sigma3=4;  % error variance in data
xdata=x(1:50:end)+sigma3*xn;
ydata=y(1:50:end)+sigma3*yn;
zdata=z(1:50:end)+sigma3*zn;
```

The idea is to use these experimental points along with the noisy initial conditions shown in Fig. 21.4 in order to enhance our prediction for the future state.

Given that $\mathbf{H} = \mathbf{I}$ in this example, this reduces the EKF algorithm to computing

$$\bar{\mathbf{x}}_{k+1} = \mathbf{x}_{0_{k+1}} + \mathbf{K}_{k+1}\left(\mathbf{y}_{k+1} - \mathbf{x}_{0_{k+1}}\right) \tag{21.3.4}$$

where $\mathbf{K}_{k+1}$ is the Kalman gain matrix given by

$$\mathbf{K}_{k+1} = \mathbf{P}_{k+1}(\mathbf{P}_{k+1} + \mathbf{R})^{-1} \tag{21.3.5}$$

and $\mathbf{R}$ is the noise covariance matrix. The error covariance of the updated state vector is given by

$$\bar{\mathbf{P}}_{k+1} = (\mathbf{I} - \mathbf{K}_{k+1})\mathbf{P}_{k+1} . \tag{21.3.6}$$

Note that the matrices involved are 3×3 matrices and the innovation is simply given by $\mathbf{y}_{k+1} - \mathbf{x}_{0_{k+1}}$. From (21.2.14), and using the fact that the dynamics propagates in an error-free fashion ($\mathbf{q}_1 = 0$), then

$$\mathbf{P}_{k+1} = \mathbf{J}(\mathbf{f})\mathbf{P}_k\mathbf{J}(\mathbf{f})^T \tag{21.3.7}$$

where the Jacobian for the Lorenz equation can be easily computed to give

$$\mathbf{J}(\mathbf{f}) = \begin{bmatrix} -\sigma & \sigma & 0 \\ r - z & -1 & -x \\ y & x & -b \end{bmatrix} \tag{21.3.8}$$

and the matrix $\mathbf{P}_k$ measures the error in estimating the initial state of the system at time t_k. This error is determined by (21.3.2) and the parameter σ_2.

Everything is in place then to implement the data assimilation procedure. The following is an algorithmic outline of what needs to occurs:

(i) Determine the sources of error and how to incorporate them. The error in the data measurement determines the matrix $\mathbf{R}$ while the error in the initial condition determines the matrix $\mathbf{P}_k$ (note that we are ignoring errors generated in the dynamics themselves).

(ii)   Compute the Jacobian at time t_k using the best estimate for the state vector $\mathbf{x}_0(t_k)$ and combine it with the computation of $\mathbf{P}_k$ in order to compute $\mathbf{P}_{k+1}$.

(iii) With $\mathbf{P}_{k+1}$ and $\mathbf{R}$, compute the Kalman gain matrix $\mathbf{K}_{k+1}$.

(iv)   Compute the new state of the system using the innovation vector and the Kalman gain matrix.

(v)   Use the new state of the system to again project to another time into the future where observational data is once again available.

The Lorenz equation is a fairly trivial example to consider. Moreover, our treatment of the data assimilation will be for the simplest case possible. In this example, the error variance for both the initial conditions and data measurements will both be unity so that $\sigma_2 = \sigma_3 = 1$, respectively. The Kalman gain matrix will then be given as in the one-dimensional case: $K = \sigma_2/(\sigma_2 + \sigma_3)$. A code for simulating this system and making adjustments based upon the data measurements is as follows:

```
x_da=[];   % data assimilation solution
for j=1:length(tdata)-1   %  step through every t=0.5
  tspan=0:0.01:0.5;   % time between data collection
  [tspan,xsol]=ode45('lor_rhs',tspan,xic,[],sigma,b,r);

  xic0=[xsol(end,1); xsol(end,2); xsol(end,3)]   % model estimate
  xdat=[xdata(j+1); ydata(j+1); zdata(j+1)] % data estimate
  K=sigma2/(sigma2+sigma3);   % Kalman gain
  xic=xic0+(K*[xdat-xic0])   %  adjusted state vector

  x_da=[x_da; xsol(1:end-1,:)];   % store the data
end
x_da=[x_da; xsol(end,:)];   % store last data time
```

In this simulation, the vector **data-xic0** is the innovation vector that is weighted according to the Kalman gain matrix. Figure 21.6 shows the results of this simulation for one representative realization of the error vectors. In the top panels, the nondata-assimilated computation is shown showing that the simulation solution diverges from the true solution around $t \approx 5$ (see also Fig. 21.4). The error between the true solution and the model solution is shown in the right panel. Note that the error is quite large around $t \approx 5$, thus making any prediction beyond this time fairly useless. In the bottom panel, the data assimilated solution is demonstrated (bold line) and compared to the true dynamics (line). The data assimilated solution is nearly indistinguishable from the true dynamics. The experimental observations are shown by circles at every half-unit of time. The error between the data assimilated solution and true dynamics is shown in the right panel. Note that the error remains quite small in comparison to the direct simulation from noisy initial data. Regardless, there is a build up of error that will eventually grow large as the data assimilated

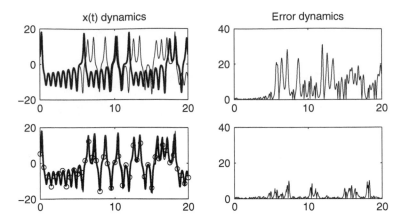

Figure 21.6: Comparison of the model dynamics using a direct numerical simulation of the noisy initial conditions (top panels) with the data assimilated solution with noisy initial conditions and noisy data measurements (bottom panels). The direct simulation (bold line) fails to predict the true dynamics (line) beyond $t \approx 5$ (top panel). Indeed, the error in this case grows quite large at this time (top right panel). When making use of the data assimilation technique (bottom panels) and the data measurements (circles), the solution stays close to the true solution for a much longer time with much smaller error (bottom right panel). Thus data assimilation can greatly extend the time window under which the model can be useful.

solution also diverges from the true dynamics. Ultimately, the data assimilated solution allows for significant extension of the time window where the model prediction is valid.

Data assimilation, in general, is much more sophisticated than what has been applied here to a simple 3×3 system. Indeed, for highly complex systems, the model error in the dynamics plays a fundamental role characterizing the behavior in addition to measurement and initial condition errors. How one chooses to not only treat this error, but how to sample from the dynamics in time, gives rise to a great variety of mathematical techniques for enhancing the data assimilation method [83, 84, 85, 86]. Such data assimilation methods are at the heart of cutting-edge technology, for instance, in weather prediction and/or climate modeling. The hope here was simply to illustrate the basic ideas of this tremendously powerful methodology.

Equation-Free Modeling

Multi-scale phenomena, and their associated complex systems, are an increasingly important class of problems that need to be addressed from a mathematical point of view. As illustrated previously in the chapter on wavelets, some mathematical progress has been made in separating differing length-scales by more sophisticated (wavelet) transforms. Alternatively, one can think about the difficulty in simulating both fine-scale time and spatial features alongside coarse time and spatial features. The recently developed equation-free modeling concept allows for efficient simulations of complex, multi-scale physical systems without requiring the user to specify the system-level (coarse scale) equations [88, 89, 90], thus providing a novel approach to computing such systems.

22.1 Multi-Scale Physics: An Equation-Free Approach

It is often the case that in many systems of interest we know, or can derive, the microscopic description from some governing first principles. For instance, in atomic systems the Schrödinger equation can play the role of describing single atoms and their interactions. Similarly, $\mathbf{F} = m\mathbf{a}$ is the starting point of understanding many classical systems of interacting particles. Gas dynamics can also be considered both at the microscopic level of atoms moving rapidly and interacting with each other, or as a macroscopic collection of these atoms which produce a temperature, pressure and volume. In this case, the gas can be modeled with kinetic theory at the atomic level, or by something like the ideal gas law at the macroscopically observable level.

The dichotomy in characteristic length-scales between micro- and macro-physics is readily apparent in the above examples. Ultimately in constructing an appropriate mathematical description of the system, physically justifiable evolution equations, preferably derived from first principles, are critical for predictive analysis of a given system. Often it is the case that the microscopic interactions and dynamics are known and an attempt is made to systematically construct the macroscopic (observable) scale of interest from asymptotic or averaging arguments of the fine-scale behavior. However, for sufficiently complex systems, this may be an impossible task and instead, *qualitative* models are constructed.

As an example of such micro-to-macro-scale systems and dynamics, consider the myriad of interesting patterns found in nature that are clearly generated from a complex interaction at small scales in order to form coherent patterns on the macroscopic scales of interest. These patterns range from the development of clouds, which ultimately involves a large number of interacting water molecules, to sand patterns, which involve the interactions of a tremendously large number of individual sand particles, to chemical reactions, which involve the interaction of two distinct species of chemical concentrates, to the formation of patterns on sea shells and amoebas, which are functions of the microscopic growth and chemicals released during the growth process. In many of these cases, qualitative models can be constructed, such as the ubiquitous reaction–diffusion systems common in pattern forming systems. However, rarely are these models connected back to the microscale which actually generated the macroscopic evolution equations.

To connect from the microscale to the macroscale, sophisticated mathematical machinery is often brought to bear on the problem. This will be illustrated in the context of a many body quantum mechanical system where a mean field approach will be pursued. This will allow for a self-consistent, asymptotically justified macroscopic evolution model. A similar approach can be taken in fluid mechanics: deriving the Navier–Stokes equations from the kinetic equations and interaction at the molecular level of a fluid. Interestingly enough, the Navier–Stokes equations where known long before its justification from the atomic level description, thus showing the power, importance and necessity of macroscopic descriptions.

Microscopic to macroscopic example: Bose–Einstein Condensation

To consider a specific model that allows the connection from the microscopic scale to the macroscopic scale, we examine mean field theory of many particle quantum mechanics with the particular application of Bose–Einstein Condensations (BECs) trapped in a standing light wave [93]. The classical derivation given here is included to illustrate how the local model and its nonlocal perturbation are related. The inherent complexity of the dynamics of N pairwise interacting particles in quantum mechanics often leads to the consideration of such simplified mean field descriptions. These descriptions are a blend of symmetry restrictions on the particle wavefunction [94] and functional form assumptions on the interaction potential [94, 95, 96].

The dynamics of N identical pairwise interacting quantum particles is governed by the time-dependent, N-body Schrödinger equation

$$i\hbar\frac{\partial\Psi}{\partial t}=-\frac{\hbar^2}{2m}\Delta^N\Psi+\sum_{i,j=1,i\neq j}^{N}W(\mathbf{x}_i-\mathbf{x}_j)\Psi+\sum_{i=1}^{N}V(\mathbf{x_i})\Psi, \tag{22.1.1}$$

where $\mathbf{x_i}=(x_{i1},x_{i2},x_{i3})$, $\Psi=\Psi(\mathbf{x}_1,\mathbf{x}_2,\mathbf{x}_3,...,\mathbf{x}_N,t)$ is the wavefunction of the N-particle system, $\Delta^N=\left(\nabla^N\right)^2=\sum_{i=1}^{N}\left(\partial_{x_{i1}}^2+\partial_{x_{i2}}^2+\partial_{x_{i3}}^2\right)$ is the kinetic energy or Laplacian operator for

N-particles, $W(\mathbf{x_i} - \mathbf{x_j})$ is the symmetric interaction potential between the ith and jth particle, and $V(\mathbf{x}_i)$ is an external potential acting on the ith particle. Also, $\hbar$ is Planck's constant divided by 2π and m is the mass of the particles under consideration.

One way to arrive at a mean field description is by using the Lagrangian reduction technique [97], which exploits the Hamiltonian structure of Eq. (22.1.1). The Lagrangian of Eq. (22.1.1) is given by [94]

$$
L = \int_{-\infty}^{\infty} \left\{ i\frac{\hbar}{2} \left(\Psi \frac{\partial \Psi^*}{\partial t} - \Psi^* \frac{\partial \Psi}{\partial t} \right) + \frac{\hbar^2}{2m} |\nabla^N \Psi|^2 \right.
$$
$$
\left. + \sum_{i=1}^{N} \left(\sum_{j \neq i}^{N} W(\mathbf{x}_i - \mathbf{x}_j) + V(\mathbf{x}_i) \right) |\Psi|^2 \right\} d\mathbf{x}_1 \cdots d\mathbf{x}_N. \tag{22.1.2}
$$

The Hartree–Fock approximation (as used in [94]) for bosonic particles uses the separated wavefunction ansatz

$$
\Psi = \psi_1(\mathbf{x}_1, t)\psi_2(\mathbf{x}_2, t) \cdots \psi_N(\mathbf{x}_N, t) \tag{22.1.3}
$$

where each one-particle wavefunction $\psi(\mathbf{x}_i)$ is assumed to be normalized so that $\langle \psi(\mathbf{x}_i)|\psi(\mathbf{x}_i)\rangle^2 = 1$. Since identical particles are being considered,

$$
\psi_1 = \psi_2 = \ldots = \psi_N = \psi, \tag{22.1.4}
$$

enforcing total symmetry of the wavefunction. Note that for the case of BECs, assumption (22.1.4) is approximate if the temperature is not identically zero.

Integrating Eq. (22.1.2) using (22.1.3) and (22.1.4) and taking the variational derivative with respect to $\psi(\mathbf{x}_i)$ results in the Euler–Lagrange equation [97]

$$
i\hbar \frac{\partial \psi(\mathbf{x}, t)}{\partial t} = -\frac{\hbar^2}{2m} \Delta \psi(\mathbf{x}, t) + V(\mathbf{x})\psi(\mathbf{x}, t) + (N-1)\psi(\mathbf{x}, t) \int_{-\infty}^{\infty} W(\mathbf{x}-\mathbf{y})|\psi(\mathbf{y}, t)|^2 d\mathbf{y}. \tag{22.1.5}
$$

Here, $\mathbf{x} = \mathbf{x}_i$, and Δ is the one-particle Laplacian in three dimensions. The Euler–Lagrange equation (22.1.5) is identical for all $\psi(\mathbf{x}_i, t)$. Equation (22.1.5) describes the nonlinear, nonlocal, mean field dynamics of the wavefunction $\psi(\mathbf{x}, t)$ under the standard assumptions (22.1.3) and (22.1.4) of Hartree–Fock theory [94]. The coefficient of $\psi(\mathbf{x}, t)$ in the last term in Eq. (22.1.5) represents the effective potential acting on $\psi(\mathbf{x}, t)$ due to the presence of the other particles.

At this point, it is common to make an assumption about the functional form of the interaction potential $W(\mathbf{x} - \mathbf{y})$. This is done to render Eq. (22.1.5) analytically and numerically tractable. Although the qualitative features of this functional form may be available, for instance from experiment, its quantitative details are rarely known. One convenient assumption in the case of short-range potential interactions is $W(\mathbf{x} - \mathbf{y}) = \kappa\delta(\mathbf{x} - \mathbf{y})$ where δ is the Dirac delta function. This leads to the Gross–Pitaevskii [95, 96] mean field description:

$$
i\hbar \frac{\partial \psi}{\partial t} = -\frac{\hbar^2}{2m} \Delta \psi + \beta|\psi|^2\psi + V(\mathbf{x})\psi, \tag{22.1.6}
$$

where $\beta = (N-1)\kappa$ reflects whether the interaction is repulsive ($\beta > 0$) or attractive ($\beta < 0$). The above string of assumptions is difficult to physically justify. Nevertheless, Lieb and Seiringer [98] show that Eq. (22.1.6) is the correct asymptotic description in the dilute-gas limit. In this limit, Eqs. (25.3.5) and (25.3.6) are asymptotically equivalent. Thus, although the non-local equation (25.3.5) is in no way a more valid model than the local equation (22.1.6), it can be interpreted as a perturbation to the local equation (22.1.6).

Equation-free macroscopic modeling

In the case of the above mean field model, or the Navier–Stokes equation, a systematic approach can be taken to constructing the macroscopic governing equations. However, for sufficiently complex systems, this may not be possible, either because there is no tractable analytic formalism, or the governing microscopic equations do not lend themselves to an asymptotic/averaging analysis. In this case, only the microscopic equations remain known and at the macroscopic level one must resort to qualitative models which make no connection to the first principles of the system. Alternatively, the *equation-free* modeling technique allows one to use the microscopic description to produce macroscopic observables along with their corresponding macroscopic time evolution without an explicit set of macroscopic equations. Thus we would like the following:

- A system level (coarse or macroscopic) description of the system.
- The macroscopic description should be "simple" (i.e. of significantly lower dimension than the microscopic evolution).
- The macroscopic description should be practically predictive.
- The macroscopic description should be computationally tractable, unlike the complex microscopic system.

Ultimately, the problem formulation is a multi-scale (multi-resolution) type problem that has fine-scale (microscopic) behavior which is only of marginal interest in comparison to the coarse (macroscopic) behavior.

Mathematically, the problem formulation results in an attempt to construct the macroscopic dynamical evolution, for instance, as a system of differential equations:

$$\frac{d\mathbf{x}}{dt} = f(\mathbf{x}) \tag{22.1.7}$$

where the behavior is determined by $f(\mathbf{x})$. Such a system can be easily solved computationally by using a standard time-stepping algorithm. Specifically, consider the Euler time-stepping algorithm which simply makes use of a slope formula and the definition of derivative:

$$\mathbf{x}_{n+1} = \mathbf{x}_n + \Delta t f(\mathbf{x}_n). \tag{22.1.8}$$

Provided we know $f(\mathbf{x})$, the solution can easily be advanced in time by time steps Δt. But here is the point: *we don't know $f(\mathbf{x})$!* Regardless, the Euler method essentially provides a template, or protocol, for specifying *where* to perform function evaluations. A couple of key observations should be stated at this point:

- An explicit representation of the function $f(\mathbf{x})$ is not needed, rather only the discrete values at $f(\mathbf{x}_n)$ are necessary.

- The function $f(\mathbf{x}_n)$ could be evaluated through a variety of techniques including a table lookup, simulation or from experiment itself.

- The sequence $\mathbf{x}_n$ is not known *a priori*, but must be determined as part of the computational task.

- The protocol is determined from an underlying algorithm associated with the macroscopic evolution equations.

Thus our objective is in the construction of $f(\mathbf{x}_n)$ by any means at discrete points. Indeed, the above arguments lead us to consider the following algorithm (to be explored in detail in the next section) for evolving on the macroscopic scale.

1 Estimate the initial state of the system $\mathbf{x}(t = 0)$.

2 Simulate the microscale physics for a short time in order to estimate $d\mathbf{x}/dt$.

3 Take a "big" mesoscale step into the future, i.e. a time-step much larger than the microscale system allows, but on the time-step scale of the macrosystem.

4 Repeat.

This algorithm is the shell of what the equation-free approach does. Indeed, the purpose is simply to use the microscale to estimate, not specify, the macroscopic evolution $f(\mathbf{x})$. Key to this algorithm will be our ability to go between the micro- and macro-scales. The process of going from micro to macro is known as *restriction* whereas the macro to macro is termed *lifting*. Both of these concepts and operations will be considered in detail in the next section. Figure 22.1 demonstrates

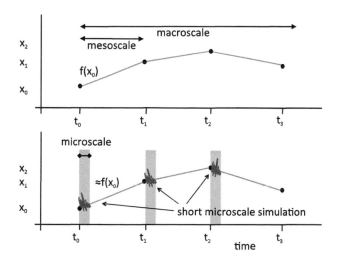

Figure 22.1: Graphical depiction of the algorithm used in the equation-free modeling methodology. In the top figure, the standard Euler stepping scheme is displayed which predicts the future through the quantity $f(\mathbf{x}_n)$. The bottom figure shows the equation-free technique which estimates $\approx f(\mathbf{x}_n)$ through short bursts of simulations at the microscopic level.

graphically the process undertaken in the above algorithm. In the top plot, the standard Euler algorithm is considered where the function $f(\mathbf{x})$ is known. The bottom plot shows the equation-free process by which a burst of microscale simulation generates a numerical approximation to $f(\mathbf{x})$ giving the necessary projection to the future.

22.2 Lifting and Restricting in Equation-Free Computing

We are interested in the macroscopic behavior of large complex systems. This will be the scale of interest in our given application. Two assumptions will be made: only the microscopic (fine-scale) equations of motion are available, and the macroscopic (coarse-scale) equations are unknown. The microscopic variables are often not directly represented in the macroscopic formulation. For instance, for a gas simulation, the individual gas particle velocities and distributions are of critical importance in the computation of the fine-scale equations of motion and state. On the macroscopic level, however, the interest is the gas pressure, volume and temperature, for instance. Thus a clear connection needs to be made between the microscopic quantities of interest and how they project to the macroscopic variables, i.e. this is the so-called restriction step.

Restriction: Micro- to macro-scale

To formulate this in a more concrete mathematical framework, the microscopic and macroscopic variables and their dynamics need to be specified. We begin with the microscopic scale for which we know the evolution equations. For simplicity, assume we can explicitly construct the microscopic evolution as a large system of differential equations

$$\frac{d\mathbf{u}}{dt} = f(\mathbf{u}), \quad \mathbf{u} \in \mathbb{R}^m, \quad m \gg 1. \qquad (22.2.1)$$

The variable $u(t)$ represents the detailed (fine-scale) microscopic solution of the system of interest. In this case, $f(u)$ is explicitly known. The fundamental premise is this: the degree of freedom of the microscopic equations is so large, $m \gg 1$, that it renders solving (22.2.1) impractical or too computationally expensive for large times. Perhaps one can think of $m \to \infty$ in order to reinforce the high dimensionality of the complex system being considered.

Regardless of time constraints and computational expense, the system of differential equations can be formally time-stepped into the future with a wide variety of algorithms. The generic representation for such a time-stepper is in the one-step map:

$$\mathbf{u}(s + t) = T_d^t \mathbf{u}(s) \qquad (22.2.2)$$

where the time-stepper takes a step t into the future with the map T_d^t. The little d denotes that this is the *detailed* stepper. This represents a time-t map that can be, in theory, stepped discretely into the future with a specified accuracy. For the forward Euler scheme, the time-stepper is trivially given by $T_d^t \mathbf{u}(s) = \mathbf{u}(s) + tf(\mathbf{u}(s))$. This has a local error of $O(t^2)$ and a global error of

$O(t)$. However, any time-stepping routine could be used depending upon your accuracy needs. Note that since we know $f(\mathbf{u})$, then the time-stepping map T^t can be computed explicitly.

The fine-scale model (22.2.1) could be in practice a variety of things: a molecular dynamics simulation, a quantum mechanical N-body calculation, or even a finely discretized partial differential equation. Alternatively, the fine-scale evolution can come from a stochastic system simulation via Monte Carlo or Brownian dynamics methods. The key now is to bring the microscopic simulations to the coarse (macroscopic) level in some self-consistent way. Essentially, we must decide how to represent the microscopic variables at a macroscopic level. This would often involve *statistical methods* for bringing up (restricting) the dynamics from micro- to macroscales. Alternatively, one could also think about using SVD/PCA/POD to extract out the principal components of the microscale to project onto the macroscopic system. Mathematically then, we must choose a set of variables at the macroscopic level:

$$\mathbf{U} \in \mathbb{R}^M \tag{22.2.3}$$

where

$$M \ll m. \tag{22.2.4}$$

Here $\mathbf{U}$ is the system of variables representing the macroscopic behavior that is, in some sense, of greatly lower dimension than the fine-scale dynamics. Note that there is some ambiguity here. For instance, if the dynamics of $\mathbf{U}$ is represented by a partial differential equation, it is then mathematically an infinite dimensional system. However, what we would be considering for the equations of $\mathbf{U}$ is its discretized counterpart which should be, in theory, of much lower dimension (complexity) than the microscopic system.

The choice of macroscopic variables determines the *restriction* operator, i.e. how the microscopic variables are restricted to project to the macroscopic level. The restriction operator can be denoted by $\mathcal{M}$ so that its effect is given by

$$\mathbf{U} = \mathcal{M}\mathbf{u}. \tag{22.2.5}$$

It is this step that either averages over the fine scales (in both space and time) or applies some low-rank approximation which may come from, for instance, the SVD/PCA/POD formalism. Other methods include the maximum likelihood inference model and the cumulative probability distribution for generating the restriction operator.

Lifting: Macro- to micro-scale

If the microscopic system can be successfully modeled at the macroscopic scale by the set of variables $\mathbf{U}$, the remaining statistical quantities at the micro-scale should be well approximated by functionals of the selected ones used for the restriction operation. To illustrate this, consider the transformation of the microscopic dynamics to a new set of variables

$$\mathbf{u} \mapsto \mathbf{v} \tag{22.2.6}$$

which gives a system of differential equations

$$\frac{d\mathbf{v}}{dt} = g(\mathbf{v}_1, \mathbf{v}_2) \tag{22.2.7}$$

where $\mathbf{v} = (\mathbf{v}_1, \mathbf{v}_2)$ is partitioned into M and $m - M$ components. Further, the compatible partition g is $(g_1, g_2/\epsilon)$ with $\epsilon \ll 1$. This gives the system of equations

$$\frac{d\mathbf{v}_1}{dt} = g_1(\mathbf{v}_1, \mathbf{v}_2) \quad \in \mathbb{R}^M \tag{22.2.8a}$$

$$\frac{d\mathbf{v}_2}{dt} = \frac{1}{\epsilon} g_2(\mathbf{v}_1, \mathbf{v}_2) \quad \in \mathbb{R}^{m-M}. \tag{22.2.8b}$$

This formalism gives a multi-scale or multi-resolution mathematical architecture that can be used to construct a self-consistent fast-/slow-scale analysis. Specifically, the variable $\mathbf{v}_2$ represents the fast-scale dynamics of the micro-scale. Since it is a fast scale, it should reach its steady state behavior rather quickly (on some slow manifold). Thus $\mathbf{v}_2$ closes to $\mathbf{v}_1$ so that $\mathbf{v}_2 = r(\mathbf{v}_1)$ where $g_2(\mathbf{v}_1, r(\mathbf{v}_1)) = 0$. The variable $\mathbf{v}_1$ can then be identified with the macroscopic variable $\mathbf{U}$ so that

$$\frac{d\mathbf{U}}{dt} = \frac{d\mathbf{v}_1}{dt} = g_1(\mathbf{v}_1, r(\mathbf{v}_1)) = F(\mathbf{U}). \tag{22.2.9}$$

The only way for this to make sense is for us to be able to systematically construct the unknown functions g, r and F, i.e. they are not known in closed form, but rather constructed computationally from the fine-scale simulations and the restriction process. This is what gives it the name *equation-free* since the macroscopic governing equations $F(\mathbf{U})$ are not known.

If the construction of $F(\mathbf{U})$ can be performed, then Eq. (22.2.9) can be stepped forward in time with

$$\mathbf{U}(t + \tau) = T_c^\tau \mathbf{U}(t) \tag{22.2.10}$$

where the time-stepper takes a step τ into the future with the map T_c^τ. The little c denotes that this is the *coarse* stepper. As before, this represents a time-τ map that can be, in theory, stepped discretely into the future with a specified accuracy. The parameter $\tau > 0$ is the time horizon of the coarse time-stepper. The length of time τ appropriate for this problem is such that it renders the computation of the complex system tractable. Specifically, it must be chosen to be small enough on the macroscale so that good accuracy can be achieved at the coarse scale, yet it must be chosen big enough so that microscale simulations can be short enough to be tractable.

The coarse time-stepper relies on a variety of factors, including the ability of the algorithm to project back the macroscopic details to the microscopic system. This project from coarse to fine detail is called the *lifting* operation and it is denoted by μ. The lifting operation is philosophically equivalent to the inverse of the *restriction* operation $\mathcal{M}$. Thus for consistency:

$$\mathcal{M}\mu = \mathbf{I} \tag{22.2.11}$$

where $\mathbf{I}$ is the identify operator. Thus the combination of lifting and restriction allows flexibility to move in and out of the micro- and macro-scales.

Equation-free algorithm

With the mathematical concept of lifting and restriction in place, an algorithm can be presented for performing complex, high-dimensional, multi-scale modeling. The following is the outline of the algorithm that would be used assuming a macroscopic condition $U(t)$ is specified at some moment in time.

1 **Lift** Transform the initial data on the coarse scale to one (or more) fine, consistent microscopic realizations $\mathbf{u}(s, t)$, i.e. use the lifting operator to produce $\mathbf{u}(0, \mathbf{t}) = \mu \mathbf{U}$.

2 **Evolve** The microscopic simulator can then be applied to advance the fine-scale initial data a short, macroscopic time τ. Thus the following evolution in time is constructed: $\mathbf{u}(\tau, t) = T_d^\tau \mathbf{u}(0, t)$.

3 **Restrict** Obtain the restriction of $\mathbf{u}$ so that the coarse time-stepper evolution can be constructed from $T_c^\tau \mathbf{U}(t) = \mathcal{M}\mathbf{u}(\tau, t)$.

The algorithm is then repeated to continue advancing the solution into the future. Effectively, you are replacing the full evolution by a multi-scale approach where *short bursts of the fine-scale simulations are used to generate large time-steps at the coarse level.*

The above algorithm shows that the coarse stepper is thus defined by

$$T_c^\tau = \mathcal{M} T_d^\tau \mu. \qquad (22.2.12)$$

which is a combination of repeated lifting, fine-scale evolution and restriction. No equations are ever specified at the large (coarse or macroscopic) scale, nor do they need to be specified. It is equation-free.

A few things are worth noting in the above formulation. The first is that the choice of lifting μ determines the initial condition for $\mathbf{v}_2$ at time t in Eq. (22.2.8b). For a simulation time $s \gg \epsilon$, and with the assumption that $\mathbf{v}_2$ approximately reduces to a function of $\mathbf{v}_1$ given by $g_2 \approx 0$ (independent of the initial state), the equation quickly approaches the slow manifold for any lifting operator μ. Thus after some initial transient, the dynamics is well approximated by Eq. (22.2.9) and $\mathbf{U}(t + s) \approx \mathcal{M}\mathbf{u}(s, t)$. More importantly, the derivates of $\mathcal{M}\mathbf{u}(s, t)$ will also be a good approximation to the derivatives of $\mathbf{U}(t + s)$ after the transient time. *We will thus use the derivatives of $\mathcal{M}\mathbf{u}(s, t)$ to approximate the derivatives of $\mathbf{U}(t + s)$ and thereby create the coarse time-stepper T_c^τ.*

Coarse projection integration

The final step in the equation-free algorithm is to make use of the short-time, fine-scale simulations to take large time-steps on the macroscopic time-scale. Thus a high degree of computational efficiency can be readily achieved. Mathematically speaking, we would like to simulate the macroscopic equation (22.2.9) for times $t \gg \tau$. Coarse projection integration will allow us to take the *short-time* coarse time-stepper and perform *long-time* (system level) tasks.

To understand the coarse projection integration, consider the following two time-scales

Δt large time-step (slow coarse dynamics)

δt small time-step (fast, transient, microscale dynamics).

Further let the following scaling hold

$$\delta t \leq \tau \ll \Delta t. \tag{22.2.13}$$

The step δt is often thought of as the numerical time-step of the microscopic evolution (22.2.1). By setting $t_k = k\delta t$ and defining $\mathbf{U(t)}$ as $\mathbf{U}^N \approx \mathbf{U}(N\Delta t)$, then we can consider the action of the microscopic time-stepper

$$\mathbf{u}^{k,N} = T_d^{t_k}\mathbf{u}^{0,N} \tag{22.2.14}$$

where $\mathbf{u}^{0,N} = \mu\mathbf{U}^N$. As long as the number of small time-steps, k, remains relatively small so that $t_k = k\delta t = O(\tau)$, then this step can be afforded computationally.

The coarse approximations to these fine steps are computed using the restriction operator so that

$$\mathbf{U}^{k,N} = \mathcal{M}\mathbf{u}^{k,N} \approx \mathbf{U}(N\Delta t + t_k) \tag{22.2.15}$$

where for consistency $\mathbf{U}^{0,N} = \mathbf{U}^N$. This formula does not allow for the computation of $\mathbf{U}^{N+1}$ since the $t_k \ll \Delta t$. However, the value of $\mathbf{U}^{N+1}$ can be extrapolated by using a coarse scheme such as

$$\mathbf{U}^{N+1} = \mathbf{U}^{k,N} + (\Delta t - t_k)\tilde{F}(\mathbf{U}^{k,N}) \tag{22.2.16}$$

where $\tilde{F}$ approximates F. For $t_k \geq \tau$, the fast modes have in principle died out so that the differences of $\mathbf{U}^{k,N}$ are useful for approximating derivatives of the exact solution. Specifically, the following approximation is used:

$$\tilde{F}(\mathbf{U}^{k,N}) = \frac{\mathbf{U}^{k+1,N} - \mathbf{U}^{k,N}}{\delta t} \approx \frac{d\mathbf{U}(N\Delta t + t_k)}{dt} = F(\mathbf{U}(N\Delta t + t_k)) \approx F(\mathbf{U}^{k,N}). \tag{22.2.17}$$

This method of extrapolation is called *projective forward Euler*, and it is the simplest case of the class of methods known as projective integration methods. This is the explicit *outer* integrator that is built around the *inner* integrator associated with the fine-scale simulations.

Thus the coarse-scale time evolution as modeled by (22.2.9) is performed by approximations of F achieved by the fine-scale simulations. The entire process is represented in Fig. 22.2 where the interaction of the fine-scale, course-scale, projective stepping along with the lifting/restriction operations are identified. It is quite a remarkable process that potentially can allow for highly improved, and computationally tractable, methodologies for handling multi-scale problems. Part of the connection here to data analysis is in the projection and lifting operators: many of the techniques we have considered could be used to perform these critical operations.

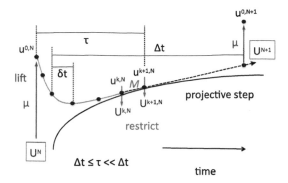

Figure 22.2: Graphical depiction of the algorithm used in the equation-free modeling methodology. The micro-to-macro and macro-to-micro scales are conducted through the restriction and lifting steps, respectively. This is used as the basis for constructing the projective numerical step which allows for large time-steps in the multi-scale simulations. The macroscopic evolution is donated by $\mathbf{U}^N$ whereas the time-step in the microscopic evolution is denoted by $\mathbf{u}_{k,N}$.

22.3   Equation-Free Space–Time Dynamics

The discussion in the previous section highlighted the ability of the equation-free algorithm to advance in time a numerical solution on a coarse scale, via an Euler projective time-step, given simulations on the microscopic time-scale. In this section, we extend this idea to include both time- and space-evolution dynamics. Specifically, we consider the case when the microscopic equations of motion both occur on fast scales in relation to the macroscopic observations of interest.

Mathematically, the analytic framework considered on the microscale is now modified to

$$\frac{d\mathbf{u}}{dt} = f(\mathbf{u}), \quad \mathbf{u} \in \mathbb{R}^m, \quad m \gg 1 \quad + \text{boundary conditions} \tag{22.3.1}$$

where the variable $u(t)$ once again represents the detailed (fine-scale) microscopic solution of the system of interest. In this case, $f(u)$ is explicitly known to be a function of both space and time. As before, the fundamental premise is this: the degree of freedom of the microscopic equations is so large, $m \gg 1$, that it renders solving (22.3.1) impractical or too computationally expensive for large times. Perhaps one can think of $m \to \infty$ in order to reinforce the high dimensionality of the complex system being considered.

At the course (macroscopic) level, the equations of motion are now governed by a partial differential equation of time and space so that

$$\frac{\partial \mathbf{U}}{\partial t} = L\left(t, \mathbf{x}, \mathbf{U}, \frac{\partial \mathbf{U}}{\partial \mathbf{x}}, \frac{\partial^2 \mathbf{U}}{\partial \mathbf{x}^2}, \cdots, \frac{\partial^p \mathbf{U}}{\partial \mathbf{x}^p}\right) \tag{22.3.2}$$

where $\mathbf{U}(t, \mathbf{x})$ is the macroscopic variable of interest and L is a general integro-differential operator of degree p in space $\mathbf{x}$. The operator L is in general, nonlinear. Further, it is unknown given the premise that it is too complicated to achieve an averaging or asymptotic scheme capable of deriving it from the microscopic description. The goal once again is then to *approximate L through microscopic simulations versus actually explicitly constructing it.*

To illustrate how to proceed, consider the scenario for which we know the operator L. Specifically, suppose that the operator L gives the diffusion equation at the macroscopic level so that

$$\frac{\partial U}{\partial t} = \frac{\partial^2 U}{\partial x^2} \tag{22.3.3}$$

where the system considered is single degree of freedom in time and space, i.e. $\mathbf{U}(t, \mathbf{x}) = U(t, x)$. Standard finite difference methods can be used to advance the solution of (22.3.3). Indeed, Tables 4.1–4.2 explicitly demonstrate the finite difference calculations achieved from Taylor series expansions. Thus if we approximate the solution by a finite number of points, then

$$U_n(t) = U(x_n, t) \tag{22.3.4}$$

and Eq. (22.3.3) discretizes to

$$\frac{\partial U_n}{\partial t} = \frac{U_{n+1} - 2U_n + U_{n-1}}{\Delta x^2} \tag{22.3.5}$$

where Δx is the size of the discretization in the macroscopic variable. If further we discretize time so that

$$U_n^m = U(x_n, t_m) \tag{22.3.6}$$

then this gives a complete finite difference discretization of the original diffusion equation. By using an Euler time-stepping algorithm

$$\frac{\partial U_n}{\partial t} = \frac{U_n^{m+1} - U_n^m}{\Delta t} \tag{22.3.7}$$

the center-difference, forward Euler discretization of the equation results in

$$U_n^{m+1} = U_n^m + \frac{\Delta t}{\Delta x^2} \left(U_{n+1}^m - 2U_n^m + U_{n-1}^m \right) \tag{22.3.8}$$

This gives a stencil for propagating the solution forward in macroscopic time with a macroscopic (coarse) description in space. Figure 22.3 shows the basic stencil used in a graphical format. The solution of the partial differential equation reduces to iterating on the grid points in order to advance the solution in time on a coarse scale.

The point of the above discretization is the following: by definition, discretization lays down a grid on which the coarse solution values are evaluated, namely the points in space and time $U_n^m = U(t_m, x_n)$. From the macroscopic perspective, the evolution of the coarse scale can be accurately calculated or computed from this gridded data. Thus once the grid is determined, the space–time evolution is characterized by evolving the governing equations (22.3.2). Of course,

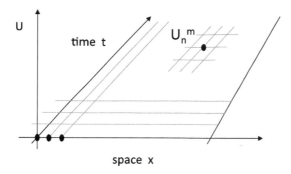

Figure 22.3: Graphical depiction of the coarse-scale stencil used to propagate the solution into the future with a finite difference approach. Each location of the coarse scale is denoted by $U_n^m = U(t_m, x_n)$.

these equations are not known, and that is the point. Thus the microscopic evolution needs to be tied into this analysis in order to make further progress.

To guide the discussion consider Fig. 22.4. This shows the key features of the spatial component of the equation-free scheme. In particular, at each coarse grid point, a small (microscopic) neighborhood of the point is selected in order to perform simulations of the fine-scale system in both time and space. Two key parameters are considered:

$$H \quad \text{macroscopic space resolution} \tag{22.3.9a}$$

$$h \quad \text{microscopic space resolution} \tag{22.3.9b}$$

where $h \ll H$. Just like the previous time-stepping scheme, in space micro-sections of the spatial variable are used to construct interpolated values for the macroscopic description in space.

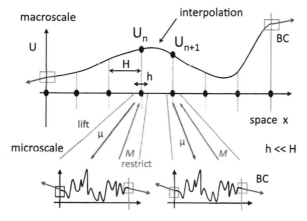

Figure 22.4: The equation-free method applied to the evolution in space and time. At each coarse time-step, microscopic simulations are performed at the stencil points of the coarse grid. By lifting (μ) and restriction ($\mathcal{M}$), a self-consistent method can be developed for going from micro-to-macro and macro-to-micro scales. Boundary conditions (BC) must be carefully considered at both the macroscopic and microscopic scales.

Note that this is unlike the *extrapolation* of the time-stepping produced from the projective Euler method. In either case, both space and time coarse steps are produced from microscale simulations. Key to this is the restriction and lifting steps necessary for going from micro-to-macro and macro-to-micro scales, respectively. Thus the same concepts apply in space as they do in time, leading to the space–time equation-free algorithm:

1   The initial condition in each microscale box is lifted in a self-consistent way from the macro-to-micro scales. Boundary conditions are computed to the microscale from the coarse scale.

2   Microscopic boundary conditions are applied to the microscale computational boxes.

3   The microscale is evolved over a microscopic time-scale until the fast transients die out and we are on the slow manifold.

4   The simulation results are restricted back to the macroscale in order to predict the projective time-step. The projective step is an *extrapolation* in time.

5   After advancing a coarse time-step, interpolation is used to fill in the spatial points in between the coarse grid.

6   Repeat the process.

This is the basic skeleton of the scheme. Note that as a tremendous advantage, *the scheme is ideally suited for parallel computing*. Indeed, each microscale computation can be performed on a separate processor. The method thus has the possibility of greatly extending your computational ability for a large, complex system in space and time, again, without ever explicitly knowing the macroscopic evolution equations.

Complex Dynamical Systems: Combining Dimensionality Reduction, Compressive Sensing and Machine Learning

The data-driven techniques developed in this book have been largely applied in isolation, i.e. they have been separated from each other and the other methods in the text. However, to fully exploit these methods, modern applications in complex systems should aim to use as many techniques as necessary to elucidate the behavior of a given system. Efficient strategies for managing data and integrating it into the underlying dynamics are leading to paradigm shifts in modeling techniques including data-assimilation techniques using reduced-order modeling methods (PCA, POD, dynamic mode-decomposition and equation-free modeling) with the two extremely successful and recent mathematical strategies for handling data: the so-called *compressive sensing* (CS) and *machine learning* (ML) methods. In the former, sparsity in the data and/or representation of the data plays a defining role. In the latter, statistical learning is accomplished by computing patterns or principal components (dimensionality reduction) of the data. Our objective is to integrate CS and ML together with dimensionality reduction methods in order to enhance our modeling and predictive capabilities of complex dynamical systems. These two techniques (CS, ML) are quite novel in the context of dynamical systems and data assimilation, and we demonstrate their usefulness in applications to complex dynamical systems.

 23.1   Combining Data Methods for Complex Systems

To begin thinking about combining our various data-driven methods, we will first consider a given complex system and its underlying dynamical evolution. Specifically, let us begin by assuming that the system under consideration has the following nonlinear partial differential equation

$$\mathbf{U}_t = N(\mathbf{U}, \mathbf{U}_x, \mathbf{U}_{xx}, \cdots, x, t, \beta) \qquad (23.1.1)$$

where $\mathbf{U}$ is a vector of physically relevant quantities and the subscripts t and x denote partial differentiation. The function $N(\cdot)$ captures the space–time governing equations specific to the system being considered. In general, this can be a complicated, nonlinear function of the quantity $\mathbf{U}$, its derivatives and both space and time. Along with this PDE are some prescribed boundary conditions and initial conditions:

$$\mathbf{U}(x, 0) = \mathbf{U}_0(x). \qquad (23.1.2)$$

The parameter β is included as a bifurcation parameter for which the solutions of the governing PDE change markedly for different ranges of β.

Of course, techniques for solving such a system have been considered extensively throughout the second part of this book. Indeed, provided that N is well behaved, i.e. continuous and differentiable, for instance, then a numerical solution to the prescribed problem can be found uniquely by discretizing the PDE into an n-dimensional system of ODEs where $n \gg 1$. For many complex dynamical systems, the PDE discretization process can easily yield a million or billion (or more) degree of freedom system, thus straining computational resources. For such high dimensional systems, it may be computationally expensive or even intractable to reconstruct the full state from sparse samples using compressive sensing. However, the high dimensional system is often governed by dynamics on a low dimensional attractor. This makes it possible to use compressive sensing to reconstruct the amplitudes of the few most important modes in POD space, which is considerably easier because of the low dimension. From these amplitudes, the full state may then be reconstructed.

Data assimilation methods can be applied to understanding the dynamical system of interest. In particular, measurements of the system can be made in order to inform the characterization of the system. However, measurements are certainly not practical on the order of the million or billion (or more) degree of freedom system generated through discretization. Instead, a limited number of measurements are taken, let's say m of them where $m \ll n$, in order to inform our dynamics. Thus the measurements can be considered *sparse* and this ultimately becomes a compressive sensing problem. Suppose that $\hat{\mathbf{U}}(x, t)$ are m observations at a given time t of the state of the system. The objective is to use these m-dimensional sparse observations in order to best reconstruct the full n-dimensional state vector $\mathbf{U}$. Or better yet, can the sparse measurements be used to directly construct a low-rank approximation of the dynamics? But in order to reconstruct this state vector, the underlying dynamics must be known to be sparse in some basis, leading to the construction of a library of low-rank approximations to the dynamics of (23.1.1).

Library of dynamics: Machine learning

Fundamental to understanding how to combine the low-rank dynamical approximations with compressive sensing is the building of a library of vectors containing the low-rank approximations of (23.1.1). Such low-rank approximations are generated for various values of the bifurcation parameter β and a dynamical reconstruction can be made, for instance, using the techniques of Section 18.3. Thus the goal is to make m sparse measurements in order to identify the modes necessary to solve (23.1.1) using the low-rank approximation

$$\mathbf{U}(x, t) = \sum_{k=1}^{M} a_k(t)\phi_k(x) \tag{23.1.3}$$

where the $\phi_k(x)$ are library elements and $M \ll n$. What remains is to construct library modes $\phi_k(x)$ and identify which are *active* given the sparse measurement of the state $\hat{\mathbf{U}}(x, t)$.

To construct the library, simulations of the full system, or dense measurements/observations of an experimental system, can be conducted over a prescribed period of time with a fixed value of β. Snapshots of the simulation data are taken at a given interval. Thus a data matrix is generated from the evolution of the full state of the system:

$$\mathbf{A}_{\beta_k} = \begin{bmatrix} | & | & & | \\ \mathbf{U}(t, x) & \mathbf{U}(t + \Delta t, x) & \cdots & \mathbf{U}(t + m\Delta t, x) \\ | & | & & | \end{bmatrix} \tag{23.1.4}$$

where m is the number of snapshots taken at an interval of Δt. Note, Δt here is not the time-discretization step used for numerically integration (23.1.1). Here β_k denotes the specific value of β used in (23.1.1). The goal is to simulate over the various values of β that produce nontrivial dynamics. Each unique dynamical regime is denoted by $\beta_1, \beta_2, \cdots, \beta_K$.

Once the data matrix (23.1.4) is constructed, its principal components, or POD modes, are identified through the singular value decomposition. For a given β_k, these modes (vectors) are computed to be

$$\phi_1(x, \beta_k), \phi_2(x, \beta_k), \cdots, \phi_{m_k}(x, \beta_k) \tag{23.1.5}$$

where m_k is determined by a cut-off criterion for modal energy (see Section 18.3). For instance, one could specify that the modes that comprise 99.9% of the energy of the data matrix be retained, thus giving the value of m_k. Recall that these POD modes are orthogonal to each other and ordered in decreasing order of *importance*. Figure 23.1 depicts the basic idea in collecting the low-rank approximations for the various dynamical regimes of the PDE (23.1.1) as a function of the bifurcation parameter β.

With these modes identified, a library can be constructed containing all the low-rank approximations of the data. Thus the library ψ_L is composed of the following vectors:

$$\psi_L = \begin{bmatrix} | & & | & | & & | & & | & & | \\ \phi_1(x, \beta_1) & \cdots & \phi_{m_1}(x, \beta_1) & \phi_1(x, \beta_2) & \cdots & \phi_{m_2}(x, \beta_2) & \cdots & \phi_1(x, \beta_K) & \cdots & \phi_{m_K}(x, \beta_K) \\ | & & | & | & & | & & | & & | \end{bmatrix}. \tag{23.1.6}$$

In principle, the library contains the representative low-rank modes for *all possible dynamical behavior of the governing system*. Let us assume that there are a total number of p library elements (basis vectors). Thus it becomes somewhat of a look-up table for the dynamics. Or more precisely, this is the supervised *machine learning* portion of the analysis. Of importance is that these modes

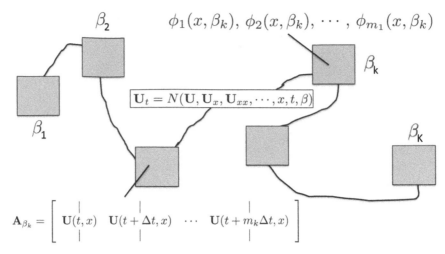

Figure 23.1: Depiction of the library building procedure. The governing PDE (23.1.1) is simulated for the various values of the bifurcation parameter β that generate interesting dynamics. For each unique region of dynamical behavior β_k (represented abstractly by the shaded boxed regions), snapshots of the dynamics are used to construct a data matrix modeled by some low-rank approximation of modes $\phi_1(x, \beta_k), \phi_2(x, \beta_k), \cdots, \phi_{m_k}(x, \beta_k)$. The goal is to simulate the system at each interesting region of parameter space for the POD library construction. The system can then be allowed to wander from one parameter regime to another (denoted abstractly by the bolded black line) and modes are *switched* in the reconstruction process as we go from β_j to β_k.

represent the dynamics in the best sparse basis. It should be noted that the library modes are *not* orthogonal. Rather, groups of library POD modes are orthogonal when selected for a given β_k.

Compressive sensing with POD library

With the library (23.1.6) constructed, our attention is once again turned towards the low-rank POD dynamic reconstruction technique

$$\mathbf{U}(x, t) = \sum_{k=1}^{K} \sum_{m=1}^{m_k} a_{km}(t)\phi_m(x, \beta_k) = \psi_L \mathbf{a} \tag{23.1.7}$$

where now the PDE solution is represented in the p library elements constructed for the various values of β. If measurements of the full n-dimensional PDE system were available, then the POD library expansion (23.1.7) could be completely prescribed.

The idea at this point is to use the sparse data measurements, which are both more efficient and realistic in practice, to inform the POD library expansion (23.1.7). In particular, since at any given time the dynamics should belong to only a subset of the modes, since the dynamics are occurring for a given β_k, then the majority of coefficients $a_{km}(t)$ should be zero, thus ensuring sparsity of the dynamics. This then provides an ideal framework for the implementation of compressive sensing.

Suppose a number of *spatial* measurements are made on the dynamical system at a number of prescribed locations. Then the following relation holds (see Section 17.3)

$$\hat{\mathbf{U}} = \Phi\mathbf{U} \qquad (23.1.8)$$

where Φ is a permutation of the identity matrix which specifies the measurement locations. Since it is assumed in this argument that $\mathbf{U}$ and $\hat{\mathbf{U}}$ are $n \times 1$ and $m \times 1$, respectively, the matrix Φ is $m \times n$. Recall that because of sparse sampling $m \ll n$.

Using (23.1.8) and (23.1.7) results in an underdetermined linear system. Here there are m equations (constraints) and p unknowns (modal coefficients) so that $m \ll p$:

$$\hat{\mathbf{U}} = (\Phi\psi_L)\mathbf{a}. \qquad (23.1.9)$$

Our objective is to determine the vector $\mathbf{a}$ since we have the sparse measurements to construct $\hat{\mathbf{U}}$, we know the measurement locations to construct Φ, and we have already constructed the library ψ_L. This is an underdetermined system that can be solved in a variety of ways, but our aim is to use the L^1 norm error in order to promote sparsity in the solution vector $\mathbf{a}$ (see Section 17.3). Thus an L^1 convex optimization procedure is performed to evaluate $\mathbf{a}$. This works because our library elements are the sparse basis modes that represent the dynamics of the PDE (23.1.1).

Algorithm for combining machine learning and compressive sensing

The library construction and sparse sensing combine together to give a highly efficient way to reconstruct the dynamics of the full PDE (23.1.1) using a limited number of sensors and a machine learned data base. The procedure is summarized as follows:

- Run extensive simulations of the governing PDE (23.1.1) at values of the bifurcation parameter β that produce signature dynamics in the system.
- For each signature value β_k ($k = 1, 2, \cdots, K$), sample the low-rank dynamics (slow manifold) through a snapshot based method. Alternatively, other dimensionality reduction techniques can be used to produce the correct low-rank encodings such as dynamic mode decomposition or equation-free modeling. Retain only the principal components (POD modes) that capture a prescribed high percentage of the data.
- Construct a library ψ_L of the p POD vectors sampled at the various β_k.
- Building the library is a one-time cost. Once done, use sparse measurements $\hat{\mathbf{U}}$ and a compressive sensing scheme (convex L^1 optimization) to reconstruct the best-fit (in an L^1 sense) sparse evaluation of the full state of the system $\mathbf{U}$ at any given time t by evaluating $\mathbf{a}$.
- Either the sparse sampling can be used continuously to produce a sparse representation of the state of the system in terms of the library modes ψ_L, or occasional sparse sampling can be used to generate the correct low-rank POD dynamics of $\mathbf{a}$.

The above algorithm will be demonstrated in two example systems that follow in the next two sections. Although not mentioned yet, there are a great number of significant advantages of such a scheme for representing complex dynamical systems: (i) Once the library is constructed from the extensive simulations, future prediction of the system can be done quite a bit more efficiently since the correct POD modes for any dynamical regime β_k have already been computed, (ii) the algorithm works equally well with experimental data and/or in an equation-free context and (iii) given the low-rank space in which the algorithm always attempts to work, it is ideal for the implementation of control strategies. Indeed, control theory can only be practically applied in real time to low dimensional systems, and the framework of the complex dynamical system has now been placed perfectly in the context where control theory can make an impact.

23.2 Implementing a Dynamical Systems Library

To illustrate the algorithm developed in the last section, a toy model is constructed that has a fairly rich bifurcation structure. The steps of the algorithm laid out previously will be followed in order to develop a library of the low-rank dynamics of the system. Once achieved, sparse sensing will be applied to (i) identify the correct β_k regime where the dynamics is occurring, and (ii) use the POD modes to then reconstruct the low dimensional dynamics.

The complex Ginzburg–Landau equation

As a model system, consider the complex Ginzburg–Landau model of mathematical physics. Here it is modified to include both quintic terms and a fourth-order diffusion term much like the Swift–Hohenberg equation. The model is the following:

$$iU_t + \left(\frac{1}{2} - i\tau\right) U_{xx} - i\kappa\, U_{xxxx} + (1 - i\mu)|U|^2 U + (v - i\sigma)|U|^4 U - i\gamma\, U = 0 \qquad (23.2.1)$$

where $U(x,t)$ is a complex function of time and space and each subscript t and x denotes differentiation with respect to that variable.

Interesting solutions to this governing equation abound. However, we will consider a limited number of parameter regimes, each of which displays a different characteristic dynamics. In this study, we will consider a vector of the parameters:

$$\beta = (\tau, \kappa, \mu, v, \sigma, \gamma). \qquad (23.2.2)$$

In particular, six parameter regimes are considered that illustrate different dynamical behaviors. Table 23.1 gives a breakdown of the different parameters used and a comment on the type of behavior exhibited.

Using the computational methods outlined in the second part of this book, a spectral method is used to simulate the evolution of (23.2.1) for the various parameter regimes highlighted in Table 23.1. Figure 23.2 is 2×3 set of subplots illustrating the low-rank behavior produced in the simulations. As is common in many complex dynamical systems, especially those of a dissipative nature, low dimensional attractors are embedded in the high dimensional space. These figures simply show the low dimensional structures which are spontaneously formed from generic,

Table 23.1 Parameter regimes β_k considered for the complex Ginzburg–Landau equation (23.2.1) along with a brief description of the dynamics (see Fig. 23.2). The low-rank approximations of these parameter regimes are used to construct the elements of the library ψ_L

	τ	κ	μ	ν	σ	γ	Description
β_1	−0.3	−0.05	1.45	0	−0.1	−0.5	three-hump, localized solution
β_2	−0.3	−0.05	1.4	0	−0.1	−0.5	localized solution with small side lobes
β_3	0.08	0	0.66	−0.1	−0.1	−0.1	breather
β_4	0.125	0	1	−0.6	−0.1	−0.1	exploding (periodic) dissipative soliton
β_5	0.08	−0.05	0.6	−0.1	−0.1	−0.1	fat bump
β_6	0.08	−0.05	0.5	−0.1	−0.1	−0.1	dissipative soliton

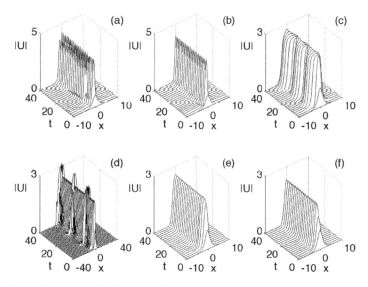

Figure 23.2: Evolution dynamics of (23.2.1) for the six parameter regimes given in Table 23.1: (a) β_1, (b) β_2, (c) β_3, (d) β_4, (e) β_5 and (f) β_6. All parameter regimes exhibit stable, low-dimensional attractors.

localized initial data. The low dimensional structures allow for the low-rank approximations used in the library.

The MATLAB code for simulating (23.2.1) can be constructed as follows:

```
n=512; L=20;  % domain size and discretization points
tspan=linspace(0,60,61);  % time domain

tau=0.08;  % tau  %% PARAMETERS for beta_k
ka=-0.05;  % kappa
beta=0.5;  % beta
```

Continued

```
nu=-0.1;    % nu
s=-0.1;     % sigma
g=-0.1;     % gamma


x2=linspace(-L/2,L/2,n+1); x=x2(1:n);   % grid
k=(2*pi/L)*[0:n/2-1 -n/2:-1].';   % wavenumbers


u=sech(1.0*t).'; ut=fft(u);    % initial conditions
[t,utsol]=ode45('cqgle_rhs',zspan,ut,[],k,tau,ka,beta,nu,s,g);
for j=1:length(t)
    usol(j,:)=ifft(utsol(j,:)); % solution matrix
end
```

The above code calls the right-hand side function **cqgle_rhs.m** which contains the partial differential equation itself.

```
function rhs=cqgle_rhs(z,ut,dummy,k,tau,ka,beta,nu,s,g)
u=ifft(ut);
rhs= -(i/2+tau)*(k.^2).*ut + ka*(k.^4).*ut ...
    + (i + beta)*fft( (abs(u).^2).*u ) ...
    + (i*nu + s)*fft( (abs(u).^4).*u ) ...
    + g*ut;
```

This code is sufficient to simulate (23.2.1) for the various dynamical regimes listed in Table 23.1. The simulation results have already been demonstrated in Fig. 23.2.

Library construction

The simulation results allow for the construction of the library elements of ψ_L. For each dynamical regime demonstrated in Fig. 23.2 corresponding to β_j where $j = 1, 2, \cdots, 6$, the POD modes (SVD) are computed for the evolution once the initial transient has died away. In particular, for the simulations illustrated in Fig. 23.2, the data have been sampled every $\Delta t = 1$ from time $t \in [0, 60]$. Given the initial transient response, the snapshots from $t \in [20, 60]$ are used to construct the low-rank approximations of the data matrix. In MATLAB, this can be simply done with the following commands

```
[U,S,V]=svd(usol(20:end,:));
sig=diag(S);
```

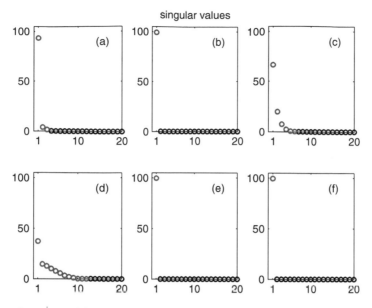

Figure 23.3: Singular values of the SVD performed on the evolution data of (23.2.1) shown in Fig. 23.2 corresponding to regions (a)–(f) respectively. The sampling occurs for every $\Delta t = 1$ in the interval $t \in [20, 60]$. Note the low dimensional structure of each dynamic regime. Specifically, the magenta circles represent the modes that comprise 99% of data. For the breather and the exploding dissipative soliton, a larger number of modes are necessary to represent the dynamics. The collection of modes associated with the dominant (magenta) modes is used in the library ψ_L.

where the singular values are then written into the vector **sig**. This vector is plotted in Fig. 23.3 for each dynamical regime, clearly demonstrating the low-rank nature of the dynamics of each regime β_k. This also allows us to determine a threshold for mode selection. Specifically, only the first m_k modes are kept where these modes capture 99% of the dynamics (data matrix).

Figure 23.4 shows the library elements (modes) that comprise the library ψ_L at the six different values of β_k. In many of the regimes, a single dominant mode is sufficient to characterize the

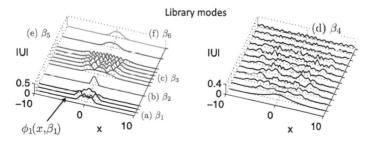

Figure 23.4: Library ψ_L constructed from the dominant modes found in Fig. 23.3. The groupings, identified by their β_k value, are associated with the different dynamical regimes (a)–(f) in Fig. 23.2. Note that the modes of the exploding dissipative solution have been including separately in the right panel as there are 12 modes required to capture 99% of the dynamics in this regime.

parameter space since a steady state attractor is exhibited. For the more dynamically interesting behaviors in Fig. 23.2, such as the breather and exploding dissipative soliton, more modes are required to capture the dynamics. The total number of modes in the library is $p = 24$.

Compressive sensing and reconstruction of dynamics

The objective is now to highlight the role of compressive sensing in the identification and reconstruction of the dynamical system using the library ψ_L. Specifically, suppose that for a given system, the governing equation (23.2.1) represents the underlying evolution of the system. Moreover, the bifurcation parameter $\beta = \beta(t)$ is time-dependent so that the dynamics switches from one attractor to another as β changes in time (see Fig. 23.1 for a cartoon of the basic idea). Specifically, consider the following example

$$\beta = \begin{cases} \beta_1 & t \in [0, 100) \\ \beta_2 & t \in [100, 200) \\ \beta_3 & t \in [200, 300]. \end{cases} \tag{23.2.3}$$

The evolution dynamics for this case is illustrated in the left panel of Fig. 23.5. The goal is to use sparse measurements to sample/observe/measure the dynamics of the underlying system and use the library to reconstruct the dynamics with the appropriate low-rank model.

To be more precise about the example to be used here, consider the relation between the number of measurements, number of library elements and the original size of the system:

$$m = 3 \text{ measurements at } x_0 = 0, x_1 = 0.5, x_2 = 1$$
$$p = 24 \text{ number of library elements}$$
$$n = 128 \text{ dimension of original system.}$$

Note that $m \ll p \ll n$ as it should be. Moreover, our measurement locations will be located at $x = 0, 0.5$ and 1 for this case. For the example here, assume that the measurements take place at $t = 50, 150$ and 250. The goal will be to take a single time-snapshot measurement at three spatial

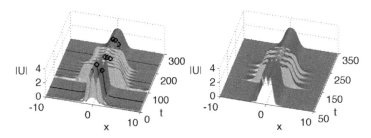

Figure 23.5: Full evolution dynamics (left panel) and the low-rank POD dynamics reconstruction using compressive sensing (right panel). The black lines on the left panel at $t = 50, 150$ and 250 represent the sampling times while the three black circles represent the three sparse measurement locations. From the three samples, the right panel is reconstructed by identifying the correct POD modes (see Fig. 23.6).

locations and use them to discover the modes of importance for doing a POD reconstruction of the dynamics, for instance.

To accomplish this task, the matrix Φ in (23.1.8) must be constructed. For the three measurements proposed here, the matrix takes the form:

$$\Phi = \begin{bmatrix} 0 & \cdots & 0 & 1 & 0 & \cdots & & & & & & & 0 \\ 0 & \cdots & & & & \cdots & 0 & 1 & 0 & \cdots & & & 0 \\ 0 & \cdots & & & & & & \cdots & 0 & 1 & 0 & \cdots & 0 \end{bmatrix} \qquad (23.2.4)$$

where the nonzero entries are the spatial locations on the numerical grid where $x_0 = 0$, $x_1 = 0.5$ and $x_2 = 1$, respectively. Thus the matrix $\Phi\psi_L$ in the under-determined system (23.1.9) is computed to be a 3×24 matrix.

For the system with three measurements on a domain $x \in [-10, 10]$ discretized with $n = 128$ points, the approximate positions for $x_0 = 0$, $x_1 \approx 0.5$ and $x_2 \approx 1$ are given by

```
n1=65; n2=68; n3=72; % measurement locations
phi=zeros(3,128);
phi(1,n1)=1; phi(2,n2)=1; phi(3,n3)=1; % matrix phi
```

where the 65th, 68th and 72nd grid points are appropriate for where we want to measure ($x = 0$, $x \approx 0.5$ and $x \approx 1$). Note that sensor location is critical in the compressive sensing process. Indeed, for best performance the sensors should be placed at locations where maxima and minima are observed in the POD library modes [91, 92]. Performance can drop significantly using poor sampling locations.

The following code constructs the $\Phi\psi_L$ matrix (3×24 or more generally $m \times p$) in (23.1.9) with the three sensor locations ($x = 0$, $x \approx 0.5$ and $x \approx 1$) at the time slice corresponding to $t = 50$ in Fig. 23.5. This figure represents both the time slice and the specific sensor locations. Once the measurements have been made, the vector $\hat{U}$ is also determined and then only the vector $\mathbf{a}$ of library mode coefficients needs to be determined. They are determined through L^1 optimization in order to promote sparsity (compressive sensing). As in Section 17.3, a convex optimization package is used that can be directly implemented with MATLAB: http://cvxr.com/cvx/. The following code reconstructs the sparse solution vector $\mathbf{a}$ given the three measurements.

```
PhiPsi=phi*abs(psiL);
Uhat=[abs(umaster(21,n1)); abs(umaster(21,n2)); abs(umaster(21,n3))];

p=24;
cvx_begin;
variable a(p);
    minimize( norm(a,1) );
    subject to
    PhiPsi*a == Uhat;
cvx_end;
```

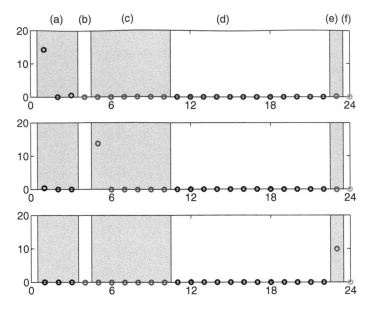

Figure 23.6: Results of the compressive sensing reconstruction procedure of the underdetermined system (23.1.9) for the three sparse measurements $\hat{U}$ at $x_0 = 0, 0.5, 1$ for the times $t = 50, 150, 250$ (top, middle and bottom panels, respectively). The value a_n of each of the 24 modal coefficients is given showing that the sparse sensing scheme correctly identifies the correct parameter regimes of β_1 (top panel), β_3 (middle panel) and β_5 (bottom panel). The a_n are color coded according to the library elements depicted in Fig. 23.4. For ease of viewing, the different β_k regimes are separated by shaded/nonshaded regions and are further identified at the top with Fig. 23.2 using the parameter regimes (a)–(f).

The resulting vector **a** is the solution that determines what modal content is identified in the measurement process.

Figure 23.6 shows the results of the three measurements $x_0 = 0, x_1 = 0.5, x_2 = 1$ for the times $t = 50, 150, 250$. In particular, the vector **a** is constructed at each of these time locations. As has been conjectured, the sparse sensing identified the modes of the parameter regime β_1, β_3 and β_5 to be active at the sampling times. This is indeed the case and allows for a reconstruction of the dynamics through a POD low-rank dynamical projection (see Section 18.3). For this example, ideal sampling locations were chosen based upon considering the library modes of ψ_L. If poor choices are made in the spatial sampling locations, i.e. not where maxima and minima are observed in the POD library modes [91, 92], then modes are often misidentified or from mixed β_k (see Fig. 23.7). This poses additional difficulties in the method, thus suggesting that sensor placement should be carefully considered.

The right panel of Fig. 23.5 shows the reconstruction dynamics in the POD basis determined through the compressive sensing scheme. In particular, since we sample in time at $t = 50, 150$ and 250, the reconstruction is only performed at these specific times. To perform the low-dimensional reconstruction (23.1.3) with the POD library elements: (i) the modes corresponding to the specific β_k are extracted, (ii) the compressive sensing measurement of the a_n are projected on to

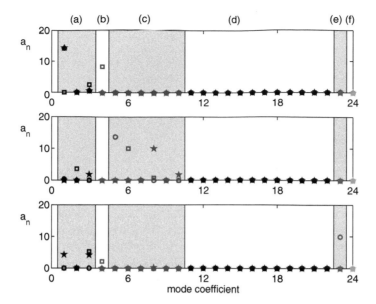

Figure 23.7: Comparison of mode identification for the three sparse measurements $\hat{U}$ at $x_0 = 0, 0.5, 1$ (circles), $x_0 = 0, 1, 2$ (stars) and $x_0 = 0, 2, 4$ (squares). The circles represent the ideal sampling location whereas the two other locations mis-identify the active modes of the system. However, since the modes are not orthogonal in the library ψ_L, this does not mean that reconstructions cannot still be relatively accurate.

these modes, (iii) the modes are evolved according to the POD-Galerkin projection technique of Section 18.3, and (iv) the full time–space evolution is reconstructed by using the spatial modes from the library ψ_L in conjunction with their time dynamics $a_n(t)$.

As an example of the above procedure, the following MATLAB code is used in the reconstruction from the measurements at time $t = 150$. The modes identified were those corresponding to β_3, thus the six modes numbered 5 through 11 (see Fig. 23.6).

```
t=[150:2:250];    % evolve to next measurement
psiA=psiL(:,5:10); % library POD modes

for j=1:6 % compute 2nd/4th POD derivatives
  pA2x(:,j)=ifft(-(k.^2).*fft(psiA(:,j)));
  pA4x(:,j)=ifft( (k.^4).*fft(psiA(:,j)));
end
lhs=inv((psiA.')*psiA); % compute inner products

a_ic=a(5:10);  % initial conditions from L^1 optimization
[t,a_sol]=ode45('a_rhs',t,a_ic,[],psiA,pA2x,pA4x,lhs,tau,ka,be,nu,si,ga);

for j=1:length(z) % build solution matrix
    u_sol(j,:)=( psiA*(a_sol(j,:).') ).';
end
```

where the right-hand side function **a_rhs.m** is given by

```
function a_rhs=a_rhs(z,a,dummy,psiA,pA2x,pA4x,lhs,tau,ka,be,nu,si,ga)
A1=(psiA*a); A2=(psiA2x*a); A3=(psiA4x*a);

rhs=(i*0.5+tau)*A2 + ka*A3 + (i+be)*(abs(A1).^2).*A1 ...
    +(i*nu+si)*(abs(A1).^4).*A1 + ga*A1;
rhs2=(psiA.')*rhs;

a_rhs=lhs*(rhs2);
```

This code provides a generic architecture for simulating the PDE dynamics on the low dimensional manifold spanned by the POD modes. Figure 23.5 shows the comparison of the low-rank dynamical approximation (in this case a 3×3 system of ODEs followed by a 6 × 6 and a 1 × 1) to the full PDE. The agreement is quite remarkable given that only three observations were used to reconstruct the original high dimensional data. In the next section, a much more sophisticated problem will be considered and more careful consideration of sensor location will be addressed. Regardless, this example serves to illustrate the power of combining low dimensional POD approximations with machine learning and compressive sensing.

23.3   Flow Around a Cylinder: A Prototypical Example

The previous section provided a proof-of-concept example of the integration of dimensionality reduction with machine learning and compressive sensing. In this section, a second, more complex example will be given of a classic problem of applied mathematics: the flow around a cylinder. Figure 23.8 demonstrates the basic phenomenon exhibited for the flow around a cylinder as a function of the Reynolds number *Re*.

To be more precise about the integration of CS and ML with the dynamics of flow around a cylinder, consider a standard numerical scheme for solving PDEs by discretization. That is, turn the fluid system into a system of ordinary differential equations (ODE), then solve the latter equations with standard time-stepping tools such as Runge–Kutta. For complex dynamical systems, such as the flow around a cylinder, the immediate difficulty is dimensionality: the dimension of the ODE depends on the number of points in the grid, which can be prohibitively large. This is due to the fact that an accurate PDE solution generally requires a large number of grid points, making the numerical solution computationally expensive. But a well-known observation is that in many systems the dynamics, including those of the fluid and pressure fields, computed in the high dimensional ODE, are observed to be low dimensional.

To this end, the POD method can be used for characterizing the flow around the cylinder. Thus the dynamics is projected from the high dimensional discretized system to a low dimensional system that can be solved with a significant computational improvement, and moreover, is often amenable to analytic analysis or control.

(a) laminar flow

(b) stagnation point

(c) von Kármán vortices

(d) turbulent flow

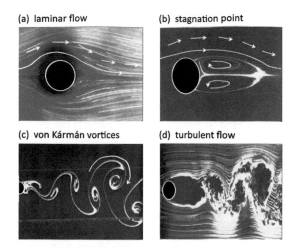

Figure 23.8: Fluid flow past a cylinder for increasing Reynolds number. The four pictures correspond to $Re =$ (a) 9.6, (b) 41, (c) 140, and (d) 10000. Note the transition from laminar (steady state) flow to a recirculating steady state flow, to von Kármán vortex shedding and finally to fully turbulent flow [modified from M. Van Dyke, *An Album of Fluid Motion,* Parabolic Press (1982) with permission from Prof. Hassan Nagib, Illinois Institute of Technology].

The integration of the CS and ML schemes exploits the underlying low dimensional (POD modes) dynamics observed in simulation and/or experiment. As has been stated, CS takes advantage of sparsity by considering a basis wherein the dynamics is sparse. Here, the sparse basis elements are the POD modes computed from full PDE simulations of the system. By characterizing various dynamical parameter regimes of the system, a library of sparse POD modes can be constructed, i.e. a supervised ML algorithm is effectively enacted. Sparse measurements of either the pressure field on the cylinder or flow/pressure field around the cylinder can then be used with the CS algorithm to identify the correct POD modes for the given measurements, thus allowing for reconstruction of the dynamics. In our example of flow around a cylinder [92], the sparse POD basis elements are constructed from different Reynolds numbers where the dynamical behavior changes significantly. A limited number of pressure field measurements are all that is necessary to reconstruct the pressure field, flow field and determine the current Reynolds number.

Governing flow equations and boundary conditions

To demonstrate the integration of machine learning and compressive sensing on complex dynamical systems, we study the flow of an incompressible fluid around a cylinder in a channel. This system is described by the Navier–Stokes equation:

$$\frac{\partial u}{\partial t} + u \cdot \nabla u + \nabla p - Re \nabla^2 u = 0$$
$$\nabla \cdot u = 0 \tag{23.3.1}$$

where $u(x, t) \in R^2$ represents the 2D velocity of the fluid, and $p(x, t) \in R$ is the corresponding pressure. The boundary condition are as follows:

1   Constant flow of $u = (1, 0)^T$ at $x = -15$, i.e. the entry of the channel.

2   Constant pressure of $p = 0$ at $x = 25$, i.e. the end of the channel.

3   Neumann boundary conditions, i.e. $\partial u/\partial \mathbf{n} = 0$ on the boundary of the channel.

4   No-slip boundary conditions on the cylinder.

This system is chosen since it is a well-known and classic problem of applied mathematics. Further, it has been demonstrated to exhibit low dimensional behavior as the parameter Re, the Reynolds number, is varied.

Simulations show the same behavior as the experimental observations in Fig. 23.8. Specifically, as the Reynolds number is increased, the laminar flow develops two circulating regions along with a stagnation point behind the cylinder. Further increasing the Reynolds number induces the onset of periodic, von Kármán vortex shedding. Finally, further increasing the flow rate beyond a critical level produces fully developed turbulence in the wake of the cylinder. As might be expected, the first three dynamical regimes produce low dimensional representations of the fluid motion. But remarkably, even in the fully turbulent regime, snapshot based methods for $O(1)$ samples of time show the fluid flow to be relatively low dimensional (20–30 modes).

Numerical computations and library construction

The numerical solution of the incompressible Navier–Stokes equation is typically computed using, for instance, high-order finite element methods. The finite elements are especially important for handling the boundary conditions imposed by the cylinder in the computational domain. The computational mesh used is ideally nonuniform in order to efficiently provide refinement around the cylinder and to the right of the cylinder where the most interesting fluid flow occurs. In contrast, no refinement is required far to the left, top and bottom of the computational domain.

The simulations used for library construction are generated for the Reynolds numbers $Re = 10$, 40, 150, 300, 1000 and 10 000, with equivalent and arbitrary initial conditions. The dynamics were simulated until a steady state or periodic dynamics was observed for the non turbulent dynamics. Upon achieving the steady or periodic state, the dynamics was sampled through a series of snapshots whose time interval between snapshots was much faster than the periodicity observed in the dynamics, for instance. Figure 23.9 show a typical snapshot of the pressure profiles taken once the system was in the equilibrium or time-periodic state.

For the POD mode selection, the dimension of the projected, low-rank system must be specified. This is chosen according to the data by choosing a threshold of, for instance, 99% for the energy we want to conserve with the projection. Our objective is to select the minimal number of POD modes needed to do so for each Reynolds numbers. For the Reynolds number $Re = 10$, only a single POD mode is necessary to capture the dynamics, whereas for $Re = 1000$, approximately 10 modes are required. Once all the POD modes are computed in the preprocessing stage, i.e. for the various Reynolds numbers, we place them as the column vectors in the library matrix ψ_L. The library encapsulates all known behaviors of the system, or at least those explored via the computational simulations. Success of the library-based method is based upon appropriately sampling

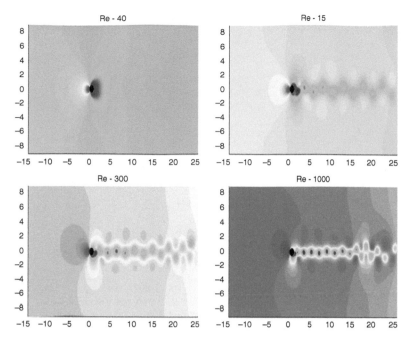

Figure 23.9: Typical pressure field measurement for Reynolds numbers $Re =$40, 150, 300, 1000 (top left to bottom right). Such snapshots separated in time by Δt are used in the data matrix for producing the POD modes for each Reynolds number. [From I. Bright, G. Lin and J. N. Kutz, Compressive sensing and machine learning strategies for characterizing the flow around a cylinder with limited pressure measurements, Physics of Fluids (submitted).]

all relevant dynamics and parameter ranges under which the system will be considered. This is equivalent to constructing the matrix with 24 library library elements in the CQGLE equation of the last section.

Figure 23.10 shows the three most dominant POD modes for the pressure fields on the cylinder for $Re = 40, 150, 300, 1000$. These are part of the library matrix ψ_L and are used as the sparse modes in the compressive sensing reconstruction of the pressure and flow fields. In addition to the pressure modes around the cylinder, their corresponding pressure and streamline POD modes in the flow field are also kept for reconstruction of the entire pressure/flow field dynamics on and off the cylinder.

Compressive sensing: Cylinder pressure measurements

To demonstrate the ability of the compressive sensing to characterize the fluid flow, the method is used to compute the Reynolds number of the flow field from a sparse number of pressure measurements on the cylinder as the flow field is changed in time. The location of the pressure measurements is important in proper use of the technique [91, 92]. This Reynolds number reconstruction is applied to a numerically generating flow where the Reynolds number varies slowly

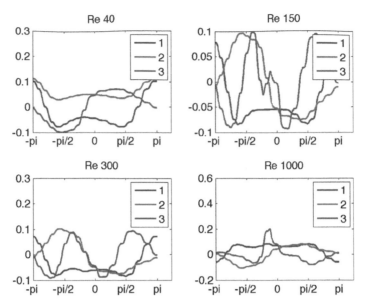

Figure 23.10: The first three dominant POD modes for the pressure field around the cylinder for $Re = 40$, 150, 300, 1000 (top left to bottom right). The compressive sensing scheme identifies which modes are active in a given flow measurement while zeroing out the other modes. [From I. Bright, G. Lin and J. N. Kutz, Compressive sensing and machine learning strategies for characterizing the flow around a cylinder with limited pressure measurements, Physics of Fluids (submitted).]

in time from one Reynolds regime to another. Specifically, a flow is generated that starts with Reynolds number 40 and which jumps successively to 150, 300 and 1000. The change in Reynolds number occurs after a steady state/periodic state is achieved. The reconstruction of the Reynolds number is applied to every snapshot through the compressive sensing scheme.

Recall that the reconstruction of the data from a single snapshot is performed by solving an L^1 optimization problem for an underdetermined system, thus promoting sparsity in the POD selection and obtaining the coefficient vector **a**. Each entry in **a** corresponds to the weighting of a single POD mode from our library ψ_L. Since the optimization in L_1 promotes sparsity, only those coefficients from POD modes associated with the measured flow remain nonzero, or ideally so. The classification of the Reynolds number is done by summing the absolute value of the coefficient that correspond to each Reynolds number. To account for the large number of coefficients that we allocated for the higher Reynolds number (which may more than 20 POD modes, rather than a single coefficient for Reynolds number 40) we divide by the square root of the number of POD modes allocated in **a** for each Reynolds number. The classified method is the one that has the largest magnitude after this process.

Finally, to visualize the entire sparse-sensing and reconstruction process more carefully, Fig. 23.11 shows both the Reynolds number reconstruction for the time-changing flow field along with the pressure field and flow field reconstructions at select locations in time. Note that the compressive sending scheme along with the supervised machine learning library provide an effective method for characterizing the flow strictly through sparse measurements.

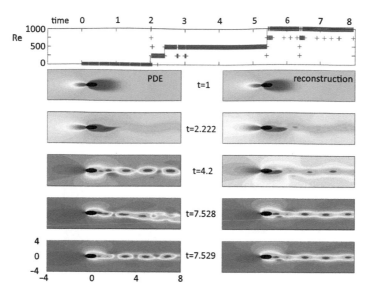

Figure 23.11: Pressure field and streamline reconstruction from 15 pressure measurements around the cylinder at a given instance in time. Note that the compressive sensing in combination with the supervised machine learning strategy for building ψ_L works well for characterizing the flow around a cylinder. [From I. Bright, G. Lin and J. N. Kutz, Compressive sensing and machine learning strategies for characterizing the flow around a cylinder with limited pressure measurements, Physics of Fluids (submitted).]

In concluding this chapter, it is important to note that modern techniques of data analysis are providing transformative algorithms and methods for exploring complex systems. In the context of exploring high dimensional dynamical systems, a variety of techniques have successfully been applied to developing low dimensional models capable of greatly improving the analysis, control and predictability of a given model. The two promising data-analysis techniques of machine learning and compressive (sparse) sensing have hitherto been unexplored in the context of dynamical systems theory. When used in conjunction, ML and CS provide an ideal framework for using sparse sensors in a complex dynamical system for reconstruction of the dynamics and evolution of the system on an optimal low dimensional manifold.

The combination of dimensionality reduction, compressive sensing and machine learning advocated here can be applied in an exceptionally broad context. Indeed, such algorithmic strategies can be used to enhance computation and efficiently identify measurement locations in a given system by *remembering* the key characteristics of the system. An additionally appealing aspect of the methodology presented is that the low-rank nature of the approximations used allow one to construct control algorithms that wrap around the dimensionality reduction, CS and ML infrastructure. For instance, in flight control, the identification of the dominant dynamical modes of the system through a sparse set of measurements would only be the first step in applying a control algorithm for stabilizing a constant lift, for instance. Thus the proposed method would serve as the internal portion of the control algorithm.

PART IV

Scientific Applications

The purpose of this part is to be able to solve realistic and difficult problems which arise in a variety of physical settings. This part outlines a few physical problems which give rise to phenomena of great general interest. The techniques developed in these chapters on finite difference, spectral and finite element methods provide all the necessary tools for solving these problems. Thus each problem isn't simply a mathematical problem, but rather a way to describe phenomena which may be observed in various physically interesting systems.

Applications of Differential Equations and Boundary Value Problems

24

24.1   Neuroscience and the Hodgkin–Huxley Model

In 1952, Alan Hodgkin and Andrew Huxley published a series of five papers detailing the mechanisms underlying the experimentally observed dynamics and propagation of voltage spikes in the giant squid axon (see Fig. 24.1). This seminal work eventually led them to receive the 1963 Nobel Prize in Physiology and Medicine. Not only were the experimental techniques and characterization of the ion channels unparalleled at the time, but Hodgkin and Huxley went further to develop the underpinnings of modern theoretical/computational neuroscience [100, 101, 102]. Given its transformative role in neuroscience, the Hodgkin–Huxley model is regarded as one of the great achievements of twentieth-century biophysics.

At its core, the Hodgkin–Huxley model framed how the action potential in neurons behaves like some analogous electric circuit. This formulation was based upon their observations of the dynamics of ion conductances responsible for generating the nerve action potential. In physiology, an action potential is a short-lasting event in which the electrical membrane potential of a cell rapidly rises and falls, following a consistent, repeatable trajectory. Action potentials occur in several types of animal cells, called excitable cells, which include neurons, muscle cells, and endocrine cells, as well as in some plant cells.

Driving the dynamics was the fact that there was a potential difference in voltage between the outside and inside of a given cell. A typical neuron has a resting potential of −70 mV. The cell membrane separating these two regions consists of a lipid bilayer. The cell membrane also has channels, either gated or nongated, across which ions can migrate from one region to the other. Nongated channels are always open, whereas gated channels can open and close with the probability of opening often depending upon the membrane potential. Such gated channels, which often select for a single ion, are called *voltage-gated channels*.

The Hodgkin–Huxley model builds a dynamic model of the action potential based upon the ion flow across the membrane. The principal ions involved in the dynamics are sodium ions

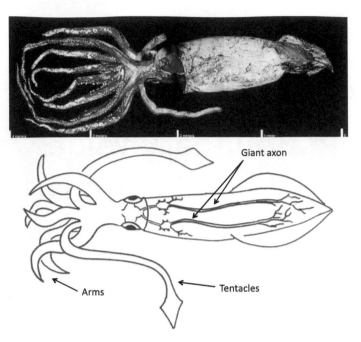

Figure 24.1: Picture and diagram of a giant squid with the giant axon denoted along the interior of the body. The giant axon was ideal for experiments given its length and girth. The top photograph is courtesy of NASA.

(Na$^+$), potassium ions (K$^+$), and chloride anions (Cl$^-$). Although the Cl$^-$ does not flow across the membrane, it does play a fundamental role in the charge balance and voltage dynamics of the sodium and potassium. In animal cells, two types of action potential exist: a voltage-gated sodium channel and a voltage-gated calcium channel. Sodium-based action potentials usually last for under one millisecond, whereas calcium-based action potentials may last for 100 milliseconds or longer. In some types of neurons, slow calcium spikes provide the driving force for a long burst of rapidly emitted sodium spikes. In cardiac muscle cells, on the other hand, an initial fast sodium spike provides a primer to provoke the rapid onset of a calcium spike, which then produces muscle contraction.

Accounting for the channel gating variables along with the membrane potential results in the 4×4 systems of nonlinear differential equations originally derived by Hodgkin and Huxley [100, 101, 102]:

$$C_M \frac{dV}{dt} = -\bar{g}_K N^4 (V - V_K) - \bar{g}_{Na} M^3 H (V - V_{Na}) - \bar{g}_L (V - V_L) + I_0 \tag{24.1.1a}$$

$$\frac{dM}{dt} = \alpha_M (1 - M) - \beta_M M \tag{24.1.1b}$$

$$\frac{dN}{dt} = \alpha_N (1 - N) - \beta_N N \tag{24.1.1c}$$

$$\frac{dH}{dt} = \alpha_H (1 - H) - \beta_H H \tag{24.1.1d}$$

where $V(t)$ is the voltage of the action potential that is driven by the voltage-gated variables for the potassium ($N(t)$) and the sodium ($M(t)$ and $H(t)$). The specific functions in the equations were derived to fit experimental data for the squid axon and are given by

$$\alpha_M = 0.1\frac{25 - V}{\exp((25 - V)/10) - 1} \tag{24.1.2a}$$

$$\beta_M = 4\exp\left(\frac{-V}{18}\right) \tag{24.1.2b}$$

$$\alpha_H = 0.07\exp\left(\frac{-V}{20}\right) \tag{24.1.2c}$$

$$\beta_H = \frac{1}{\exp((30 - V)/10) + 1} \tag{24.1.2d}$$

$$\alpha_N = 0.01\frac{10 - V}{\exp((10 - V)/10) - 1} \tag{24.1.2e}$$

$$\beta_N = 0.125\exp\left(\frac{-V}{80}\right) \tag{24.1.2f}$$

with the parameters in the first equation given by $\bar{g}_{Na} = 120$, $\bar{g}_K = 36$, $\bar{g}_L = 0.3$, $V_{Na} = 115$, $V_K = -12$ and $V_L = 10.6$. Here I_0 is some externally applied current to the cell. One of the characteristic behaviors observed in simulating this equation is the production of spiking behavior in the action potential which agrees with experimental observations in the giant squid axon.

FitzHugh–Nagumo reduction

The Hodgkin–Huxley model, it can be argued, marks the start of the field of computational neuroscience. Once it was established that neuron dynamics had sound theoretical underpinnings, the field quickly expanding and continues to be a dynamic and exciting field of study today, especially as it pertains to characterizing the interactions of large groups of neurons responsible for decision making, functionality and data/signal-processing in sensory systems.

The basic Hodgkin–Huxley model already hinted at a number of potential simplifications that could be taken advantage of. In particular, there was clear separation of some of the time-scales involved in the voltage-gated ion flow. In particular, sodium-gated channels typically are very fast in comparison to calcium-gated channels. Thus on the *fast* scale, calcium-gating looks almost constant. Such arguments can be used to derive simpler versions of the Hodgkin–Huxley model. Perhaps the most famous of these is the FitzHugh–Nagumo model which is a two-component system that can be written in the form [100, 101]:

$$\frac{dV}{dt} = V(a - V)(V - 1) - W + I_0 \tag{24.1.3a}$$

$$\frac{dW}{dt} = bV - cW \tag{24.1.3b}$$

where $V(t)$ is the voltage dynamics and $W(t)$ models the refractory period of the neuron. The parameter a satisfies $0 < a < 1$ and determines much of the resulting dynamics. One of the most attractive features of this model is that standard phase-plane analysis techniques can be applied and a geometrical interpretation can be applied to the underlying dynamics. Although only qualitative in nature, the FitzHugh–Nagumo model has allowed for significant advancements in our qualitative understanding of neuron dynamics.

FitzHugh–Nagumo and propagation

Finally, we consider propagation effects in the axon itself. Up to this point, no propagation effects have been considered. One typical way to consider the effects of diffusion is to think of the voltage $V(t)$ as diffusing to neighboring regions of space and potentially causing these neighboring regions to become excitable and spike. The simplest nonlinear propagation model is given by the FitzHugh–Nagumo model with diffusion. This is a modification of (24.1.3) to include diffusion [100, 101]:

$$\frac{dV}{dt} = D\frac{d^2V}{dx^2} + V(a - V)(V - 1) - W + I_0 \qquad (24.1.4a)$$

$$\frac{dW}{dt} = bV - cW \qquad (24.1.4b)$$

where the parameter D measures the strength of the diffusion.

Projects

(a) Calculate the time dynamics of the Hodgkin–Huxley model for the default parameters given by the model.

(b) Apply an outside current $I_0(t)$, specifically a step function of finite duration, to see how this drives spiking behavior. Apply the step function periodically to see periodic spiking of the axon.

(c) Repeat (a) and (b) for the FitzHugh–Nagumo model.

(d) Although you can compute the fixed points and their stability for the FitzHugh–Nagumo model by hand, verify the analytic calculations for the location of the fixed points and determine their stability computationally.

(e) For fixed values of parameters in the FitzHugh–Nagumo model, perform a convergence study of the time-stepping method by controlling the error tolerance, TOL, in ODE45 and ODE23:

```
TOL=1e-4;
OPTIONS = odeset('RelTol',TOL,'AbsTol',TOL);
[T,Y] = ODE45('F',TSPAN,Y0,OPTIONS);
```

Show that indeed the schemes are fourth order and second order, respectively, by running the computation across a given time domain and adjusting the tolerance. In particular, plot on a log-log scale the average step-size (*x*-axis) using the *diff* and *mean* command versus the tolerance (*y*-axis) for a large number of tolerance values. What are the slopes of these lines? Note that the local error should be $O(\Delta t^5)$ and $O(\Delta t^3)$, respectively. What are the local errors for ODE113 and ODE15s?

(f)   Assume a Gaussian initial pulse for the spatial version of the FitzHugh–Nagumo equation (24.1.4) and with $D = 0.1$ and investigate the dynamics for differing values of *a*. Assume periodic boundary conditions and solve using second-order finite differencing for the derivative. Explore the parameter space so as to produce spiking behavior and relaxation oscillations.

(g)   Repeat (f) with implementation of the fast Fourier transform for computing the spatial derivatives. Compare methods (f) and (g) in terms of both computational time and convergence.

(h)   Try to construct initial conditions for the FitzHugh–Nagumo equation (24.1.4) with $D = 0.1$ so that only a right traveling wave is created. (Hint: Make use of the refractory field which can be used to suppress propagation.)

(i)   Make a movie of the spike wave dynamics and their propagation down the axon.

(j)   Use these numerical models as a control algorithm for the construction of a giant squid made from spare parts in your garage. Attempt to pick up some of the smaller dogs in your neighborhood (see Fig. 24.1).

24.2   Celestial Mechanics and the Three-Body Problem

The movement of the heavenly bodies has occupied a central and enduring place in the history of science. From the earliest theoretical framework of the *Doctrine of the Perfect Circle* in the second century A.D. by Claudius Ptolemy to our modern understanding of celestial mechanics, gravitation has taken a pivotal role in nearly two thousand years of scientific and mathematical development. Indeed, some of the most prominent scientist in history have been involved in the development of the theoretical underpinnings of celestial mechanics and the notion of gravity, including Copernicus, Galileo, Kepler and Newton (Fig. 24.2). The two contemporaries, Galileo and Kepler, produced conjectures and observations that would lay the foundation for Newton's law of universal gravitation. Galileo built on the ideas of Copernicus and forcefully asserted that the Earth revolved around the Sun and further challenged the notion of the perfection of the heavenly spheres. Of course, his recalcitrant attitude towards the church did not help his career very much.

Kepler, a contemporary of Galileo's, further advanced the notion of planetary motion by asserting that (i) planets have elliptical orbits with the Sun as a focus, (ii) a radius vector from the Sun to a planet sweeps out equal areas in equal times, and (iii) the period of an orbit squared is

Figure 24.2: Giants of celestial mechanics. (Left) Nicolaus Copernicus (1473–1543, artist unknown) proposed a heliocentric model to replace the dominant Ptolemic viewpoint. (Second from left) Galileo Galilei (1564–1642, portrait by Giusto Sustermans) invented the first telescope and was able to strongly argue for the Copernican model. Additionally, he studied and documented the first observations concerning uniformly accelerated bodies. (Second from right) Johannes Kepler (1571–1630, artist unknown) made observations leading to three key findings about planetary orbits, including their elliptical trajectories. (Right) Isaac Newton (1642–1727, portrait by Sir Godfrey Kneller), perhaps the most celebrated physicist in history, placed the work of Copernicus, Galileo and Kepler on a mathematical footing with the invention of calculus and the notion of a force called gravity.

proportional to the semi-major axis of the ellipse cubed. Such conjectures where able to improve the accuracy of predicting planetary motion by two orders of magnitude over both the Ptolemic and Copernican systems.

The law of universal gravitation was proposed on the heels of the seminal work of Galileo and Kepler. In order to propose his theory of forces, Sir Isaac Newton formulated a new mathematical framework: calculus. This led to the much celebrated law of motion $\mathbf{F} = m\mathbf{a}$ where $\mathbf{F}$ is a given force vector and $\mathbf{a}$ is the acceleration vector, which is the second derivative of position. In the context of gravitation, Newton proposed the attractive gravitational law

$$F = \frac{GMm}{r^2} \tag{24.2.1}$$

where G is a universal constant, M and m are the masses of the two interacting bodies and r is the distance between them. With this law of attraction, the so-called Kepler's law could be written

$$\frac{d^2\mathbf{r}}{dt^2} = -\frac{\mu}{r^3}\mathbf{r} \tag{24.2.2}$$

where $\mu = G(m + M)$ and the dynamics of the smaller mass m is written in terms of a frame of reference with the larger mass M fixed at the origin. Thus $\mathbf{r}$ measures the distance from mass M to mass m. Equation (24.2.2) can be solved exactly to first, mathematically confirm the three assertions and/or observations made by Kepler, and second, to predict the dynamics of any two-mass system. This was known as the *two-body* problem and its solutions where conic sections in polar form, yielding straight lines, parabolas, ellipses and hyperbolas as solution orbits. Surprisingly enough, after more than 300 years of gravitation, this is the only gravitation problem for which we have a closed form solution. There are reasons for this as will be illustrated shortly.

The three-body problem

One can imagine the scientific excitement that came from Newton's theory of gravitation. No doubt, after the tremendous success of the two-body problem, there must have been some anticipation that solving a problem involving three masses would be more difficult, but perhaps a tenable proposition. As history has shown, the three-body problem has defied our ability to provide a closed form solution aside from some perturbative solutions in restricted asymptotic regimes.

We consider the restricted three-body problem for which two of the masses m_1 and m_2 are much larger than m_3 so that $m_1, m_2 \gg m_3$ (see Fig. 24.3). Further, it is assumed that we are in a frame of reference such that the two masses m_1 and m_2 are located at the points $(0,0)$ and $(-1,0)$, respectively. Thus we consider the motion of m_3 under the gravitational pull of m_1 and m_2. This could, for instance, model an Earth–Moon–rocket system.

The governing equations are

$$x' = u$$

$$u' = 2v + x - \frac{\mu(-1 + x + \mu)}{\left(y^2 + z^2 + (-1 + x + \mu)^2\right)^{3/2}} - \frac{(1-\mu)(x+\mu)}{\left(y^2 + z^2 + (x+\mu)^2\right)^{3/2}}$$

$$y' = v$$

$$v' = -2u + y - \frac{\mu y}{\left(y^2 + z^2 + (-1 + x + \mu)^2\right)^{3/2}} - \frac{(1-\mu)y}{\left(y^2 + z^2 + (x+\mu)^2\right)^{3/2}}$$

$$z' = w$$

$$w' = -\frac{\mu z}{\left(y^2 + z^2 + (-1 + x + \mu)^2\right)^{3/2}} - \frac{(1-\mu)z}{\left(y^2 + z^2 + (x+\mu)^2\right)^{3/2}}$$

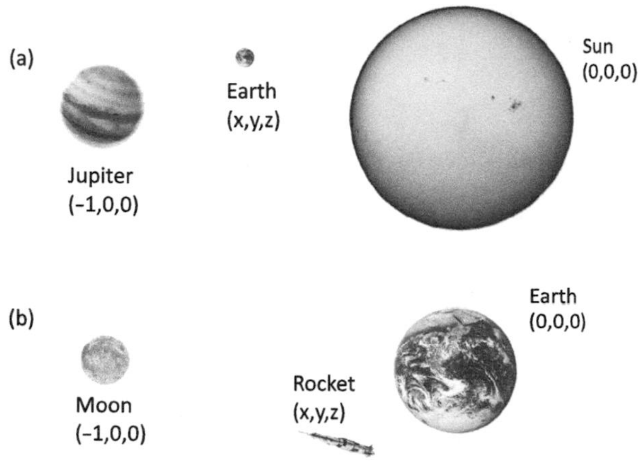

Figure 24.3: Two classic examples of the three-body problem: (a) the Sun–Jupiter–Earth system and (b) the Earth–Moon–rocket/satellite problem.

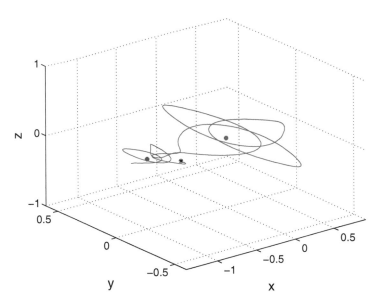

Figure 24.4: Trajectory of a satellite in a three-body system. In this case, the satellite performs an orbit transfer from m_1 to m_2 and back again. For this example, the following parameters where used with $\mu = 0.1$ and with initial conditions $(x, u, y, v, z, w) = (-1.2, 0, 0, 0, 0, 0)$. The evolution was for $t \in [0, 12]$. The red dot is mass m_1, the magenta dot is mass m_2, and the red line is the trajectory following the mass m_3 denoted by a black star.

where the position of mass m_3 is given by $\vec{r} = (x \ y \ z)$ and its velocity is $\vec{v} = (u \ v \ w)$. The parameter $0 < \mu < 1$ measures the relative masses of the smaller to large mass, i.e. m_2/m_1. An example trajectory for the three-body problem is shown in Fig. 24.4.

We can further restrict the problem so that m_3 is in the same plane (x–y plane) as masses m_1 and m_2. Thus we throw out the z and w dynamics and only consider x, y, u and v. For this case:

(a)   Using a root solving routing (and analytic methods), find the five critical (Lagrange) points.

(b)   Find the stability of the critical point by linearizing around each point. MATLAB can be used to help find the eigenvalues.

(c)   Evolve the governing nonlinear ODEs and verify the stability calculations of (b) above.

(d)   Show that the system is chaotic by demonstrating sensitivity to initial conditions, i.e. measure the separation in time of two nearly identical initial conditions.

(e)   By exploring with MATLAB, show the following possible behaviors:
   ▪ Stable evolution around m_1
   ▪ Stable evolution around m_2
   ▪ Transfer of orbit from m_1 to m_2

- Periodic and stable evolution near the two Lagrange points where $y \neq 0$
- Escape from the m_1, m_2 system.

(f) Explore the change in behavior as a function of the parameter μ, i.e. as the dominance, or lack thereof, of one of the masses changes.

24.3 Atmospheric Motion and the Lorenz Equations

Understanding climate patterns and providing accurate weather predictions are of central importance in the fields of atmospheric sciences and oceanography. And despite decades of study and large meteorological models, it would seem that weather forecasts are still quite inaccurate, and mostly untrustworthy for more than one week into the future. Ultimately, there is a fundamental mathematical reason for this. Much like the three-body problem of celestial mechanics, even a relatively simple idealized model can demonstrate the source of the underlying difficulties in predicting weather.

The first scientist to predict that weather and climate dynamics where indeed a *global* phenomenon was Sir Gilbert Walker. In 1904 at the age of 36, he was posted in India as the Director General of Observations. Of pressing interest to him was the ability to predict monsoon failures and the ensuing droughts that would accompany them as this had a profound impact on the people of India. In an attempt to understand these seemingly random failures, he looked for correlations in weather and climate across the globe and found that the random failure of monsoons in India often coincided with low pressure over Tahiti, high pressure over Darwin, Australia, and relaxed trade winds over the Pacific Ocean. Indeed, his observations showed that pressure highs in Tahiti were directly correlated to pressure lows in Darwin and vice versa. This is now called the *southern oscillation*. In years when the southern oscillation was weak, the following observations were made: there was heavy rainfall in the central Pacific, droughts in India, warm waters in south-west Canada and cold winters in south-east United States. This led to his conjecture that weather was indeed a global phenomenon. He was largely ignored for decades, and even ridiculed for his suggestion that climatic conditions over such widely separated regions of the globe could be linked.

Sixty-five years after Walker's claims, Jacob Bjerknes in 1969 advanced a conceptual mathematical model tying together the well documented El Niño phenomenon to the southern oscillation: the El Niño southern oscillation (ENSO). A schematic of the model is illustrated in Fig. 24.5 where the normal Walker circulation can be broken if the easterly trade winds slacken. In this case, warm water from the central Pacific is pushed back to the eastern Pacific leading to both the El Niño phenomenon and the see-sawing of high pressures between Darwin, Australia and Tahiti.

More than a schematic as shown in Fig. 24.5, the theory proposed by Bjerknes was a mathematical description of the circulation and its potential breakdown. The model makes a variety of simplifying assumptions. First on the list of assumptions is that the velocity of the surface wind u_a

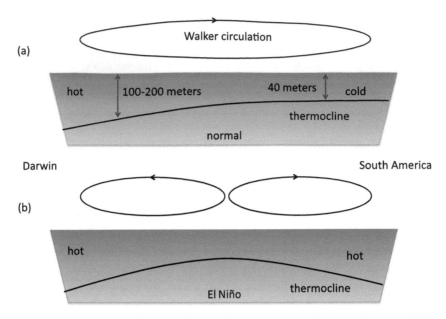

Figure 24.5: Schematic of the southern oscillation phenomenon and El Niño. The normal Walker circulation pattern is illustrated in (a) where a large circulation occurs from the South American continent to Australia due to the easterly trade winds. If the trade winds slacken, warm water moves back east and warms the easter Pacific resulting in El Niño. Note that the breakage of the normal Walker circulation pattern leads to a see-saw in pressure highs between Darwin and Tahiti.

is driven by the temperature difference $T_e - T_w$, i.e. hot air moves to cold regions. In differential form, this can be stated mathematically as

$$\frac{du_a}{dt} = b(T_e - T_w) + r(u_0 - u_a) \tag{24.3.1}$$

where u_0 measures the velocity of the normal easterly winds, r is the rate at which the u_a relaxes to u_0, and b is the rate at which the ocean influences the atmosphere. But since the ocean temperature and velocity change slowly in relation to the atmospheric temperature and velocity, the following approximation holds: $du_a/dt \approx 0$. Thus

$$u_a = \frac{b}{r}(T_e - T_w) + u_0 \tag{24.3.2}$$

gives an approximation to the air velocity at the surface.

The air velocity interacts with the ocean through a stress coupling at the water–air interface. This stress coupling is responsible for generating the ocean circulation velocity near the surface of $u = U$. The simplest model for this coupling is given by

$$\frac{dU}{dt} = du_a - cU \tag{24.3.3}$$

where d measures the strength of the coupling and c is the damping (relaxation) back to the zero velocity state $U = 0$. Note that if the coupling is $d = 0$, then the solution is given by $U = U(0) \exp(-ct)$ so that $U \to 0$ as $t \to 0$. Thus u_a is responsible for driving the current. Substituting the value of u_a previously found into this expression yields the velocity dynamics

$$\frac{dU}{dt} = B(T_e - T_w) - C(U - U_0) \tag{24.3.4}$$

where $B = db/r$ and $U_0 = du_0/c$. This is the approximate governing equation for the air–sea interaction dynamics.

Ocean temperature advection

The equation governing the evolution of the temperature is a standard advection equation which is not derived here. The spatial advection of temperature is given by the partial differential equation

$$\frac{\partial T}{\partial t} + u\frac{\partial T}{\partial x} + w\frac{\partial T}{\partial z} \tag{24.3.5}$$

where $T = T(x, z, t)$ gives the temperature distribution as a function of time and the horizontal and vertical location. The temperature partial differential equation can be approximated using a finite difference scheme for the spatial derivatives. Thus we take

$$\frac{\partial T}{\partial x} \approx \frac{T(x + \Delta x) - T(x - \Delta x)}{2\Delta x}$$

$$\frac{\partial T}{\partial z} \approx \frac{T(z + \Delta z) - T(z - \Delta z)}{2\Delta z}$$

where Δz, $\Delta x \ll 1$ in order for the approximation to be somewhat accurate. The key now is to divide the ocean into four (east and west and top/bottom) finite difference boxes as illustrated in Fig. 24.6. In particular, there is assumed to be an east ocean box and a west ocean box. In the east ocean box, the following approximate relations hold:

$$u\frac{\partial T}{\partial x} \to u\frac{T(x + \Delta x) - T(x - \Delta x)}{2\Delta x} = \frac{U}{2\Delta x}(T_e - T_w)$$

$$w\frac{\partial T}{\partial z} \to w\frac{T(z + \Delta z) - T(z - \Delta z)}{2\Delta z} = -\frac{W}{2\Delta z}(T_e - T_d).$$

Inserting these approximate expressions into the original temperature advection equations yields the temperature equation

$$\frac{\partial T_e}{\partial t} + \frac{U}{2\Delta x}(T_e - T_w) - \frac{W}{2\Delta z}(T_e - T_d) = 0. \tag{24.3.6}$$

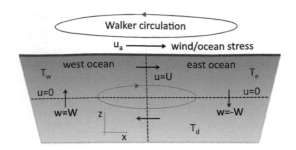

Figure 24.6: Mathematical construction of the southern oscillation scenario. The parameter u_a measures the wind speed from the Walker circulation that interacts via sheer stress with the ocean in the purple top layer. The resulting ocean velocity in the top layer is assumed to be $u = U$. The temperature in the west ocean, east ocean and deep ocean is given by T_w, T_e and T_d, respectively. These temperature gradients drive the circulation dynamics.

Applying the same approximation in the west ocean box yields a similar formula aside from a change in sign for $w = W$. This yields the west ocean box equation

$$\frac{\partial T_w}{\partial t} + \frac{U}{2\Delta x}(T_e - T_w) + \frac{W}{2\Delta z}(T_w - T_d) = 0 \,. \tag{24.3.7}$$

Thus a significant approximation occurred by assuming that the entire Pacific Ocean could be divided into two finite difference cells.

The Lorenz equations

Finally, in order to enforce mass conservation, we must require $W\Delta x = U\Delta z$ so that $W = U\Delta z/\Delta x$. Inserting this relation into the formulas above for the east and west ocean temperatures and considering in addition the air–sea interaction dynamics of (24.3.4) gives the governing ENSO dynamics

$$x' = \sigma y - \rho(x - x_0) \tag{24.3.8a}$$
$$y' = x - xz - y \tag{24.3.8b}$$
$$z' = xy - z \tag{24.3.8c}$$

where the prime denotes time differentiation and the following rescalings are used: $x = U/(A2\Delta x)$, $y = (T_e - T_w)/(2T_0)$, $z = 1 - (T_e + T_w)/(2T_0)$, $t \to At$, $x_0 = U_0/(A2\Delta x)$, $\sigma = 2BT_0/(\Delta x A^2)$ and $\rho = c/A$. A final rescaling with $x_0 = 0$ can move the ENSO dynamics into the more commonly accepted form

$$x' = -\sigma x + \sigma y \tag{24.3.9a}$$
$$y' = rx - xz - y \tag{24.3.9b}$$
$$z' = xy - bz. \tag{24.3.9c}$$

These are the famed *Lorenz equations* that first arose in the study of the ENSO phenomenon at the end of the 1960s. Reasonable values of parameters for Earth's atmosphere give $\sigma = 8/3$ and $\sigma = 10$. The objective is then to see how the solutions change with the parameter r which relates the temperate difference in the layers of the atmosphere.

(a) Using a root solving routine and/or analytic methods, find the critical value of the parameter r where the number of critical points changes from one to three.

(b) Find the stability of the critical point by linearizing around each point. MATLAB can be used to help find the eigenvalues.

(c) Evolve the governing nonlinear ODEs and verify the stability calculations of (b) above.

(d) Determine the critical value $r = r_c$ numerically at which the system becomes chaotic by demonstrating sensitivity to initial conditions, i.e. measure the separation in time of two nearly identical initial conditions. Show that for $r < r_c$, the system is not chaotic while for $r > r_c$ the system is chaotic.

(e) By exploring with MATLAB, also consider the periodically forced Lorenz equations as a function of γ and ω:

$$x' = \sigma y - \rho(x - x_0) + \gamma \cos(\omega t) \tag{24.3.10a}$$
$$y' = x - xz - y \tag{24.3.10b}$$
$$z' = xy - z. \tag{24.3.10c}$$

Use the PLOT3 command to observe the dynamics parametrically in time as a function of $x(t), y(t)$ and $z(t)$.

24.4 Quantum Mechanics

One of the most celebrated physics developments of the twentieth century was the theoretical development of quantum mechanics. The genesis of quantum mechanics, and a quantum description of natural phenomena, began with Max Planck and his postulation that energy is radiated and absorbed in discrete quanta (or energy elements), thus providing a theoretical framework for understanding experimental observations of blackbody radiation. In his theory, each quantum of energy was proportional to its frequency (ν) so that

$$E = h\nu \tag{24.4.1}$$

where h is Planck's constant. This is known as Planck's law. Einstein used his hypothesis to explain the photoelectric effect, for which he won his Nobel Prize.

Shortly after the work of Planck and Einstein, the foundations of quantum mechanics were laid down in Europe by some of the most prominent physicists of the twentieth century. Figure 24.7 is a picture from the 1927 conference on quantum mechanics in Brussels, Belgium. It would be difficult to find a conference or workshop with a higher ratio of Nobel Prize winners than

Figure 24.7: The Quantum Mechanics Dream Team at the 1927 Solvay Conference on Quantum Mechanics. The photograph is by Benjamin Couprie, Institut International de Physique Solvay, Brussels, Belgium. Front Row: Irving Langmuir, Max Planck, Marie Curie, Hendrik Lorentz, Albert Einstein, Paul Langevin, Charles Guye, Charles Thomson, Rees Wilson, Owen Richardson. Middle Row: Peter Debye, Martin Knudsen, William Bragg, Hendrik Kramers, Paul Dirac, Arthur Compton, Louis de Broglie, Max Born, Niels Bohr. Back Row: Auguste Piccard, Émile Henriot, Paul Ehrenfest, Édouard Herzen, Théophile de Donder, Erwin Schrödinger, Jules-Émile Verschaffelt, Wolfgang Pauli, Werner Heisenberg, Ralph Howard Fowler, Léon Brillouin. Thanks to this group, you have that sweet iPhone.

this conference. And only a few years later, the scientific community of Europe would begin to fracture under the rise of Hitler and the imminence of World War II.

Quantum mechanics has had enormous success in explaining, among other things, the individual behavior of the subatomic particles that make up all forms of matter (electrons, protons, neutrons, photons and others), how individual atoms combine covalently to form molecules, the basic operation of lasers, transistors and electron microscopes, etc. In semiconductors, the working mechanism for resonant tunneling in a diode device, based on the phenomenon of quantum tunneling through potential barriers, has led to the entire industry of quantum-electronically based devices, such as the iPhone and your laptop. Ultimately, quantum mechanics has gone from a theoretical construct with many interesting philosophical implications to the greatest engineering design and manufacturing tool of the modern, high-tech world.

In what we develop here, we will follow Schrödinger's formulation of the problem. De Broglie had already introduced the idea of wave-like behavior to atomic particles. However, the standard wave equation requires two initial conditions in order to correctly pose the initial value problem.

But this violated *Heisenberg's uncertainty principle* which stated that the momentum and position of a particle could not be known simultaneously. Indeed, if the exact position was known, then no information was known about its momentum. On the other hand, if the exact momentum was known, then nothing could be known about its position. This led Schrödinger to formulate the following wave equation

$$i\hbar \frac{\partial \psi}{\partial t} = -\frac{\hbar^2}{2m} \frac{\partial^2 \psi}{\partial t^2} \tag{24.4.2}$$

where $\hbar = h/2\pi$, m is the particle mass, and $\psi(x, t)$ is the wavefunction that represents the probability function for finding the particle at a location x at a time t. As a consequence of this probabilistic interpretation, $\int |\psi|^2 dx = 1$. This equation is known as Schrödinger's equation. It is a wave equation that requires only a single initial condition, thus it retains wave-like behavior without violating the Heinsenberg uncertainty principle.

In most cases, the particle is also subject to an external potential field. Thus the probability density evolution in a one-dimensional trapping potential, in this case assumed to be the harmonic potential, is governed by the partial differential equation:

$$i\hbar \psi_t + \frac{\hbar^2}{2m} \psi_{xx} + V(x)\psi = 0, \tag{24.4.3}$$

where ψ is the probability density and $V(x) = kx^2/2$ is the harmonic confining potential. A typical solution technique for this problem is to assume a solution of the form

$$\psi = \sum_1^N a_n \phi_n(x) \exp\left(i\frac{E_n}{2\hbar}t\right) \tag{24.4.4}$$

which is called an eigenfunction expansion solution (ϕ_n = eigenfunction, E_n = eigenvalue). Plugging in this solution ansatz to Eq. (24.4.3) gives the boundary value problem:

$$\frac{d^2 \phi_n}{dx^2} - \left[Kx^2 - \varepsilon_n\right] \phi_n = 0 \tag{24.4.5}$$

where we expect the solution $\phi_n(x) \to 0$ as $x \to \pm\infty$ and E_n is the quantum energy. Note here that $K = km/\hbar^2$ and $\varepsilon = Em/\hbar^2$. In what follows, take $K = 1$ and always normalize so that $\int_\infty^\infty |\phi_n|^2 dx = 1$.

(a) Calculate the first five *normalized* eigenfunctions (ϕ_n) and eigenvalues (ε_n) using a shooting scheme.

(b) Calculate the first five *normalized* eigenfunctions (ϕ_n) and eigenvalues (ε_n) using a direct method. Be sure to use forward- and backward-differencing for the boundary conditions. (Hint: $3 + \Delta x\sqrt{KL^2 - E} \approx 3$.)

(c) There has been suggestions that in some cases, nonlinearity plays a role such that

$$\frac{d^2 \phi_n}{dx^2} - \left[\gamma |\phi|^2 Kx^2 - \varepsilon_n\right] \phi_n = 0. \tag{24.4.6}$$

Depending upon the sign of γ, the probability density is focused or defocused. Find the first three *normalized* modes for $\gamma = \pm 0.2$ using shooting.

(d) For a fixed value of the energy (take, for instance, ε_1), perform a convergence study of the shooting method by controlling the error tolerance, TOL, in ODE45 and ODE23:

```
TOL=1e-4;
OPTIONS = odeset('RelTol',TOL,'AbsTol',TOL);
[T,Y] = ODE45('F',TSPAN,Y0,OPTIONS);
```

Show that indeed the schemes are fourth order and second order, respectively, by running the computation accross the computational domain and adjusting the tolerance. In particular, plot on a log-log scale the average step-size (*x*-axis) using the *diff* and *mean* command versus the tolerance (*y*-axis) for a large number of tolerance values. What are the slopes of these lines? Note that the local error should be $O(\Delta t^5)$ and $O(\Delta t^3)$, respectively. What are the local errors for ODE113 and ODE15s?

(e) Compare your solutions with the exact Gauss–Hermite polynomial solutions for this problem.

(f) For a double-well potential

$$V(x) = \begin{cases} -1 & 1 < x < 2 \text{ and } -2 < x < -1 \\ 0 & \text{otherwise,} \end{cases} \qquad (24.4.7)$$

calculate the symmetric ground state ϕ_1 and the first, antisymmetric state ϕ_2. Note that $|\varepsilon_1 - \varepsilon_2| \ll 1$, thus the difficulty in this problem. Make a 3D plot $(x, t, |\psi|)$ of the solution (24.4.4) for the initial condition $\psi = \phi_1 + \phi_2$ which shows the tunneling between potential wells. Note that for plotting purposes, take $\hbar = 1$ and $m = 1$.

 ## 24.5   Electromagnetic Waveguides

The propagation of electromagnetic energy in a one-dimensional optical waveguide is governed by the partial differential equation:

$$2ikU_z + U_{xx} + k^2 \Delta^2 n(x)U = 0, \qquad (24.5.1)$$

where U is the envelope of the electromagnetic field and the equation has been nondimensionalized such that the unit length is the waveguide core radius $a = 10$ μm. Here, the dimensionless wavenumber is $k = 2\pi n_0 a / \lambda_0$ where $n_0 = 1.46$ is the cladding index of refraction and $\lambda_0 = 1.55$ μm is the free-space wavelength. The parameter $\Delta^2 = (n_{core}^2 - n_0^2)/n_0^2$ measures the difference between the peak value of the index in the core $n_{core} = 1.48$ and the cladding n_0. Note that z measures the distance traveled in the waveguide and x is the transverse dimension.

A typical solution technique for this problem is to assume a solution of the form

$$U = \sum_{1}^{N} A_n \psi_n(x) \exp\left(\frac{i}{2k}\beta_n z\right) \tag{24.5.2}$$

which is called an eigenfunction expansion solution (ψ_n = eigenfunction, β_n = eigenvalue). Plugging in this solution ansatz to Eq. (24.5.1) gives the boundary value problem

$$\frac{d^2 \psi_n}{dx^2} + \left[k^2 \Delta^2 n(x) - \beta_n\right] \psi_n = 0 \tag{24.5.3}$$

where we expect the solution $\psi_n(x) \to 0$ as $x \to \pm\infty$ and β_n is called the propagation constant. The function $n(x)$ gives the index of refraction profile of the fiber and is typically manufactured to enhance the performance of a given application. For ideal profiles,

$$n(x) = \begin{cases} 1 - |x|^\alpha & 0 \le |x| \le 1 \\ 0 & |x| > 1. \end{cases} \tag{24.5.4}$$

Two cases of particular interest are for $\alpha = 2$ and $\alpha = 10$.

(a) Calculate the first five *normalized* eigenfunctions (ψ_n) and eigenvalues (β_n) for these two cases using a shooting scheme. (Note: normalization $\int_{-\infty}^{\infty} |\psi_n|^2 dx = 1$.)

(b) Calculate the first five *normalized* eigenfunctions (ψ_n) and eigenvalues (β_n) for these two cases using a direct solve scheme. (Note: normalization $\int_{\infty}^{\infty} |\psi_n|^2 dx = 1$.)

(c) For high-intensity pulses, the index of refraction depends upon the intensity of the pulse itself. The propagating modes are thus found from

$$\frac{d^2 \psi}{dx^2} + \left[\gamma|\psi|^2 + k^2 \Delta^2 n(x) - \beta\right]\psi = 0. \tag{24.5.5}$$

Depending upon the sign of γ, the waveguide leads to focusing or defocusing of the electromagnetic field. Find the first three *normalized* modes for $\gamma = \pm 0.2$ using shooting.

(d) For the case $\alpha = 2$ and for a fixed value of the propagation constant (take, for instance, β_1), perform a convergence study of the shooting method by controlling the error tolerance, TOL, in ODE45 and ODE23:

```
TOL=1e-4;
OPTIONS = odeset('RelTol',TOL,'AbsTol',TOL);
[T,Y] = ODE45('F',TSPAN,Y0,OPTIONS);
```

Show that indeed the schemes are fourth order and second order, respectively, by running the computation accross the computational domain and adjusting the tolerance. In partic-ular, plot on a log-log scale the average step-size (*x*-axis) using the *diff* and *mean* command versus the tolerance (*y*-axis) for a large number of tolerance values. What are the slopes of these lines? Note that the local error should be $O(\Delta t^5)$ and $O(\Delta t^3)$, respectively. What are the local errors for ODE113 and ODE15s?

25 Applications of Partial Differential Equations

25.1 The Wave Equation

One of the classic equations of mathematical physics and differential equations is the *wave equation*. It has long been the prototypical equation for modeling and understanding the underlying behavior of linear wave propagation and phenomena in a myriad of fields including electrodynamics, water waves and acoustics, to name a few application areas. And although we have a number of analytic techniques available to solve the wave equation in closed form, it is a great test-bed for the application of the numerical techniques developed here for solving differential and partial differential equations.

To begin, the classic derivation of the wave equation will be given. It is based entirely on Newton's law $\mathbf{F} = m\mathbf{a}$. The derivation will be given for the displacement of an elastic string in space and time $U(x, t)$ that is pinned at two end points: $U(0, t) = 0$ and $U(L, t) = 0$. Moreover, the string will be assumed to be of constant density (mass per unit length) and the effects of gravity can be ignored.

Figure 25.1 shows the basic physical configuration associated with the vibrating string. If we consider any given location x on the string, then the string is assumed to vibrate vertically at that point, thus no translations of the string are allowed. A simple consequence of this argument, which holds for sufficiently small displacements, is that the sum of the horizontal forces is zero while the sum of the vertical is not. Mathematically, this is given as

$$\text{vertical direction:} \quad \sum \mathbf{F}_y = m\mathbf{a}_y \tag{25.1.1a}$$

$$\text{horizontal direction:} \quad \sum \mathbf{F}_x = 0 \tag{25.1.1b}$$

where the subscripts $\mathbf{F}_y$ and $\mathbf{F}_x$ are the forces in the vertical and horizontal directions, respectively.

Consider then the specific spatial interval from x to $x + \Delta x$ as depicted in Fig. 25.1. For Δx infinitesimally small, the ideas of calculus dictate the derivation of the governing equations. In particular, the tension (forces) on the string at x and $x + \Delta x$ are given by the vectors T_- and

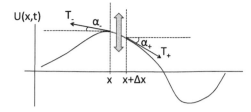

Figure 25.1: Forces associated with the deflection of a vibrating elastic string. At the points x and $x + \Delta x$, the string is subject to the tension (force) vectors $T_{\pm}$. The horizontal forces must balance to keep the string from translating whereas the vertical forces are not balanced and produce acceleration and motion of the string.

T_+, respectively. The vertical and horizontal components of these forces can be easily computed using trigonometry rules so that

$$\text{vertical direction:} \quad T_- \sin \alpha_- - T_+ \sin \alpha_+ = -\rho \Delta x \frac{d^2 U}{dt^2} \tag{25.1.2a}$$

$$\text{horizontal direction:} \quad T_+ \cos \alpha_+ - T_- \cos \alpha_- = 0 \tag{25.1.2b}$$

where ρ is the constant density of the string so that $\rho \Delta x$ is the mass of the infinitesimally small piece of string considered. Note that the second condition ensures that there is no acceleration of the string in the horizontal direction. Further, the sign is negative in front of the second derivative term in the vertical direction since the acceleration is in the negative direction as indicated in the figure.

From (25.1.2b), we have that $T_+ \cos \alpha_+ = T_- \cos \alpha_- = T$. Dividing (25.1.2a) by the constant T yields

$$\frac{T_- \sin \alpha_-}{T} - \frac{T_+ \sin \alpha_+}{T} = -\frac{\rho \Delta x}{T} \frac{d^2 U}{dt^2}$$

$$\frac{T_- \sin \alpha_-}{T_- \cos \alpha_-} - \frac{T_+ \sin \alpha_+}{T_+ \cos \alpha_+} = -\frac{\rho \Delta x}{T} \frac{d^2 U}{dt^2}$$

$$\tan \alpha_- - \tan \alpha_+ = -\frac{\rho \Delta x}{T} \frac{d^2 U}{dt^2}. \tag{25.1.3}$$

The important thing to note at this point in the derivation is that $\tan \alpha_{\pm}$ is simply the slope at x and $x + \Delta x$, respectively. Dividing through by a factor of $-\Delta x$ and replacing the tangent with its slope yields the equation

$$\frac{1}{\Delta x} \left[\frac{\partial U(x + \Delta x, t)}{\partial x} - \frac{\partial U(x, t)}{\partial x} \right] = \frac{\rho}{T} \frac{\partial^2 U(x, t)}{\partial t^2}. \tag{25.1.4}$$

As $\Delta x \to 0$, the definition of derivative applies and the left-hand side is the derivative of the derivative, i.e. the second derivative. This gives the wave equation

$$\frac{\partial^2 U}{\partial t^2} = c^2 \frac{\partial^2 U}{\partial x^2} \tag{25.1.5}$$

where $c^2 = T/\rho$ is a positive quantity.

Interestingly enough, the derivation can also be applied to a vibrating string where the density varies in x. This gives rise to the nonconstant coefficient version of the wave equation:

$$\frac{\partial^2 U}{\partial t^2} = \frac{\partial}{\partial x}\left(c^2(x)\frac{\partial U}{\partial x}\right). \tag{25.1.6}$$

This makes it more difficult to solve, but only moderately so, as will be seen in the projects associated with the wave equation.

Initial conditions

In addition to the governing equations (25.1.5) or (25.1.5) both boundary and initial conditions must be specified. Much like differential equations, the number of initial conditions to be specified depends upon the highest derivative in time. The wave equation has two derivatives in time and thus requires two initial conditions:

$$U(x,0) = f(x) \tag{25.1.7a}$$

$$\frac{\partial U(0,x)}{\partial t} = g(x). \tag{25.1.7b}$$

Thus once $f(x)$ and $g(x)$ are specified, the system can be evolved forward in time.

Boundary conditions

A variety of boundary conditions can be applied, including pinned where $U = 0$ at the boundary, no-flux where $\partial U/\partial x = 0$ at the boundary, periodic where $U(0, t) = U(L, t)$ at the boundary or some combination of pinned and no-flux. In the algorithms developed here, consider then the following possibilities:

$$\text{pinned: } U(0, t) = U(L, t) = 0 \tag{25.1.8a}$$

$$\text{no-flux: } \frac{\partial U(0, t)}{\partial x} = \frac{\partial U(L, t)}{\partial x} = 0 \tag{25.1.8b}$$

$$\text{mixed: } U(0, t) = 0 \quad \text{and} \quad \frac{\partial U(L, t)}{\partial x} = 0 \tag{25.1.8c}$$

$$\text{periodic: } U(0, t) = U(L, t). \tag{25.1.8d}$$

Each of these gives rise to a different differentiation matrix $\mathbf{A}$ for computing two derivatives in a finite difference scheme.

Projects

(a) Given $f(x) = \exp(-x.^2)$ and $g(x) = 0$ as initial conditions, compute the solution numerically on the domain $x \in [0, 40]$ using a second-order accurate finite difference scheme in space with time-stepping dictated by **ode45**. Be sure to first rewrite the system as a set of first-order equations. Solve the system for all four boundary types given above and be sure to simulate the system long enough to see the waves interact with the boundaries several times.

(b) Repeat experiment (a), but now implement fourth-order accurate schemes for computing the derivatives. In this case, the no-flux conditions can use second-order accurate schemes for computing the effects of the boundary on the differentiation matrix.

(c) Using periodic boundary conditions, simulate the system using the fast Fourier transform method.

(d) Do a convergence study of the second-order, fourth-order and spectral methods using a high-resolution simulation of the spectral method as the *true solution*. See how the solution converges to this true solution as a function of Δx. Verify that the differentiation matrices are indeed second-order and fourth-order accurate. Further, determine the approximate order of accuracy of the FFT method.

(e) What happens when the density changes as a function of the space variable x? Consider $c^2(x) = \cos(10\pi x/L)$ and solve (25.1.6) with pinned boundaries and fourth-order accuracy. Note how the waves slow down and speed up depending upon the local value of c^2.

(f) Given the Gaussian initial condition for the field, determine an appropriate initial time derivative condition $g(x)$ such that the wave travels only right or only left.

(g) For electromagnetic phenomena, the intensity of the electromagnetic field can be brought to such levels that the material properties (index of refraction) change with the intensity of the field itself. This gives a governing wave equation:

$$\frac{\partial^2 U}{\partial t^2} = c^2 \frac{\partial^2 U}{\partial x^2} + \beta |U|^2 U \tag{25.1.9}$$

where $|U|^2$ is the intensity of the field. Solve this equation for β either positive or negative. As β is increased, note the effect it has on the wave propagation.

 25.2 Mode-Locked Lasers

The invention of the laser in 1960 is a historical landmark of scientiþc innovation. In the decades since, the laser has evolved from a fundamental science phenomenon to a ubiquitous, cheap and readily available commercial commodity. A large variety of laser technologies and applications exist from the commercial, industrial, medical and academic arenas. In many of these applications, the laser is required to produce ultrashort pulses (e.g. tens to hundreds of femtoseconds), which are localized light waves, with nearly uniform amplitudes. The generation of such localized light pulses is the subject of mode-locking theory.

As with all electromagnetic phenomena, the propagation of an optical field in a given medium is governed by Maxwell's equations:

$$\nabla \times \mathbf{E} = -\frac{\partial \mathbf{B}}{\partial t} \tag{25.2.1a}$$

$$\nabla \times \mathbf{H} = \mathbf{J}_f + \frac{\partial \mathbf{D}}{\partial t} \tag{25.2.1b}$$

$$\nabla \cdot \mathbf{D} = \rho_f \tag{25.2.1c}$$

$$\nabla \cdot \mathbf{B} = 0. \tag{25.2.1d}$$

Here the electromagnetic field is denoted by the vector $\mathbf{E}(x, y, z, t)$ with the corresponding magnetic field, electromagnetic and magnetic flux densities being denoted by $\mathbf{H}(x, y, z, t)$, $\mathbf{D}(x, y, z, t)$ and $\mathbf{B}(x, y, z, t)$, respectively. In the absence of free charges, which is the case of interest here, the current density and free charge density are both zero so that $\mathbf{J}_f = 0$ and $\rho_f = 0$.

The flux densities $\mathbf{D}$ and $\mathbf{B}$ characterize the constitutive laws of a given medium. Therefore, any specific material of interest is characterized by the constitutive laws and their relationship to the electric and magnetic fields:

$$\mathbf{D} = \epsilon_0 \mathbf{E} + \mathbf{P} = \epsilon \mathbf{E} \tag{25.2.2a}$$

$$\mathbf{B} = \mu_0 \mathbf{H} + \mathbf{M} \tag{25.2.2b}$$

where ϵ_0 and μ_0 are the free space permittivity and free space permeability, respectively, and $\mathbf{P}$ and $\mathbf{M}$ are the induced electric and magnetic polarizations. At optical frequencies, $\mathbf{M} = 0$. The nonlocal, nonlinear response of the medium to the electric field is captured by the induced polarization vector $\mathbf{P}$.

In what follows, interest will be given solely to the electric field and the induced polarization. Expressing Maxwell's equations in terms of these two fundamental quantities $\mathbf{E}$ and $\mathbf{P}$ can be easily accomplished by taking the curl of (25.2.1a) and using (25.2.1b), (25.2.2a) and (25.2.2b). This yields the electric field evolution

$$\nabla^2 \mathbf{E} - \nabla(\nabla \cdot \mathbf{E}) - \frac{1}{c^2} \frac{\partial^2 \mathbf{E}}{\partial t^2} = \mu_0 \frac{\partial^2 \mathbf{P}(\mathbf{E})}{\partial t^2}. \tag{25.2.3}$$

For the quasi-one-dimensional case of fiber propagation, the above expression for the electric field reduces to

$$\frac{\partial^2 E}{\partial x^2} - \frac{1}{c^2} \frac{\partial^2 E}{\partial t^2} = \mu_0 \frac{\partial^2 P(E)}{\partial t^2} \tag{25.2.4}$$

where now the electric field is expressed as the scalar quantity $E(x, t)$. Note that we could formally handle the transverse field structure as well in an asymptotic way, but for the purposes of this exercise, this is not necessary.

Only the linear propagation is considered at first. The nonlinearity will be added afterwards. Thus one considers the following linear function with linear polarization response function:

$$\frac{\partial^2 E}{\partial x^2} - \frac{1}{c^2}\frac{\partial^2 E}{\partial t^2} = \frac{1}{c^2}\frac{\partial^2}{\partial t^2}\left(\int_{-\infty}^{t} \chi^{(1)}(t-\tau)E(\tau,x)d\tau\right) \tag{25.2.5}$$

where $\chi^{(1)}$ is the linear response to the electric field that includes the time response and causality. It should be recalled that $c^2 = 1/(\epsilon_0\mu_0)$. The left-hand side of the equation is the standard wave equation that can be solved analytically. It generically yields waves propagating left and right in the media at speed c. Here, we are interested in waves moving only in a single direction down the fiber.

The center-frequency expansion is an asymptotic approach that assumes the electromagnetic field, to leading order, is at a single frequency. A slowly varying envelope equation is then derived for the evolution of an envelope equation that contains many cycles of the electric field. Thus at leading order, one can think of this approach as considering a delta function in frequency and a plane wave in time.

The center-frequency expansion begins by assuming

$$E(x,t) = Q(x,t)e^{i(kx-\omega t)} + c.c. \tag{25.2.6}$$

where c.c. denotes complex conjugate and k and ω are the wavenumber and optical frequency, respectively. These are both large quantities so that $k, \omega \gg 1$. Indeed, in this asymptotic expansion the small parameter is given by $\epsilon = 1/k \ll 1$. Note that if Q is constant, than this assumption amounts to assuming a plane wave solution, i.e. a delta function in frequency.

Inserting (25.2.6) into (25.2.5) yields the following:

$$e^{i(kx-\omega t)}\left[Q_{xx} + 2ikQ_x - k^2 Q\right] - \frac{1}{c^2}e^{i(kx-\omega t)}\left[Q_{tt} + 2i\omega Q_t - \omega^2 Q\right]$$
$$= \frac{1}{c^2}\frac{\partial^2}{\partial t^2}\left(\int_{-\infty}^{t} \chi^{(1)}(t-\tau)Q(\tau,x)e^{i(kx-\omega\tau)}d\tau\right) \tag{25.2.7}$$

where subscripts denote partial differentiation. In the integral on the right-hand side, a change of variables is made so that $\sigma = t - \tau$ and

$$e^{i(kx-\omega t)}\left[Q_{xx} + 2ikQ_x - k^2 Q\right] - \frac{1}{c^2}e^{i(kx-\omega t)}\left[Q_{tt} + 2i\omega Q_t - \omega^2 Q\right]$$
$$= \frac{1}{c^2}\frac{\partial^2}{\partial t^2}\left(e^{i(kx-\omega t)}\int_{0}^{\infty} \chi^{(1)}(\sigma)Q(t-\sigma,x)d\sigma\right). \tag{25.2.8}$$

Using the following Taylor expansion

$$Q(t-\sigma,x) = Q(t,x) - \sigma Q_t(x,t) + \frac{1}{2}\sigma^2 Q_{tt}(x,t) + \cdots \tag{25.2.9}$$

in (25.2.8) results finally in the following equation for $Q(x,t)$:

$$Q_{xx} + 2ikQ_x - k^2 Q + \frac{1}{c^2}\left[\omega^2(1+\hat{\chi})Q + i\left[(\omega^2\hat{\chi})_\omega + 2\omega\right]Q_t\right.$$
$$\left. - 1/2\left[(\omega^2\hat{\chi})_{\omega\omega} + 2\right]Q_{tt} + \cdots\right] = 0 \tag{25.2.10}$$

where $\hat{\chi} = \int_0^\infty \chi^{(1)}(\sigma)e^{i\omega\sigma}\,d\sigma$ is the Fourier transform of the linear susceptibility function $\chi^{(1)}$. It is at this point that the asymptotic balancing of terms begins. In particular, the leading order balances with the largest terms of size k^2 and ω^2 so that

$$k^2 = \frac{\omega^2}{c^2}(1 + \hat{\chi}). \tag{25.2.11}$$

For now, we assume that the linear susceptibility is real. An imaginary part will lead to attenuation. This will be considered later in the context of the laser cavity, but not now in the envelope approach.

Once the dominant balance (25.2.11) has been established, it is easy to show that the following two relations hold:

$$2kk' = \frac{1}{c^2}\left[(\omega^2\hat{\chi})_\omega + 2\omega\right] \tag{25.2.12a}$$

$$kk'' + (k')^2 = \frac{1}{2c^2}\left[(\omega^2\hat{\chi})_{\omega\omega} + 2\right] \tag{25.2.12b}$$

where the primes denote differentiation with respect to ω. Such relations can be found for higher-order terms in the Taylor expansion of the integral. Thus the equation for $Q(x,t)$ reduces to

$$2ik\left(Q_x + k'Q_t\right) + Q_{xx} - \left[kk'' + (k')^2\right]Q_{tt} + \sum_{n=3}^\infty \beta_n \partial_t^{(n)}Q = 0. \tag{25.2.13}$$

Noting that $k' = dk/d\omega$ is in fact the definition of the inverse group velocity in wave propagation problems, we can move into the group-velocity frame of reference by defining the new variables:

$$T = t - k'x \tag{25.2.14a}$$

$$Z = x. \tag{25.2.14b}$$

This yields the governing linear equations

$$iQ_Z - \frac{k''}{2}Q_{TT} + \sum_{n=3}^\infty \beta_n \partial_T^{(n)}Q = 0 \tag{25.2.15}$$

where the β_n are the coefficients of the higher order dispersive terms (β_n for $n > 2$). If we ignore the higher order dispersion, i.e. assume $\beta_n = 0$, then the resulting equation is just the linear Schrödinger equation we expect. Note that the Z and T act as the *time* and *space* variables, respectively, in the moving coordinate system.

The nonlinearity will now be included with the linear susceptibility response. At present, only an instantaneous response will be introduced so that the governing equation (25.2.5) is modified to

$$\frac{\partial^2 E}{\partial x^2} - \frac{1}{c^2}\frac{\partial^2 E}{\partial t^2} = \frac{1}{c^2}\frac{\partial^2}{\partial t^2}\left[\int_{-\infty}^t \chi^{(1)}(t-\tau)E(\tau,x)\,d\tau + \chi^{(3)}E^3\right] \tag{25.2.16}$$

where $\chi^{(3)}$ is the nonlinear (cubic) susceptibility. Note that it is assumed that the propagation is in a centro-symmetric material so that all $\chi^{(2)} = 0$. Both the center-frequency and short-pulse asymptotics proceed to balance the cubic response with the dispersive effects derived in Eqs. (25.2.15) and (25.2.16).

In the center-frequency asymptotics, the cubic term can be carried through the derivation assuming the ansatz (25.2.6) and following the steps (25.2.7) through (25.2.14). The leading order contribution to (25.2.10) becomes

$$\frac{\omega^2}{c^2}\chi^{(3)}|Q|^2 Q + \cdots \tag{25.2.17}$$

where the dots represent higher order terms. Note that because of the ansatz approximation (25.2.6), there are no derivatives on the cubic term. Thus the effects of the two derivatives applied to the nonlinear term in (25.2.16) produce at leading order the derivative of the plane wave ansatz (25.2.6), i.e. it simply produces a factor of ω^2 in front of the nonlinearity. This yields the governing linear equation

$$iQ_Z - \frac{k''}{2}Q_{TT} + \sum_{n=3}^{\infty}\beta_n \partial_T^{(n)}Q + \alpha|Q|^2 Q = 0 \tag{25.2.18}$$

where $\alpha = \omega^2\chi^{(3)}/2kc^2$.

With appropriate normalization and only including the second-order chromatic dispersion, the standard nonlinear Schrödinger (NLS) equation is obtained:

$$iQ_Z \pm \frac{1}{2}Q_{TT} + |Q|^2 Q = 0. \tag{25.2.19}$$

Such a description holds if the envelope $Q(Z, T)$ contains many cycles of the electric field so that the asymptotic ordering based upon $1/k \ll 1$ holds. If only a few cycles of the electric field are underneath the envelope, than the asymptotic ordering that occurs in (25.2.10) cannot be applied and the reduction to the NLS description is suspect. Note that the dispersion can be either positive (anomalous dispersion) or negative (normal dispersion).

Saturating gain dynamics and attenuation

In addition to the two dominant effects of dispersion and self-phase modulation as characterized by the NLS equation (25.2.19), a given laser cavity must also include the gain and loss dynamics. The loss is not only due to the attenuation in splicing the laser components together, but also from inclusion of an output coupler in the laser cavity that taps off a certain portion of the laser cavity energy every round trip. Thus gain must be applied to the laser system. A simple model that accounts for both gain saturation that is bandwidth limited and attenuation is the following:

$$Q_Z = -\Gamma Q + g(Z)\left(1 + \tau\partial_T^2\right)Q \tag{25.2.20}$$

where

$$g(Z) = \frac{2g_0}{1 + \|Q\|^2/E_0}. \tag{25.2.21}$$

Here Γ measures the attenuation rate of the cavity, the parameter τ measures the bandwidth of the gain (assumed to be parabolic in shape around the center-frequency), g_0 is the gain pumping strength, and E_0 is the cavity saturation energy. Note that $\|Q\|^2 = \int_{-\infty}^{\infty} |Q|^2 dT$ is the total cavity energy. Thus the gain dynamics $g(Z)$ provides the saturation behavior that is characteristic of any physically realistic gain media. Specifically, as the cavity energy grows, the effective gain $g(Z)$ decreases since the number of atoms in the inverted population state becomes depleted.

Intensity discrimination

In addition to the gain and loss dynamics, one other physically important effect must be considered, namely the intensity discrimination (or saturable absorption) necessary for laser mode-locking. Essentially, there must be some physical mechanism in the cavity that favors high intensities over low intensities of the electric field, thus allowing for an intensity selection mechanism. Physically, such an effect has been achieved in a wide variety of physical scenarios including the nonlinear polarization laser, a laser mode-locked via nonlinear interferometry, or quantum saturable absorbers. A phenomenological approach to considering such phenomena can be captured by the simple equation

$$Q_Z = \beta |Q|^2 Q - \sigma |Q|^4 Q \qquad (25.2.22)$$

where β models a cubic gain and σ is a quintic saturation effect. Normally, the cubic gain would dominate and the quintic saturation is only proposed to prevent solutions from blowing up. Thus typically $\sigma \ll \beta$.

The Haus master mode-locking model

The late Hermann Haus of the Massachusetts Institute of Technology proposed that the dominant laser cavity effects should be included in a single phenomenological description of the electric field evolution [58]. As such, he proposed the master mode-locking model that includes the chromatic dispersion and nonlinearity of (25.2.19), the saturating gain and loss of (25.2.21), and the intensity discrimination of (25.2.22). Averaged together, they produce the Ginzburg–Landau-like equation

$$iQ_Z \pm \frac{1}{2}Q_{TT} + \alpha |Q|^2 Q = i \left[g(Z) \left(1 + \tau \partial_T^2 \right) Q - \Gamma Q + \beta |Q|^2 Q - \sigma |Q|^4 Q \right] \qquad (25.2.23)$$

where the gain $g(Z)$ is given by (25.2.21).

Mode-locking with waveguide arrays

Another model of mode-locking is based on a more quantitative approach to the saturable absorption mechanism. The intensity discrimination in the specific laser considered here is provided by the nonlinear mode-coupling in the waveguides as described in the previous section. When placed within an optical fiber cavity, the pulse shaping mechanism of the waveguide array leads to stable and robust mode-locking. The resulting approximate evolution dynamics describing the waveguide array mode-locking is given by [99]

$$i\frac{\partial Q}{\partial Z} \pm \frac{1}{2}\frac{\partial^2 Q}{\partial T^2} + \alpha|Q|^2 Q + CV + i\gamma_0 Q - ig(Z)\left(1 + \tau\frac{\partial^2}{\partial T^2}\right)Q = 0 \qquad (25.2.24a)$$

$$i\frac{\partial V}{\partial Z} + C(W + Q) + i\gamma_1 V = 0 \qquad (25.2.24b)$$

$$i\frac{\partial W}{\partial Z} + CV + i\gamma_2 W = 0, \qquad (25.2.24c)$$

where the gain $g(Z)$ is again given by (25.2.21) and the $V(Z, T)$ and $W(Z, T)$ fields model the electromagnetic energy in the neighboring channels of the waveguide array. Note that the equations governing these neighboring fields are ordinary differential equations. Figure 25.2 demonstrate the prototypical mode-locking behavior as a function of the gain parameter g_0.

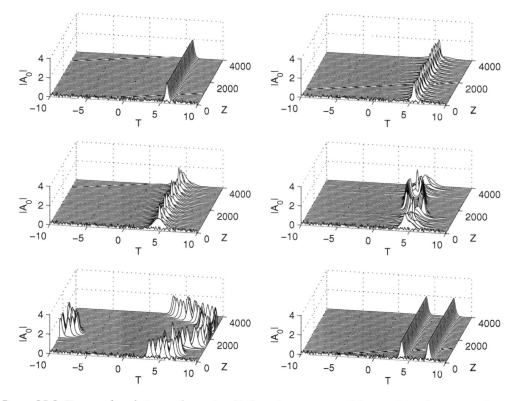

Figure 25.2: Temporal evolution and associated bifurcation structure of the transition from one pulse per round-trip to two pulses per round-trip. The corresponding values of gain are $g_0 = 2.3, 2.35, 2.5, 2.55, 2.7$ and 2.75. For the lowest gain value only a single pulse is present. The pulse then becomes a periodic breather before undergoing a "chaotic" transition between a breather and two-pulse solution. Above a critical value ($g_0 \approx 2.75$), two pulse are stabilized. [From J. N. Kutz and B. Sandstede, Theory of passive harmonic mode-locking using waveguide arrays, Optics Express **16**, 636–650 (2008) ©OSA.]

Project and application

Consider the mode-locking equations (25.2.23) and (25.2.24) with the gain given by (25.2.21). Use the following parameter values for (25.2.23) $(E_0, \tau, \alpha, \Gamma, \beta, \sigma) = (1, 0.1, 1, 1, 0.5, 0.1)$ with a negative sign of dispersion, and consider (25.2.24) with a positive sign of dispersion and $(E_0, \tau, C, \alpha, \gamma_0, \gamma_1, \gamma_2) = (1, 0.1, 5, 8, 0, 0, 10)$.

Initial conditions

For both models, assume white-noise initial conditions at the initial time $Q(0, T)$. Recall that Z is the time-like variable and T is the space-like variable in this moving coordinate frame.

Boundary conditions

The cavity is assumed to be very long in relation to the width of the pulse. Therefore, it is usually assumed that the localized mode-locked pulses are essentially in an infinite domain. To approximate this, one can simply use a large domain and an FFT based solution method. Be sure to pick a domain large enough so that boundary effects do not play a role.

(a)   For both governing models (25.2.23) and (25.2.24), explore the dynamics as a function of increasing gain g_0. In particular, determine the ranges of stability for which stable one-pulse mode-locking occurs.

(b)   Carefully explore the region near the transition regions where the one-pulse solution undergoes instability to a two-pulse solution. Specifically, identify the region where periodic oscillations occur via a Hopf bifurcation and determine if there are any chaotic regions of behavior near the transition point.

(c)   Determine a parameter regime for positive dispersion in (25.2.23) where stable mode-locking can occur. Key parameters to adjust for this model are α, β and σ. Additionally, determine the mode-locking regime of stability for (25.2.24) with negative dispersion. The key parameters to adjust for this model are α and C.

 25.3   Bose–Einstein Condensates

To consider a specific model that allows the connection from the microscopic scale to the macroscopic scale, we examine mean field theory of many particle quantum mechanics with the particular application of BECs trapped in a standing light wave [93]. The classical derivation given here is included to illustrate how the local model and its nonlocal perturbation are related. The inherent complexity of the dynamics of N pairwise interacting particles in quantum mechanics often leads to the consideration of such simplified mean field descriptions. These descriptions are a blend of symmetry restrictions on the particle wavefunction [94] and functional form assumptions on the interaction potential [94, 95, 96].

The dynamics of N identical pairwise interacting quantum particles is governed by the time-dependent, N-body Schrödinger equation

$$i\hbar \frac{\partial \Psi}{\partial t} = -\frac{\hbar^2}{2m} \Delta^N \Psi + \sum_{i,j=1, i\neq j}^{N} W(\mathbf{x}_i - \mathbf{x}_j) \Psi + \sum_{i=1}^{N} V(\mathbf{x}_i) \Psi, \qquad (25.3.1)$$

where $\mathbf{x_i} = (x_{i1}, x_{i2}, x_{i3})$, $\Psi = \Psi(\mathbf{x}_1, \mathbf{x}_2, \mathbf{x}_3, \ldots, \mathbf{x}_N, t)$ is the wavefunction of the N-particle system, $\Delta^N = \left(\nabla^N \right)^2 = \sum_{i=1}^{N} \left(\partial_{x_{i1}}^2 + \partial_{x_{i2}}^2 + \partial_{x_{i3}}^2 \right)$ is the kinetic energy or Laplacian operator for N particles, $W(\mathbf{x_i} - \mathbf{x_j})$ is the symmetric interaction potential between the ith and jth particle, and $V(\mathbf{x}_i)$ is an external potential acting on the ith particle. Also, h is Planck's constant divided by 2π and m is the mass of the particles under consideration.

One way to arrive at a mean field description is by using the Lagrangian reduction technique [97], which exploits the Hamiltonian structure of Eq. (25.3.1). The Lagrangian of Eq. (25.3.1) is given by [94]

$$L = \int_{-\infty}^{\infty} \left\{ i\frac{\hbar}{2} \left(\Psi \frac{\partial \Psi^*}{\partial t} - \Psi^* \frac{\partial \Psi}{\partial t} \right) + \frac{\hbar^2}{2m} \left| \nabla^N \Psi \right|^2 \right.$$
$$\left. + \sum_{i=1}^{N} \left(\sum_{j\neq i}^{N} W(\mathbf{x}_i - \mathbf{x}_j) + V(\mathbf{x}_i) \right) |\Psi|^2 \right\} d\mathbf{x}_1 \cdots d\mathbf{x}_N. \qquad (25.3.2)$$

The Hartree–Fock approximation (as used in [94]) for bosonic particles uses the separated wavefunction ansatz

$$\Psi = \psi_1(\mathbf{x}_1, t)\psi_2(\mathbf{x}_2, t) \cdots \psi_N(\mathbf{x}_N, t) \qquad (25.3.3)$$

where each one-particle wavefunction $\psi(\mathbf{x}_i)$ is assumed to be normalized so that $\langle \psi(\mathbf{x}_i) | \psi(\mathbf{x}_i) \rangle^2 = 1$. Since identical particles are being considered,

$$\psi_1 = \psi_2 = \cdots = \psi_N = \psi, \qquad (25.3.4)$$

enforcing total symmetry of the wavefunction. Note that for the case of BECs, assumption (25.3.3) is approximate if the temperature is not identically zero.

Integrating Eq. (25.3.2) using (25.3.3) and (25.3.4) and taking the variational derivative with respect to $\psi(\mathbf{x}_i)$ results in the Euler–Lagrange equation [97]

$$i\hbar \frac{\partial \psi(\mathbf{x}, t)}{\partial t} = -\frac{\hbar^2}{2m} \Delta \psi(\mathbf{x}, t) + V(\mathbf{x})\psi(\mathbf{x}, t)$$
$$+ (N-1)\psi(\mathbf{x}, t) \int_{-\infty}^{\infty} W(\mathbf{x} - \mathbf{y}) |\psi(\mathbf{y}, t)|^2 d\mathbf{y}. \qquad (25.3.5)$$

Here, $\mathbf{x} = \mathbf{x}_i$, and Δ is the one-particle Laplacian in three dimensions. The Euler–Lagrange equation (25.3.5) is identical for all $\psi(\mathbf{x}_i, t)$. Equation (25.3.5) describes the nonlinear, nonlocal, mean field dynamics of the wavefunction $\psi(\mathbf{x}, t)$ under the standard assumptions (25.3.3) and (25.3.4) of Hartree–Fock theory [94]. The coefficient of $\psi(\mathbf{x}, t)$ in the last term in Eq. (25.3.5) represents the effective potential acting on $\psi(\mathbf{x}, t)$ due to the presence of the other particles.

At this point, it is common to make an assumption on the functional form of the interaction potential $W(\mathbf{x} - \mathbf{y})$. This is done to render Eq. (25.3.5) analytically and numerically tractable. Although the qualitative features of this functional form may be available, for instance from experiment, its quantitative details are rarely known. One convenient assumption in the case of short-range potential interactions is $W(\mathbf{x} - \mathbf{y}) = \kappa\delta(\mathbf{x} - \mathbf{y})$ where δ is the Dirac delta function. This leads to the Gross–Pitaevskii [95, 96] mean field description:

$$i\hbar\frac{\partial\psi}{\partial t} = -\frac{\hbar^2}{2m}\Delta\psi + \beta|\psi|^2\psi + V(\mathbf{x})\psi, \tag{25.3.6}$$

where $\beta = (N - 1)\kappa$ reflects whether the interaction is repulsive ($\beta > 0$) or attractive ($\beta < 0$). The above string of assumptions is difficult to physically justify. Nevertheless, Lieb and Seiringer [98] show that Eq. (25.3.6) is the correct asymptotic description in the dilute-gas limit. In this limit, Eqs. (25.3.5) and (25.3.6) are asymptotically equivalent. Thus, although the nonlocal Eq. (25.3.5) is in no way a more valid model than the local equation (25.3.6), it can be interpreted as a perturbation to the local equation (25.3.6).

Project and application

In what follows, you will be asked to consider the dynamics of a BEC in one, two and three dimensions. The potential in (25.3.6) determines much of the allowed or enforced dynamics in the system.

One-dimensional condensate

In this initial investigation, the BEC cavity is assumed to be trapped in a cigar shaped trap where the longitudinal direction is orders of magnitude longer than the transverse direction. Therefore, it is usually assumed that the localized mode-locked pulses are essentially in an infinite domain. To approximate this, one can simply use a large domain and an FFT based solution method. Be sure to pick a domain large enough so that boundary effects do not play a role. The governing equation (25.3.6) is then reduced to nondimensional form

$$i\frac{\partial\psi}{\partial t} + \frac{1}{2}\frac{\partial^2\psi}{\partial x^2} + \alpha|\psi|^2\psi - \left[V_0\sin^2(\omega(x - \bar{x})) + V_1 x^2\psi\right] = 0, \tag{25.3.7}$$

where a periodic (sinusoidal) and harmonic (parabolic) potential have been assumed to be acting on the system.

(a) Investigate the dynamics of the system as a function of the parameters V_0 and V_1 and with $\alpha = 1$. Assume a localized initial condition, something like a Gaussian that can be generically imposed off-center ($x \neq 0$) from the harmonic trap (see, for instance, Fig. 25.3). Also investigate the three scenarios where the width of the pulse is much wider, about the same width and narrower than the period of the oscillatory potential. Consider the stability of the BEC and the amplitude–width dynamics as the condensate propagates in time and slides along the combination of periodic and harmonic traps.

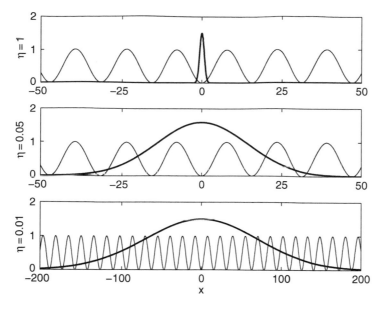

Figure 25.3: Localized condensate (bold line) and applied periodic potential (light line) for various ratios of period of the potential to a Gaussian input $\psi = 1.5\sqrt{\eta} \exp\left(-\eta^2 x^2\right)$. Here we choose $\omega = 0.2$ with the value of η given on the left of each panel. [From N. H. Berry and J. N. Kutz, Dynamics of Bose–Einstein condensates under the influence of periodic and harmonic potentials, Physical Review E **75**, 036214 (2007) ©APS.]

(b) For the case of a purely periodic potential, $V_1 = 0$ and $V_0 \neq 0$, consider whether standing wave solutions or periodic solutions are permissible. Assume an initial periodic sinusoidal solution and investigate both an attractive ($\alpha = 1$) or repulsive condensate ($\alpha = -1$).

Two- and three-dimensional condensate lattices

An interesting BEC configuration to consider is when a periodic lattice potential is imposed on the system. In two dimensions, this gives the governing equations

$$i\frac{\partial \psi}{\partial t} + \frac{1}{2}\frac{\partial^2 \psi}{\partial x^2} + \frac{1}{2}\frac{\partial^2 \psi}{\partial y^2} + \alpha|\psi|^2\psi - \left[A_1 \sin^2(\omega_1 x) + B_1\right]\left[A_2 \sin^2(\omega_2 y) + B_2\right] = 0, \quad \text{(25.3.8)}$$

where A_i, ω_i and B_i completely characterize the nature of the periodic potential.

(c) For a symmetric lattice with $A_1 = A_2$, $B_1 = B_2$ and $\omega_1 = \omega_2$, consider the dynamics of both localized and periodic solutions on the lattice for both attractive and repulsive condensates. Are there relatively stable BEC configurations that allow for a long-time stable configuration of the BEC (see, for instance, Fig. 25.4)? Consider solutions that are localized in the individual troughs of the periodic potential as well as those that are localized on the peaks of the potential.

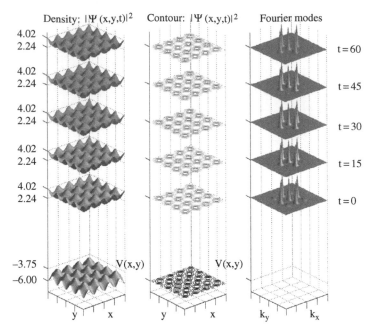

Figure 25.4: Evolution of stable two-dimensional BEC on a periodic lattice. [From B. Deconinck, B. Frigyik and J. N. Kutz, Stability of exact solutions of the defocusing nonlinear Schrödinger equation with periodic potential in two dimensions, Physics Letters A **283**, 177–184 (2001) ©Elsevier.]

(d) Explore the dynamics of the lattice when symmetry is broken in the *x*- and *y*-directions.

(e) Generalize (25.3.8) to three dimensions and consider (c) and (d) in the context of a three-dimensional lattice structure. Use the **isosurface** and **slice** plotting routines to demonstrate the evolution dynamics.

 ## 25.4 Advection–Diffusion and Atmospheric Dynamics

The shallow-water wave equations are of fundamental interest in several contexts. In one sense, the ocean can be thought of as a shallow-water description over the surface of the Earth. Thus the circulation and movement of currents can be studied. Second, the atmosphere can be thought of as a relatively thin layer of fluid (gas) above the surface of the Earth. Again the circulation and atmospheric dynamics are of general interest. The shallow-water approximation relies on a separation of scale: the height of the fluid (or gas) must be much less than the characteristic horizontal scales. The physical setting for shallow-water modeling is illustrated in Fig. 25.5. In this figure, the characteristic height is given by the parameter D while the characteristic horizontal fluctuations are given by the parameter L. For the shallow-water approximation to hold, we must have

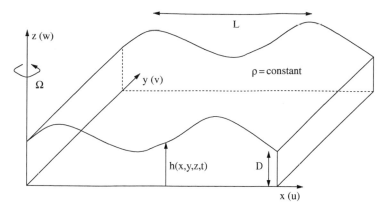

Figure 25.5: Physical setting of shallow-atmosphere or shallow-water equations with constant density ρ and scale separation $D/L \ll 1$. The vorticity is measured by the parameter Ω.

$$\delta = \frac{D}{L} \ll 1. \qquad (25.4.1)$$

This gives the necessary separation of scales for reducing the Navier–Stokes equations to the shallow-water description.

The motion of the layer of fluid is described by its velocity field

$$\mathbf{v} = \begin{pmatrix} u \\ v \\ w \end{pmatrix} \qquad (25.4.2)$$

where u, v, and w are the velocities in the x-, y-, and z-directions, respectively. An alternative way to describe the motion of the fluid is through the quantity known as the vorticity. Roughly speaking the vorticity is a vector which measures the twisting of the fluid, i.e. $\boldsymbol{\Omega} = (\omega_x \quad \omega_y \quad \omega_z)^{\mathrm{T}}$. Since with shallow water we are primarily interested in the vorticity or fluid rotation in the x–y plane, we define the vorticity of interest to be

$$\omega_z = \omega = \frac{\partial v}{\partial x} - \frac{\partial u}{\partial y}. \qquad (25.4.3)$$

This quantity will characterize the evolution of the fluid in the shallow-water approximation.

Conservation of mass

The equations of motion are a consequence of a few basic physical principles, one of those being conservation of mass. The implications of conservation of mass are that the time rate of change of mass in a volume must equal the net inflow/outflow through the boundaries of the volume. Consider a volume from Fig. 25.5 which is bounded between $x \in [x_1, x_2]$ and $y \in [y_1, y_2]$. The mass in this volume is given by

$$\text{mass} = \int_{x_1}^{x_2} \int_{y_1}^{y_2} \rho(x,y)h(x,y,t)dxdy \tag{25.4.4}$$

where $\rho(x,y)$ is the fluid density and $h(x,y,t)$ is the surface height. We will assume the density ρ is constant in what follows. The conservation of mass in integral form is then expressed as

$$\frac{\partial}{\partial t} \int_{x_1}^{x_2} \int_{y_1}^{y_2} h(x,y,t)dxdy$$

$$+ \int_{y_1}^{y_2} \left[u(x_2,y,t)h(x_2,y,t) - u(x_1,y,t)h(x_1,y,t) \right] dy$$

$$+ \int_{x_1}^{x_2} \left[v(x,y_2,t)h(x,y_2,t) - v(x,y_1,t)h(x,y_1,t) \right] dx = 0, \tag{25.4.5}$$

where the density has been divided out. Here the first term measures the rate of change of mass while the second and third terms measure the flux of mass across the x boundaries and y boundaries, respectively. Note that from the fundamental theorem of calculus, we can rewrite the second and third terms:

$$\int_{y_1}^{y_2} \left[u(x_2,y,t)h(x_2,y,t) - u(x_1,y,t)h(x_1,y,t) \right] dy = \int_{x_1}^{x_2} \int_{y_1}^{y_2} \frac{\partial}{\partial x}(uh)dxdy \tag{25.4.6a}$$

$$\int_{x_1}^{x_2} \left[v(x,y_2,t)h(x,y_2,t) - v(x,y_1,t)h(x,y_1,t) \right] dx = \int_{x_1}^{x_2} \int_{y_1}^{y_2} \frac{\partial}{\partial y}(vh)dxdy. \tag{25.4.6b}$$

Replacing these new expressions in (25.4.5) shows all terms to have a double integral over the volume. The integrand must then be identically zero for conservation of mass. This results in the expression

$$\frac{\partial h}{\partial t} + \frac{\partial}{\partial x}(hu) + \frac{\partial}{\partial y}(hv) = 0. \tag{25.4.7}$$

Thus a fundamental relationship is established from a simple first-principles argument. The second and third equations of motion for the shallow-water approximation result from the conservation of momentum in the x- and y-directions. These equations may also be derived directly from the Navier–Stokes equations or conservation laws.

Shallow-water equations

The conservation of mass and momentum generates the following three governing equations for the shallow-water description

$$\frac{\partial h}{\partial t} + \frac{\partial}{\partial x}(hu) + \frac{\partial}{\partial y}(hv) = 0 \tag{25.4.8a}$$

$$\frac{\partial}{\partial t}(hu) + \frac{\partial}{\partial x}\left(hu^2 + \frac{1}{2}gh^2 \right) + \frac{\partial}{\partial y}(huv) = fhv \tag{25.4.8b}$$

$$\frac{\partial}{\partial t}(hv) + \frac{\partial}{\partial y}\left(hv^2 + \frac{1}{2}gh^2 \right) + \frac{\partial}{\partial x}(huv) = -fhu. \tag{25.4.8c}$$

Various approximations are now used to reduce the governing equations to a more manageable form. The first is to assume that at leading order the fluid height $h(x, y, t)$ is constant. The conservation of mass equation (25.4.8a) then reduces to

$$\frac{\partial u}{\partial x} + \frac{\partial v}{\partial y} = 0, \tag{25.4.9}$$

which is referred to as the *incompressible flow* condition. Thus under this assumption, the fluid cannot be compressed.

Under the assumption of a constant height $h(x, y, t)$, the remaining two equations reduce to

$$\frac{\partial u}{\partial t} + 2u\frac{\partial u}{\partial x} + \frac{\partial}{\partial y}(uv) = fv \tag{25.4.10a}$$

$$\frac{\partial v}{\partial t} + 2v\frac{\partial v}{\partial y} + \frac{\partial}{\partial x}(uv) = -fu. \tag{25.4.10b}$$

To simplify further, take the y derivative of the first equation and the x derivative of the second equation. The new equations are

$$\frac{\partial^2 u}{\partial t \partial y} + 2\frac{\partial u}{\partial y}\frac{\partial u}{\partial x} + 2u\frac{\partial^2 u}{\partial x \partial y} + \frac{\partial^2}{\partial y^2}(uv) = f\frac{\partial v}{\partial y} \tag{25.4.11a}$$

$$\frac{\partial^2 v}{\partial t \partial x} + 2\frac{\partial v}{\partial x}\frac{\partial v}{\partial y} + 2v\frac{\partial^2 v}{\partial x \partial y} + \frac{\partial^2}{\partial x^2}(uv) = -f\frac{\partial u}{\partial x}. \tag{25.4.11b}$$

Subtracting the first equation from the second gives the following reductions

$$\frac{\partial}{\partial t}\left(\frac{\partial v}{\partial x} - \frac{\partial u}{\partial y}\right) - 2\frac{\partial u}{\partial y}\frac{\partial u}{\partial x} - 2u\frac{\partial^2 u}{\partial x \partial y} - \frac{\partial^2}{\partial y^2}(uv)$$

$$+ 2\frac{\partial v}{\partial x}\frac{\partial v}{\partial y} + 2v\frac{\partial^2 v}{\partial x \partial y} + \frac{\partial^2}{\partial x^2}(uv) = -f\left(\frac{\partial u}{\partial x} + \frac{\partial v}{\partial y}\right)$$

$$\frac{\partial}{\partial t}\left(\frac{\partial v}{\partial x} - \frac{\partial u}{\partial y}\right) + 2\frac{\partial u}{\partial x}\left(\frac{\partial v}{\partial x} - \frac{\partial u}{\partial y}\right) + 2\frac{\partial v}{\partial y}\left(\frac{\partial v}{\partial x} - \frac{\partial u}{\partial y}\right) \tag{25.4.12}$$

$$+ u\left(\frac{\partial^2 v}{\partial x^2} - \frac{\partial^2 v}{\partial y^2} - 2\frac{\partial^2 u}{\partial x \partial y}\right) + v\left(\frac{\partial^2 u}{\partial x^2} - \frac{\partial^2 u}{\partial y^2} + 2\frac{\partial^2 v}{\partial x \partial y}\right) = -f\left(\frac{\partial u}{\partial x} + \frac{\partial v}{\partial y}\right)$$

$$\frac{\partial}{\partial t}\left(\frac{\partial v}{\partial x} - \frac{\partial u}{\partial y}\right) + u\frac{\partial}{\partial x}\left(\frac{\partial v}{\partial x} - \frac{\partial u}{\partial y}\right) + v\frac{\partial}{\partial y}\left(\frac{\partial v}{\partial x} - \frac{\partial u}{\partial y}\right) - u\frac{\partial}{\partial y}\left(\frac{\partial u}{\partial x} + \frac{\partial v}{\partial y}\right)$$

$$+ v\frac{\partial}{\partial x}\left(\frac{\partial u}{\partial x} + \frac{\partial v}{\partial y}\right) + 2\left(\frac{\partial v}{\partial x} - \frac{\partial u}{\partial y}\right)\left(\frac{\partial u}{\partial x} + \frac{\partial v}{\partial y}\right) = -f\left(\frac{\partial u}{\partial x} + \frac{\partial v}{\partial y}\right).$$

The final equation is reduced greatly by recalling the definition of the vorticity (25.4.3) and using the imcompressibility condition (25.4.9). The governing equations then reduce to

$$\frac{\partial \omega}{\partial t} + u\frac{\partial \omega}{\partial x} + v\frac{\partial \omega}{\partial y} = 0. \tag{25.4.13}$$

This gives the governing evolution of a shallow-water fluid in the absence of diffusion.

The streamfunction

It is typical in many fluid dynamics problems to work with the quantity known as the streamfunction. The streamfunction $\psi(x, y, t)$ is defined as follows:

$$u = -\frac{\partial \psi}{\partial y} \qquad v = \frac{\partial \psi}{\partial x}. \tag{25.4.14}$$

Thus the streamfunction is specified up to an arbitrary constant. Note that the streamfunction automatically satisfies the imcompressibility condition since

$$\frac{\partial u}{\partial x} + \frac{\partial v}{\partial y} = -\frac{\partial^2 \psi}{\partial x \partial y} + \frac{\partial^2 \psi}{\partial x \partial y} = 0. \tag{25.4.15}$$

In terms of the vorticity, the streamfunction is related as follows:

$$\omega = \frac{\partial v}{\partial x} - \frac{\partial u}{\partial y} = \nabla^2 \psi. \tag{25.4.16}$$

This gives a second equation of motion which must be considered in solving the shallow-water equations.

Advection–diffusion

The advection of a fluid is governed by the evolution (25.4.13). In the presence of frictional forces, modification of this governing equation occurs. Specifically, the motion in the shallow-water limit is given by

$$\frac{\partial \omega}{\partial t} + [\psi, \omega] = \nu\nabla^2\omega \tag{25.4.17a}$$

$$\nabla^2\psi = \omega \tag{25.4.17b}$$

where

$$[\psi, \omega] = \frac{\partial \psi}{\partial x}\frac{\partial \omega}{\partial y} - \frac{\partial \psi}{\partial y}\frac{\partial \omega}{\partial x} \tag{25.4.18}$$

and $\nabla^2 = \partial_x^2 + \partial_y^2$ is the two-dimensional Laplacian. The diffusion component which is proportional to ν measures the frictional forces present in the fluid motion.

The advection–diffusion equations have the characteristic behavior of the three partial differential equation classifications: parabolic, elliptic and hyperbolic:

$$\text{parabolic:} \quad \frac{\partial \omega}{\partial t} = \nu \nabla^2 \omega \tag{25.4.19a}$$

$$\text{elliptic:} \quad \nabla^2 \psi = \omega \tag{25.4.19b}$$

$$\text{hyperbolic:} \quad \frac{\partial \omega}{\partial t} + [\psi, \omega] = 0. \tag{25.4.19c}$$

Two things need to be solved for as a function of time:

$$\psi(x, y, t) \quad \text{streamfunction} \tag{25.4.20a}$$

$$\omega(x, y, t) \quad \text{vorticity.} \tag{25.4.20b}$$

We are given the initial vorticity $\omega_0(x, y)$ and periodic boundary conditions. The solution procedure is as follows:

1 **Elliptic solve** Solve the elliptic problem $\nabla^2 \psi = \omega_0$ to find the streamfunction at time zero $\psi(x, y, t = 0) = \psi_0$.

2 **Time-stepping** Given now ω_0 and ψ_0, solve the advection–diffusion problem by time-stepping with a given method. The Euler method is illustrated below

$$\omega(x, y, t + \Delta t) = \omega(x, y, t) + \Delta t \left(\nu \nabla^2 \omega(x, y, t) - [\psi(x, y, t), \omega(x, y, t)] \right).$$

This advances the solution Δt into the future.

3 **Loop** With the updated value of $\omega(x, y, \Delta t)$, we can repeat the process by again solving for $\psi(x, y, \Delta t)$ and updating the vorticity once again.

This gives the basic algorithmic structure which must be followed in order to generate the solution for the vorticity and streamfunction as a function of time. It only remains to discretize the problem and solve.

Project and application

The time evolution of the vorticity $\omega(x, y, t)$ and streamfunction $\psi(x, y, t)$ are given by the governing equation:

$$\omega_t + [\psi, \omega] = \nu \nabla^2 \omega \tag{25.4.21}$$

where $[\psi, \omega] = \psi_x \omega_y - \psi_y \omega_x$, $\nabla^2 = \partial_x^2 + \partial_y^2$, and the streamfunction satisfies

$$\nabla^2 \psi = \omega. \tag{25.4.22}$$

Initial conditions

Assume a Gaussian shaped mound of initial vorticity for $\omega(x, y, 0)$. In particular, assume that the vorticity is elliptical with a ratio of 4:1 or more between the width of the Gaussian in the x- and y-directions. I'll let you pick the initial amplitude (one is always a good start).

Diffusion

In most applications, the diffusion is a small parameter. This fact helps the numerical stability considerably. Here, take $\nu = 0.001$.

Boundary conditions

Assume periodic boundary conditions for both the vorticity and streamfunction. Also, I'll let you experiment with the size of your domain. One of the restrictions is that the initial Gaussian lump of vorticity should be well-contained within your spatial domains.

Numerical integration procedure

Discretize (second-order) the vorticity equation and use ODE23 to step forward in time.

(a) Solve these equations where for the streamline ($\nabla^2 \psi = \omega$) use a fast Fourier transform.

(b) Solve these equations where for the streamline ($\nabla^2 \psi = \omega$) use the following methods:
- A/b
- LU decomposition
- BICGSTAB
- GMRES.

  Compare all of these methods with your FFT routine developed in part (a) (check out the CPUTIME command for MATLAB). In particular, keep track of the computationl speed of each method. Also, for BICGSTAB and GMRES, for the first few times solving the streamfunction equations, keep track of the residual and number of iterations needed to converge to the solution. Note that you should adjust the tolerance settings in BICGSTAB and GMRES to be consistent with your accuracy in the time-stepping. Experiment with the tolerance to see how much more quickly these iteration schemes converge.

(c) Try out these initial conditions with your favorite/fastest solver on the streamfunction equations.
- Two oppositely "charged" Gaussian vorticies next to each other, i.e. one with positive amplitude, the other with negative amplitude.
- Two same "charged" Gaussian vorticies next to each other.
- Two pairs of oppositely "charged" vorticies which can be made to collide with each other.
- A random assortment (in position, strength, charge, ellipticity, etc.) of vorticies on the periodic domain. Try 10–15 vorticies and watch what happens.

(d) Make a 2D movie of the dynamics. Color and coolness are key here. (MATLAB command: movie, getframe). I would very much like to see everyone's movies.

25.5   **Introduction to Reaction–Diffusion Systems**

To begin a discussion of the need for generic reaction–diffusion equations, we consider a set of simplified models relating to *predator-prey* population dynamics. These models consider the interaction of two species: predators and their prey. It should be obvious that such species will have a significant impact on one another. In particular, if there is an abundance of prey, then the predator population will grow due to the surplus of food. Alternatively, if the prey population is low, then the predators may die off due to starvation.

To model the interaction between these species, we begin by considering the predators and prey in the absence of any interaction. Thus the prey population (denoted by $x(t)$) is governed by

$$\frac{dx}{dt} = ax \tag{25.5.1}$$

where $a > 0$ is a net growth constant. The solution to this simple differential equation is $x(t) = x(0)\exp(at)$ so that the population grows without bound. We have assumed here that the food supply is essentially unlimited for the prey so that the unlimited growth makes sense since there is nothing to kill off the population.

Likewise, the predators can be modeled in the absence of their prey. In this case, the population (denoted by $y(t)$) is governed by

$$\frac{dy}{dt} = -cy \tag{25.5.2}$$

where $c > 0$ is a net decay constant. The reason for the decay is that the population starves off since there is no food (prey) to eat.

We now try to model the interaction. Essentially, the interaction must account for the fact the predators eat the prey. Such an interaction term can result in the following system:

$$\frac{dx}{dt} = ax - \alpha xy \tag{25.5.3a}$$

$$\frac{dy}{dt} = -cx + \alpha xy, \tag{25.5.3b}$$

where $\alpha > 0$ is the interaction constant. Note that α acts as a decay to the prey population since the predators will eat them, and as a growth term to the predators since they now have a food supply. These nonlinear and autonomous equations are known as the *Lotka–Volterra equations*. There are two fundamental limitations of this model: the interaction is only heuristic in nature and there is no spatial dependence. Thus the validity of this simple modeling is certainly questionable.

Spatial dependence

One way to model the dispersion of a species in a given domain is by assuming the dispersion is governed by a diffusion process. If in addition we assume that the prey population can saturate at a given level, then the governing population equations are

$$\frac{\partial x}{\partial t} = a\left(x - \frac{x^2}{k}\right) - \alpha xy + D_1\nabla^2 x \tag{25.5.4a}$$

$$\frac{\partial y}{\partial t} = -cx + \alpha xy + D_2\nabla^2 y, \tag{25.5.4b}$$

which are known as the *modified Lotka–Volterra* equations. They include the species interaction with saturation, i.e. the *reaction* terms, and spatial spreading through diffusion, i.e. the *diffusion* term. Thus it is a simple reaction–diffusion equation.

Along with the governing equations, boundary conditions must be specified. A variety of conditions may be imposed; these include periodic boundaries, clamped boundaries such that x and y are known at the boundary, flux boundaries in which $\partial x/\partial n$ and $\partial y/\partial n$ are known at the boundaries, or some combination of flux and clamped. The boundaries are significant in determining the ultimate behavior in the system.

Spiral waves

One of the many phenomena which can be observed in reaction–diffusion systems is spiral waves. An excellent system for studying this phenomenon is the FitzHugh–Nagumo model which provides a heuristic description of an excitable nerve potential:

$$\frac{\partial u}{\partial t} = u(a - u)(1 - u) - v + D\nabla^2 u \tag{25.5.5a}$$

$$\frac{\partial v}{\partial t} = bu - \gamma v, \tag{25.5.5b}$$

where a, D, b and γ are tunable parameters.

A basic understanding of the spiral wave phenomenon can be achieved by considering this problem in the absence of diffusion. Thus the reaction terms alone give

$$\frac{\partial u}{\partial t} = u(a - u)(1 - u) - v \tag{25.5.6a}$$

$$\frac{\partial v}{\partial t} = bu - \gamma v. \tag{25.5.6b}$$

This reduces to a system of differential equations for which the fixed points can be considered. Fixed points occur when $\partial u/\partial t = \partial v/\partial t = 0$. The three fixed points for this system are given by

$$(u, v) = (0, 0) \tag{25.5.7a}$$

$$(u, v) = (u_\pm, (a - u_\pm)(1 - u_\pm)u_\pm) \tag{25.5.7b}$$

where $u_\pm = [(a + 1) \pm ((a + 1)^2 - 4(a - b/\gamma))^{1/2}]/2$.

The stability of these three fixed points can be found by linearization [7]. In particular, consider the behavior near the steady state solution $u = v = 0$. Thus let

$$u = 0 + \tilde{u} \tag{25.5.8a}$$

$$v = 0 + \tilde{v} \tag{25.5.8b}$$

where $\tilde{u}, \tilde{v} \ll 1$. Plugging in and discarding higher order terms gives the linearized equations

$$\frac{\partial \tilde{u}}{\partial t} = a\tilde{u} - \tilde{v} \tag{25.5.9a}$$

$$\frac{\partial \tilde{v}}{\partial t} = b\tilde{u} - \gamma\tilde{v}. \tag{25.5.9b}$$

This can be written as the linear system

$$\frac{d\mathbf{x}}{dt} = \begin{pmatrix} a & -1 \\ b & -\gamma \end{pmatrix} \mathbf{x}. \tag{25.5.10}$$

Assuming a solution of the form $\mathbf{x} = \mathbf{v}\exp(\lambda t)$ results in the eigenvalue problem

$$\begin{pmatrix} a - \lambda & -1 \\ b & -\gamma - \lambda \end{pmatrix} \mathbf{v} = 0 \tag{25.5.11}$$

which has the eigenvalues

$$\lambda_{\pm} = \frac{1}{2}\left[(a - \gamma) \pm \sqrt{(a - \gamma)^2 + 4(b - a\gamma)}\right]. \tag{25.5.12}$$

In the case where $b - a\gamma > 0$, the eigenvalues are purely real with one positive and one negative eigenvalue, i.e. it is a saddle node. Thus the steady state solution in this case is unstable.

Further, if the condition $b - a\gamma > 0$ holds, then the two remaining fixed points occur for $u_- < 0$ and $u_+ > 0$. The stability of these points may also be found. For the parameter restrictions considered here, these two remaining points are found to be stable upon linearization.

The question which can then naturally arise: if the two stable solutions $u_\pm$ are connected, how will the front between the two stable solutions evolve? The scenario is depicted in one dimension in Fig. 25.6 where the two stable $u_\pm$ branches are connected through the unstable $u = 0$

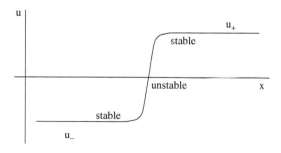

Figure 25.6: Front solution which connects the two stable branches of solutions $u_\pm$ through the unstable solution $u = 0$.

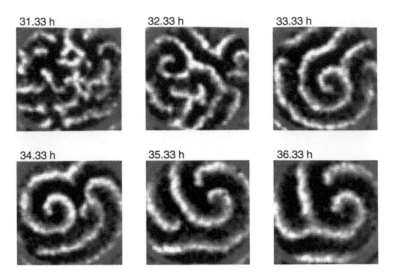

Figure 25.7: Experimental observations of typical spatio-temporal evolution of spiral waves in the early stage of a primary culture of dissociated rat ventricular cells. Few spiral cores survive at the end. [From S.-J. Woo, J. H. Hong, T. Y. Kim, B. W. Bae and K. J. Lee, Spiral wave drift and complex-oscillatory spiral waves caused by heterogeneities in two-dimensional in vitro cardiac tissues, New Journal of Physics **10**, 015005 (2008) ©IOP.]

solution. Not just an interesting mathematical phenomena, spiral waves are also exhibited in nature. Figure 25.7 illustrates an experimental observation in which spiral waves are exhibited in rat ventricular cells. Spiral waves are seen to be naturally generated in this and many other systems and are of natural interest for analytic study.

The $\lambda-\omega$ reaction–diffusion system

The general reaction–diffusion system can be written in the vector form

$$\frac{\partial \mathbf{u}}{\partial t} = \mathbf{f}(\mathbf{u}) + D\nabla^2 \mathbf{u} \tag{25.5.13}$$

where the diffusion is proportional to the parameter D and the reaction terms are given by $\mathbf{f}(\mathbf{u})$. The specific case we consider is the $\lambda-\omega$ system which takes the form

$$\frac{\partial}{\partial t}\begin{pmatrix} u \\ v \end{pmatrix} = \begin{pmatrix} \lambda(A) & -\omega(A) \\ \omega(A) & \lambda(A) \end{pmatrix}\begin{pmatrix} u \\ v \end{pmatrix} + D\nabla^2\begin{pmatrix} u \\ v \end{pmatrix} \tag{25.5.14}$$

where $A = u^2 + v^2$. Thus the nonlinearity is cubic in nature. This allows for the possibility of supporting three steady-state solutions as in the FitzHugh–Nagumo model which led to spiral wave behavior. The spirals to be investigated in this case can have one, two, three or more arms. An example of a spiral wave initial condition is given by

$$u_0(x, y) = \tanh |\mathbf{r}| \cos (m\theta - |\mathbf{r}|) \tag{25.5.15a}$$

$$v_0(x, y) = \tanh |\mathbf{r}| \sin (m\theta - |\mathbf{r}|) \tag{25.5.15b}$$

where $|\mathbf{r}| = \sqrt{x^2 + y^2}$ and $\theta = x + iy$. The parameter m determines the number of arms on the spiral. The stability of the spiral evolution under perturbation and in different parameter regimes is essential in understanding the underlying dynamics of the system. A wide variety of boundary conditions can also be applied. Namely, periodic, clamped, no-flux or mixed. In each case, the boundaries have significant impact on the resulting evolution as the boundary effects creep into the middle of the computational domain.

Project and application

Consider the $\lambda - \omega$ reaction–diffusion system

$$U_t = \lambda(A)U - \omega(A)V + D_1 \nabla^2 U \tag{25.5.16a}$$

$$V_t = \omega(A)U + \lambda(A)V + D_2 \nabla^2 V \tag{25.5.16b}$$

where $A^2 = U^2 + V^2$ and $\nabla^2 = \partial_x^2 + \partial_y^2$.

Boundary conditions

Consider the two boundary conditions in the x- and y-directions:

- Periodic
- No flux: $\partial U/\partial n = \partial V/\partial n = 0$ on the boundaries.

Numerical integration procedure

The following numerical integration procedures are to be investigated and compared.

- For the periodic boundaries, transform the right-hand side with FFTs
- For the no flux boundaries, use the Chebychev polynomials.

You can advance the solution in time using ode45 (or any one of the built-in ODE solvers from the MATLAB suite).

Initial conditions

Start with spiral initial conditions in U and V.

```
[X,Y]=meshgrid(x,y);
m=1; % number of spirals
u=tanh(sqrt(X.^2+Y.^2)).*cos(m*angle(X+i*Y)-(sqrt(X.^2+Y.^2)));
v=tanh(sqrt(X.^2+Y.^2)).*sin(m*angle(X+i*Y)-(sqrt(X.^2+Y.^2)));
```

Consider the specific $\lambda-\omega$ system:

$$\lambda(A) = 1 - A^2 \tag{25.5.17a}$$

$$\omega(A) = -\beta A^2. \tag{25.5.17b}$$

Look to construct one- and two-armed spirals for this system. Also investigate when the solutions become unstable and "chaotic" in nature. Investigate the system for all three boundary conditions. Note $\beta > 0$ and further consider the diffusion to be not too large, but big enough to kill off Gibbs phenomena at the boundary, i.e. $D_1 = D_2 = 0.1$.

25.6 Steady State Flow Over an Airfoil

The derivation for the steady state flow over an airfoil is similar to that of the vorticity and streamfunction dynamics in shallow water. The physical setting of interest now is geared towards the behavior of fluids around airfoils. In this case, a separation of scales is achieved by assuming the airfoil (wing) is much longer in length than its thickness or width. The scale separation once again allows for a two-dimensional study of the problem. Additionally, since we are considering the steady state flow, the time dependence of the problem will be eliminated. The resulting problem will be of the elliptic type. This will give the steady state streamfunction and flow over the wing.

The physical setting for the airfoil modeling is illustrated in Fig. 25.8. In this figure, the characteristic wing width is given by the parameter D while the characteristic wing length is given by the parameter L. For a quasi-two-dimensional description to hold, we must have

$$\delta = \frac{D}{L} \ll 1. \tag{25.6.1}$$

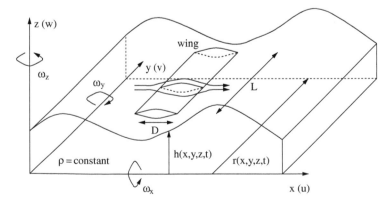

Figure 25.8: Physical setting of an airfoil in a constant density fluid ρ with scale separation $D/L \ll 1$.

This gives the necessary separation of scales for reducing the Navier–Stokes equations to the shallow-water description. In particular, the shallow-water reduction which led to the vorticity-streamfunction equations can be pursued here. The governing equations which result from mass conservation and momentum conservation in the x and z planes give

$$\frac{\partial r}{\partial t} + \frac{\partial}{\partial x}(ru) + \frac{\partial}{\partial z}(rw) = 0 \qquad (25.6.2a)$$

$$\frac{\partial}{\partial t}(ru) + \frac{\partial}{\partial x}\left(ru^2 + \frac{1}{2}gr^2\right) + \frac{\partial}{\partial z}(ruw) = 0 \qquad (25.6.2b)$$

$$\frac{\partial}{\partial t}(rw) + \frac{\partial}{\partial z}\left(rw^2 + \frac{1}{2}gr^2\right) + \frac{\partial}{\partial x}(ruw) = 0 \qquad (25.6.2c)$$

where we have ignored the Coriolis parameter.

At leading order, we assume $r(x, y, t) \approx$ constant so that the conservation of mass equation reduces to

$$\frac{\partial u}{\partial x} + \frac{\partial w}{\partial z} = 0, \qquad (25.6.3)$$

which is again referred to as the *imcompressible flow* condition. Under the assumption of a constant width $r(x, z, t)$, the remaining two momentum equations reduce further. The final reduction comes from taking the z derivative of the first equation and the x derivative of the second equation and performing the algebra associated with (25.4.10) through (25.4.13) which gives

$$\frac{\partial \omega}{\partial t} + [\psi, \omega] = \nu \nabla^2 \omega \qquad (25.6.4a)$$

$$\nabla^2 \psi = \omega \qquad (25.6.4b)$$

where

$$[\psi, \omega] = \frac{\partial \psi}{\partial x}\frac{\partial \omega}{\partial z} - \frac{\partial \psi}{\partial z}\frac{\partial \omega}{\partial x} \qquad (25.6.5)$$

and $\nabla^2 = \partial_x^2 + \partial_z^2$ is the two-dimensional Laplacian. The diffusion component which is proportional to ν measures the frictional forces present in the fluid motion. Note that we have once again introduced the streamfunction $\psi(x, z, t)$ which is defined as

$$u = -\frac{\partial \psi}{\partial z} \qquad w = \frac{\partial \psi}{\partial x}. \qquad (25.6.6)$$

Thus the streamfunction is specified up to an arbitrary constant. Note that the streamfunction automatically satisfies the imcompressibility condition since

$$\frac{\partial u}{\partial x} + \frac{\partial w}{\partial z} = -\frac{\partial^2 \psi}{\partial x\partial z} + \frac{\partial^2 \psi}{\partial x\partial z} = 0. \qquad (25.6.7)$$

In terms of the vorticity, the streamfunction is related as follows:

$$\omega = \omega_y = \frac{\partial w}{\partial x} - \frac{\partial u}{\partial z} = \nabla^2 \psi. \qquad (25.6.8)$$

This gives a second equation of motion (25.6.4a) which must be considered in solving the airfoil equations.

Steady state flow

The behavior of the fluid as it propagates over the airfoil can change drastically depending upon the density and velocity of the fluid. The typical measure of the combined fluid density and velocity is given by a quantity known as the *Reynolds number R*. Essentially, as the Reynolds number increases, then the fluid is either moving at higher velocities or has a lower density. The experimental fluid flow around a cylinder, which can be thought of as an airfoil, is depicted in Fig. 23.8. The four pictures correspond to $R = 9.6, 41.0, 140$ and 10 000. Note the transition from laminar (steady state) flow to a recirculating steady state flow, to Kárman vortex shedding and finally to fully turbulent flow. The same type of phenomenon occurs also in three dimensions.

The general turbulent behavior is difficult to capture computationally. However, the steady state behavior can be well understood by considering the equations of motion in the steady state limit. In particular, at large times in the steady state flow the diffusion term in the advection–diffusion equation (25.6.4a) causes the vorticity to diffuse to zero, i.e. as $t \to \infty$ then $\omega \to 0$. The streamfunction equation (25.6.4b) then becomes Poisson's equation

$$\nabla^2 \psi = 0. \tag{25.6.9}$$

Thus an elliptic solve is the only thing required to find the steady state behavior.

Boundary conditions

To capture the steady state flow which is governed by the elliptic problem, we need to develop boundary conditions for both the computational domain and airfoil. Figure 25.9 illustrates the computational domain which may be implemented to capture the steady state flow over an airfoil. Five boundary conditions are specified to determine the steady state flow over the airfoil. These five boundary conditions on the domain $x \in [-L, L]$ and $y \in [-L, L]$ are as follows:

1 $x = -L$: The incoming flux flow it imposed at this location. Assuming the flow enters so that $u(x = -L) = u_0$ and $w(x = -L) = 0$ gives

$$u = u_0 = -\frac{\partial \psi}{\partial z} \quad \to \quad \frac{\partial \psi}{\partial z} = -u_0. \tag{25.6.10}$$

2 $x = L$: The outgoing flux flow should be indentical to that imposed for 1. Thus (25.6.10) should apply.

3 $z = -L$: Assume that far above or below the airfoil, the fluid flow remains parallel to the x-direction. Thus (25.6.10) still applies. However, we also have that

$$w = 0 = \frac{\partial \psi}{\partial x} \quad \to \quad \frac{\partial \psi}{\partial x} = 0. \tag{25.6.11}$$

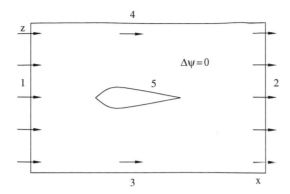

Figure 25.9: Computational domain for calculating the steady state flow over an airfoil. Note the five boundaries which must be specified to reach a solution.

4    $z = L$: The boundary conditions at the top of the domain will mirror those at the bottom.

5   Airfoil: The steady state flow cannot peneratrate into the airfoil. Thus a boundary condition is required on the wing itself. Physically, a *no-slip* boundary condition is applied. Thus the fluid is stationary on the domain of the wing:

$$u = w = 0 \quad \rightarrow \quad \frac{\partial \psi}{\partial x} = \frac{\partial \psi}{\partial z} = 0. \tag{25.6.12}$$

Although the rectangular computational domain is easily handled with finite difference techniques, the airfoil itself presents a significant challenge. In particular, the shape is not regular and the finite difference technique is only equipped to handle rectangular type domains. Thus an alternative method should be used. Specifically, finite elements can be considered since it allows for an unstructured domain and can handle any complicated wing shape. Realistically, any attempt with finite difference or spectral techniques on a complicated domain is more trouble than its worth.

26 Applications of Data Analysis

26.1 Analyzing Music Scores and the Gábor Transform

In this section, you will analyze a portion of Handel's *Messiah* with time–frequency analysis. To get started, you can use the following commands (note that Handel is so highly regarded, that MATLAB has a portion of his music already in MATLAB!):

```
load handel
v = y'/2;
plot((1:length(v))/Fs,v);
xlabel('Time [sec]');
ylabel('Amplitude');
title('Signal v(n)');
```

This code gives you a plot of the portion of music you will analyze. To play this back in MATLAB, you can use the following commands:

```
p8 = audioplayer(v,Fs);
playblocking(p8);
```

This project is rather open ended in the sense that you should explore the time–frequency signature of this 9 second piece of classic work. Things you should think about doing:

1   Through use of Gábor filtering, produce spectrograms of the piece of work.

2   Explore the window width of the Gábor transform and how it effects the spectrogram.

3 Explore the spectrogram and the idea of oversampling (i.e. using very small translations of the Gábor window) versus potential undersampling (i.e. using very course/large translations of the Gábor window).

4 Use different Gábor windows. Perhaps you can start with the Gaussian window, and look to see how the results are affected with the Mexican hat wavelet and a step-function (Shannon) window.

This is an opportunity for you to have a creative and open ended experience with MATLAB and data analysis.

Mary had a little lamb

Find two musical instruments: perhaps a recorder and a piano. With these two instruments, play the song *Mary had a little lamb*. Be sure to play it at the same cadence so as to make a good comparison between the instruments. You can record and save the music files using almost any smart phone. Typically, these will be saved in the WAV format. Thus you will have two music files **music1.wav** and **music2.wav**. To import and convert them, use the following commands for both pieces. Note that it is assumed that both music pieces are 15 seconds long, but you should use whatever recording time you actually used. Additionally note that basically both pieces are converted to a vector representing the music, thus you can easily edit the music by modifying the vector.

```
tr_piano=15;   % record time in seconds
y=wavread('music1'); Fs=length(y)/tr_piano;
plot((1:length(y))/Fs,y);
xlabel('Time [sec]'); ylabel('Amplitude');
title('Mary had a little lamb (piano)');   drawnow
p8 = audioplayer(y,Fs); playblocking(p8);

figure(2)
tr_rec=15;   % record time in seconds
y=wavread('music2'); Fs=length(y)/tr_rec;
plot((1:length(y))/Fs,y);
xlabel('Time [sec]'); ylabel('Amplitude');
title('Mary had a little lamb (recorder)');
p8 = audioplayer(y,Fs); playblocking(p8);
```

1 Through use of the Gábor filtering, reproduce the music score for this simple piece. See Fig. 26.1 which has the music scale in hertz (note: to get a good clean score, you may want to filter out overtones, see below).

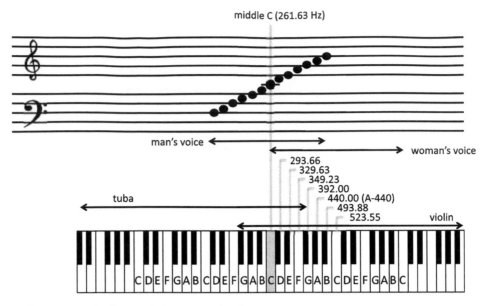

Figure 26.1: Music scale along with frequency of each note in Hz.

2   What is the difference between a recorder and piano? Can you see the difference in the time–frequency analysis? Note that many people talk about the difference of instruments being related to the *timbre* of an instrument. The timbre is related to the overtones generated by the instrument for a center frequency. Thus if you are playing a note at frequency ω_0, an instrument will generate overtones at $2\omega_0$, $3\omega_0$, $\cdots$ and so forth. Note that the piccolo is an instrument that plays almost pure tones while most other instruments lay out a clear pattern of overtones, thus allowing them to sound different and unique.

 ## 26.2   Image Denoising through Filtering and Diffusion

Denoising is one of the most common image processing objectives in image analysis. In this example, we will take an image and add noise both globally and locally and try to recover the original image. Consider loading in a given image

```
A=imread('photo','jpeg');
Abw2=rgb2gray(A);
Abw=double(Abw2);
[nx,ny]=size(Abw);
```

Figure 26.2: Original images in color (left) and black and white (right) along with their noisy (global) counterparts.

The loaded JPEG image is saved both as a color and black-and-white image. The size of each is determined from the **size** command which gives the number of pixels in the horizontal and vertical direction. In the example here, the uploaded photo of Erice, Italy, is 800×600 pixels.

Two things will be done to these images:

(i) White noise will be added to the picture in a global way (see Fig. 26.2). The following code will add such white noise and plot the original images in color and black-and-white along with their noisy counterparts.

```
noise2=randn(nx,ny); noise3=randn(nx,ny,3);
u2=uint8(Abw+50*noise2);
u3=uint8(double(A)+50*randn(nx,ny,3));

subplot(2,2,1), imshow(A)
subplot(2,2,2), imshow(Abw2)
subplot(2,2,3), imshow(u3)
subplot(2,2,4), imshow(u2)
```

Figure 26.3: Original images in color (left) and black and white (right) along with their noisy (local) counterparts.

The noise added is simply in the spatial domain with a randomly distributed, normal variable.

(ii) White noise will be added to the picture in a local way (see Fig. 26.3). The following code will add such white noise to a small patch of the picture and plot the original images in color and black and white along with their noisy counterparts.

```
Abw_spot=Abw;
Abw_spot(200-40:1:200+40,400-40:1:400+40) ...
  =Abw(200-40:1:200+40,400-40:1:400+40) ...
  +100*noise2(200-40:1:200+40,400-40:1:400+40);

Abw_spot2=double(A);
Abw_spot2(200-40:1:200+40,400-40:1:400+40,:) ...
  =double(A(200-40:1:200+40,400-40:1:400+40,:)) ...
  +100*noise3(200-40:1:200+40,400-40:1:400+40,:);
```

Continued

```
A_n=uint8(Abw_spot);
A_n2=uint8(Abw_spot2);

figure(2)
subplot(2,2,2),  imshow(Abw2);
subplot(2,2,4),  imshow(A_n);
subplot(2,2,1),  imshow(A);
subplot(2,2,3),  imshow(A_n2);
```

The objective now will be to reconstruct, as best as possible, the original quality images.

Task 1

Use filtering to try to clean up the corrupted images with global noise. Filter both the black-and-white image and the three levels of the RGB data of the color picture. Investigate the results as a function of filter type and filter width.

Task 2

Use diffusion to locally try to remove the white noise patch. You could potentially use both diffusion and filtering to get the best image quality. Again, you now have three diffusion equations to solve for the three levels of the color picture.

26.3   Oscillating Mass and Dimensionality Reduction

Perform the experiment outlined in Sections 15.3 and 15.4. To do this, you will need a spring, a mass and three smart phones or video cameras to record your results. The experiment is an attempt to illustrate various aspects of the PCA and its practical usefulness and the effects of noise on the PCA algorithms.

- **(Test 1) Ideal case** Consider a small displacement of the mass in the z-direction and the ensuing oscillations. In this case, the entire motion is in the z-directions with simple harmonic motion being observed.

- **(Test 2) Noisy case** Repeat the ideal case experiment, but this time, introduce camera shake into the video recording. This should make it more difficult to extract the simple harmonic motion. But if the shake isn't too bad, the dynamics will still be extracted with the PCA algorithms.

- **(Test 3) Horizontal displacement** In this case, the mass is released off-center so as to produce motion in the $x-y$ plane as well as the z-direction. Thus there is both a pendulum motion and a simple harmonic oscillations. See what the PCA tells us about the system.

- **(Test 4) Horizontal displacement and rotation** In this case, the mass is released off-center and rotates so as to produce motion in the $x-y$ plane, rotation, as well as the z-direction. Thus there is both a pendulum motion and simple harmonic oscillations. See what the PCA tells us about the system.

In order to extract out the mass movement from the video frames, the following MATLAB code is needed. This code is a generic way to read in movie files to MATLAB for post-processing.

```
obj=mmreader('matlab_test.mov')

vidFrames = read(obj);
numFrames = get(obj,'numberOfFrames');

for k = 1 : numFrames
    mov(k).cdata = vidFrames(:,:,:,k);
    mov(k).colormap = [];
end

for j=1:numFrames
  X=frame2im(mov(j));
  imshow(X); drawnow
end
```

The multimedia reader command **mmreader** simply characterizes the file type and its attributes. The bulk of time in this is the **read** command which actually uploads the movie frames into the MATLAB desktop for processing. Once the frames are extracted, they can again be converted to double precision numbers for mathematical processing. In this case, the position of the mass is to be determined from each frame. This basic shell of code is enough to begin the process of extracting the spring–mass system information.

Explore the PCA method on this problem and see what you find. Note that you will need to generically trim each movie clip down to the same size as well as align them in time. Precise alignment is not necessary to get good results using a PCA analysis.

26.4   Music Genre Identification

Music genres are instantly recognizable to us, whether it be jazz, classical, blues, rap, rock, etc. One can always ask how the brain classifies such information and how it makes a decision based upon hearing a new piece of music. The objective of this project is to attempt to write a code that can classify a given piece of music by sampling a five second clip.

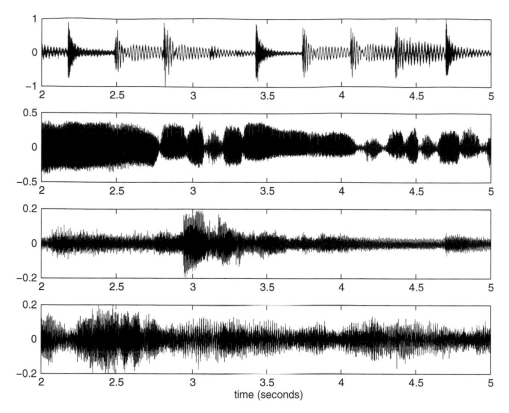

Figure 26.4: Instantly recognizable, these four pieces of music are (in order of top to bottom): Dr. Dre's *Nuthin' but a 'G' thang* (The Chronic), John Coltrane's *A Love Supreme* (A Love Supreme), Led Zeppelin's *Over The Hills and Far Away* (Houses of the Holy), and Mozart's *Kyrie* (Requiem). Illustrated is a three-second clip from time two seconds to five seconds of each of these songs.

As an example, consider Fig. 26.4. Four classic pieces of music are demonstrated spanning genres of rap, jazz, classic rock and classical. Specifically, a three-second sample is given of Dr. Dre's *Nuthin' but a 'G' thang* (The Chronic), John Coltrane's *A Love Supreme* (A Love Supreme), Led Zeppelin's *Over The Hills and Far Away* (Houses of the Holy), and Mozart's *Kyrie* (Requiem). Each has a different signature, thus begging the question whether a computer could distinguish between genres based upon such a characterization of the music.

- **(Test 1) Band classification** Consider three different bands of your choosing and of different genres. For instance, one could pick Michael Jackson, Soundgarden, and Beethoven. By taking five-second clips from a variety of each of their music, i.e. building training sets, see if you can build a statistical testing algorithm capable of accurately identifying "new" five-second clips of music from the three chosen bands.

- **(Test 2) The case for Seattle** Repeat the above experiment, but with three bands from within the same genre. This makes the testing and separation much more challenging. For instance, one could focus on the late 1990s Seattle grunge bands: Soundgarden, Alice in Chains, and Pearl Jam. What is your accuracy in correctly classifying a five-second sound clip? Compare this with the first experiment with bands of different genres.

- **(Test 3) Genre classification** One could also use the above algorithms to simply broadly classify songs as jazz, rock, classical, etc. In this case, the training sets should be various bands within each genre. For instance, classic rock bands could be classified using sound clips from Zep, AC/DC, Floyd, etc. while classical could be classified using Mozart, Beethoven, Bach, etc. Perhaps you can limit your results to three genres, for instance, rock, jazz, classical.

Warning and Notes You will probably want to SVD the spectrogram of songs versus the songs themselves. Interestingly, this will give you the dominant spectrogram *modes* associated with a given band. Moreover, you may want to resample your data (i.e. take every other point) in order to keep the data sizes more manageable. Regardless, you will need lots of processing time.

References

1 A. Greenbaum, *Iterative Methods for Solving Linear Systems* (SIAM, 1997).

2 J. R. Shewchuk, "An Introduction to the Conjugate Gradient Method Without the Agonizing Pain (Edition 1 1/4)" (1994). http://www.cs.berkeley.edu/~jrs/jrspapers.html#cg; http://www.cs.cmu.edu/~quake-papers/painless-conjugate-gradient.pdf.

3 L. Sirovich and M. Kirby, "Low-dimensional procedure for the characterization of human faces," Journal of the Optical Society of America A **4**, 519524 (1987).

4 M. Kirby and L. Sirovich, "Application of the Karhunen–Loeve procedure for the characterization of human faces," IEEE Transactions on Pattern Analysis and Machine Intelligence **12**, 103108 (1990).

5 M. Turk and A. Pentland, "Eigenfaces for recognition," Journal of Cognitive Neuroscience **3**, 7186 (1991).

6 J. C. Lagarias, J. A. Reeds, M. H. Wright, and P. E. Wright, "Convergence properties of the Nelder–Mead simplex method in low dimensions," SIAM Journal of Optimization **9**, 112–147 (1998).

7 W. E. Boyce and R. C. DiPrima, *Elementary Differential Equations and Boundary Value Problems*, 7th ed. (Wiley, 2001).

8 R. Finney, F. Giordano, G. Thomas, and M. Weir, *Calculus and Analytic Geometry*, 10th ed. (Prentice Hall, 2000).

9 C. W. Gear, *Numerical Initial Value Problems in Ordinary Differential Equations* (Prentice Hall, 1971).

10 J. D. Lambert, *Computational Methods in Ordinary Differential Equations* (Wiley, 1973).

11 R. L. Burden and J. D. Faires, *Numerical Analysis* (Brooks/Cole, 1997).

12 D. G. Luenberger and Y. Ye, *Linear and Nonlinear Programming*, 3rd ed. (Springer, 2008).

13 S. M. Cox and P. C. Matthews, "Exponential time differencing for stiff systems," Journal of Computational Physics **176**, 430–455 (2002).

14 A.-K. Kassam and L. N. Trefethen, "Fourth-order time-stepping for stiff PDEs," SIAM Journal of Scientific Computing **26**, 12141233 (2005).

15 W. H. Press, S. A. Teukolsky, W. T. Vetterling, and B. P. Flannery, *Numerical Recipes*, 2nd ed. (Cambridge, 1992).

16 R. Courant, K. O. Friedrichs, and H. Lewy, "Über die partiellen differenzengleichungen der mathematischen physik," Mathematische Annalen **100**, 32–74 (1928).

17 J. C. Strikwerda, *Finite Difference Schemes and Partial Differential Equations* (Chapman & Hall, 1989).

18 N. L. Trefethen, *Spectral Methods in MATLAB* (SIAM, 2000).

19 G. Strang, *Introduction to Applied Mathematics* (Wellesley-Cambridge Press, 1986).

20 B. Deconinck and J. N. Kutz, "Computing spectra of linear operators using the Floquet–Fourier–Hill method," Journal of Computational Physics **219**, 296–321 (2006).

21 G. W. Hill, "On the part of the lunar perigee which is a function of the mean motions of the sun and the moon," Acta Mathematica **8**, 1–36 (1886).

22 A. H. Nayfeh and D. T. Mook, *Nonlinear Oscillations* (Wiley, 1995).

23 J. Guckenheimer and P. Holmes, *Nonlinear Oscillations, Dynamical Systems, and Bifurcations of Vector Fields* (Springer, 1983).

24 I. M. Gelfand and S. V. Fomin, *Calculus of Variations* (Prentice-Hall, 1963).

25 M. R. Spiegel, *Probability and Statistics* (Schaum's Outline Series, McGraw-Hill, 1975).

26 S. Ross, *Introduction to Probability Models*, 4th ed. (Academic Press, 1989).

27 L. Debonath, *Wavelet Transforms and Their Applications* (Birkhäuser, 2002).

28 A. Haar, "Zur Theorie der orthogonalen funktionen-systeme," Mathematische Annalen **69**, 331–371 (1910).

29 T. F. Chan and J. Shen, *Image Processing and Analysis* (SIAM, 2005).

30 L. N. Trefethen and D. Bau, *Numerical Linear Algebra* (SIAM, 1997).

31 J. Shlens, "A tutorial on principal component analysis," http://www.snl.salk.edu/~shlens/pca.pdf.

32 P. Holmes, J. L. Lumley, and G. Berkooz, *Turbulence, Coherent Structures, Dynamical Systems, and Symmetry* (Cambridge, 1996).

33 J. P. Cusumano, M. T. Sharkady, and B. W. Kimble, "Dynamics of a flexible beam impact oscillator," Philosophical Transactions of the Royal Society of London **347**, 421–438 (1994).

34 B. F. Feeny and R. Kappagantu, "On the physical interpretation of proper orthogonal modes in vibrations," Journal of Sound and Vibration **211**, 607–616 (1998).

35 T. Kubow, M. Garcia, W. Schwind, R. Full, and D. E. Koditschek, "A principal components analysis of cockroach steady state leg motion," (in preparation).

36 R. Ruotolo and C. Surace, "Damage assessment of multiple cracked beams: numerical results and experimental validation." Journal of Sound and Vibration **206**, 567–588 (1997).

37 K. L. Briggman, H. D. I. Abarbanel, and W. B. Kristan Jr., "Optical imaging of neuronal populations during decision-making," Science **307**, 896 (2005).

38 A. Chatterjee, "An introduction to the proper orthogonal decomposition," Current Science **78**, 808–817 (2000).

39 J. Tenenbaum, V. de Silva, and J. C. Langford, "A global geometric framework for nonlinear dimensionality reduction," Science **290**, 2319–2323 (2000).

40 S. Roweis and L. K. Saul, "Linear embedding nonlinear dimensionality reduction by locally linear embedding," Science **290**, 2323–2326 (2000).

41 K. Q. Weinberger and L. K. Saul, "Unsupervised learning of image manifolds by semidenite programming," International Journal of Computer Vision **70**, 77–90 (2006).

42 E. J. Candés, X. Li, Y. Ma, and J. Wright, "Robust principal component analysis?" Journal of ACM **58**, 1–37 (2011).

43 Z. Lin, M. Chen, L. Wu, and Y. Ma, "The augmented Lagrange multiplier method for exact recovery of corrupted low-rank matrices." UIUC Technical Report UILU-ENG-09-2215, arXiv:1009.5055v2, October 26, 2009.

44 Z. Lin, A. Ganesh, J. Wright, M. Chen, L. Wu, and Y. Ma, "Fast convex optimization algorithms for exact recovery of a corrupted low-rank matrix." UIUC Technical Report UILU-ENG-09-2214, August 2009.

45 J. Cai, E. J. Candes, and Z. Shen, "A singular value thresholding algorithm for matrix completion," SIAM Journal of Optimization **20**, 1956–1982 (2010).

46 X. Yuan and J. Yang, "Sparse and low-rank matrix decomposition via alternating direction methods," www.optimization-online.org (2009).

47 A. Hyvärinen and E. Oja, "Independent component analysis: Algorithms and applications," http://mlsp.cs.cmu.edu/courses/fall2012/lectures/ICA_Hyvarinen.pdf.

48 H. Farid and E. H. Adelson, "Separating reflections from images by use of independent component analysis," Journal of the Optical Society of America A **16**, 2136–2145 (1999).

49 R. A. Fisher, "The use of multiple measurements in taxonomic problems," Annals of Eugenics **7**, 179–188 (1936).

50 C. Moler, "Magic reconstruction: Compressed sensing," Mathworks News & Notes, Cleves Corner (2010).

51 E. Candés and M. Wakin, "An introduction to compressive sampling," IEEE Signal Processing Magazine, pp. 21–30, March 2008.

52 R. Baraniuk, "Compressive sensing," IEEE Signal Processing Magazine, pp. 118–124, July 2007.

53 D. Donoho, "Compressed sensing," IEEE Transactions on Information Theory **52**, 1289–1306 (2006).

54 E. Candés, *L1-magic toolbox*, from E. Candés' homepage. http://www-stat.stanford.edu/~candes/l1magic/.

55 M. Duarte, M. Davenport, D. Takhar, J. Laska, T. Sun, K. Kelly, and R. Baraniuk, "Single pixel imaging via compressive sampling," IEEE Signal Processing Magazine **25**, 83–91 March 2008.

56 M. Duarte, M. Wakin, and R. Baraniuk, "Wavelet-domain compressive signal reconstruction using a hidden Markov tree model," International Conference on Acoustics, Speech, and Signal Processing (ICASSP), 2008, Las Vegas, NV. pp. 5137–5140.

57 E. Ding, E. Shlizerman, and J. N. Kutz, "Modeling multipulsing transition in ring cavity lasers with proper orthogonal decomposition," Physical Review A **82**, 023823 (2010).

58 H. A. Haus, "Mode-locking of lasers," IEEE Journal of Selected Topics in Quantum Electronics **6**, 1173 (2000).

59 J. N. Kutz, "Mode-locked soliton lasers," SIAM Review **48**, 629 (2006).

60 S. Namiki, E. P. Ippen, H. A. Haus, and C. X. Yu, "Energy rate equations for mode-locked lasers," Journal of the Optical Society of America B **14**, 2099 (1997).

61 J. P. Gordon and H. A. Haus, "Random walk of coherently amplified solitons in optical fiber transmission," Optics Letters **11**, 665–667 (1986).

62 C. W. Rowley and J. E. Marsden, "Reconstruction equations and the Karhunen–Loeve expansion," Physica D **142**, 1–19 (2000).

63 A. Dhooge, W. Govaerts, and Yu. A. Kuznetsov, "MATCONT: A MATLAB package for numerical bifurcation analysis of ODEs," ACM Transactions on Mathematical Software **29**, 141 (2003).

64 P. J. Schmid, "Dynamic mode decomposition of numerical and experimental data," Journal of Fluid Dynamics **656**, 5–28 (2010).

65 P. J. Schmid, L. Li, M. P. Juniper, and O. Pust, "Applications of the dynamic mode decomposition," Theoretical and Computational Fluid Dynamics **25**, 249–259 (2011).

66 C. W. Rowley, I. Mezić, S. Bagheri, P. Schlatter, and D. S. Henningson, "Spectral analysis of nonlinear flows," Journal of Fluid Mechanics **641**, 115–127 (2009).

67 K. K. Chen, J. H. Tu, and C. W. Rowley, "Variants of dynamic mode decomposition: Boundary conditions, Koopman, and Fourier analyses," Journal of Nonlinear Science (submitted). DOI 10.1007/s00332-012-9130-9.

68 C. Penland, "Random forcing and forecasting using principal oscillations pattern analysis," Monthly Weather Review **117**, 2165–2185 (1989).

69 C. Penland and L. Matrosova, "Expected and actual errors of linear inverse model forecasts," Monthly Weather Review **129**, 1740–1745 (2001).

70 M. A. Alexander, L. Matrosova, C. Penland, J. D. Scott, and P. Chang, "Forecasting Pacific SSTs: Linear inverse model predictions of the PDO," Journal of Climate **21**, 385–402 (2008).

71 I. Mezić and A. Banaszuk, "Comparison of systems with complex behavior," Physica D **197**, 101–133 (2004).

72 I. Mezić, "Spectral properties of dynamical systems, model reduction and decompositions," Nonlinear Dynamics **41**, 309–325 (2005).

73 M. Williams, P. Schmid, and J. N. Kutz, "Hybrid reduced order integration with the proper orthogonal decomposition and dynamic mode decomposition," SIAM Journal of Multiscale Modeling and Simulation (2013).

74 J. A. Lee and M. Verleysen, *Nonlinear Dimensionality Reduction* (Springer, 2007).

75 B. Sonday, A. Singer, C. W. Gear, and I. G. Kevrekidis, "Manifold learning techniques and model reduction applied to dissipative PDEs," SIAM Journal of Applied Dynamical Systems, 1–20 (2010).

76 M. Ilak and C. W. Rowley, "Modeling of transitional channel flow using balanced proper orthogonal decomposition," Physics of Fluids **20**, 034103 (2008).

77 M. Rathinam and L. Petzold, "A new look at proper orthogonal decomposition," SIAM Journal on Numerical Analysis **41**, 1893–1925 (2004).

78 C. Rowley, "Model reduction for fluids using balanced proper orthogonal decomposition," International Journal of Bifurcation and Chaos **15**, 997–1013 (2005).

79 M. Ilak and C. W. Rowley, "Reduced-order modeling of channel flow using traveling POD and balanced POD," in Third AIAA Flow Control Conference, 2006.

80 X. Chen and S. Akella, "A dual-weighted trust-region adaptive POD 4-D Var applied to a finite volume shallow water equations model on the sphere," International Journal of Numerical Methods in Fluids **68**, 377–402 (2012).

81 P. del Sastre and R. Bermejo, "The POD technique for computing bifurcation diagrams: A comparison among different models in fluids," in *Numerical Mathematics and Advanced Applications*, A. B. de Castro, D. Gomez, P. Quintela, and P. Salgado, eds., 880–888 (Springer, 2006).

82 J. R. Holton and G. J. Hakim, *An Introduction to Dynamic Meteorology*, 5th ed., Ch. 13.6 (Academic Press, 2013).

83 M. Ghil, S. E. Cohn, and A. Dulcher, "Sequential estimation of data assimilation and initialization," in *The Interaction Between Objective Analysis and Initialization*, D. Williamson, ed., Publ. Meteorol. **127**, 83–97, Proceedings of the 14th Stanstead Seminar, McGill University, Montreal.

84 R. N. Miller, M. Ghil, and F. Gauthiez, "Advanced data assimilation for strongly nonlinear dynamical systems," Journal of Atmospheric Sciences **51**, 1037–1056 (1994).

85 P. Gauthier, "Chaos and quadri-dimensional data assimilation: a study based on the Lorenz model," Tellus **44**, 2–17 (1992).

86 G. Evensen, "Advanced data assimilation for strongly nonlinear dynamics," Monthly Weather Review **125**, 1342–1354 (1997).

87 A. H. Jazwinski, *Stochastic Processes and Filtering Theory* (Academic Press, 1970).

88 I. G. Kevrekidis, C. W. Gear, and G. Hummer, "Equation free: The computer-aided analysis of complex multiscale systems," American Institute of Chemical Engineering **50**, 1346–1355 (2004).

89 B. A. Cipra, "Equation-free computing gets something from (almost) nothing," SIAM News **38** (2005).

90   I. G. Kevrekidis, C. W. Gear, J. M. Hyman, P. G. Kevrekidis, O. Runborg, and C. Theodoropoulos, "Equation-free, course-grained multiscale computation: enabling microscopic simulators to perform system-level analysis," Communications in Mathematical Science **1**, 715–762 (2003).

91   B. Yildirim, C. Chryssostomidis, and G. E. Karniadakis, "Efficient sensor placement for ocean measurements using low-dimensional concepts," Ocean Modelling **27**, 160–173 (2009).

92   I. Bright, G. Lin, and J. N. Kutz, "Compressive sensing and machine learning strategies for characterizing the flow around a cylinder with limited pressure measurements," Physics of Fluids (submitted).

93   B. Deconinck and J. N. Kutz, "Singular instability of exact stationary solutions of the nonlocal Gross–Pitaevskii equation," Physics Letters A **319**, 97–103 (2003).

94   G. Baym, *Lectures in Quantum Mechanics* (Addison-Wesley, 1990).

95   E. P. Gross, "Structure of a quantized vortex in boson systems," Nuovo Cimento **20**, 454 (1961).

96   L. P. Pitaevskii, "Vortex lines in an imperfect Bose gas," Soviet Physics JETP **13**, 451 (1961).

97   G. B. Whitham, *Linear and Nonlinear Waves* (Wiley, 1974).

98   E. H. Lieb and R. Seiringer, "Proof of Bose–Einstein condensation for dilute trapped gases," Physical Review Letters **88**, 170409 (2002).

99   J. N. Kutz and B. Sandstede, "Theory of passive harmonic mode-locking using waveguide arrays," Optics Express **16**, 636–650 (2008).

100 J. Keener and J. Sneyd, *Mathematical Physiology* (Springer, 1998).

101 G. B. Ermentrout and D. H. Terman, *Mathematical Foundations of Neuroscience* (Springer, 2010).

102 P. Dayan and L. F. Abbott, *Theoretical Neuroscience* (MIT Press, 2001).

Index of MATLAB Commands

Index